Computer and Information Sciences II

Erol Gelenbe · Ricardo Lent
Georgia Sakellari
Editors

Computer and Information Sciences II

26th International Symposium on Computer and Information Sciences

Springer

Erol Gelenbe
Department of Electrical
and Electronics Engineering
Imperial College
London
UK
e-mail: e.gelenbe@imperial.ac.uk

Ricardo Lent
Imperial College
London
UK
e-mail: r.lent@imperial.ac.uk

Georgia Sakellari
University of East London
London
UK

ISBN 978-1-4471-2154-1 e-ISBN 978-1-4471-2155-8
DOI 10.1007/978-1-4471-2155-8
Springer London Dordrecht Heidelberg New York

British Library Cataloguing in Publication Data
A catalogue record for this book is available from the British Library

Library of Congress Control Number: 2011938586

Cover design: eStudio Calamar S.L.

Printed on acid-free paper

Springer is part of Springer Science+Business Media (www.springer.com)

Preface

The 26th Annual International Symposium on Computer and Information Science was held in London at the Royal Society, on 26–28 September 2011.

The symposium included presentations of the papers in these proceedings which were selected from some 140 submission through a refereeing process carried out by both the programme committee members and additional referees. In addition there were several other invited papers and special sessions.

Special thanks go to both the programme committee members and to the referees, for the diligent efforts that they have made to select the best of the submitted papers, and then for their generous help to the authors so that they could improve the final versions through constructive suggestions and comments.

The topics included in these proceedings cover timely topics such as Web Computing and Search, Data Engineering, Green ICT which aims at saving energy in information processing and communication systems, Wireless Networks, Computer Networks, Discovery Science, Computer Vision and Image Processing, Machine Learning, Agent Based Systems, Security and Safety, Modelling and Simulation, Architecture and Systems, and Algorithms.

As computer science and engineering has matured, we see that the emphasis of much of the research in this conference turns towards the interaction between basic scientific problems and the important areas of application. In a way, this leads to a more exciting proceedings volume with broader appeal both for industry and for the research community.

Erol Gelenbe
Ricardo Lent
Georgia Sakellari

Contents

Part IX Security and Safety

Part X Modelling and Simulation

Part XI Architecture and Systems

Part XII Maching Learning

Search Computing: Addressing Complex Search on the Web

Stefano Ceri and Marco Brambilla

Abstract Web search is considered a playground for a few giants—such as Google, Yahoo! and Bing—that relegate the other players to market niches. However, Web search is far from satisfying all information needs, because many search queries are complex, require information integration, and go beyond what can be offered by a single Web page; on these queries, generalized search engines do not perform well enough. This paper addresses a new paradigm for search-driven data integration, called Search Computing, based on combining data extraction from distinct sources and data integration by means of specialized integration engines. The Search Computing project has the ambitious goal of lowering the technological barrier required for building complex search applications, thereby enabling the development of many new applications which will cover relevant search needs.

Keywords Search computing · Software engineering · Search engine · Conceptual models · Ranking · Information retrieval

1 Introduction

Searching for information is perhaps the most important application of today's computing systems. In the new century, all the World's citizens have become accustomed to thinking of the Web as the source for answering their information

S. Ceri (✉) · M. Brambilla
Dipartimento di Elettronica ed Informazione,
Politecnico di Milano, V. Ponzio 34/5,
20133 Milan, Italy
e-mail: ceri@elet.polimi.it

M. Brambilla
e-mail: mbrambil@elet.polimi.it

E. Gelenbe et al. (eds.), *Computer and Information Sciences II*,
DOI: 10.1007/978-1-4471-2155-8_1,

needs, and search engines as their Web interface. If any Web page in the world stores the answer to our information need, then we expect the search engine to link that page and describe it through a snippet appearing in the first page of the search results. A few search engine companies are able to meet such expectations, and completely cover the search engine market.

However, offering a link to a Web page does not cover all information needs. The notion of "best page" for solving a given problem is typically inadequate when the query is complex and requires composing information spanning over multiple pages. Even simple problems, such as "which theatre offers a at least-three-stars action movie in London close to a good Italian restaurant", can only be solved by searching the Web multiple times, e.g. by extracting a list of the recent action movies filtered by ranking, then looking for movie theatres, then looking for Italian restaurants close to them. While doing so, the user is applying ranking while extracting movies, then Italian restaurants based on proximity with the theatres showing them; the best way for building the matching is to use the search engine itself, by entering movie's names as keywords, and then theatre's locations to search for restaurants, and looking at them upon city maps. While the search engine hints to useful information, the user's brain is the fundamental platform for information integration.

Complex queries are supported in certain domains, such as travels, hotel booking, and book purchasing, by specialized, domain-specific search integrators. It is instructive to assess how travel assistants solve the problem: they offer a few, predefined queries to build the itinerary, then offer additional services (e.g., car rentals, hotels, local events, insurance) so as to complete the plan around the itinerary. Thus, they perform specialized steps of integration and then let the user enrich the solution incrementally and interactively, with customized interfaces. Search integrators outperform general search engines in many domains, which in turn exhibit regular query patterns and are associated with substantial business.

The search computing project (SeCo), funded by the European Research Council as an advanced IDEAS grant, aims at building concepts, algorithms, tools, and technologies to support complex Web queries. The project, described in [1, 2], is now in the third of a five-year lifespan (Nov. 2008–2013); it proposes a new paradigm for solving complex queries based on combining data extraction from distinct sources and data integration by means of specialized integration engines. The search computing project has the ambitious goal of lowering the technological barrier required for building complex search applications, thereby enabling the development of many new applications which will cover relevant search needs.

Search computing covers many research directions, which are all required in order for providing an overall solution to complex search. The core of the project is the technology for search service integration, which requires both theoretical investigation and engineering of efficient technological solutions. The core theory concerns the development of result integration methods that not only denote "top-k optimality", but also the need of dealing with proximity, approximation, and uncertainty. Such a theory is supported by an open, extensible and scalable architecture for computing queries over data services, designed so as to incorporate the project's results by adding new operations, by encoding new join methods, and by injecting new features dealing with incremental evaluation and adaptivity.

A number of further research dimensions complement such core. Formulation of a complex query and browsing over solutions is a complex cognitive task, whose intrinsic difficulty has to be lowered as much as possible so as to meet usability requirements. Therefore, we are investing a consistent effort in the development of user-friendly interfaces which are targeted at assisting users in expressing their needs and then browsing on results. Solving a complex problem requires supporting users in the interactive and incremental design of their queries, thereby assisting search as a long-term process for exploring the solution space; result differences can be better appreciated by visualizing results (e.g., through maps or timelines).

The project success depends as well on the ability of registering new sources and making them available for solving complex problems; therefore, we have designed abstractions, architectural solutions, and model-driven design tools for service registration and for application development, aiming at assisting service publishing, application design, and query execution tuning. While the current description of Web resources is very simple, so as to enable an equally simple description of Web interactions, we aim at linking the service description to ontological sources, so as to enable high-level expressive interfaces covering the gap from high-level interactions to query expression.

Figure 1 shows an overview of the Search Computing framework, constituted by several sub-frameworks. The *service mart framework* provides the scaffolding for wrapping and registering data sources in service marts. The *user framework* provides functionality and storage for registering users, with different roles and capabilities. The *query framework* supports the management and storage of queries as first class citizens: a query can be executed, saved, modified, and published for other users to see. The *service invocation framework* masks the technical issues involved in the interaction with the service mart, e.g., the Web service protocol and data caching issues.

The core of the framework aims at executing multi-domain queries. The *query manager* takes care of splitting the query into sub-queries (e.g., "Which are the at-least-three-stars action movies?"; "Which theatres show them?"; "Where are Italian restaurants nearby those theatres?") and bounding them to the respective relevant data sources registered in the service mart repository; starting from this mapping, the *query planner* produces an optimized query execution plan, which dictates the sequence of steps for executing the query. Finally, the execution engine actually executes the query plan, by submitting the service calls to designated services through the *service invocation framework*, building the query results by combining the outputs produced by service calls, computing the global ranking of query results, and producing the query result outputs in an order that reflects their global relevance (Fig. 1).

This paper is organized as follows: Sect. 2 describes service registration, Sect. 3 describes the query engine and the process of query planning and

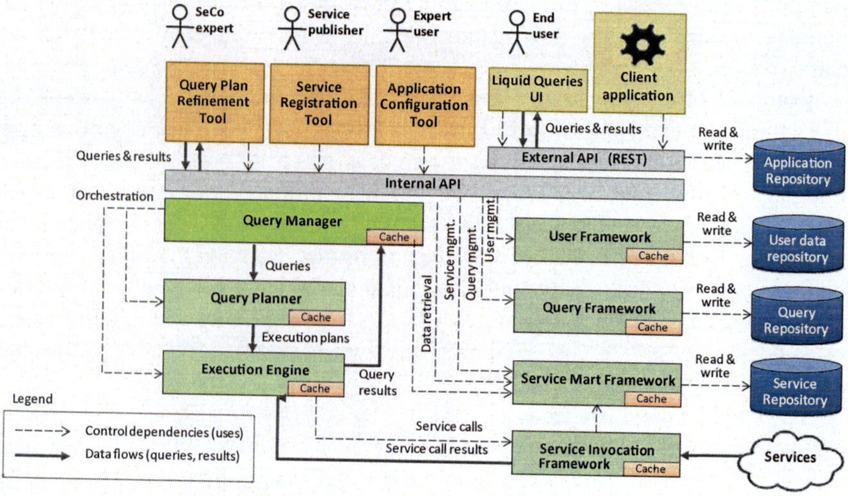

Fig. 1 Overview of the search computing framework

execution, Sect. 4 describes the liquid query interface, Sect. 5 describes the exploration and visualization activities, Sect. 6 describes the process of application development.

2 Service Registration

Service registration is an essential aspect of the Search computing project; the process is very critical because it must satisfy two conflicting requirements. On one side, services must be described with enough details about their interfaces and deployment so as to support their composition and invocation by means of fully automatic processes. On the other side, the actual mapping of services to real-world objects and facts must be exposed, so as to enable the construction of high-level user interfaces covering the semantic gap between user interaction and service selection.

The service model used for registration must describe not only the object or fact exposed by a service, but also the logic that a specific service performs while accessing an object, so that an interpretation system can select the specific service which best matches the user's requirements expressed in an informal or semi-formal way. Moreover, the model must support processes that aim at recognizing, at service registration time, when services describe the "same" objects or properties through "different" notations (e.g., names or types), so as to support matching processes. The scope of service registration in SeCo is therefore quite broad, as it must cover aspects ranging from performance indicators up to the semantic description of services and of their parameters.

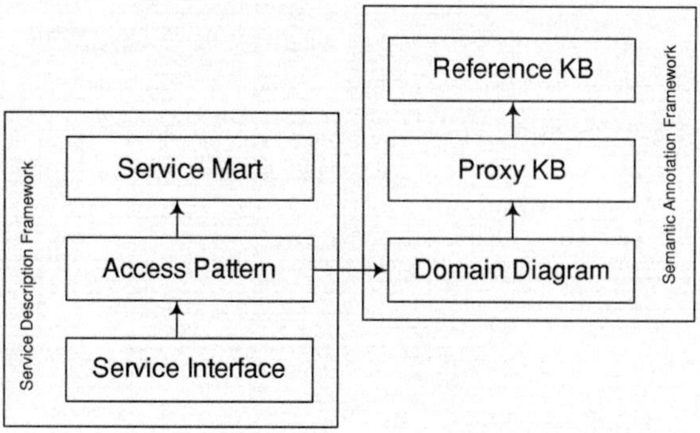

Fig. 2 Overview of the service description and annotation frameworks

An overview of the approach is described in Fig. 2: during registration, each data service becomes known in terms of: the concepts that it describes (conceptual view), its access pattern, i.e. the input–output parameters (logical view), and the actual supported interaction protocol, with a variety of quality parameters (physical view); these three views describe the *Service Description Framework (SDF)* at different levels of abstractions.

In addition, information about services being used constitutes a *Service Annotation Framework (SAF),* which helps the query system in understanding the query and selecting the appropriate services. Access pattern information from all the services is used to create the Domain Diagram (DD), a simple ER model of the concepts which are used by services; these may reference the Knowledge Base Proxy (KBP), which in turn refers to one or more Knowledge Bases (KB). In SeCo, we use general-purpose knowledge bases, such as Yago, or domain-specific knowledge bases, such as the Gene Ontology. The Proxy represents a subset of the knowledge base summarizing the concepts that are most useful for query analysis.

The *Service Description Framework* adopted so far in SeCo for service description [3] uses a multi-level modeling approach, consisting of conceptual, logical, and physical layers. The *conceptual level* is a very simple model which characterizes real world entities, called *Service Marts (SM)*, structurally defined by means of attributes, and their relationships. The *logical level* describes the access to the conceptual entities in terms of data retrieval patterns (called *Access Patterns, AP*) described by input and output attributes. Finally, the *physical level* represents the mappings of these patterns to concrete Web *Service Interfaces (SI)*, which incorporate the details about the endpoint and the protocol to be used for the invocation, together with some basic statistics on the service behavior that can be used for granting some given levels of quality of service (QoS). Figure 3 provides an example of presents the concept Movie, registered as a service mart, together with two associated access patterns and service interfaces, with the respective attribute mappings (notice that mappings are shown only for the first access pattern for clarity).

A three-layered architecture fulfills the following needs: (1) abstracting the conceptual properties of objects from the large amount of services that access

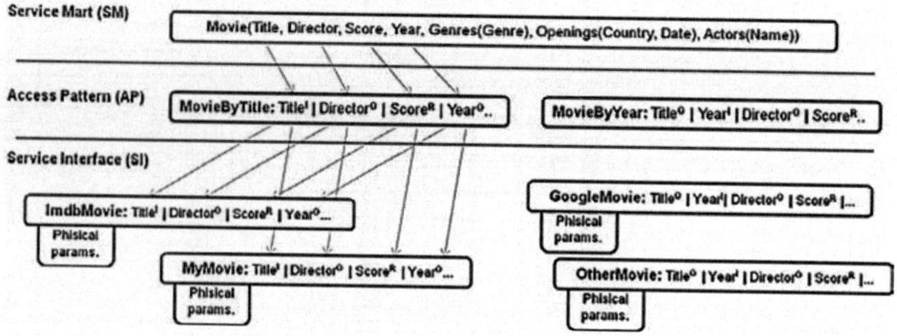

Fig. 3 Multi-level description of Web services about "movies"

them; (2) applying the separation of concerns principle to the service description task, by granting the independence of concept definitions, access methods, and concrete service descriptions.

3 Query Execution

The execution of multi-domain queries upon search services is assigned to the query process, which deals with the problem of scheduling service calls, taking into account their invocation constraints, pursuing some optimization objectives at compile-time, and then dealing at run-time with their actual, possibly unexpected, behaviour during the execution [4, 5]. Figure 4 summarizes the architecture of the query execution component and shows the inputs, outputs, and interplay between its components.

The system processes multi-domain queries. Independently of the higher-level languages and representations in which queries are formulated at the user interface level, any query is, from the engine's perspective, the specification of a collection of services to be invoked and a set of conjunctive conditions over their results. SeCoQL, the declarative textual language chosen to represent abstract queries, serves well as matching point between different components, is easily generated by the UI modules and easily parsed by the underlying modules, and is compact and readable enough to be convenient also for expert users and developers. SeCoQL has a declarative SQL-like syntax, in which the query result is defined as the concatenation of the tuples qualifying from the services listed in the FROM clause by means of the evaluation of the predicates listed in the WHERE clause and projected according to the list of attributes of the SELECT clause.

A compile-time analysis of the SeCoQL query performs a cost-driven optimization of the scheduling of service invocations, producing logical plans. The input of this stage is a SeCoQL query in which the Service Interfaces to be invoked have already been chosen by higher layers, and the join types (pipe vs parallel) are therefore already fixed. The planner exploits the remaining available degrees

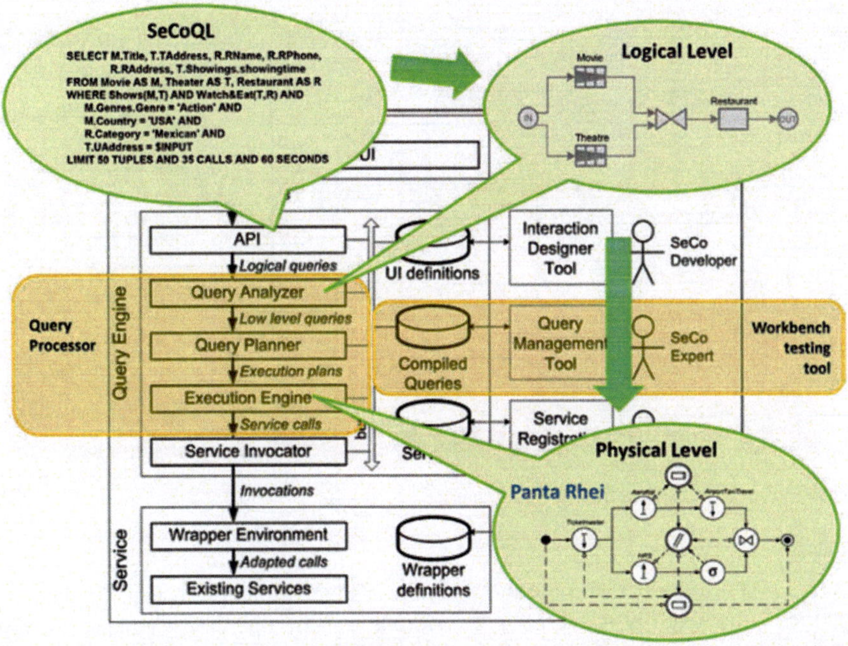

Fig. 4 Overview of the query execution flow

of freedom to decide the topology, the number and sequence of service invocations, and the join strategies. The output is a *logical plan,* i.e., a specification of a workflow with quantitative estimates of the size of partial results and of the number of invocations to be performed on each service in order to produce these results.

Logical plans are then translated into *physical plans* that are directly executable by the query engine. These plans are expressed in Panta Rhei, a unit-based language with support for parallelism, stateless and stateful execution, and backwards and forward control. Panta Rhei was designed to bridge the gap from the compile-time analysis performed by the query planner at the logical level to the run-time enactment of the query. It was designed with the objective of providing a clear specification of the engine behaviour and also enabling runtime adaptivity in the form of reactions to events that do not match the expectations of the user or the assumptions made by the system at compile-time. Distribution, parallelization, and replication issues have also been considered.

The query engine is implemented as an interpreter of Panta Rhei plans. The execution of a query is based on the simple assumption that any query consists of either (a) a simple invocation of a service interface, or (b) the combination of the results of two subqueries. In the former case, the engine supports the invocation of service interfaces that wrap data sources of many different kinds. As for the latter case, the results of the sub-queries can only be joined in series (pipe join) or in

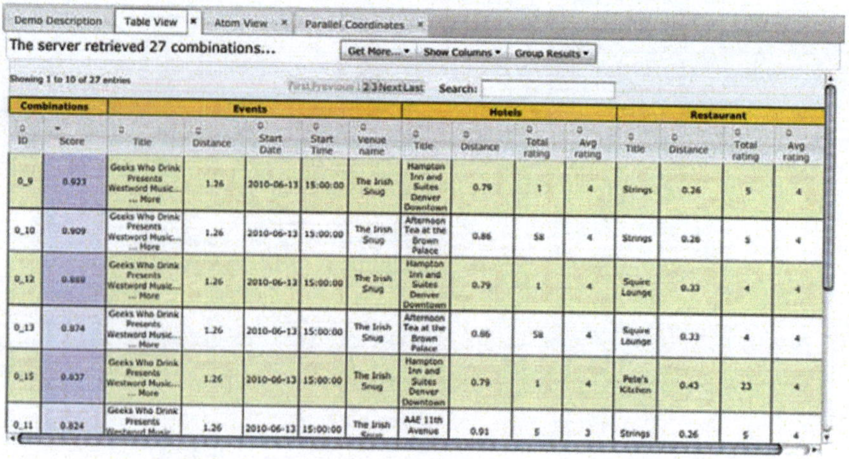

Fig. 5 Tabular representation of liquid queries

parallel (parallel join), and the interleaving of invocations performed on the subqueries is handled by interpreting the signals sent by operators dedicated to this task, called strategy units. Also, the execution of a query can be monitored in all its stages by means of a workbench tool.

4 Liquid Query Paradigm

The Liquid Query interface is used to present composite answers produced by joins over results of registered search services [6]. Google Squared and Google Fusion Tables are the closest approaches to Liquid Query, but the emphasis on search service federation gives to our tabular structures an additional dimension which emphasizes the source of results and provides a more powerful representation.

The *tabular format,* shown in Fig. 5, presents the result set in terms of projected attributes (i.e., shown columns), ordering attributes, clustering attributes, grouping attribute, and expansions. A liquid result instance is shown as a row in the tabular representation of the liquid result schema. A liquid result page is a set of liquid result instances that are shown altogether in the user interface. Notice that users can dynamically change the order within the list of selected services and within the lists of their projection attributes and ordering attributes. Figure 5 depicts the Liquid Result page for the running example.

When browsing the result set, the end user is expected to perform a set of interaction primitives that refine or change the shape of the query, the content of the result set, and the visualization format of the results. Such primitives may need to access the server-side (Remote Query Interaction Primitives), or could impact

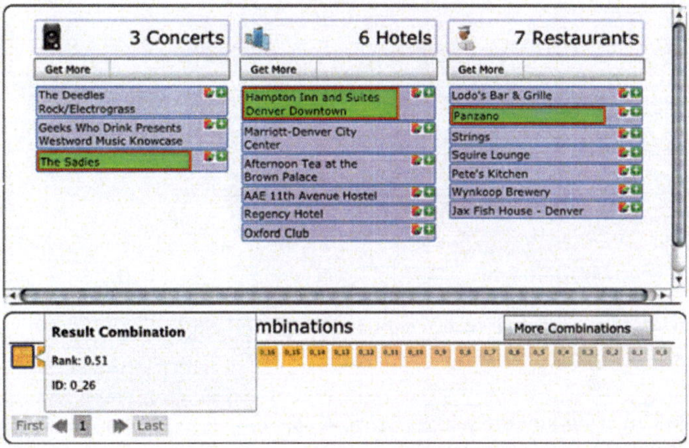

Fig. 6 Atom view of liquid queries

only the client-side version of the extracted result set (Local Resultset Interaction Primitives). The exploration primitives include:

- *Expand* (*target service, selected tuples*) The operation expands the result schema by adding one new target service and joining it with selected tuples. The expansion causes a set of exact queries to the expanded service interface, on the values selected by the user. If the expansion requires additional inputs, a dialog box could be shown to the user for submitting the needed parameters. As an example, an expand operation could get other events scheduled in the same days as some specific events selected from the initial result set. The interface will ask the user to provide the missing parameters (e.g., the event type) and will produce a new service result, with the new events matching the selected one.
- *MoreAll* The operation loads additional tuples from all the selected services in the currently specified query (excluding extensions); this command is typically executed as reaction to a user interaction asking for more information about the information need as a whole.
- *MoreOne* (*service*) The operation asks for additional tuples from a specific service interface in the currently defined query. This command is typically executed as reaction to a user interaction asking more information on a specific domain (e.g., more movies or restaurants), leaving the others unchanged.

To improve over the tabular representation, we implemented a result visualization called *atom view*, which forms the basis for the exploration of the resource graph and the iterative expansion of the query. Instead of showing the combinations as table rows, the atom view adopts the visualization strategy shown in Fig. 6: items are shown in separate lists, one for each entity involved in the query: in the example of Fig. 6 (*central part*), the set of hotels is shown side by side with the list of events. The lists are completely independent and show the top-k items retrieved by the query for each entity. In the atom view the unified global ranking

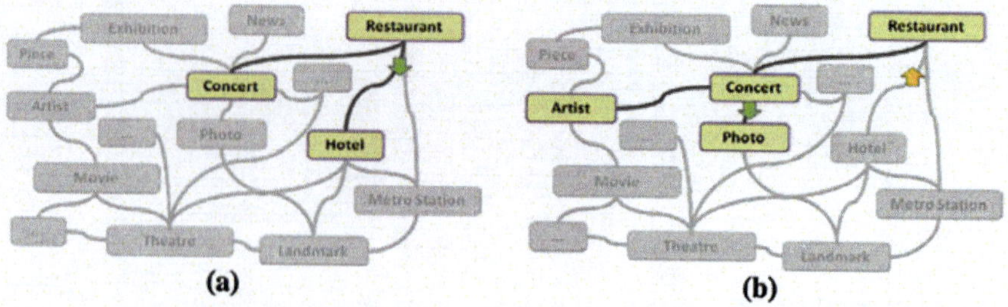

Fig. 7 Moving "forward" and "backward" on the SRF

is displayed in a dedicated widget (Fig. 6, *bottom part*), where the joined combinations of objects are represented (and ordered) by their global ranking score: relative importance of combinations is represented by the ordering of items, while absolute value of the ranking score is shown through appropriate visual clues (like colour gradient or object size). By moving the mouse over the list of combinations, the user can select one and automatically highlight the items in the various object lists that contribute to the combination; vice versa, he can select one element from a list and thus highlight the combinations comprising it. The advantage of this view is the good simultaneous visibility of the items extracted, of their combinations and of the global ranking.

5 Exploration and Visualization

The exploration activities is shown in Fig. 7 as a navigation of the Data Dictionary in the Service Annotation Framework [7]. Users express their queries directly upon the concepts that are known to the system, such as hotels, restaurants, or movie shows; moreover, users are aware of the connections between the concepts, and therefore they can, e.g., select a show and then relate it to several other concepts: the performing artist, the close-by restaurants, the transportation and parking facilities, other shows being played in the same night in town, and so on.

The exploratory query interaction paradigm proceeds as follows. The user starts by selecting one of the available objects, and submits an "object query" to extract a subset of object instances. For example, a user could choose a "*concert*" object, and ask for "jazz" concerts in a "club" in the "Village" area of "New York", or choose a "*restaurant*" object, and ask for "vegetarian" restaurants in the "south end" of "the city" in a given "price range". Object queries are conjunctions of selection predicates: each predicate is represented in the user interface as a widget for inputting the corresponding selection parameters, e.g. a drop-down list for choosing a value among the available restaurant types or city districts, or a slider bar for setting price ranges. Queries results are ranked and the user interface also lets the user specify his ranking preferences, e.g. "distance", "quality", "price range" or a combination of them.

At this point, the user can select the most relevant object instances and continue with the exploration by choosing the next concept among the various ones that can be connected to the selected objects; the system will then retrieve connected object instances and form a "combination" with the objects retrieved at the preceding steps. Result combinations at any given step of the search process are ranked according to the global ranking criterion. Continuing in the above example, after selecting three or four "jazz clubs", the user is offered to add to this plan several additional options: restaurants, exhibitions, movies, music shows, dancing places, and other amenities. If they select a "restaurant", the connection to "concert" maps concerts to the restaurant instances that are close to the concert instances, and then the connection to "restaurants" adds them to the solution. If users further select "metro", they are asked to indicate the starting point of their ride, and the search obtains for every selected object instance at the previous interaction the best metro station and ride, with the detail of trains, line changes, and expected time.

At any stage, users can "move forward" in the exploration, by adding a new object to the query, starting from the connections available in the data dictionary and from the objects that have been previously extracted. Forward path exploration can be applied to all previously extracted objects/combinations or to a subset only, manually "checked" by the user. Users can also "move backwards" (backtrack) in the exploration, by excluding one of the objects from the query, or by "unchecking" some of their previous manual selections of relevant object instances. For example, a user may decide that the bus ride is too inconvenient, prefer to use a car instead, and then explore parking opportunities for the selected restaurants. Figure 7a, b show how nodes can be added to and removed from a given query.

Visualization of results can be optimized according to the type of their attributes. Types are classified as:

- *Interval* Quantitative attributes measured relative to an arbitrary interval (e.g., Celsius degrees, latitude and longitude, date, GPA). In this class, two important subclasses are further distinguished for visualization purposes:
  - *Geographic points and addresses* They admit domain specific operations like the computation of distance, the visualization on maps, the determination of routes, etc.
  - *Time points* They admit the representation on time scales and calendars, at different domain-specific granularity.
- *Ratio* Quantitative attributes measured as the ratio with a known magnitude unit (e.g., most physical properties).
- *Nominal* Categorical labels without notion of ordering (e.g., music genre). They can be visualized by means of textual labels (for example within tables). In case of a low number of categories, they can be represented through visual clues, for example different shapes or colors.

- *Ordinal* Data values that admit order, but not size comparison (e.g., quality levels).

The process of generating the visualization aims at producing a representation that maximizes the understandability of the result set, by considering the type and semantics of data and the functions of multi-domain search [8]. The visualization maps the result set onto a presentation space by considering the types of attributes that describe the properties of combinations and the visualization families that are best usable for rendering the data dimensions.

In order to identify the primary visualization dimension, one can assume that an ordering of data type exists and that it guides the selection of the primary attributes determining the visualization space. Such ordering depends on the capacity of the data types to "delimit" a visualization space where single objects and combinations included in the result set can be conveniently positioned. For example, if a given object O has a geo-referenced attribute, then its instances can be represented on a map by using that attribute as a primary visualization; similarly, objects with attributes representing temporal events can be placed on a timeline. Representing a combination then amounts to finding suitable representations for the majority of its objects, by highlighting them upon a given visualization space, and then relating together the object instances of a combination through orthogonal visual mechanisms. Once placed on the visualization space, an object is succinctly represented by some of its attributes (e.g., identifiers); typically the local ranking of the object can also be visually represented (e.g., through conventional shapes, or colors). Other attributes are omitted from the visualization, and can be accessed through secondary visualization methods, such as pop-up windows.

For example, the results of queries about geo-localized can use maps. When the user searches for upcoming events (concert, sport matches, movie shows, and so on) close to a specified location, considering also availability of close-by hotels, he may be offered a map as primary visualization, together with visual clues that indicate the object's rankings. The user controls the progression of search, can backtrack, and can turn to totally different solutions, as if he were operating on the Liquid Query framework (Fig. 8).

6 Search Computing Tools

Building and configuring search computing applications may become a non-trivial task for a designer. To ease his work, the proper set of development, configuration and monitoring tools are needed. The tools should support all the development phases necessary to support all the aspects of a multi-domain search application, according to sensible development process guidelines [8]. The development process is structured into service registration and application configuration; the latter, in turn is composed by query specification and user interface definition. *Service registration* comprises several substeps, corresponding to the different abstraction

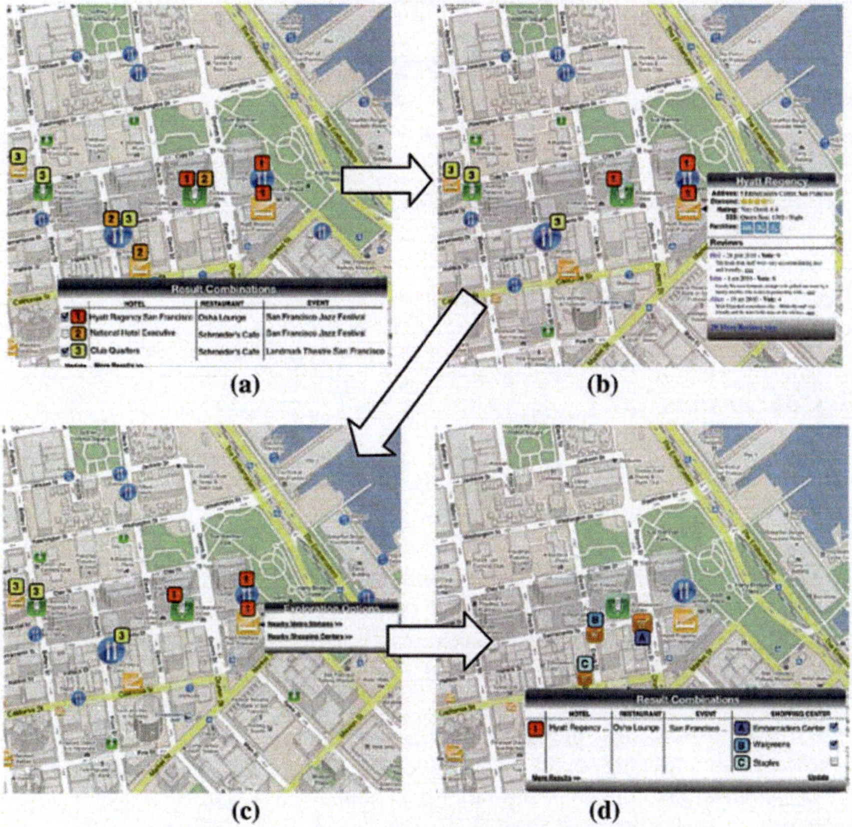

Fig. 8 Result Exploration in the entertainment planning scenario. **a** Visualization and selection of ranked combinations **b** display of details of a specific object (hotel Hyatt Regency), including non geo-referenced objects, such as reviews **c** menu of exploration options relevant for the selected hotel **d** exploration additional geo-referenced objects, such as shopping centers (now included in the combinations)

levels that describe the resources: Service Mart (SM), Access Pattern (AP), and Service Interface (SI); as well as their annotations in the Data Dictionary (DD) and Knowledge Proxy (KP). *Application configuration* consists in configuring a search computing application for a vertical domain; in particular, this task includes *query specification,* which is performed with a mashup-based approach [9], illustrated in Fig. 9, and *user interface configuration*, i.e. choosing the interfaces of query submission forms and the display style of the result set, including the choices on the initial display of the result set (e.g., sorting, grouping, and clustering attributes or the size of the result list) and on specific visual representations (e.g., maps or timelines).

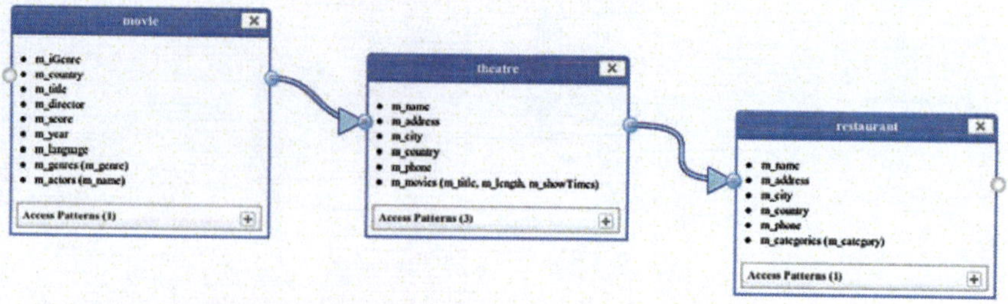

Fig. 9 Tool user interface for query configuration

7 Conclusions

While currently search in the Web is dominated by giant companies and monolithic systems, we believe that a new generation of search computing systems will soon be developed, with a much more composite software organization, addressing the needs of complex queries over a fragmented market. Generic search systems are already dominated by domain-specific vertical search systems, e.g. with travels and electronic bookstores. When the threshold complexity of building such verticals will be lowered, a variety of new market sectors will become more profitable.

We also expect that the future will support easy-to-use software systems interfaces for combining sources; application developers will act as brokers of new search applications built by assembling data sources, exposed as search services. We envision a growing market of specialized, localized, and sophisticated search applications, addressing the *long tail of search needs* (e.g., the "gourmet suggestions" about slow-food offers in given geographic regions). In this vision, large communities of service providers and brokers (e.g., like *ProgrammableWeb.com* and *Mashape.com* for mashups) could be empowered by support design environment and tools for executing search service compositions and orchestrations.

Thanks to the lowering of programming barriers one could expect an ultimate user's empowerment, whereby end users could compose data sources at will from predefined data source registries and collections; e.g., the interaction could be built progressively, by starting from simple, menu-driven interfaces where the user is just asked to select the concepts and the orchestration is then inferred. Demos providing some evidence of the feasibility of this approach were recently presented at WWW [10] at ACM-Sigmod [11]. We also envision that, with suitable ontological support and possibly within a narrow domain, queries could be generated from keywords, as with conventional search engines.

Acknowledgments This research is part of the Search Computing (SeCo) project, funded by the European Research Council (ERC), under the 2008 Call for "IDEAS Advanced Grants", a program dedicated to the support of frontier research. The project lasts from November 2008 to October 2013.

References

1. Ceri, S., Brambilla, M. (eds.): Search Computing Challenges and Directions, vol. 5950. Springer, Heidelberg (2010)
2. Ceri, S., Brambilla, M. (eds.): Search Computing Trends and Developments, vol. 6585. Springer, Heidelberg (2011)
3. Campi, A., Ceri, S., Maesani, A., Ronchi, S.: Designing Service Marts for Engineering Search Computing Applications. International Web Engineering (ICWE) Conference, Vienna, Austria (2010)
4. Braga, D., Corcoglioniti F., Grossniklaus, M., Vadacca, F.: Efficient computation of search computing queries, in [2]
5. Braga, D., Grossniklaus, M., Paton, N.W.: Run-time adaptivity for search computing, in [2]
6. Bozzon, A., Brambilla, M., Ceri, S., Fraternali, P.: Liquid query: multi-domain exploratory search on the web. WWW 2010, Raleigh, USA, ACM, April 2010
7. Bozzon, A., Brambilla, M., Ceri, S., Fraternali, P.: Information exploration in search computing, in [2]
8. Bozzon, A., Brambilla, M., Ceri, S., Fraternali, P., Matera M., Catarci, T.: Visualization of multi-domain ranked data, in [2]
9. Braga, D., Ceri, S., Daniel, F., Martinenghi, D.: Mashing up search services. IEEE. Internet. Comput. **12**(5), 16–23 (2008)
10. Brambilla M., Bozzon A., Ceri S., Fraternali P., Vadacca S.: Exploratory search in multi-domain information spaces with Liquid Query, WWW Conference, Demo session, 31 Mar 2011
11. Brambilla M., Bozzon A., Ceri S., Corcoglioniti F., Fraternali P., Vadacca S.: Search Computing: Multi-domain search on ranked data. ACM-Sigmod Conference, Demo session, June 2011

Part I
Web Computing and Search

Hypergraph-Theoretic Partitioning Models for Parallel Web Crawling

Ata Turk, B. Barla Cambazoglu and Cevdet Aykanat

Abstract Parallel web crawling is an important technique employed by large-scale search engines for content acquisition. A commonly used inter-processor coordination scheme in parallel crawling systems is the link exchange scheme, where discovered links are communicated between processors. This scheme can attain the coverage and quality level of a serial crawler while avoiding redundant crawling of pages by different processors. The main problem in the exchange scheme is the high inter-processor communication overhead. In this work, we propose a hypergraph model that reduces the communication overhead associated with link exchange operations in parallel web crawling systems by intelligent assignment of sites to processors. Our hypergraph model can correctly capture and minimize the number of network messages exchanged between crawlers. We evaluate the performance of our models on four benchmark datasets. Compared to the traditional hash-based assignment approach, significant performance improvements are observed in reducing the inter-processor communication overhead.

1 Introduction

In order to maintain the accuracy of the information presented to their users, search engines build very large document repositories, trying to cover the entire Web while maintaining the freshness of constructed repositories. Web crawlers are among the most important software employed in forming these repositories. A web crawler is mainly responsible for locating, fetching, and storing web pages by following the link structure between them.

A. Turk · C. Aykanat (✉)
Computer Engineering Department, Bilkent University, Ankara, Turkey
e-mail: aykanat@cs.bilkent.edu.tr

B. B. Cambazoglu
Yahoo! Research, Barcelona, Spain

E. Gelenbe et al. (eds.), *Computer and Information Sciences II*,
DOI: 10.1007/978-1-4471-2155-8_2,

There are many challenges in sequential web crawling [1]. Efficient web crawling requires the use of distributed systems having high network bandwidth, large processing power, memory, and disk capacities [2, 3]. Consequently, in state-of-the-art search engines, the web crawling problem is addressed by parallel crawlers running on distributed memory parallel architectures, where each processor hosts a separate crawler. In [4], a taxonomy of parallel crawlers is provided, based on the coordination among processors. As explained in [4], the exchange scheme is one of the best inter-processor coordination schemes for parallel crawling. In the exchange scheme, the retrieval task for a link is assigned to the processor that has the responsibility of storing the page pointed by the link.

The hash-based task assignment scheme [4–7] has been widely used in parallel web crawling systems that utilize the exchange coordination scheme. In [7], the hashes are computed over the whole URLs to assign pages to crawlers, whereas, in [4–6], the hashes are computed over the host component of URLs. Compared to the page-hash-based scheme, the site-hash-based scheme has the advantage of reducing the number of inter-processor links since pages within the same site are more likely to link each other.

The above-mentioned page-to-processor assignment schemes implicitly address the load balancing problem, but they do not capture the cost of inter-processor communication. In [8], a greedy constructive algorithm is proposed to integrate the cost of communication into the execution cost of a crawling task on a processor in a heterogeneous system. In [9], we proposed a graph-partitioning-based (GP-based) model for page-to-processor assignment. In this model, the partitioning constraint addresses the load balancing problem while the partitioning objective of minimizing the cutsize defined over the edges that span more than one parts corresponds to minimizing the total volume of inter-processor communication. In parallel crawling, the messages exchanged are relatively small in size (around 50 bytes), but many. Hence, the latency overhead, which can be expressed in terms of the number of messages exchanged between processors, is a better indicator of the communication overhead. In [10], a GP-based page-to-processor assignment scheme is proposed to improve the partitioning strategy of a geographically distributed crawler by utilizing geographical information (server location and geographic scope of web pages).

The state-of-the-art parallel crawling systems generally have a distributed architecture, and crawling agents are connected through a wide area network. Depending on the characteristics of the underlying communication network, the message latency overhead in link exchanges between processors may become a bottleneck. We try to minimize this latency overhead (number of network messages) in the exchange scheme via a hypergraph-partitioning-based assignment model that minimizes the total message count while enforcing two load balancing constraints. The first constraint enforces a balance on the retrieval and storage task loads of processors while the second constraint enforces a balance on the number of issued page download requests.

2 Preliminaries

We represent the Web as a two-tuple $(\mathcal{P},\mathcal{L})$, where $\mathcal{P}$ and $\mathcal{L}$ indicate the set of pages and the set of links between pages, respectively. Without loss of generality, multiple links from a page p_i to page p_j in $\mathcal{P}$ are assumed to be coalesced into a single link ℓ_{ij} in $\mathcal{L}$. For efficient crawling, our models utilize the web structure obtained in the previous crawling session to provide a better page-to-processor mapping for the following crawling session.

A hypergraph $\mathcal{H}=(\mathcal{V},\mathcal{N})$ is defined as a set of vertices $\mathcal{V}$ and a set of nets (hyperedges) $\mathcal{N}$, where each net connects a number of distinct vertices. The vertices connected by a net n_i are said to be its pins and denoted as $Pins(n_i)$. A cost $c(n_j)$ is assigned as the cost of a net $n_j \in \mathcal{N}$. Multiple weights $w^1(v_i), w^2(v_i), \ldots, w^R(v_i)$ may be associated with a vertex $v_i \in \mathcal{V}$.

$\Pi=\{\mathcal{V}_1,\mathcal{V}_2,\ldots,\mathcal{V}_K\}$ is said to be a K-way partition of a given hypergraph $\mathcal{H}$ if vertex parts (i.e., every $\mathcal{V}_k$, for $k\in\{1,\ldots,K\}$) are mutually disjoint and their union is exhaustive. The K-way hypergraph partitioning (HP) problem can be defined as finding a K-way vertex partition Π that optimizes a partitioning objective defined over the nets that connect more than one part while satisfying a given partitioning constraint. The partitioning constraint is to satisfy multiple balance criteria on part weights, i.e.,

$$W^r(\mathcal{V}_k) \le W^r_{\text{avg}}(1+\epsilon^r), \text{for } k=1 \text{ to } K, r=1 \text{ to } R. \tag{1}$$

For the rth constraint, weight $W^r(\mathcal{V}_k)$ of a part $\mathcal{V}_k$ is defined as the sum of the weights $w^r(v_i)$ of vertices in $\mathcal{V}_k$, W^r_{avg} is the weight that each part should have in the case of perfect balancing, and ϵ^r is the maximum imbalance ratio allowed.

In a partition Π of hypergraph $\mathcal{H}$, a net is said to connect a part if it has at least one pin in that part. The connectivity set $\Lambda(n_j)$ of a net n_j is the set of parts connected by n_j. The connectivity $\lambda(n_j)=|\Lambda(n_j)|$ of a net n_j is the number of parts connected by n_j. A net n_j is said to be cut if it connects more than one part (i.e., $\lambda(n_j)>1$) and uncut otherwise. In our particular problem, the partitioning objective is to minimize the connectivity-1 metric

$$Cutsize(\Pi) = \sum_{n_j \in \mathcal{N}_{\text{cut}}} c(n_j)(\lambda(n_j)-1), \tag{2}$$

defined over the set $\mathcal{N}_{\text{cut}}$ of cut nets. The HP problem is known to be NP-hard [11, 12]. However, there are successful HP tools (e.g., hMeTiS [13] and PaToH [14]).

3 Site-Based Partitioning Model

In our site-based HP model, we represent the link structure $(\mathcal{P},\mathcal{L})$ by a site hypergraph $\tilde{\mathcal{H}}=(\tilde{\mathcal{V}},\tilde{\mathcal{N}})$. We assume that a set $\mathcal{S}=\{S_1,S_2,\ldots\}$ of sites is given,

Table 1 Properties of the datasets

Dataset	Number of Pages	Sites	Links	% of intra-site links	Site-based hypergraph Number of Vertices	Nets	Pins
google[a]	913,569	15,819	4,480,218	87.42	15,819	86,552	277,805
in-2004 [15, 16]	1,382,908	4,380	16,917,053	95.92	4,380	205,106	555,195
de-fr [17][b]	8,212,782	38,741	39,330,882	75.06	38,741	2,091,986	4,968,036
indochina [15, 17]	7,414,866	18,984	194,109,311	92.80	18,984	250,931	799,783

[a] Google contest, available at http://www.google.com/programming-contest.
[b] Crawled with Larbin: multi-purpose web crawler, available at http://larbin.sourceforge.net.

where sites are mutually disjoint and exhaustive page subsets of $\mathcal{P}$. All pages belonging to a site S_i are represented by a single vertex $v_i \in \tilde{\mathcal{V}}$. The weights $w^1(v_i)$ and $w^2(v_i)$ of each vertex v_i are set equal to the total size of pages (in bytes) and the number of pages hosted by site S_i, respectively.

A K-way partition is applied on $\tilde{\mathcal{H}}$ for a parallel system that contains K crawlers/processors. In a K-way partition $\tilde{\Pi} = (\tilde{\mathcal{V}}_1, \tilde{\mathcal{V}}_2, \ldots, \tilde{\mathcal{V}}_K)$ of $\tilde{\mathcal{H}}$, each vertex part $\tilde{\mathcal{V}}_k$ corresponds to a subset $\mathcal{S}_k$ of the set $\mathcal{S}$ of sites, where pages of each site $S_i \in \mathcal{S}_k$ are to be retrieved and stored by processor P_k.

In the site-based hypergraph $\tilde{\mathcal{H}}$, vertices are in site granularity, whereas nets are in page granularity. For each page p_j having at least one outgoing inter-site link, there exists a net n_j with cost $c(n_j) = 1$. Vertex v_k is a pin of net n_j if and only if site S_k contains page p_j or a page p_i pointed by link ℓ_{ji}. That is,

$$Pins(n_j) = \{v_k : p_j \in S_k\} \cup \{v_k : \ell_{ji} \in \mathcal{L} \wedge p_i \in S_k\}. \tag{3}$$

Pages of a site that does not have outgoing links to page(s) of any other site do not incur nets in $\tilde{\mathcal{H}}$.

4 Experimental Results

To validate the applicability of the proposed model, we run simulations over a set of precrawled page collections. These collections are converted into hypergraphs as described in Sect. 3. Properties of the test datasets and respective site-based hypergraphs are displayed in Table 1. Since the page size information is not available for the in-2004, indochina, and de-fr datasets, unit page sizes are assumed for vertex weighting in hypergraphs corresponding to these datasets. As seen in the table, in all datasets, more than 75% of the links remain among the pages belonging to the same site.

The state-of-the-art HP tool PaToH [14] is used to partition constructed hypergraphs. The imbalance tolerance is set to 5% for both weight constraints. Due to the randomized nature of PaToH, experiments are repeated 8 times, and average values are reported.

Table 2 Load imbalance values and message counts

		Load imbalance (%)		Message count (10^3)	
Dataset	(K)	Hash-based	HP-based	Hash-based	HP-based
	4	14.57	4.99	24.0	8.4
	8	20.79	4.99	40.0	9.3
google	16	22.34	4.99	56.4	10.9
	32	70.18	4.99	71.7	12.3
	64	131.46	4.79	84.7	13.7
	4	11.69	4.84	5.3	1.4
	8	25.00	4.29	8.9	1.9
in-2004	16	40.58	8.20	12.8	2.5
	32	73.47	9.54	16.6	3.1
	64	142.86	10.80	19.9	3.6
	4	12.75	5.00	52.6	9.5
	8	34.51	5.00	86.8	10.9
de-fr	16	58.62	5.00	128.8	10.7
	32	81.95	10.33	177.8	12.4
	64	203.39	91.61	231.9	14.1
	4	3.71	5.00	18.4	4.1
	8	9.91	5.00	33.7	5.2
indochina	16	21.77	5.00	54.5	6.4
	32	41.97	5.00	80.0	7.6
	64	51.91	8.62	109.8	8.1

Table 2 displays the performance of the proposed site-based HP model against the site-hash-based model in load balancing and reducing the number of messages. In terms of load balance, only computational load imbalance values related to the page download request counts of processors are reported. Storage imbalance values are not reported since actual storage requirements are not used due to their unavailability in three out of four datasets. The communication overhead due to the link exchange operation is reported in terms of the number of messages exchanged between crawlers. We use K values of 4, 8, 16, 32, and 64. As seen in Table 2, the HP model performs significantly better than the site-hash-based model in balancing the page request loads of processors. We observe that, the load balancing performance of the site-hash-based model drastically deteriorates with increasing K values. This is basically due to the high variation in the sizes of the sites in the datasets used. This experimental observation shows the need for intelligent algorithms instead of hash-based algorithms even solely for load balancing purposes in site-based assignment, especially for large K values.

As also seen in Table 2, the HP model performs significantly better than the hash-based model in reducing the message latency overhead as well. More specifically, on average, the HP model produces partitions with 4.9, 4.9, 7.3, and 8.7 times fewer number of messages than the partitions produced by the site-hash-based model in google, in-2004, de-fr, and indochina datasets, respectively. In general, the performance ratio between the HP and hash-based models increases with increasing number of processors. For example, for the largest number of processors ($K = 64$), the site-based HP model produces partitions which incur 6.2, 5.5, 16.5, and 13.6 times fewer number of messages than the site-hash-based model in google, in-2004, de-fr, and indochina datasets, respectively.

These experimental findings confirm the need for intelligent algorithms such as HP models in order to maintain the message latency overhead at acceptable levels for large K values.

5 Conclusion

In this paper, we proposed a model for minimizing the communication overhead of parallel crawlers that work in the exchange mode. The model provides considerable improvement in terms of the number of network messages exchanged between the crawlers, relative to the hash-based assignment approach.

References

1. Lee, H.-T., Leonard, D., Wang, X., Loguinov, D.: IRLbot: scaling to 6 billion pages and beyond. In: Proceedings of the 17th International Conference on World Wide Web, pp. 427–436 (2008)
2. Baeza-Yates, R., Castillo, C., Junqueira, F., Plachouras, V., Silvestri, F.: Challenges in distributed information retrieval. In: International Conference on Data Engineering, pp. 6–20 (2007)
3. Cambazoglu, B.B., Plachouras, V., Junqueira, F., Telloli, L.: On the feasibility of geographically distributed web crawling. In: Proceedings of the 3rd International Conference on Scalable Information Systems, pp. 1–10 (2008)
4. Cho, J., Garcia-Molina, H.: Parallel crawlers. In: Proceedings of the 11th Int'l Conference on World Wide Web, pp. 124–135 (2002)
5. Edwards, J., McCurley, K., Tomlin, J.: An adaptive model for optimizing performance of an incremental web crawler. In: Proceedings of the 10th International Conference on World Wide Web, pp. 106–113 (2001)
6. Heydon, A., Najork, M.: Mercator: a scalable, extensible web crawler. World Wide Web **2**(4), 219–229 (1999)
7. Shkapenyuk, V., Suel, T.: Design and implementation of a high-performance distributed web crawler. In: Proceedings of the 18th International Conference on Data Engineering, pp. 357–368 (2002)
8. Teng, S.-H., Lu, Q., Eichstaedt, M., Ford, D., Lehman, T.: Collaborative web crawling: information gathering/processing over Internet. In: Proceedings of the 32nd Annual Hawaii International Conference on System Sciences (1999)
9. Cambazoglu B, B., Turk, A., Aykanat, C.: Data-parallel web crawling models. Lect. Notes. Comput. Sci. **3280**, 801–809 (2004)
10. Exposto, J., Macedo, J., Pina, A., Alves, A., Rufino, J.: Efficient partitioning strategies for distributed web crawling. Lect. Notes. Comput. Sci. **5200**, 544–553 (2008)
11. Berge, C.: Graphs and Hypergraphs. North-Holland Publishing Company, New York (1973)
12. Lengauer, T.: Combinatorial Algorithms for Integrated Circuit Layout. Wiley, UK (1990)

13. Karypis, G., Kumar, V.: Multilevel k-way hypergraph partitioning. In: Proceedings of the 36th annual ACM/IEEE Design Automation Conference, pp. 343–348 (1999)
14. Çatalyürek, U.V., Aykanat, C.: PaToH: a multilevel hypergraph partitioning tool, version 3.0. Technical report, Bilkent University. Department of Computer Engineering (1999)
15. Boldi, P., Codenotti, B., Santini, M., Vigna, S.: UbiCrawler: a scalable fully distributed web crawler. Softw. Pract. Experience **34**(8), 711–726 (2004)
16. Boldi, P., Vigna, S.: The WebGraph framework I: compression techniques. In: Proceedings of the 13th International Conference on World Wide Web, pp. 595–602 (2004)
17. Jean-Loup, G., Latapy, M., Viennot, L.: Efficient and simple encodings for the web graph. In: Proceedings of the 3rd International Conference on Advances in Web-Age Information Management, pp. 328–337 (2002)

Web Service Discovery: A Service Oriented, Peer-to-Peer Approach with Categorization

Mustafa Onur Özorhan and Nihan Kesim Cicekli

Abstract This paper discusses automated methods to achieve web service advertisement and discovery, and presents efficient search and matching techniques based on OWL-S. The service discovery and matchmaking is performed via a centralized peer-to-peer web service repository. The repository has the ability to run on a software cloud, which improves the availability and scalability of the service discovery. An OWL-S based unified ontology—Suggested Upper Merged Ontology—is used in web service annotation. User-agents generate query specification using the system ontology, to provide semantic unification between the client and the system during service discovery. Query matching is performed via complex Hilbert Spaces composed of conceptual planes and categorical similarities for each web service.

Keywords Peer-to-peer web service discovery · Categorization · Cloud · Ranking

1 Introduction

Service oriented technologies revolutionized the modern world of computing with the paradigm shift they brought. However, to be able to use a Web Service, a service user must have access to the descriptor of the service. Then again, the service user should be able to access the Web Service descriptions of multiple

M. O. Özorhan
Central Bank of Turkey, Ankara, Turkey
e-mail: onur.ozorhan@tcmb.gov.tr

N. K. Cicekli (✉)
Department of Computer Engineering, Middle East Technical University, Ankara, Turkey
e-mail: nihan@ceng.metu.edu.tr

E. Gelenbe et al. (eds.), *Computer and Information Sciences II*,
DOI: 10.1007/978-1-4471-2155-8_3,

Web Services from a preferably single, but pre-defined location. In general, pre-defined locations containing service descriptions are called service repositories. There are mainly three types of service repositories, (i) centralized service registries, (ii) peer-to-peer service registries and (iii) service search engines [1]. Web services are traditionally described with the use of the WSDL [2], which unfortunately cannot adequately represent their actual semantics. Most prominent ontology—based approaches in Web services description are OWL-S [3], WSMO and SAWSDL [4]. Using ontological concepts for service description in OWL-S files, the meaning of the functionality of a service can be passed to a software-agent, eliminating human intervention in many processes such as service discovery and service composition. For a software agent to be able to properly execute a web service, first a selection from a set of web services should be made. This process is called *Web Service Discovery*. The main problems faced in Web Service Discovery are the availability of the Service Repositories, availability of the Web Services and semantically deficient Web Services.

This paper solves these problems by a semi-decentralized, peer-to-peer service repository, which runs on a cloud computing architecture on the service repository side. The provided architecture provides solutions to the service semanticity related problems with semi-automatic methods for WSDL to OWL-S conversion, and annotation via a unified ontology which merges a wide range of ontologies together in a single ontology, namely SUMO [5]. WSDL is selected as the source for service descriptor conversion, since it is the current standard for web service description, and is widely used in the industry. The contributions of this paper can be summarized as follows: 1) Web service similarity computation based on categories and concepts in Hilbert Spaces. 2) Automated semantic web service discovery, with web service ranking. 3) Web service semanticization via web service discovery queries.

The rest of the paper is organized as follows. Section 2 shows how we use Hilbert Spaces in our work. Section 3 outlines the system architecture, and describes the modules participating in our solution. Section 4 describes the methodology proposed in Semi Automated Service Publishing technique. In Sect. 5, the algorithms used in query processing and service ranking are described. Section 6 concludes the paper, and provides a pathway for the future work.

2 Hilbert Spaces

Hilbert Spaces are complex multi-dimensional spaces with applications in various disciplines including mathematics, physics and engineering. In Mathematics, Hilbert Spaces are used for Functional Analysis, which mostly deals with infinitely dimensional, topological vector spaces. In our system, Hilbert Spaces are used in a similar manner in order to represent web services and queries in infinite dimensional spaces with ontologies. Hilbert Spaces are selected in our system for three main reasons: 1) Hilbert Spaces are infinitely dimensional. 2) New dimensions can be added to a Hilbert Space in time, since data points are represented with infinite

Fig. 1 Three dimensional view of a concept

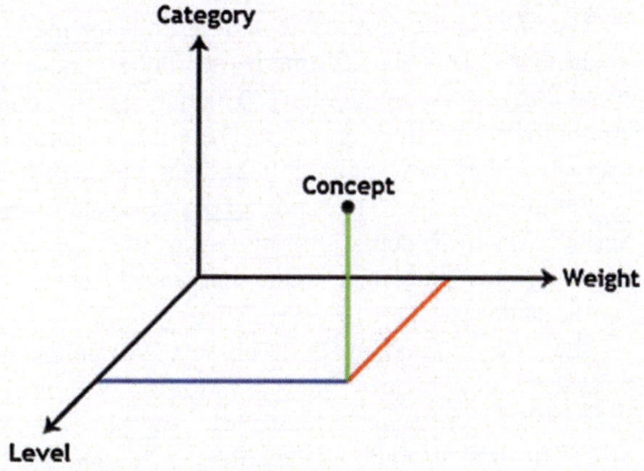

coordinates. 3) Hilbert Spaces support the Skew Coordinate system, with which we can mathematically model the relationships between categories.

Our system uses the unified ontology SUMO, which is created by merging ontologies of multiple categories. Our idea is to represent the concepts and categories of SUMO in a space. Firstly, the concepts in the ontology are clustered by their categories. Each category is represented with a dimension in the Hilbert Space. Each dimension carries concepts from their own category. Since an infinite number of dimensions are available in the Hilbert Space, additional dimensions can be added to the space, if the ontology is modified, and new conceptual categories are provided. Our Hilbert Space also contains weight and level dimensions, as shown in Fig. 1.

Both of these dimensions are orthogonal to the remaining dimensions in the complex space. The weight dimension is used to store the weight information about the concepts added to the space; and the level dimension is used to store the hierarchical levels of the concepts.

Another property of Hilbert Spaces is that, they do not need to be square summable. This means that only countably finite axes that are orthogonal are enough to form an orthonormal basis for the entire space [6]. Hilbert Spaces can use coordinates that intersect with different angles. In our space, similar categories have less than 90° intersection angles, and dissimilar categories have more than 90° intersection angles. Since each dimension interacts with any other dimension in the space, the similarities between each dimension are computed by the algorithm described later (Sect. 4).

3 System Architecture

The system is built on a cloud computing architecture. It is composed of three main components: *Service Discovery Client* (*SDC*), *Service Publisher Peer* (*SPP*) and *Service Repository Cloud* (*SRC*).

The *SDC* is the part of the system built for the use of human user agents. This client helps users specify queries to find a certain web service, or a web service providing certain functionality. The queries are composed of one or more criteria combined with a filter operator. The criteria can be about the structure of the web service, or they can be about the semantic properties of the web service. The query is submitted to *SRC*. The repository is searched for matching services, and a set of Service Match Records are returned to the user. Service Match Records contain Service Name, Publisher Name and a URL where more details about the service can be obtained.

The *SPP* is used by the web service publishers. With *SPP*, a web service publisher can publish services to the SRC. With the embedded Service Repository Peer (SRP) component, a publisher can opt to function as a relay station for the *SRC*. The *SPP* also allows the publisher to annotate the OWL-S Service Profile of the service with the concepts from *SRC*'s unified ontology SUMO.

The *SRC* is the main component of the proposed system, and consists of a set of modules responsible for service publishing, discovery and presentation. The *SRC* runs on a cloud computing architecture, and is hosted at Google App Engine. When a web service publisher publishes a web service, the OWL-S service descriptor files and additional semantic information (i.e. the name of the service, the main category of the service) are transferred to the *SRC*. When a query is received, the Service Discovery Manager searches the web service space in the Data Store for matching services, ranks them for similarity, and sends the responses to the *SDC*.

Service oriented computing defines an architectural style whose goal is to achieve loose coupling among interacting software entities [7]. In this respect all of the services provided by the system are built as web services themselves. API users can search for services using the interfaces of the Service Discovery Manager to receive programmatically readable data.

4 Semi-Automated Service Publishing

In the present web service environment, services are prevalently described by WSDL files. The main problem regarding WSDL files as service descriptors is their lack of semantic information. OWL-S files on the other hand enable the expression of semantic properties for the service. The problem can be overcome by converting the WSDL files to OWL-S syntax. The WSDL2OWL-S [8] library is used for a preliminary WSDL to OWL-S conversion. The Service Profile file generated by the system is presented to the user for annotation in an easy to use editor that allows further OWL-S annotation. The annotatable nodes currently include the Input, Output, Precondition and Effect types. The publisher can also specify a main category for the service to be published.

The SUMO ontology contains links to WordNet ontology. Our system uses both of these ontologies for service publishing and annotation purposes. The categorization of the semantic entities is made based on the root nodes in mid-level ontologies, just after ontology import procedure is completed. A recursive algorithm traverses all the nodes in the data store to label each node with a category plane. Each semantic node in the SRC is assigned numeric level information, to be used in Service Match ranking.

When a service is uploaded to our system, a Hilbert Space is generated for the service, and the space is used for service and query match score computation during web service discovery. There are an undefined number of dimensions in the created complex space for each uploaded service, limited by the categories defining the service. The lifecycle of the Hilbert Space for the uploaded web services are discussed below.

For SRC, there are two types of data that needs to be processed: (i) conceptual data (i.e. type information of IOPE and Service Main Category) and (ii) semantic, non-conceptual data (i.e. id information of IOPE, Service Name and Service Description). Usually the data points accumulated by the IOPE types are not enough to accurately annotate a service, due to their low number. Therefore, the categorical spaces are enriched with concepts by other concepts that in one way relate to the already present concepts. The weight of a data point decreases as relationship links are followed.

There are multiple dimensions in the complex space computed for a web service, and all of these dimensions contribute to the service discovery procedure. Our system computes a dimension significance score for each dimension, and includes this score in the discovery process to make sure that the data points generated in the categories that are important to the web service has more impact in a service discovery [9].

There are multiple dimensions and category planes in the complex category space generated for the web services. However, an explicit relationship is not present between these categories. This introduces a problem since inter-category plane relationships play an important role on web service discovery at most times. The algorithm for the computation of the category plane similarity score is similar to Google's Page Rank [10] algorithm, in which links between two entities bond the entities together. There are two types of links: (i) incoming links and (ii) outgoing links. If a concept A has a relationship to a concept B, the relationship is an outgoing link for concept A and an incoming link for concept B. For each category, the incoming and outgoing links with any other category are counted, and the categories with most links in between are evaluated to be more similar.

5 Automated Semantic Service Discovery

The query for a service discovery can be generated by: (i) a user-agent or (ii) a software-agent. The data in a query needs to be: a number of query criterion and a filter operator. A sample query is illustrated in Table 1.

Table 1 Example query generated by user agent

Join operator	Parameter	Parameter
AND(	Input = Price|	Output = Ticket)
OR(	Input = Movie Name|	Output = Ticket)

Table 2 Query criteria importance order

Query criteria
Input = Output > Precondition = Effect > Service Name > Full Text

The criteria in a query can be of different importance levels. The SRC processes the given query to find the individual criteria, and sorts the criteria in the order of importance. The importance of the query criteria is shown in Table 2.

In order to create a main category for the query, the criteria containing the Input, Output, Precondition and Effect variables are examined, and the category of the conceptual data points rooting from these are evaluated. The first data necessary for the matching algorithm is created when a web service is published to the SRC: a permanent conceptual space is created for the given service, and stored for future use in service discovery. The second data necessary for the matching algorithm is created when the query is sent to the system: a temporary conceptual space is created for the given query. These conceptual spaces, dimension significance values and category plane similarities are used by the service match ranking algorithm to obtain the most similar services [9]. When certain conditions are satisfied, the SRC might forward the service discovery request of an agent to a SRP. A query forward might mean one of the following three: (i) there is not a reasonable number of available services in the cloud to make an accurate service ranking, (ii) there is an adequate number of services in the cloud, however none of the services are suitable enough for the received query or (iii) there is a SRP specialized to the main category of the query, which might contribute better results to the service discovery. In these cases, the SRC returns a forward message, bundled with the Service Match Record list to the query owner agent. The agent is free to follow or discard the forward.

6 Conclusions

This paper proposes a software system which allows web service publishers to convert their WSDL web service descriptions to OWL-S web service descriptions, annotate their web service descriptions, and publish the descriptions to a semi-decentralized peer-to-peer network using a client application or a software API. Publishing and discovery operations are performed via a single, unified ontology. The proposed approach is different from existing systems in architectural and semantic areas. Architecturally, it is highly scalable and service oriented. Semantically, the web service similarities are computed with a novel approach, the

Hilbert Space approach, based on categories and concepts. Web service semanticity is enhanced via web service discovery queries, which is again a new approach.

A future work might be allowing transparent structural and semantic composition for the provided queries. With this advancement, our system can transparently compose web services in the repository to create a web service that matches the complex solid described by the query Hilbert Space. The composition can simply be modeled as the unification of solids to create a bigger multi-dimensional object.

References

1. Verma, K.: METEOR-S WSDI: a scalable P2P infrastructure of registries for semantic publication and discovery of web services. Inf. Technol. Manag. **6**(1), 17–39 (2005)
2. Christiensen, E., Curbera, F., Meredith, G., Weerawarana, S.: Web Services Description Language 1.1. http://www.w3.org/TR/wsdl
3. W3C, OWL-S: Semantic Markup for Web Services. http://www.w3.org/Submission/OWL-S
4. Pantazoglou, M., Tsalgatidou, A.: A P2P platform for socially intelligent web service publication and discovery. The third international multi-conference on computing in the global information technology, pp. 271–276 (2008)
5. Pease, A.: Suggested Upper Merged Ontology. http://www.ontologyportal.org/
6. Hilbert, D., von Neumann, J., Nordheim, L.: Über die grundlagen der quantenmechanik. MathematischeAnnalen **98**(1), 1–30 (1928)
7. Amoretti, M., Zanichelli, F., Conte, G.: Enabling peer-to-peer web service architectures with JXTA-SOAP. In: Proceedings of IADIS International Conference e-Society, pp. 75–90. Algarve, Portugal (2008)
8. CMU Atlas Project Group, WSDL2OWL-S. http://www.daml.ri.cmu.edu/wsdl2owls/
9. Ozorhan, M.O.: A service oriented peer to peer web service discovery mechanism with categorization. http://etd.lib.metu.edu.tr/upload/12611634/index.pdf
10. Page, L., Brin, S., Motwani, R., Winograd, T.: The pagerank citation ranking: bringing order to the web. Stanford InfoLab, 66 (1999)

Finding High Order Dependencies in Data

Rosa Meo and Leonardo D'Ambrosi

Abstract We propose *DepMiner*, a method implementing a simple but effective model for the evaluation of the high-order dependencies in a set S of observations. S can be either ordered—thus forming a sequence of events—or not. DepMiner is based on Δ, a measure of the degree of surprise of S based on the departure of the probability of S from a referential probability estimated in the condition of maximum entropy. The method is powerful: at the same time it detects significant positive dependencies as well as negative ones suitable to identify rare events. The system returns the patterns ranked by Δ; they are guaranteed to be statistically significant and their number results reduced in comparison with other methods.

1 Introduction

In statistics, machine learning and data mining the problem of the determination of set of variables whose values are correlated represents an important knowledge in many fields such as in feature selection, database design, market basket analysis, information retrieval, machine translation, biology, etc. Often in the scientific literature, the study of the dependence between variables is limited to pairs [1] but finding correlations among more than two variables is essential for many commercial and sociological studies (collaborations and interaction networks), bio-medical (interaction among drugs and proteins) and scientific domains.

Furthermore, many observations have a special meaning if considered as sequentially ordered: examples span from the nucleotide sequences of the

R. Meo (✉)
University of Torino, Torino, Italy
e-mail: meo@di.unito.it

L. D'Ambrosi
Regional Agency for Health Care Services—A.Re.S.S, Piemonte, Italy

E. Gelenbe et al. (eds.), *Computer and Information Sciences II*,
DOI: 10.1007/978-1-4471-2155-8_4,

organisms genome to the series of events generated in a network of interactions (like a social network, a hypertext or a computer network), to the analysis of web server logs or the detection of intrusion attempts in information systems.

Formally, a set of observations $I = \{i_1, i_2, \ldots, i_k\}$ is named *itemset* where all the elements or items of the set occur together in the same example (or transaction). Here the term transaction denotes a group of events occurring at the same time-stamp and sharing the same order number. Any single observation or item i_j represents a pair $(Variable, Value)$ that means that the *Variable* describing one of the attributes of the observation assumes the indicated value. The set of items I is defined as a non-empty subset of elements from a language $\mathcal{L}$, the item collection. In this manner, an itemset represents the set of items that occur together and have the same order number.

A sequence S is an ordered list of itemsets $S = <I_1, I_2, \ldots, I_n>$. A subsequence S' of S, constituted by $m<n$ itemsets, is denoted by $S' \subset S$ and is constituted by a subset of the itemsets of S in which the precedence relationship between any pair of itemsets is maintained.

In this paper we deal with both itemsets and sequences of itemsets and we are able to model both of them with the same probabilistic model. We treat an itemset as a composite event and we represent it in our model by the joint probability that all its items occur together. Our model is based on the probability of occurrence of events. We model the probability of the sequences by means of the probability of the events represented by the sequence members. For the itemsets we record the frequency with each all the items occur together; for sequences, we record the frequency with which each of the event sequence occurs in the correct order. In order to unify the two concepts of itemsets and sequences we use the common term *pattern*. In practical cases, often it happens that the set of returned patterns is large and much of the information is redundant because many of the patterns are returned together with some of their subsets. The reduction of the redundancy in the result set answers to two major challenges: reducing the overwhelming size of the mining results and eliminating redundancy in the information content and the overlapping between the patterns.

Deciding which patterns are redundant is not straightforward. It might depend on the applications. For instance, the inclusion of patterns with some common elements could be acceptable because the itemsets might have different meaning or occur in different situations. Instead, the inclusion in the result of both subsets and their super-sets is not acceptable if the super-sets do not add new information to the information carried by the subsets. In literature, redundant itemsets are detected in many different and sometimes opposite ways. Calders and Goethals [2] allows the reduction of the number of the itemsets with a loss-less compression of the result because all the information lacking from the result can be restored by application of the concept of closed itemsets. Zhang et al. [3] considers the correlation among the items as strong only when all the items are considered together and when their subsets have instead a weak correlation. On the opposite side, Duan and Street [4] considers interesting an itemset if all its subsets are closely related to all other subsets.

In data mining there exist computationally efficient methods to discover significant dependencies and correlations among the items of a frequent pattern [5–7]. In itemsets mining, in order to determine the dependencies in patterns composed by more than 2 items, either they make the multi-way independence assumption or they evaluate the contribution to the overall itemset of each variable separately [3, 5, 6]. The difficulty stems from the fact that there is not an easy way to determine a referential probability of an itemset I that represents a condition of independence among the subsets if we do not suppose independence among all the single variables in I. Gallo et al. [5] ranks the frequent itemsets according to the unlikelihood that they appear under the hypothesis that all the items in the itemset are independent. But the multi-way independence condition gives a problem: according to this definition of independence, if a dependence already exists in a subset of I, this dependence is "inherited" from the subset to I and to all the super-sets of I [8]. Thus we do not have a way to distinguish if an intrinsic dependence exists in an itemset I in addition to the dependencies inherited from its subsets.

We can solve the problem in terms of quantity of information that an itemset provides. Chakrabarti et al. [9] proposes the use of Minimum Description length to identify the interesting sequences. Similarly, we are interested in identifying the surprising patterns whose probability, as derived by the number of their observations, has a great departure w.r.t. a referential, estimated probability as determined only by its subsets. We are interested in patterns that add any information to their subsets while we want to discard those itemsets that can be foreseen given the observation of their subsets. We proposed in [10] a solution based on the maximum entropy. The entropy of an itemset I is computed by an estimation of the probability of I computed on the basis of the probability of its subsets. The probability of I at which the entropy is maximum (denoted by $P_E(I)$) corresponds to the probability that the itemset I would have in the condition in which it carries the maximum amount of information in addition to its subsets. The interest measure that we proposed for an itemset I is the departure of the probability of I w.r.t. the referential value computed at maximum entropy: $\Delta(I) = P(I) - P_E(I)$. The higher the departure between the two probabilities, the less the itemset can be correctly foreseen from the observation of its subsets. This departure identifies a dependence between the items and tells us that this dependence is not due to the subsets only. As a consequence the itemset is non-redundant.

$\Delta(I)$ decreases with the increase in the cardinality of itemsets and it is not suitable for the comparison of itemsets of different cardinality. For this purpose in this paper we propose Δ_n, a version of Δ normalized w.r.t. the probability of the itemset:

$$\Delta_n(I) = \frac{P(I) - P_E(I)}{P(I)}$$

Similarly it happens with a sequence pattern S. Its probability is computed on the basis of the occurrence of all its sequence elements (itemsets) when these occurrences respect the correct order in the sequence.

Δ_n takes both positive and negative values, in the range from $[-\infty, 1]$. Specifically, if the value is positive, it means a positive dependence, i.e., a pattern that is more frequent than expected; if the value if negative it means a negative dependence, i.e., a pattern that occurs rarer than expected. In this paper we present a method for the computation of the interesting and non redundant patterns based on the above observations. Δ_n is used as a score function to rank the patterns. In Sect. 3 we show how we succeeded to determine the significance level of Δ_n and to identify the significant patterns.

The rest of the paper is organized as follows. In Sect. 2 we summarize how Δ is computed. Section 3 shows how to determine the significance level of Δ_n. Section 4 presents an empirical evaluation study on the results of the system on some common data-sets. Finally, Sect. 5 draws the conclusions.

2 Estimation of the Referential Probability

Here we generalize our model based on Δ to generic patterns, including also sequences. Consider a pattern $W = \{w_1, w_2, \ldots, w_k\}$ where w_i might represent either a single item or an itemset in an ordered sequence. Henceforth, we will use the generic term *pattern element*. Entropy $H(W) = -\sum P(w_1^*, w_2^*, \ldots, w_k^*) \log[P(w_1^*, w_2^*, \ldots, w_k^*)]$ with w_j^* representing the element w_j either affirmed (representing an event present in an observation) or negated (absent event). Summation ranges over the probabilities of all the combinations of the k elements taken affirmed or negated. $H(W)$ is not computed by assumption that singletons are independent but taking in consideration the actual probability of occurrence of each subset of W, as observed from the database. The exclusion-inclusion principle [2] is adopted to compute the entropy of W starting from the probability of the subsets of W and it is a function of the probability of W. The estimate of the probability of W is the probability value that maximizes the entropy and corresponds to the case in which the amount of information on the presence of W is minimum given the knowledge on the presence of the subsets of W. Notice that in making the estimate we considered only the observed probabilities of the subsets. Thus, if the dependence in a pattern W is intrinsic, due to the synergy between all its elements, then the observed probability of W departs with respect to the estimate. As a result, $\Delta_n(W)$ makes emerge the intrinsic, actual dependencies, existing among all the elements in W.

3 Setting a Threshold for Δ

Another problem that we have to solve is how large must be Δ_n such that a pattern is deemed significant. We use a null model in which there are not dependencies between the variables. The null model is generated empirically via a randomization of the original data-set. Randomization is generally accepted as a way to allow a statistical test on significance of results [11]. Randomization occurs by independently shuffling the variable values among the examples. As a result, the new data-set will have the same marginal probabilities of the single variables but the dependencies between them are spoiled.

In a successive step, we compute patterns both from the real and the randomized data. Next, we compute the minimum negative value of Δ_n (denoted by UB) and the maximum positive value of Δ_n (denoted by LB) in randomized data. Then, we use UB as an upper bound for rare patterns in real data and use LB as a lower bound for the frequent patterns in real data. This is a sort of statistical test on Δ_n that accepts as dependent a pattern if its Δ_n is higher (resp. lower) than the maximum (resp. minimum) Δ_n of the patterns extracted from the randomized data.

Consider the public data-set *Mushroom* (down-loadable from the repository of UCI). After randomization, we observed the maximum value of $\Delta_n = 0.04$ and the minimum value of $\Delta_n = -0.03$. In real data, the maximum is $\Delta_n = 0.85$ and the minimum is $\Delta_n = -0.45$. Thus in *Mushroom* the positive dependencies are more abundant and marked than the negative dependencies. In Table 1 we show an output of *DepMiner*, with the ranking of patterns (itemsets) extracted from *Mushroom*. The significant itemsets (whose value of Δ_n exceeds the observed range of values in randomized data) are highlighted and shown over a gray background. Over a normal, white background, instead, are shown the other itemsets. Notice that both rarer and more frequent itemsets are interesting.

4 Experimental Evaluation

We run a set of experiments on 5 real data-sets (from FIMI and UCI Machine Learning repositories) and on 2 real data-sets coming from NASDAQ stock exchange index (from January 2001 to May 2009) and from the Italian lottery (with data on the numbers drawn from 1939). The lottery data-set is important in order to check the behavior of Δ_n on complete random data where even the marginals were uniform.

In Table 2 we include the total number of examples, *minsup* threshold, the total number of itemsets generated (N), execution times to compute Δ_n (in seconds).

We performed two experiments: the first one on the compression capability and the second one on the capability of DepMiner to determine the dependencies in contrast to methods that assume the multi-way independence condition.

1. In this experiment, to be further conservative, we compare DepMiner results at many levels. We denote as itemsets clearly non independent, the itemset whose $\Delta_n \neq 0$. Thus we include in the results both these latter ones and the itemsets whose Δ_n is acceptable by the significance test on the lower and upper bounds obtained in randomized data. We include in Table 2 three ratios: the ratio between the number of itemsets with $\Delta_n \neq 0$ and N (`Dep/N`), the ratio between the significant dependencies and N (`SDep/N`) and the ratio between non

Table 1 DepMiner output: itemsets ranking with significant itemsets highlighted over a gray background

Delta/Po	Delta	Freq.	Itemsets
0.0188	0.0075	3256	Class = poisonous, ring-number = one, bruises? = no
0.0138	0.0059	3272	stalk-surface-above-ring = smooth, ring-number = one, Class-edibility
0.0121	0.0045	3032	stalk-surface-below-ring = smooth, ring-number = one, Class = edibility
0.0055	0.0023	3376	stalk-surface-above-ring = smooth, Class = edibility, gill-size = broad
0.0007	0.0003	3216	Class = edibility, gill-size = broad, odor = none
−0.0006	−0.0002	3128	bruises? = bruises, gill-space = close, stalk-surface-above-ring = smooth
−0.0014	−0.0006	3728	veil-color = white, Class = edibility, gill-size = broad
−0.0020	−0.0008	3128	veil-color = white, ring-type = pendant, gill-size = broad
−0.0026	−0.0013	3964	gill-attach = free, stalk-above-ring = smooth, stalk-below-ring = smooth
−0.0049	−0.0021	3488	gill-attach = free, ring-number = one, Class = edibility
−0.0051	−0.0021	3368	gill-attach = free, ring-number = one, ring-type = pendant
−0.0061	−0.0022	3016	stalk-surface-above-ring = smooth, ring-type = pendant, gill-size = broad

Table 2 Experimental results

Data-set	Minsup (%)	Itemsets (N)	Dep/N (%)	SDep/N (%)	NDI/N (%)	Time(s)	γ(DM,MINI)	γ(RDM,RMINI)
Accidents	35	65,500	13.5	0.1	22.93	4294	−0.84	0.16
Chess	75	20,582	1.2	0.34	2.11	135	−0.95	−0.91
Nasdaq	0.14	242	95.8	46.69	100	107	−0.05	0.27
Kosarak	1.01	21,934	82.5	10.39	95.55	2221	−0.56	0.28
Mushroom	22.15	14,189	1.48	1.03	5.84	115	−0.94	−0.29
Retail	4.53	22,524	79.7	5.9	99.56	1322	0.02	0.55
Lottery	0.006	91,499	99.1	0	100	5804	0.81	0.77

derivable itemsets (`NDI`) obtained by the competitor method [2] and N (`NDI/N`). These ratios quantify the volume of found dependencies in data and clearly demonstrate the increased ability of DepMiner to reduce redundancies than NDI.

2. In the second experiment we compare the results of DepMiner with MINI [5], in order to determine the difference between DepMiner and another method based on the multi-way independence assumption. The last columns of Table 2 report the result of a comparison between DepMiner ranking (denoted by `DM`) and MINI. As said, our method does not consider dependencies inherited by the subsets; it coincides with the multi-way independence assumption only for patterns with cardinality 2. In order to measure the correlation between our ranking (DepMiner) and MINI's we adopted an objective measure: γ [1]. Since the methods differ in the referential probability estimate, γ quantifies the effect of this difference.

In Table 2 we also compared the two rankings computed on randomized data (RDM and RMINI). All the values reported by γ denote disagreement. The amount of discrepancies decreases (γ increases) if we move from real data to randomized data (since the high-order dependencies are spoiled during randomization). Furthermore, on complete random data (*Lottery*) the two methods agree.

5 Conclusions

We have presented *DepMiner*, a method for the extraction of significant dependencies between the values assumed by database variables. DepMiner gave good results in comparison with [5] and demonstrated a superior capability to compress results than NDI [2].

References

1. Goodman, L.A., Kruskal, W.H.: Measures of association for cross classifications. J. Amer. Stat. Ass. **49**(268), 732–764 (1954)
2. Calders, T., Goethals, B.: Non-derivable itemset mining. Data Min. Knowl. Discov. **14**(1), 171–206 (2007)
3. Zhang, X., Pan, F., Wang, W., Nobel, A.B.: Mining non-redundant high order correlations in binary data. PVLDB **1**(1), 1178–1188 (2008)
4. Duan, L., Street, W.N.: Finding maximal fully-correlated itemsets in large databases. In: Proceedings of the IEEE International Conference on Data Mining, pp. 770–775 (2009)
5. Gallo, A., Bie, T.D., Cristianini, N.: Mini: Mining informative non-redundant itemsets. In: Proceddings of PKDD Conference, pp. 438–445 (2007)
6. Xin, D., Cheng, H., Yan, X., Han, J.: Extracting redundancy-aware top-k patterns. In: Proceedings of the ACM SIGKDD Conference, pp. 444–453 (2006)
7. Omiecinski, E.: Alternative interest measures for mining associations in databases. TKDE **15**(1), 57–69 (2003)
8. Brin, S., Motwani, R., Silverstein, C.: Beyond market baskets: Generalizing association rules to correlations. In: Proceedings of the ACM SIGMOD conference, pp. 265–276 (1997)
9. Chakrabarti, S., Sarawagi, S., Dom, B.: Mining surprising patterns using temporal description length. In: Proceedings 24th International Conference on Very Large Data Bases, pp. 606–617 (1998)
10. Meo, R.: Maximum independence and mutual information. TOIT **48**(1), 318–324 (January 2002)
11. Gionis, A., Mannila, H., Mielikäinen, T., Tsaparas, P.: Assessing data mining results via swap randomization. In: Proceedings of the SIGKDD, pp. 167–176 (2006)
12. Aggarwal, C.C., Yu, P.S.: A new framework for itemset generation. In: Proceedings of the PODS, pp. 18–24 (1998)

Heuristic Algorithms for Real-Time Unsplittable Data Dissemination Problem

Mustafa Müjdat Atanak and Atakan Doğan

Abstract In real-time communication, the timely delivery of the data transfer requests needs to be guaranteed. This work formally introduces the Real-Time Unsplittable Data Dissemination Problem (RTU/DDP), which is a generalization of the unsplittable flow problem (UFP). RTU/DDP problem is NP-hard. Therefore, heuristic approaches are required to acquire *good* solutions to the problem. The problem is divided into two sub-problems: path selection and request packing. Each of these sub-problems is formally defined and heuristic algorithms are proposed for both sub-problems. The performance of these algorithms is compared with a heuristic from the literature that can solve RTU/DDP, namely Full Path Heuristic (FPH). The results and discussions of the comparisons between the proposed heuristics and FPH are presented.

Keywords Real-time · Data dissemination · Unsplittable flow problem · Computer networks · Performance evaluation

1 Introduction

For the applications with real-time requirements, the timely delivery of related data is of utmost importance. Unlike the best effort data transfer requests that need to be completed as early as possible, real-time data transfer requests come with a deadline, before which they have to be completed in order for them to be considered *successful*. Applications with real-time data transfer requests may

M. M. Atanak (✉) · A. Doğan
Department of Electrical and Electronics Engineering,
Anadolu University, Eskişehir, Turkey
e-mail: mmatanak@anadolu.edu.tr

A. Doğan
e-mail: atdogan@anadolu.edu.tr

E. Gelenbe et al. (eds.), *Computer and Information Sciences II*,
DOI: 10.1007/978-1-4471-2155-8_5,

arise in defense and surveillance [1], wireless sensor networks [2], and Grid systems [3, 4].

In this study, we will formally introduce the Real-Time Unsplittable Data Dissemination Problem (RTU/DDP), which is a generalization of the well-known Unsplittable Flow Problem. Based on [5–9] and the related studies, the network is assumed to support end-to-end guaranteed service in which a share of any link bandwidth can be reserved and then released afterwards. RTU/DDP tries to find out the routes and the amount of flow that should be assigned to each request that maximizes the number of real-time requests that are delivered successfully.

Section 2 introduces the RTU/DDP. Then, it divides the problem into two sub-problems. Sections 3 and 4 concentrate on these two sub-problems. Section 5 presents simulation results. The last section provides conclusions.

2 Problem Formulation

A data grid system is modeled by an undirected graph $G = (V, E)$, where $V = \{v_1,\ldots,v_n\}$ defines the heterogeneous machines and $E = \{e_1,\ldots,e_m\}$ denotes the links each of which connects any two machines of the system. The machines can be storage elements with limited storage space as well as network routers (or switches). Each $e_i \in E$ is associated with a bandwidth value $c_i > 0$ and a delay value $d_i > 0$. Let $R = \{r_1,\ldots,r_k\}$denote a set of real-time data transfer requests. Each request $r_i \in R$ is modeled with $<s_i,\ t_i,\ f_i,\ \delta_i>$ quadruple in which s_i is the source machine, t_i is the destination machine, f_i is the requested data item, and $\delta_i > 0$ is the deadline of the request. It is assumed that the centralized scheduler is able to reserve and use the system resources for the duration of file transfers. A request $r_i \in R$ is assumed to be *satisfied* if its file f_i is transferred from the source storage machine s_i to the destination storage machine t_i through one of the (simple) paths between these two machines before the request deadline δ_i has passed. Let $P_i = \{p_1,\ldots,p_{l_i}\}$and $l_i > 0$ define the set of paths and the number of paths for request $r_i \in R$, respectively; $P_i(e_k) = \{p_j : p_j \in P_i \text{ and } e_k \in E\}$ denote a set of paths $p_j \in P_i (j \leq l_i)$ each of which includes link $e_k \in E$ for request $r_i \in R$.

Definition 1 Given a data grid system $G = (V, E)$ and a set of real-time data transfer requests R, the RTU/DDP seeks to maximize the number of satisfied requests where each requested file is required to be transferred over a single path.

$$\text{RTU/DDP} : max \sum_{r_i \in R} \sum_{p_j \in P_i} x_{ij}$$

$$\sum_{p_j \in P_i} x_{ij} \leq 1, \text{for all } r_i \in R$$

$$\sum_{r_i \in R} \sum_{p_j \in P_i(e_k)} \pi_{ij} x_{ij} \leq c_k, \text{ for all } r_i \in R \text{ and } e_k \in E$$

$$x_{ij} \in \{0, 1\} \text{ for all } r_i \in R \text{ and } p_j \in P_i$$

where x_{ij} is 1 if request $r_i \in R$ is satisfied over path $p_j \in P_i$, and 0 otherwise; π_{ij} denotes the bandwidth demand of request $r_i \in R$ when path $p_j \in P_i$ is chosen for the request, and it is computed as follows:

$$\pi_{ij} = \frac{|f_i|}{\delta_i - \sum_{e_k \in p_j} d_k}$$

where$|f_i|$ denotes the file size.

RTU/DDP problem is a generalization to the unsplittable flow problem (UFP), which is known to be NP-hard [10]. Therefore, an *optimal* solution to RTU/DDP cannot be found in a polynomial time, unless $P = \text{NP}$. Heuristic approaches, e.g. [1, 6], are adopted to find *good* solutions to RTU/DDP. In this study, RTU/DDP is split into two sub-problems, namely path selection and request packing. The path selection determines one path for each request and passes these paths to the second phase. Request packing selects the satisfiable requests that maximize the number of satisfied requests while respecting link capacity constraints.

3 Path Selection

Minimum hop minimum delay feasible path first (MinMin/FPF). This algorithm is based on the presumption that it is best if the requests get routed over a *feasible path* with minimum number of hops incurring a small path delay.

Definition 2 A path $p_j \in P_i$ for request $r_i \in R$ is *feasible* iff the bottleneck bandwidth (*bbw*) of $p_j \in P_i$ is greater than the bandwidth demand of request $r_i \in R$ for path $p_j \in P_i$, where *bbw* is the minimum of available bandwidth values of the links on path.

MinMin/FPF is based on the Bellman-Ford shortest path algorithm, and it is inspired by [11]. The algorithm tries to find the minimum hop count path between the source and destination. If more than one path has the same hop count, the path with minimum bandwidth is favored. The path found by the algorithm is, then, checked for feasibility. If the path is not feasible, an empty path is returned. The time complexity of minimum hop Bellman Ford algorithm is $O(|V||E|)$. MinMin/FPF additionally calculates path delays in each step, which takes $|E|$ steps in the worst case. Thus, MinMin/FPF runs in $O(|V||E|^2)$ for a single request. Since MinMin/FPF should be run for each request $r_i \in R$, the time complexity of overall procedure becomes $O(|V||E|^2|R|)$.

Edge disjoint minimum hop minimum delay FPF (ED MinMin/FPF). ED MinMin/FPF tries to route every request over a *feasible path* with minimum number of hops incurring a small path delay that are largely edge disjoint. The requests are sorted based on their bandwidth requirements and they are marked as unprocessed. The main loop begins by marking every link as available. For each unprocessed request, if a minimum hop minimum delay path is found with available links, the links that are used by this path are marked as unavailable and the request is marked as processed. The loop is repeated as long as at least one unprocessed request exists. Main loop is repeated |R| times in the worst case. Each loop finds a minimum hop minimum delay path. Therefore, overall procedure is $O(|V||E|^2|R|^2)$.

4 Request Packing

Definition 3 Given $G = (V, E)$ and R, in which $r_i \in R$ is modeled with $<p_i, \pi_{ij}>$ tuple, request packing problem (RPP) seeks to maximize the number of satisfiable requests.

$$\text{RPP}: max \sum_{r_i \in R} x_i$$

$$\sum_{r_i \in R} \sum_{p_j \in P_F(e_k)} \pi_{ij} x_i \leq c_k, \text{ for all } r_i \in R \text{ and } e_k \in E$$

$$x_i \in \{0, 1\} \text{ for all } r_i \in R$$

where x_i is 1 if request $r_i \in R$ is satisfied and 0 otherwise; $P_F = \{p_i : p_i \in P_i \text{ and } 1 \leq i \leq r\}$ where $p_i \in P_i$ is a feasible path for request $r_i \in R$; $P_F(e_k)$ is the set of feasible paths which include link $e_k \in E$. RPP is a generalization to the Multidimensional 0-1 Knapsack Problem (MKP), which is NP-hard [12]. Exact algorithms based on branch and bound and dynamic programming can solve RPP, but they work in modest size problems only.

Maximum number of outgoing flows first (*MNOFF*). MNOFF first produces a *contention graph* based on the paths chosen by the respective path selection algorithm. A contention graph (a bipartite graph) is formed as follows: for each link, the requests that require bandwidth allocation on the link are determined. Assuming that all requests are sent through the network with minimum required bandwidth, the bottleneck links whose bandwidth capacity is exceeded are determined. The vertices on the left column of the contention graph are the requests and vertices on the right are the bottleneck links. The contention graph is formed by drawing an edge between request and bottleneck link if the request's path includes the link.

In each iteration, MNOFF drops the request with the maximum number of outgoing flows (the vertex with the maximum number of edges in the contention graph) and updates the contention graph. In the case of equality, MNOFF picks the request with the highest minimum bandwidth requirement. Iterations continue as long as there exists at least one edge in the contention graph.

Table 1 Parameter values in the base set

Parameter	Value in base set
NUMBER_OF_MACHINES	50
NUMBER_OF_LINKS	100 (undirected)
MAX_BANDWIDTH	1 (Gbit/s)
MAX_DELAY	10 (m s)
NUMBER_OF_REQUESTS	500
MAX_REQUEST_SIZE	10 (Gbit)
MAX_DEADLINE	100 (s)

Table 2 Base results

Path selection	Request packing	Base results
Min–Min	MNOFF	37.10
	MOFF	50.54
ED Min–Min	MNOFF	30.90
	MOFF	43.16
FPH		11.18

Finding the request with maximum number of outgoing flow is $O(|E||R|)$ and updating the contention graph takes $O(|E||R|)$ in the worst case. Since each iteration marks one request as unsatisfiable, the loop repeats $|R|$ times. Therefore, the time complexity of MNOFF algorithm is $O(|E||R|^2)$.

Maximum outgoing flows first (*MOFF*). Key difference between MOFF and MNOFF is the criterion used to choose unsatisfiable requests. That is, the MNOFF's criterion is the number of edges leaving a request vertex, while the MOFF's is the number of edges leaving a request vertex times the request minimum required bandwidth. Note that time complexity of MOFF is the same as that of MNOFF.

5 Experimental Results

In order to evaluate the performance of the algorithms, the authors have written the simulator in C++. First, a heterogeneous computing system is randomly created based on two parameters: *NUMBER_OF_MACHINES, NUMBER_OF_LINKS*. The links have bandwidths uniformly distributed between 0 and *MAX_BANDWIDTH*; and delays uniformly distributed between 0 and *MAX_DELAY*. After generating >the network topology, *NUMBER_OF_REQUESTS* requests are generated with the following parameters: *MAX_REQUEST_SIZE* and *MAX_DEADLINE*.

Using the simulator developed, a set of simulation studies were conducted. First, a base set of results was established with parameter values shown in Table 1. The performances of the proposed heuristics are compared against Full Path Heuristic (FPH) [1]. As seen from Table 2, the best performance is acquired when (MinMin/FPF, MOFF) is chosen as the (path selection, request packing) mechanism. Result is about 40% better than FPH.

Table 3 Effect of varying *MAX_BANDWIDTH*

		MAX_BANDWIDTH (Mb/s)				
Path selection	Request packing	100	500	1000	5000	10000
Min–Min	MNOFF	7.72	21.56	37.10	64.20	73.38
	MOFF	16.08	34.70	50.54	77.22	85.48
ED Min–Min	MNOFF	6.94	20.04	30.90	60.34	70.68
	MOFF	13.70	32.14	43.16	72.06	81.50
FPH		5.94	9.70	11.18	23.16	24.34

Table 4 Effect of varying *MAX_DEADLINE*

		MAX_DEADLINE				
Path selection	Request packing	10	50	100	500	1000
Min–Min	MNOFF	5.98	21.44	37.10	67.90	79.72
	MOFF	12.08	32.98	50.54	79.20	88.84
ED Min–Min	MNOFF	5.60	21.36	30.90	63.50	76.22
	MOFF	11.50	32.52	43.16	76.14	86.18
FPH		2.62	9.02	11.18	25.96	48.06

Table 3 shows the effect of varying *MAX_BANDWIDTH*. Increasing *MAX_BANDWIDTH* has positive effect on the performance. MOFF is clearly a better choice than MNOFF. MinMin/FPF shows better performance than ED MinMin/FPF on average. The performance difference between the (MinMin/FPF, MOFF) tuple and FPH increases as *MAX_BANDWIDTH* increases.

Table 4 shows the effect of varying *MAX_DEADLINE*. According to Table 4, the real-time performances increase with increasing *MAX_DEADLINE*. (Min–Min/FPF, MOFF) tuple is still the best choice as far as the performance is concerned.

6 Conclusion

This study deals with a more general case of the unsplittable flow problem, namely RTU/DDP. A formal definition of the problem is provided. Then, the problem is divided into two sub-problems: path selection and request packing. Formal definitions of both of the sub-problems as well as some heuristics to be employed in their solutions are also presented. From the results shown in previous section, it is evident that underlying protocols as well as the system parameters have significant effect on the real-time performance of the heterogeneous system. Among the simulated algorithms, the best real-time performance for the RTU/DDP is acquired when the path selection is made by MinMin/FPF and the request packing is handled with MOFF algorithms. As a future work, the authors will consider the more general splittable case, Real-Time Splittable Data Dissemination (RTS/DDP) problem, in which more than one path can be used to satisfy a given request.

Acknowledgments This material is based on work supported by Institute of Scientific and Technological Research of Turkiye under Grant No. 108E232.

References

1. Theys, M.D., Tan, M., Beck, N., Siegel, H.J., Jurczyk, M.: A mathematical model and scheduling heuristic for satisfying prioritized data requests in an oversubscribed communication network. IEEE Trans. Parallel. Distrib. Syst. **11**(9), 969–988 (2000)
2. Hea, T., Stankovic, J. A., Lub, C., Abdelzahera, T.: SPEED: a stateless protocol for real-time communication in sensor networks. In: Proceedings of International Conference on Distributed Computing Systems (ICDCs), pp. 46–55 (2003)
3. Foster, I., Kesselman, C., Lee, C., Lindell, R., Nahrstedt, K., Roy, A.: A distributed resource management architecture that supports advance reservation and co-allocation. In: Proceedings of International Workshop Quality of Services, pp. 27–36 (1999)
4. De Assunção, M.D., Buyya, R.: Performance analysis of allocation policies for interGrid resource provisioning. Inf. Softw. Technol. **51**, 42–55 (2009)
5. Orda, A.: Routing with end to end qos guarantees in broadband networks. IEEE/ACM Trans. Netw. **7**(3), 365–374 (1999)
6. Eltayeb, M.S., Doğan, A., Özgüner, F.: Concurrent scheduling: efficient heuristics for online large-scale data transfers in distributed real-time environments. IEEE Trans. Parallel. Distrib. Syst. **17**(11), 1348–1359 (2006)
7. RFC 1633 Integrated services in the internet architecture: an overview.
8. RFC 2212 Specification of guranteed quality of service.
9. RFC 2205 Resource reservation protocol (RSVP).
10. Kleinberg, J.: Approximation algorithms for disjoint paths problems, PhD Thesis, Department of EECS, MIT (1996)
11. Guerin, R., Orda, A.: Computing shortest paths for any number of hops. IEEE/ACM Trans Netw **10**(5), 613–620 (2002)
12. Freville, A.: The multidimensional 0–1 knapsack problem: an overview. Eur. J. Oper. Res. **155**, 1–21 (2004)

Automatic Categorization of Ottoman Literary Texts by Poet and Time Period

Ethem F. Can, Fazli Can, Pinar Duygulu and Mehmet Kalpakli

Abstract Millions of manuscripts and printed texts are available in the Ottoman language. The automatic categorization of Ottoman texts would make these documents much more accessible in various applications ranging from historical investigations to literary analyses. In this work, we use transcribed version of Ottoman literary texts in the Latin alphabet and show that it is possible to develop effective Automatic Text Categorization techniques that can be applied to the Ottoman language. For this purpose, we use two fundamentally different machine learning methods: Naïve Bayes and Support Vector Machines, and employ four style markers: most frequent words, token lengths, two-word collocations, and type lengths. In the experiments, we use the collected works (*divans*) of ten different poets: two poets from five different hundred-year periods ranging from the 15th to 19th century. The experimental results show that it is possible to obtain highly accurate classifications in terms of poet and time period. By using statistical analysis we are able to recommend which style marker and machine learning method are to be used in future studies.

E. F. Can · F. Can (✉) · P. Duygulu
Department of Computer Engineering, Bilkent University,
06800 Ankara, Turkey
e-mail: canf@cs.bilkent.edu.tr

E. F. Can
e-mail: efcan@cs.bilkent.edu.tr

P. Duygulu
e-mail: duygulu@cs.bilkent.edu.tr

M. Kalpakli
Department of History, Bilkent University, 06800 Ankara, Turkey
e-mail: kalpakli@bilkent.edu.tr

E. Gelenbe et al. (eds.), *Computer and Information Sciences II*,
DOI: 10.1007/978-1-4471-2155-8_6,

1 Introduction

Automatic Text Categorization (ATC) methods aim to classify natural language texts into pre-defined categories and are used in different contexts ranging from document indexing to text mining [1]. In the literature there are a variety of studies in ATC; however, studies on historical manuscripts are rare. One reason for this is the fact that old documents are scarce in the digital environment. Resources such as Ottoman Text Archive Project (OTAP) and Text Bank Project (TBP) release transcribed versions of handwritten Ottoman literary texts [2]. There are millions of pages of texts in Ottoman that are to be analyzed and classified after transcription [3]. By considering the gap in the studies for the Ottoman language, this paper is motivated to classify a text with unknown poet or time period by employing automatic text categorization methods and ultimately show the achievability of effective automatic categorization of historical Ottoman texts so that it can be employed when these documents are transcribed.

The contributions of this study are the following. We provide the first style-centered ATC study on the Ottoman language. Within the context of this language, we evaluate the performance of two different machine learning methods in ATC by using four style markers. By using statistical analysis we are able to recommend which machine learning method and style marker are to be used in future studies. The availability of huge amount of text in the Ottoman language, especially the Ottoman archives [3], confirms the practical importance and implications of our study.

2 Related Work

In ATC, style markers are used in analyzing the writing styles of authors. Holmes [4] gives a detailed overview of the stylometry studies in the literature within a historical perspective and presents a critical review of numerous style markers. Statistical methods have been used for a long time in authorship and categorization tasks and machine learning methods are used in relatively more recent works. A Bayes' theorem-based algorithm is firstly used to classify twelve disputed Federalist Papers in [5]. McCallum and Nigam [6] compare a multivariate Bernoulli model, and multinomial model. SVM (Support Vector Machines) is another machine learning method used in authorship attribution studies. Joachims makes use of SVM in the task of text classification and observes that SVM is robust and it does not require parameter tuning for the task [7]. Kucukyilmaz et al. [8] use machine learning approaches including k-nearest neighbor (k-NN), SVM, and Naïve Bayes (NB) to determine authors of chat participants by analyzing their online messaging texts. Yu [9] focuses on text classification methods in literary studies and uses NB, and SVM classifiers. In her work, the effects of common and function words are investigated.

Table 1 Ottoman literary texts used in this study

Text (no. of poems)	Century	Life span	No. of tokens	No. of types
Mihrî Hatun's divan (245)	15th	1,460–1,512	34,735	9,188
Sinan Şeyhî's divan (221)	15th	1371?–1431	27,743	10,784
Hayalî Bey's divan (619)	16th	1,500–1,557	54,338	15,727
Revânî's divan (141)	16th	1,475–1,524	24,881	8,315
Nef'î's divan (224)	17th	1,572–1,635	51,075	14,492
Neşatî's divan (186)	17th	?–1,674	23,799	7,984
Osmanzâde Tâ'ib's divan (189)	18th	1,660–1,724	19,610	8,772
Şeyh Gâlip's divan (580)	18th	1,757–1,799	59,301	18,506
Şânîzâde's Atâullah's divan (125)	19th	1771–1826	8,265	4,409
Yenişehirli Avnî's divan (425)	19th	1826–1884	54,927	18,785
Total	–	–	358,674	62,609

To the best of our knowledge there is no previous categorization study on the Ottoman language; however, there are studies on contemporary Turkish. Can and Patton [10] analyze change of writing style with time by using word lengths and most frequent words for the Turkish authors Çetin Altan and Yaşar Kemal. In another study they analyze the *Ince Memed* tetralogy of Yaşar Kemal [11].

3 Corpus and Experimental Design

In this study, we focus on Ottoman literary texts of ten poets and five consecutive centuries. Table 1 gives information about these texts. The text associated with each poet is called *divan* which is an anthology of the poet's work, as it might be selected poems or all poems of the same poet. The poets in this study are selected in such a way that they all together provide a good representation of the underlying literature. There are nine male and one female poets from five different centuries. The works of the picked poets given in Table 1 acquire almost all characteristics of the Ottoman lyric poetry [12]. In our study, the poets whose life spanned two centuries are associated with the century they died (only exception is Mihrî Hatun since she lived in the sixteenth century for a relatively short period of time).

Each document is split into blocks with k number of words, where k is taken as 200–2,000 with 200-word increments. If the number of words in the last block is smaller than the chosen block size that block is discarded. Blocking is a common approach used in stylometric studies [13].

Can and Patton [10] show that most frequent words and word lengths (in the form of token and type lengths) as style markers have remarkable performance in determining the change of writing style with time in Turkish. Because of their observations and since Turkish is the basis of the Ottoman language we use these text features in our study. We also use two-word collocations as another style marker, since phrases are one of the characteristic features of the Ottoman language and poets. Accordingly, the style markers used in the study are: Most

Frequent Words (MFW), Token Lengths (TOL), Two-word Collocations -two consecutive words- (TWC), and Type Lengths (TYL). In our study, a token is a word and a continuous string of letters. Besides, a word that involves a dash is counted as one token. Only word of length one is '*o*' (the third person and singular pronoun). Type is defined as a distinct word. For example, the following line from Yenişehirli Avnî: '*Yâhû ne kâtib ol ne mühendis ne veznedâr*' contains eight tokens and six types.

We employ two machine learning-based classifiers: Naïve Bayes (NB)- a generative classifier and Support Vector Machines (SVM)- a discriminative classifier [14, 15]. The use of fundamentally different classifiers provides us a wide test spectrum to investigate the performance of machine learning methods in ATC of Ottoman literary texts. Furthermore, NB and SVM are commonly used in similar studies. For example, Yu [9] indicates that SVM is among the best text classifiers. In the same work it is also indicated that NB is a simple but effective machine learning method and often used as a baseline.

In this study we employ the model used in [16] for NB. In SVM we employ two different kernel functions; polynomial (poly or p), and radial-basis-function (rbf) kernels. We refer to these methods as SVM-poly (or SVM-p) and SVM-rbf, respectively. These choices are motivated by the successful results obtained by them in [17]. For the construction of training and test corpora, we prefer *K*-fold cross validation in which division of data is not important compared to splitting the corpus as training and test set. In our study, we use ten for *K*. In the experiments with SVM for the polynomial kernel we run tests when the degree is set to 1, 2, 3, 4 and 5. For the radial-basis-function kernel, we set γ (width of the kernel) to 0.6, 0.8, 1.0, 1.2, and 1.4. Similar settings for SVM are used in [17] for text classification and successful results are obtained.

We conduct a two way analysis of variance (ANOVA) in order to see if the classification performances of the tested cases are significantly different from each other. When the main effects of the factors, style markers and machine learning algorithms, are statistically significantly different in explaining the variance of classification accuracy, we conduct post-hoc multiple comparisons using Scheffe's correction [18] for the levels of each factor (an abridged presentation is provided in the next section).

4 Experimental Results

4.1 Classification by Poet

In Table 2, we provide poet classification accuracies of the style markers MFW, TOL, TWC, and TYL with the machine learning methods NB, and two versions of SVM for different block sizes. The table shows that for MFW with SVM-poly, we obtain the best accuracy score when the polynomial degree is 1; similarly, we obtain the best accuracy score for SVM-rbf when γ is 1.2. In the table the values of

Table 2 Poet classification accuracies of MFW, TOL, TWC, TYL with NB, SVM-poly, and SVM-rbf for different block sizes

	MFW			TOL			TWC			TYL		
Block Size	NB	SVM-p deg = 1	SVM-rbf $\gamma = 1.2$	NB	SVM-p deg = 5	SVM-rbf $\gamma = 1.2$	NB	SVM-p deg = 1	SVM-rbf $\gamma = 1.0$	NB	SVM-p deg = 5	SVM-rbf $\gamma = 1.2$
200	63.13	73.36	74.75	30.98	28.25	26.99	34.85	34.23	35.76	37.12	34.65	35.78
400	73.67	84.50	84.97	35.35	34.35	36.1	45.00	42.34	43.63	37.89	35.78	35.32
600	81.71	87.88	88.37	43.66	40.33	41.48	49.91	48.99	49.57	46.28	40.78	37.93
800	85.21	91.00	90.49	45.69	43.52	44.90	57.07	56.44	57.58	49.65	44.34	45.42
1,000	85.47	91.08	91.42	44.77	42.49	43.55	61.96	61.22	61.50	50.00	45.35	48.23
1,200	86.55	91.32	91.32	51.32	49.16	49.80	63.72	65.77	64.73	56.93	48.29	47.82
1,400	**91.66**	91.28	91.12	51.36	47.02	48.46	64.45	69.21	69.69	59.55	50.24	48.56
1,600	89.64	91.22	91.12	50.50	48.49	48.64	68.30	68.53	69.90	55.08	49.40	46.65
1,800	88.57	92.51	92.51	54.52	**55.38**	**56.88**	69.96	69.42	68.10	**64.22**	**56.76**	**56.32**
2,000	87.05	**92.80**	**92.80**	**57.78**	52.40	55.44	**71.93**	**71.42**	**71.42**	59.17	53.63	54.53
Avg.	83.27	88.70	88.89	46.60	44.14	45.23	58.72	58.76	59.19	51.59	45.92	45.66

The parameters, polynomial degree (deg) for SVM-poly (SVM-p) and γ for SVM-rbf that yield the listed results, are also provided these parameters that provide the best performances of TOL, TWC, and TYL are also given.

For MWF for all block sizes SVM-poly and -rbf provide better results than those of NB. Both versions of SVM have similar results. For TOL for almost all block sizes NB provides slightly better results than those of SVM-poly and -rbf. Scores of SVM-rbf are slightly better than the scores of SVM-poly. For TWC all methods yield similar accuracy scores. For TYL for all block sizes NB provides a slightly better performance that those of the SVM classifiers and both versions of SVM have similar performances. From the table we can see that for MFW the difference between NB and SVM classifiers are noticeable for the other cases NB and SVM classifiers performances are mostly compatible with each other.

Statistical Analysis We do the multiple comparisons of the style markers and machine learning algorithms in poet categorization for $\rho < 0.05$ using Scheffe's method. According to comparisons, TOL and TYL are not significantly different from each other; whereas, other pairs of style markers are significantly different from each other. Considering the machine learning algorithms, the SVM classifiers with different kernels are not significantly different from each other, but they are significantly different from the NB classifier.

4.2 Classification by Time Period

In addition to classifications of texts by poet, in this study we also study classifications of texts by time period. In the corpus, there are ten *divans* from 15th to 19th centuries (two *divans* per century). In the classification of texts by time period, MFW (Most Frequent Words) provides the best classification scores (up to 94%) with the SVM classifier. TWC provides the second best performance, and TOL and TYL follow the style marker TWC. SVM mostly performs better than NB with MFW. For TOL and TYL, NB provides slightly more accurate results than SVM. The NB and SVM classifiers have almost the same performance with TWC.

Statistical Analysis As in the poet classification section, we do the multiple comparisons of the style markers and machine learning algorithms in period categorization for $\rho < 0.05$ using Scheffe's method (they are obtained by using the results as in Table 2). According to comparisons, TOL and TYL are not significantly different; whereas, other pairs of style markers are significantly different from each other. Moreover, considering the machine learning algorithms, they are significantly different from each other for combinations of all pairs.

5 Conclusion

We present the first style-centered ATC study on Ottoman literary texts particularly on collected poems (divans) of ten different Ottoman poets from five different centuries. The statistical tests show that SVM and MFW yield performances that are mostly statistically significantly different from their counterparts. Based on these observations we recommend the use the SVM classifier and MFW style marker in future related studies on this language.

The availability of huge amount of text to be digitized in the Ottoman language confirms the practical importance and implications of our results. We hope that our work and results would serve as an incentive for more research using these documents.

Acknowledgments This work is partially supported by the Scientific and Technical Research Council of Turkey (TÜBİTAK) under the grant number 109E006. Any opinions, findings and conclusions or recommendations expressed in this article belong to the authors and do not necessarily reflect those of the sponsor.

References

1. Sebastiani, F.: Machine learning in automatic text categorization. ACM Comput. Surv. **34**(1), 1–47 (October 2002)
2. Ottoman Text Archive Project. http://courses.washington.edu/otap/ (2011)
3. Başbakanlık Devlet Arşivleri, T.C.: http://www.devletarsivleri.gov.tr (2011)
4. Holmes, D.I.: Authorship attribution. Comput. Human. **28**(2), 87–106 (October 1994)
5. Merriam, T.: An experiment with the federalist papers. Comput. Human. **23**(3), 251–254 (1989)
6. McCallum, A., Nigam, K.: A comparison of event models for naive bayes text classification. In: AAAI-98 workshop on learning for text categorization (1998)
7. Joachims, T.: A statistical learning model of text classification for support vector machines. In: Proceedings of the 24th ACM SIGIR conference, 128–136 (2001)
8. Kucukyilmaz, T., Cambazoglu, B.B., Aykanat, C., Can, F.: Chat mining: Predicting user and message attributes in computer-mediated communication. Inf. Process. Manag. **44**(4), 1448–1466 (2008)
9. Yu, B.: An evaluation of text classification methods for literary study. Lit. Ling. Comp. **23**(3), 327–343 (2008)
10. Can, F., Patton, J.M.: Change of writing style with time. Comput. Human. **38**(1), 61–82 (2004)
11. Patton, J.M., Can, F.: A stylometric analysis of Yaşar Kemal's *İnce Memed tetralogy*. Comput. Human. **38**(4), 457–467 (2004)
12. Andrews, W.G., Black, N., Kalpakli, M.: Ottoman lyric poetry. University of Texas Press, Austin, Texas, USA (1997)
13. Forsyth, R.S., Holmes, D.I.: Feature-finding for text classification. Lit. Ling. Comput. **11**(4), 162–174 (June 1996)
14. Duda, R.O., Hart, P.E., Stork, D.G.: Pattern classification (2nd edn.). Wiley-Interscience, New York (2000)
15. Vapnik, V.: The nature of statistical learning theory. Springer, New York (1995)
16. Zhao, Y., Zobel, J.: Effective and scalable authorship attribution using function words. Lect. Notes Comput. Sci. **3689**, 174–189 (November 2005)
17. Joachims, T.: Text categorization with support vector machines: learning with many relevant features. In: ECML-98, 137–142 (1998)
18. Scheffe, H.: A method for judging all contrasts in the analysis of variance. Biometrica **40**, 87–104 (1953)

An Empirical Study About Search-Based Refactoring Using Alternative Multiple and Population-Based Search Techniques

Ekin Koc, Nur Ersoy, Ali Andac, Zelal Seda Camlidere, Ibrahim Cereci and Hurevren Kilic

Abstract Automated maintenance of object-oriented software system designs via refactoring is a performance demanding combinatorial optimization problem. In this study, we made an empirical comparative study to see the performances of alternative search algorithms under a quality model defined by an aggregated software fitness metric. We handled 20 different refactoring actions that realize searches on design landscape defined by combination of 24 object-oriented software metrics. The investigated algorithms include random, steepest descent, multiple first descent, multiple steepest descent, simulated annealing and artificial bee colony searches. The study is realized by using a tool called A-CMA developed in Java that accepts bytecode compiled Java codes as its input. The empiricial study showed that multiple steepest descent and population-based artificial bee colony algorithms are two most suitable approaches for the efficient solution of the search based refactoring problem.

Keywords Search-based software engineering · Combinatorial optimization · Automated refactoring · Software maintenance · Software metrics

---

E. Koc · N. Ersoy · A. Andac · Z. S. Camlidere · I. Cereci · H. Kilic (✉)
Department of Computer Engineering, Atilim University, Incek, Ankara, Turkey
e-mail: hurevren@gmail.com

E. Koc
e-mail: antimon@gmail.com

N. Ersoy
e-mail: nersoy88@gmail.com

A. Andac
e-mail: alyandac@gmail.com

Z. S. Camlidere
e-mail: zelalseda@gmail.com

I. Cereci
e-mail: icereci@atilim.edu.tr

E. Gelenbe et al. (eds.), *Computer and Information Sciences II*,
DOI: 10.1007/978-1-4471-2155-8_7,

1 Introduction

Search based optimization techniques have become popular in software engineering research, in last decade [1]. There are diverse application areas ranging from requirements engineering, quality assessment, service-oriented software engineering, to project planning where the techniques are successfully applied. Among the application areas, automated software maintenance [2–4] requires some special elaboration due to its representational and descriptive difficulties in terms of candidate solutions and objective functions. In order to cope with the design corruption [5] problem, in object-oriented software systems, maintenance programmers are required to execute periodic and systematic refactoring activities. Although one can use CASE products for the purpose, it still requires too much effort and time of maintenance programmers. Automation of this critical software engineering task which is guided by some quality model, defined by related software metrics, supports sustainability/improvement in design quality and provides highly maintainable software products. In fact, the problem is a combinatorial optimization problem in which fitness function can be characterized as an aggregate of object-oriented metric results. Once the problem is defined in terms of solution representation, fitness function and change operator, there are wide variety of candidate metaheuristic search techniques.

In this paper, we reported the results of our empirical study including comparison of 5 alternative search algorithms and Random (RND) search in the context of the search based refactoring problem. The algorithms include steepest descent (SD), multiple first descent (MFD), SA, multiple steepest descent (MSD) and artificial bee colony (ABC) searches. The goal of this research is to investigate the potential of alternative search techniques in the automated software refactoring domain. Note that, among the alternatives to the best of the authors' knowledge, the ABC search [6] has not been applied in search-based refactoring problem before. In literature, one can find alternative approaches based on different biological metaphors [7]. Throughout the experiments, all searches are executed at the design level of abstraction and realized at the design space. The general architecture of the A-CMA tool and the details of the adopted methodology are given in Sect. 2. Section 3 describes the experimental set up and the results obtained through excessive amount of runs taken for different parameter values. Finally, the conclusions are given.

2 Tool and Methodology

A-CMA consists of 3 major modules; one for carrying out modifications on a design, one for the actual searching operation and the final one for representing a solution. The Modification module is responsible for choosing applicable modifications on a current design. The static input, denoted as "Modifications" represents a set of refactoring actions that can be applied. The "Checker" module is responsible for determining possible modifications that should guarantee that an action does not change the functional properties of the design. When a modification is chosen among this set, the "Modification Applier" module carries out necessary operations on the given design. "Search Expander" module is the central building block of the operation. It is responsible for carrying out the search process on possible designs and tries to find an optimal design using several algorithms. The search "Algorithms" are static inputs as well. "Design Representation" module is responsible for providing a design abstraction for a given object oriented software. The abstraction is modifiable by other modules, easily. Finally, the "Metrics Calculation" module is responsible for providing ability to measure design quality. "Metrics" are static inputs. A-CMA produces a refactoring suggestion sheet as output of a search procedure. The suggestion sheet contains initial and final design information regarding to the found refactoring path and a step by step refactoring suggestion list for the developer. A-CMA uses a Java bytecode manipulation and analysis framework named ASM [8]. All abstract design representations of given benchmark programs has been constructed using ASM framework. In A-CMA, each refactoring action type has a checker function that can evaluate a given design to find all appropriate actions of the given type that can be applied to the design. The static analysis method consists of several condition checks on a possible refactoring action that results in a filtering over actions. A-CMA implements 20 refactoring actions: Move Up/Down Method, Move Method, Move Up/Down Field, Instantiate Method, Freeze Method, Make Class Abstract/Final/Non-Final, Inline Method, Remove Class/Interface/Method/Field, Introduce Factory, Increase/Decrease Method Security, Increase/Decrease Field Security. Evaluation of a design is better to be based on some justifiable quality model. For example, in [9], Quality Model for Object Oriented Design (QMOOD) [10] which adopts weighted sum of 7 object-oriented metrics has been used. In our case, we have implemented 24 object-oriented metrics (5 of these metrics intersects with QMOOD model) selected from various sources including [11] without any weight consideration. A-CMA works on normalized metric values. Complexity related metrics should be minimized during a search for a better design [12]. However, it is not possible to set a precise goal for many other software metrics in such way. Since the optimization process cannot rely on subjective opinions, objectives for such metrics should be regarded as "unknown". Furthermore, in order to prevent possible conflicts, those metrics that require value maximization are also treated as minimized metrics by using the multiplicative inverses of the values.

The considered minimized metrics include: The number of the fields/methods in a class; the average number of the field/method visibility of a class; the nesting level of a class; the number of the methods of a class in a package; and the number of classes/interfaces in a package. The metrics with unknown objective values include: The number of constant fields/setter/getter methods in a class; the average number of the static methods of a class; the number of interfaces a class implements; the number of children/descendants/ancestors of a class; the number of elements on which this class depends; the number of elements that depend on this class; the number of times the class is externally used as attribute type; the number of attributes in the class having another class or interface as their type; the number of times the class is externally used as parameter type; the number of parameters in the class having another class or interface as their type; nesting level of a package and the ratio of abstract classes in a package. While it is obvious that the objective value of minimized metrics should be set to 0 (assuming no negative values possible), there is a requirement to set objectives for unknown metrics. Our normalization schema uses a comparative approach in this manner. Instead of setting precise objectives, it is decided to use an optimum design set as a feedback mechanism. Therefore, the current design is compared against the norm of a given optimum design set and the distances can be used as a metric score to be minimized during search. Assuming there are k metrics, $M_1, M_2, \ldots, M_k$ and n designs in the optimum (or ideal) design set, $D_1, D_2, \ldots, D_n$ along with the current design, D_{cur} to calculate the normalized metric score for. Hence, for each metric, the Distance function is defined as:

$$Dist(\mathrm{M_i}, \mathrm{D_{cur}}) = \begin{cases} |NormVal(\mathrm{M_i}(\mathrm{D_{cur}}), \mathrm{i}) - NormVal(0, i)|, & \mathrm{M_i} \in \text{Minimized} \\ |NormVal(\mathrm{M_i}(\mathrm{D_{cur}}), \mathrm{i})|, & \mathrm{M_i} \in \text{Unknown} \end{cases} \tag{1}$$

where the NormalVal function calculates the normalized value of x against the values of *i*th metric in the optimum design set, given as:

$$NormVal(\mathrm{x}, \mathrm{i}) = Norm_{0-1}(\mathrm{M_i}(\mathrm{D_1}), \mathrm{M_i}(\mathrm{D_2}), \ldots, \mathrm{M_i}(\mathrm{D_n}))_{\mathrm{x}} \tag{2}$$

Hence, the overall metric score of D_{cur} to be minimized is defined as;

$$Eval(D_{cur}) = \sum_{i=1}^{k} Dist(M_i, D_{cur}) \tag{3}$$

3 Experimental Results

To the best of authors' knowledge, there is no widely accepted benchmark input program set that can be used for an empirical study on search based refactoring problem. Therefore, we have included 6 input programs written in Java where 5 of them being open source programs under heavy development [13–17], the last one being a student project: Beaver (a parser generator); Mango (a collections library); JFlex (lexical analyzer generator); XML-RPC (XML-based remote procedure call library); JSon (Java library for data exchange format) and Mosaic (a student project). During experiments, Java.Math, Java.Text, Java.Util, and Javax.Swing packages from the base Java library have been chosen as the members of optimum design set since they provide most of the core functionality of the library. The experiments have been carried out on 20 computers with Intel Core2DUO CPUs and 4 GBs of memory. The underlying operating system was Ubuntu-Linux (fully patched) with Sun JRE 6. A-CMA has client/server architecture to simplify work distribution amongst several computers. We took 4 different ascent values (5, 10, 15 and 20) with 30 restarts for both MFD and MSD searches; 3 different initial temperature values (1.5, 2.5, 4.0) for SA search and 7 different food source sizes (20, 40, 60, 80, 100, 120 and 200) for ABC search. There were no alternative parameter setting values for RND and SD searches. Each run was repeated 10 times for each of 6 input programs. So, the number of total runs was $10 \times 6 \times (1 + 1 + 4 + 4 + 3 + 7) = 1200$ runs.

The results include relative and absolute quality gains for different algorithms using all 6 input programs. The mean gains obtained for all 6 input programs using MFD algorithm were varying between 5.31 and 5.39. The values did not show any dramatic change against different ascent values applied in each restart. The searches were not affected too much by the depth of random bad movements in the design space while increased number of bad movements mostly resulted in quality gain decrements. Similar conclusions were drawn for MSD where mean absolute quality gains for 5, 10, 15 and 20 ascent values were 5.64, 5.60, 5.54 and 5.53, respectively. As a result, when we consider the mean absolute quality gains, MSD search outperformed MFD search for each of different ascent depth values. While the implementation of SA has been relatively straightforward, it has been difficult to decide on the ideal cooling schedule and initial temperature values. The mean absolute quality gains obtained for all 6 input programs using simulated annealing with initial temperature values 1.5, 2.5 and 4.0 were 4.80, 4.45 and 4.22, respectively. Lower initial temperature values resulted in higher mean absolute gain scores. The mean quality gain results obtained for SA search were not as good as that obtained for MFD and MSD searches. Mean absolute quality gains obtained for all 6 input programs using ABC algorithm with 200, 120, 100, 80, 60, 40 and 20 food sources were 5.76, 5.68, 5.65, 5.64, 5.50, 5.45 and 5.32, respectively. Higher mean absolute quality gains were attained for higher number of food sources. One can still expect better gain values for higher number of food sources. The ABC search algorithm was able to outperform the MSD algorithm for the highest trial of 200 food sources. In ABC implementation, we applied the discrete version to the refactoring domain [18]. Figure 1 shows the mean quality increase for each search technique for the entire set of input programs. The values are normalized against the SD performance for each program. For almost all input programs, mean normalized quality gain values were consistent for each program.

For all initial temperature values, the SA search was outperformed by almost all other techniques including baseline SD search. Except for input program Mango, the MSD and ABC searches gave mostly better results than the MFD search known to be well performing in the literature [8]. MSD outperformed SD for all mean quality gain values obtained for input programs. For the sake of reliability, we prefer higher quality gains with small standart deviations. Especially for the Beaver input program, we observed relatively high standart deviations in mean

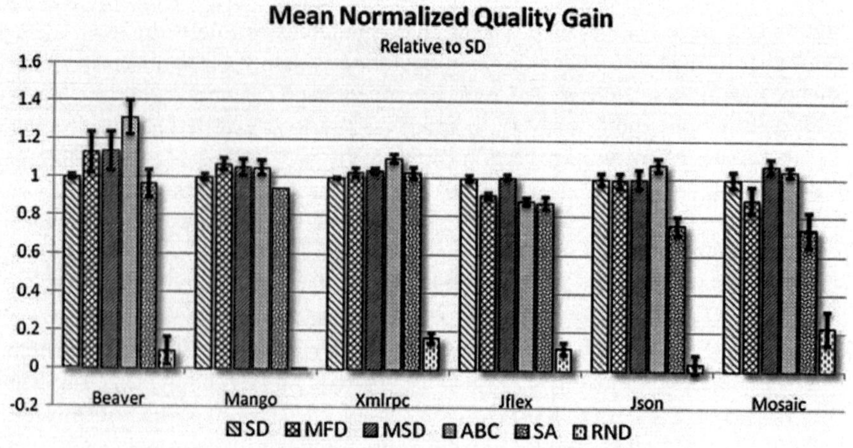

Fig. 1 Mean normalized quality gain values obtained for MFD, MSD, ABC, SA and RND searches that are calculated relative to the baseline SD search for all 6 input programs

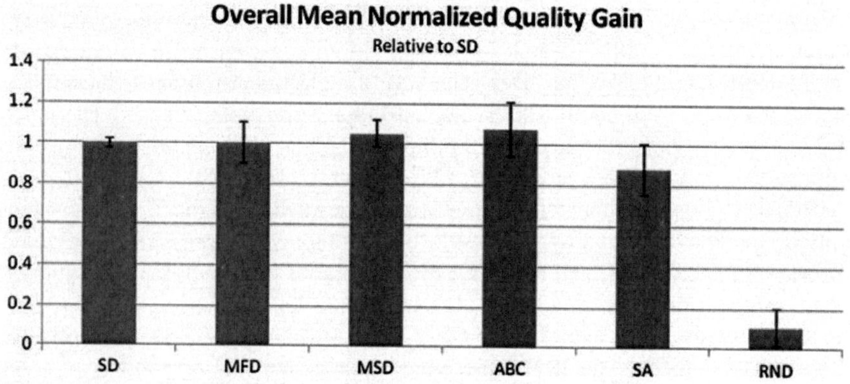

Fig. 2 Overall mean normalized quality gain values obtained for MFD, MSD, ABC, SA and RND searches that are calculated relative to the baseline SD search

absolute quality gains. When we consider the best performances attained for alternative parameter settings of each algorithm, the ABC search gave the highest mean normalized relative quality gain results for all input programs (see Fig. 2).

ABC outperformed the MSD for 3 (Beaver, XML-RPC and JSon) of 6 of the input programs in terms of mean quality gains obtained. However, high standart deviations of the gain obtained by ABC makes relatively less reliable. Therefore, we can only say that ABC and MSD results are highly competitive and do not dominate each other. The reason behind success of MSD and ABC searches is directly related with the design landscapes of the input programs defined by the aggregate software metric and available refactoring actions. Except for the input programs Json and Mosaic, the initial number of applicable refactoring instances were considerably high (being >500). The high value has been observed not only for the initial step of the search but also throughout the all search steps. Such an observation implied high branching factor of search space. As a consequence, the steepest descent technique with multiple applications MSD search performed well in most of the runs. On the other hand, success of ABC search was due to populated investigation of better designs that enables examination of alternative regions of the search space.

4 Conclusions

Based on the input programs used, quality model described and possible refactorings considered, we conclude that multiple steepest descent and artificial bee colony algorithms are two most suitable competitive approaches for the efficient solution of the search based refactoring problem. The common property of expanded search horizon of MSD and ABC makes them well performing alternatives at the cost of relatively higher execution times. MSD expands the search horizon via sequential multiple trials of full neighbors in design space while allowing bad movements in order to escape from local minima. ABC, on the other hand, expands the search horizon via memorization and improvement of more than one candidate designs, simultaneously. Our lightweight solution representation enabled us to expand search horizon efficiently which resulted in high quality designs to be found. The intensive computation requirement of the refactoring problem implies the necessity for parallel implementation of the algorithms under consideration. Also, relatively high quality results obtained via population based ABC algorithm encourage us to try parallel implementation of the population/swarm based alternative algorithms, in future.

Acknowledgements The authors would like to thank Atilim University ARGEDA department for its financial support.

References

1. Harman, M.: The current state and future of search based software engineering. In: Proceedings of Future of Software Engineering, pp. 342–357. IEEE Press, Washington (2007)
2. Bouktif, S., Antoniol, G., Merlo, E., Neteler, M.: A novel approach to optimize clone refactoring activity. In: Proceedinds of GECCO 2006, vol. 2, pp. 1885–1892, Washington (2006)
3. O'Keeffe, M., Cinneide, M.O.: Search-based software maintenance. In: Proceedings of Conference on Software Maintenance and Reengineering (CSMR'06), pp. 249–260, Italy (2006)
4. Seng, O., Stammel, J., Burkhart, D.: Search-based determination of refactorings for improving the class structure of object-oriented systems. In: Proceedings of GECCO, vol. 2, pp. 1909–1916, Washington (2006)
5. Fowler, M.: Refactoring: Improving the Design of Existing Code. Addision Wesley, Massachusetts (1999)

6. Karaboga, D., Basturk, B.: A powerful and efficient algorithm for numerical function optimization: artificial bee colony algorithm. J. Glob. Opt. **39**, 459–471 (2007)
7. Gelenbe, E., Kaptan V. Wang, Y.: Biological metaphors for agent behavior. In: LNCS- ISCIS, pp. 667−675 (2004)
8. OW2 Consortium, ASM, http://asm.ow2.org/
9. O'Keeffe, M., Cinneide, M.O.: Search based refactoring: an empirical study. J. Soft. Maint. Evol.: Res. Pract. **2**, 345–364 (2008)
10. Bansiya, J., Davis, C.G.: A hierarchical model for object-oriented design quality assessment. IEEE Trans. Softw. Eng. **28**, 4–17 (2002)
11. Chidamber, S.R., Kemerer, C.F.: A metrics suite for object oriented design. IEEE Trans. Softw. Eng. **20**, 476–493 (1994)
12. SDMETRICS tool, http://www.sdmetrics.com/
13. Beaver—a LALR Parser Generator, http://beaver.sourceforge.net
14. The Mango Library, http://www.jezuk.co.uk/cgi-bin/view/mango
15. JFlex—The Fast Scanner Generator for Java, http://jflex.de/
16. Apache XML-RPC, http://ws.apache.org/xmlrpc/
17. JSON in Java, http://www.json.org/java/index.html
18. Pan, Q., Tasgetiren, M.F., Suganthan, P.N., Chua, T.J.: A discrete artificial bee colony algorithm for the lot-streaming flow shop scheduling problem. Inf. Sci. (2010, in Press)

Part II
Data Engineering

Unsupervised Morphological Analysis Using Tries

Koray Ak and Olcay Taner Yildiz

Abstract This article presents an unsupervised morphological analysis algorithm to segment words into roots and affixes. The algorithm relies on word occurrences in a given dataset. Target languages are English, Finnish, and Turkish, but the algorithm can be used to segment any word from any language given the wordlists acquired from a corpus consisting of words and word occurrences. In each iteration, the algorithm divides words with respect to occurrences and constructs a new trie for the remaining affixes. Preliminary experimental results on three languages show that our novel algorithm performs better than most of the previous algorithms.

Keywords Unsupervised learning · Morphology · Word decomposition

1 Introduction

Natural Language Processing (NLP) is a field of computer science that investigates interactions between computers and human languages. NLP is used for both generating human readable information from computer systems and converting human language into more formal structures that a computer can understand. Well known problems of NLP are morphological analysis, part of speech tagging, wordsense disambiguation, and machine translation.

K. Ak (✉) · O. T. Yildiz
Department of Computer Science and Engineering, Işik University,
34980 Şile, Istanbul, Turkey
e-mail: koray@isikun.edu.tr; korayak@gmail.com

O. T. Yildiz
e-mail: olcaytaner@isikun.edu.tr

E. Gelenbe et al. (eds.), *Computer and Information Sciences II*,
DOI: 10.1007/978-1-4471-2155-8_8,

Morphological analysis or decomposition studies the structure of the words and identifies the morphemes (smallest meaning-bearing elements) of the language. Any word form can be expressed as a combination of morphemes. For instance, the English word "enumeration" can be decomposed as e+number+ate+ion and "interchangeable" as inter+change+able, and the Turkish word "isteyenlerle" as iste+yen+ler+le. Generally words are known as the basic units of the language but morphemes are the smallest syntactic unit and they reveal the relationship between word forms. In this respect, morphological analysis investigates the structure, formation and function of words, and attempts to formulate rules that model the language.

Morphological analysis is widely used in different areas such as speech recognition, machine translation, information retrieval, text understanding, and statistical language modeling. In many languages this task is both difficult and necessary, due to the large number of different word forms found in the text corpus. Highly inflecting languages may have thousands of different word forms of the same root, which makes the construction of an affixed lexicon hardly feasible. As an alternative to the hand-made systems, there exist algorithms that work unsupervised manner and autonomously do morphological analysis for the words in an unannotated text corpus.

In this paper, an unsupervised learning algorithm is proposed to extract information about the text corpus and the model of the language. The proposed algorithm constructs a trie that consists of characters and the occurrences of the words as nodes. The algorithm then detects roots of the given words by examining the occurrences in the path of the word. When the root is revealed, the algorithm creates a new trie from the affix parts, left after the root for each word. The algorithm continues recursively until there is no affix left to process.

The paper is organized as follows: In Sect. 2, we present previous work in the field. In Sect. 3, we present proposed algorithm. We give the results of our experiments in Sect. 4 and conclude in Sect. 5.

2 Related Work on Unsupervised Morphological Analysis

Morpho Challenge [1] is one of the competitions of the EU Network of Excellence PASCAL2 Challenge Program working on unsupervised morphological disambiguation since 2005. The objective of the challenge is to design an unsupervised machine learning algorithm that discovers which morphemes the words consist of.

In Morpho Challenge 2005, Bernhard [2] propose a method that relies on transitional probabilities of each substring of the word in the lexicon, and distinguishes stems and affixes by examining the differences in lengths and frequencies of the words.

Keshava [3] used a simple approach to gather morphemes based on finding substring words and transitional probabilities. The algorithm constructs two trees;

a forward tree where each node from top to the leaf corresponds to a word in the corpus and a backward tree to find suffix probabilities easily.

For the Turkish task in 2007, Zeman [4] proposed a paradigm based approach. All possible suffix-stem pairs are grouped into paradigms. Since all possible segmentation points are considered, the number of paradigms is huge and they are filtered. In the segmentation phase; each possible segmentation of the word is searched in the paradigms.

ParaMor [5] dominated the Morpho Challenge 2008 in all languages. Each word is examined by segmenting from every character boundary. When two or more words end in the same word-final string, ParaMor constructs a paradigm seed. Paradigms are then expanded to full candidate paradigms by adding additional suffixes.

In 2009, Monson et al. [6] proposed an improved version of ParaMor [5]. In the original version, ParaMor did not assign a numeric score to its segmentation decisions. A natural language tagger is trained to score each character boundary in each word. Using ParaMor as a source of labeled data, finite-stage tagger is trained to identify, for each character, c, in a given word, whether or not ParaMor will place a morpheme boundary immediately before c.

3 REC-TRIE

The input of the Morpho Challenge is a corpus and word list of the words with frequencies as appeared in the corpus. Since character encodings differ in the datasets, some modifications are done. English dataset consists of standard text and all words are lower-cased. Finnish dataset uses ISO Latin 1 (ISO 8859-1). The Scandinavian special letters å, ä, ö are rendered as one-byte characters. Turkish dataset is standard text and all words except the letters specific to Turkish are lower-cased. The letters specific to the Turkish language are replaced by capital letters of the standard Latin alphabet, “açıkgörüşlülügünü” is converted to “aCIkgOrUSlUlUGUnU”.

As mentioned in the first section, one of the problems in unsupervised morphological analysis is the data sparsity. Given a large dataset, most of the root words appear in the corpus. For example, in the datasets of challenge, there exist 15545 root words among 617298 words where total root count for Turkish is 23470 [7], that is 66% of the roots appeared in the dataset.

The pseudo code of our proposed algorithm (REC-TRIE) is given in Fig. 1. We simply populate word trie with words from the list that occurred more than 5 times and store the corresponding character, the number of occurrence in the corpus and the number of times that character is used in this path in this depth (Line 3). With the fact explained above, we assume the smallest most occurred word in a path is the root of the words in the path.

Once the algorithm finds root morphemes in each word, it saves the corresponding segmentation into a table and continues to the next iteration (Line 7).

```
1  Read words from Wordlist W
2  i = 1
3  Construct word Trie_i
4  Construct word Table
5  Do until no unsegmented words remain
6      For each word in the Table
7          Find boundary for unsegmented part with Trie and update Table
8          If the word is not fully segmented
9              Add unsegmented part to Trie_{i+1}
10         End If
11     End For
12 i = i + 1
13 End Do
```

Fig. 1 The pseudocode of REC-TRIE

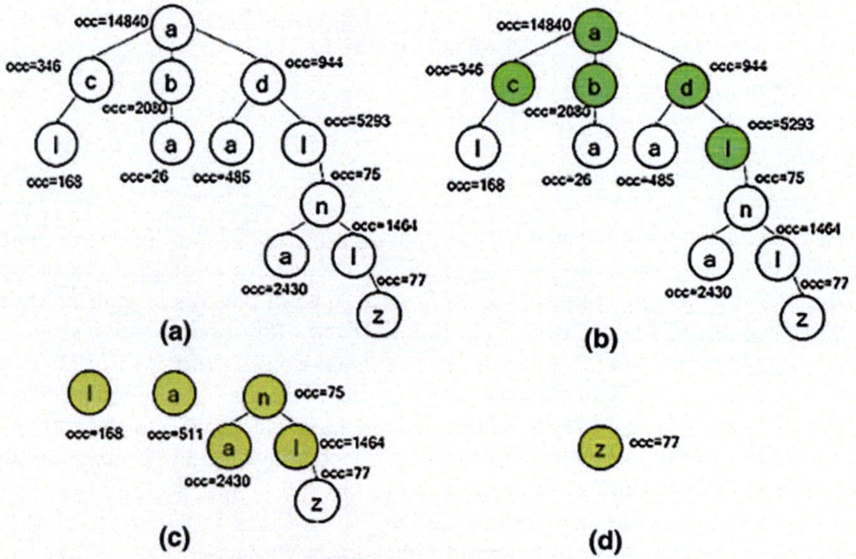

Fig. 2 Sample run from REC-TRIE. **a** A sample trie initialized. **b** After first iteration root words detected. **c** First affix trie is constructed with the extracted affix parts left from roots and first affixes are found. **d** Second iteration REC-TRIE created new affix tree and found the last affix

In each iteration, the rest part of the word is put on a new trie (Line 9) and affix boundaries are found recursively with the same method applied for root extraction. The algorithm continues until no affix candidate is left.

We present a sample run of REC-TRIE in Fig. 2. After initializing word trie the algorithm traverses each path and chooses the most occurred smallest word as root. The word "ada" is segmented as ad + a since the most occurred character is d with 944 occurrence in the path. However the words "adI", "adIn", "adIna", "adInI",

Table 1 Precision, Recall, and F-Measure of REC-TRIE compared with other algorithms in Morpho Challenge 2009 for Turkish

Author	Method	Precision (%)	Recall (%)	F-Measure (%)
Monson et al.	Paramor-Morfessor Mimic	48.07	60.39	53.53
Monson et al.	ParaMor-Morfessor Union	47.25	60.01	52.88
Monson et al.	Paramor Mimic	49.54	54.77	52.02
Our Algorithm	REC-TRIE	53.40	43.06	47.68
Lavallée and Langlais	RALI-COF	48.43	44.54	46.40
–	Morfessor CatMAP	79.38	31.88	45.49
Spiegler et al.	PROMODES 2	35.36	58.70	44.14
Spiegler et al.	PROMODES	32.22	66.42	43.39
Bernhard	MorphoNet	61.75	30.90	41.19
Can and Manandhar	2	41.39	38.13	39.70
Spiegler et al.	PROMODES committee	55.30	28.35	37.48
Golénia et al.	UNGRADE	46.67	30.16	36.64
Virpioja and Kohonen	Allomorfessor	85.89	19.53	31.82
–	Morfessor Baseline	89.68	17.78	29.67
Lavallée and Langlais	RALI-ANA	69.52	12.85	21.69
–	Letters	8.66	99.13	15.93
Can and Manandhar	1	73.03	8.89	15.86

and "adInIz" is segmented from adI since the most occurred character in these paths are I with 5293 occurrence (b). Next REC-TRIE constructs affix tries to find affix boundaries. In (c), affixes are inserted in a new trie. Note that "a" is merged from ab+a and ad+a and occurrence is summed. Again first affixes are found by selecting the most occurred character. This procedure continues until there is no affix candidate left. The final affix is found in (d) and REC-TRIE finish segmentation.

4 Experiments

Morpho Challenge gives two perl scripts to evaluate algorithms. These scripts simply compare the results against a linguistic gold standard. In the evaluation, only a subset of all words in the corpus is included. For each language, a random subset was picked, and the segmentations of these words were compared to the reference segmentations in the gold standard. The evaluation of an algorithm is based on the F-measure, which is the harmonic mean of precision and recall.

These metrics are calculated by

- Hit (H): The word is cut at the right place.
- Insertion (I): The word is cut at the wrong place.
- Deletion (D): A valid cut is ignored.

Based on these metrics; precision is the number of hits divided by the sum of the number of hits and insertions, and recall is the number of hits divided by the sum of the number of hits and deletions. So the measures are:

$$Precision = \frac{H}{(H+I)} Recall = \frac{H}{(H+D)} F - Measure = \frac{2H}{(2H+I+D)} \quad (1)$$

We have used the dataset of the Morpho Challenge 2009 and evaluate results with the perl scripts provided. Table 1 show the results of our algorithm compared with other algorithms for Turkish. Our algorithm has better F-Measure in Turkish and English than Finnish. This is due to fact that our algorithm finds roots and suffixes by traversing the trie one character at a time so found roots and suffixes are generally short. However, Finnish root words are rather long in the average due to the conservativeness of the language. Especially deletions are fairly more in Finnish since the algorithm oversegments the words. As a result, recall values for Finnish is low and pulls down the F-Measure dramatically.

5 Conclusions

We propose a novel algorithm for unsupervised morphological analysis, based on trie data structure and word occurrences. The algorithm constructs a word trie and finds root words according to the occurrences of characters in the path. After root detection is completed, remaining affix parts are used to construct affix tries. In each iteration, affix boundaries are detected and results are updated.

Although our proposed algorithm is simple, the results are encouraging. Our approach ranks 4'th in Turkish when compared with other competitors. The recall values states that we have missed some of the boundaries especially for Finnish.

The algorithm does not have any methods for prefix detection and there is no control for the irregular changes of the words or umlauts, so we should develop some strategies to cope with these situations.

References

1. Kurimo, M., Krista Lagus, S.V., Turunens, V.: Morpho challenge. http://research.ics.tkk.fi/events/morphochallenge2010/
2. Bernhard, D.: Unsupervised morphological segmentation based on segment predictability and word segments alignment. In: Proceedings of the PASCAL Challenge Workshop on Unsupervised Segmentation of Words into Morphemes (2006)
3. Keshava, S.: A simpler, intuitive approach to morpheme induction. In: Proceedings of the PASCAL Challenge Workshop on Unsupervised Segmentation of Words into Morphemes (2006)
4. Zeman, D.: Unsupervised acquiring of morphological paradigms from tokenized text. Adv. Multiling. Multimodal Inf. Retr. **5152**, 892–899 (2008)
5. Monson, C., Carbonell, J., Lavie, A., Levin, L.: Paramor and morpho challenge 2008. In: Proceedings of the 9th Cross-language evaluation forum conference on evaluating systems

for multilingual and multimodal information access. Cross-Language Evaluation Forum'08, pp. 967–974 (2009)
6. Monson, C., Hollingshead, K., Roark, B.: Probabilistic ParaMor. In: Working Notes for the CLEF 2009 Workshop, Corfu, Greece (2009)
7. Solak, A., Oflazer, K.: Design and implementation of a spelling checker for turkish. In: Literary and Linguistic Computing, vol. 8 (1993)

A Novel Approach to Morphological Disambiguation for Turkish

Onur Görgün and Olcay Taner Yildiz

Abstract In this paper, we propose a classification based approach to the morphological disambiguation for Turkish language. Due to complex morphology in Turkish, any word can get unlimited number of affixes resulting very large tag sets. The problem is defined as choosing one of parses of a word not taking the existing root word into consideration. We trained our model with well-known classifiers using WEKA toolkit and tested on a common test set. The best performance achieved is 95.61% by J48 Tree classifier.

1 Introduction

Morphological disambiguation problem is defined as the task of selecting the correct morphological parse of a word among its parses. According to the morphophonemic structure of the language and morphotactics which define the ordering of morphemes, a word may have many parses. These parses may share the same root word or may have different root words. Morphological disambiguation is considered as a preliminary step for higher level language analysis.

Turkish is one of the morphologically rich languages. Like other agglutinative languages, due to its free constituent order nature, Turkish has a large number of possible tags. There have been studies for morphological disambiguation problem in Turkish. These studies can be categorized under two main approaches: rule-

O. Görgün (✉) · O. T. Yildiz
Department of Computer Science and Engineering,
Işik University, Şile, 34980 İstanbul, Turkey
e-mail: onurg@isikun.edu.tr

O. T. Yildiz (✉)
e-mail: olcaytaner@isikun.edu.tr

E. Gelenbe et al. (eds.), *Computer and Information Sciences II*,
DOI: 10.1007/978-1-4471-2155-8_9,

based approaches and statistical approaches. In statistical approaches, a large corpus is used to train the statistical model and the trained model is tested on an unseen test corpus [1]. However, due to the large number of tags in Turkish, data sparseness is a serious problem. To cope with the data sparseness problem, morphological parses are divided into smaller parts called inflectional groups [2]. The most recent approach to the morphological disambiguation problem is presented in [3]. The methodology employed is based on ranking of the most possible parse sequences (determined by the baseline statistical model represented in [1]) with Perceptron algorithm. The very early rule-based approach to Turkish used hand-crafted rules [5]. The combination of both rule-based and machine learning approaches also exists such as [4].

In this paper we propose a classification approach to the morphological disambiguation problem. The idea behind our algorithm is as follows: Considering each set of distinct possible parses as a classification problem, one can divide the morphological disambiguation problem into multiple classification problems. Then each classification problem can be solved independently using any machine learning classifier. The inputs (features) of the classification problem are the existence of the part of speech tags in the previous two neighbor words.

The paper is organized as follows: In Sect. 2 we introduce the morphological disambiguation problem and formalize it. In Sect. 3, we will review the previous approaches to the morphological disambiguation problem in Turkish. In Sect. 4 we introduce our proposed approach. We give our experiments results in Sect. 5 and conclude in Sect. 6.

2 Morphological Disambiguation

Morphological disambiguation is the problem of selecting accurate morphological parse of a word given its possible parses. These parses are generated by a morphological analyzer [6, 7]. In morphologically rich languages like Turkish, the number of possible parses for a given word is generally more than one. Each parse is considered as a different interpretation of a single word. Each interpretation consists of a root word and sequence of inflectional and derivational suffixes. Table 1 illustrates different interpretations of the word “üzerine”.

As seen above, the first two parses share the same root but different suffix sequences. Similarly, the last two parses also share the same root, however sequence of morphemes are different. Given a parse such as

$$\text{üz} + \text{Verb} + \text{Pos} + \text{Aor} +^{\wedge} \text{DB} + \text{Adj} + \text{Zero} +^{\wedge} \text{DB} + \text{Noun} + \text{Zero} + \text{A3sg} + \text{P3sg} + \text{Dat}$$

each item is separated by “+” is a morphological feature such as Pos or Aor. Inflectional groups are identified as sequence of morphological features separated by

Table 1 Four possible parses of word "üzerine"

üzer+Noun+A3sg+P3sg+Dat
üzer+Noun+A3sg+P2sg+Dat
üz+Verb+Pos+Aor+^ DB+Adj+Zero+^ DB+Noun+Zero+A3sg+P3sg+Dat
üz+Verb+Pos+Aor+^ DB+Adj+Zero+^ DB+Noun+Zero+A3sg+P2sg+Dat

derivational boundaries (^DB). The sequence of inflectional groups forms the term tag. Root word plus tag is named as word form. So, a word form is defined as follows:

$$\text{IGroot} + \text{IG}_1 + {}^\wedge\text{DB} + \text{IG}_2 + {}^\wedge\text{DB} + \cdots + {}^\wedge\text{DB} + \text{IG}_n$$

Then the morphological disambiguation problem can be defined as follows: For a given sentence represented by a sequence of words $W = w_1^n = w_1, w_2, \ldots, w_n$, determine the sequence of parses $T = t_1^n = t_1, t_2, \ldots, t_n$, where t_i represents the correct parse of the word w_i. Using the Bayesian approach, the problem is formulated as follows:

$$\arg\max_T P(T|W) = \frac{P(T)P(W|T)}{P(W)} \tag{1}$$

where $P(W)$ is constant for all $P(W|T).P(W|T)$ is equal to 1, since given a tag sequence, there is only one possible word form corresponding to it. So the morphological disambiguation problem is simplified as the following:

$$\arg\max_T P(T|W) = \arg\max_T P(T) \tag{2}$$

3 Related Work

The baseline model described in [1] generates the most probable tag sequence for a given word sequence using Viterbi decoding. First, they break down the tag from derivation boundaries called inflectional group (IG). The problem is formulated as follows:

$$\begin{aligned} P(T) &= \prod_{i=1}^{n} P(t_i|t_{i-2}, t_{i-1}) \\ &= \prod_{i=1}^{n} ((r_i, IG_{i,1}, \ldots, IG_{i,n_i})| \\ &\quad (r_{i-2}, IG_{i-2,1}, \ldots, IG_{i-2,n_{i-2}}) \\ &\quad (r_{i-1}, IG_{i-1,1}, \ldots, IG_{i-2,n_{i-1}})) \end{aligned} \tag{3}$$

where n_i represents the number of inflectional groups associated with the ith parse and $G_{i,j}$ represents the jth inflectional group of parse i. The baseline trigram-based model is based on two basic assumptions: (1) root of the current word only depends on root of two previous words, and (2) presence of sequence of IGs in the current word depend only the last IG of two previous words. Under these assumptions, $P(T)$ is re-formulated as:

$$P(T) = \prod_{i=1}^{n}((P(r_i|r_{i-2}, r_{i-1}) \prod_{k=1}^{n_i} P(IG_{i,k}|IG_{i-2,n_{i-2}}, IG_{i-1,n_{i-1}}) \tag{4}$$

The trigram probabilities are estimated using standard n-gram probability estimation methods using morphologically disambiguated training data.

The Greedy Prepend Algorithm is a rule-based approach, based on decision lists [4]. Each pattern is formed by surface attributes of surrounding words of the current word. The decision lists are formed for each of the 126 distinct morphological features that exist. In the model, a 5-word (including word W, the first two left and two right neighbors), window is used. Greedy Prepend list reduction algorithm is used to generate the decision lists. The algorithm starts with the most general rule which covers all instances. The algorithm adds rules one by one where the best rule is determined using information gain. The algorithm stops when no improvement can be made.

The Perceptron Algorithm [3] is a combination of statistical and machine learning approaches. They use the Baseline Trigram-Based Model to generate n-best parses for each sentence. A feature set consisting of 23 features is used to disambiguate the current parse. The model also takes into account previous two words. Using the n-best parses as input to the algorithm, the algorithm makes multiple iterations over the training set to estimate parameter values. The highest scoring candidate is then selected using current parameter values. If the highest scoring candidate is different than the correct one, parameter values are updated.

4 Proposed Approach

We define the disambiguation task as identifying the correct parse from N possible parses excluding the root word. Consider the example in Table 2 for the word "üzerine". Our approach defines the classification problem as follows:

Class 1:Noun+A3sg+P3sg+Dat
Class 2:Noun+A3sg+P2sg+Dat
Class 3:Verb+Pos+Aor+^DB+Adj+Zero+^DB+Noun+Zero+A3sg+P3sg+Dat
Class 4:Verb+Pos+Aor+^DB+Adj+Zero+^DB+Noun+Zero+A3sg+P2sg+Dat

where the correct parse is Class 1. Although a word can take theoretically unlimited number of suffixes [5], the number of distinct problems (classification problem) is

Table 2 Distribution of problems with respect to the # of instances

Number of instances	Number of problems
1–10	7213
11–100	1617
101–1000	427
1001–10000	60
10001–100000	3

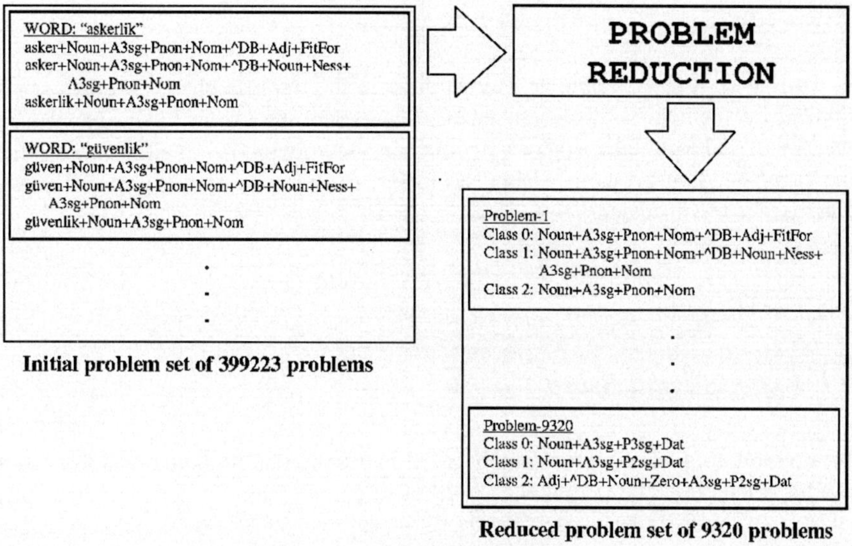

Fig. 1 Problem reduction step of the training phase. The parses of "askerlik" and "güvenlik" map to problem 1

9320 for a 1M size disambiguated train set. If *n* problems are same, only one of them is kept and others are discarded from the problem set. After this preprocessing the distribution of the problems with respect to the number of instances are given in Table 2.

Training data is processed sentence by sentence. We used 3-word window representation for each ambiguous token including it, where neighbor tokens are 2 words from left. Each neighbor is represented by a vector of 126 morphological features. Since, a neighbor may have more than one parse; each vector is formed based on the existence of morphological features. Any morphological feature that exists in any parses of neighbor words is represented by 1. Illustrations for the problem generation phase is given in Fig. 1.

Table 3 Comparison of our proposed approach (using 10 different classifiers) with three different approaches

Method	Acc. (%)	Method	Acc. (%)
NaiveBayes	93.83	Logistic Regression	94.67
Conjunctive Rule	66.25	SVM	94.98
k-NN(k=10)	95.40	LWL	94.67
J48 Tree	**95.61**	Baseline Trigram-Based Model	**95.48**
J48 Tree(no prunning)	95.09	Greedy Prepend Algorithm	**95.82**
KStar	94.36	Perceptron(23 Features)	**96.28**
NNge	90.49		

Testing is done in a similar way as done in the training phase. The test set is divided into sentences. For each sentence, we select the tokens having more than one parses and form the instance vector using 3-word window. Then, we determine the corresponding problem. Using the model of the corresponding classifier we classify the test instance.

5 Experiments

5.1 Experimental Setup

We use a training set of approximately 1M semi-automatically tagged disambiguated tokens (including end-of-sentence, end-of-title, and end-of-document markers) taken from Turkish newspapers. The training data consists of 50673 sentences where about 40% of them are morphologically ambiguous [6]. The test set consists of 958 tokens including markers mentioned above, where 42 sentences and 379 tokens are morphologically ambiguous. Performance criterion is formulated as:

$$Performance = \frac{\#\, of\ correctly\ disambiguated\ tokens}{\#\, of\ tokens} \quad (5)$$

5.2 Results

We have compared our approach with other morphological disambiguation approaches for Turkish presented in Sect. 3 We used 10 different classification algorithms in the WEKA toolkit. The results of the classification algorithms and the previous approaches are given in Table 3. According to emprical results, J48 Tree classifier has the best performance among 10 classifier and Baseline Trigram-Based Model.

Although it cannot can attain the performance of the other previous approaches, the difference is not significant for the best performer namely J48 Tree classifier.

6 Conclusion

We presented a new approach to the morphological disambiguation problem for Turkish. Our proposed approach converts original morphological disambiguation problem into multiple classification problems. Each classification problem corresponds to a set of possible parses (not including the root words) where each parse maps to a class and correct parse map to the correct class for that instance. We used 10 different classifiers from WEKA toolkit for solving the classification problems.

Our experimental results show that the best classifier is J48 Tree classifier. Although this is only better than the Baseline Trigram-Based approach among the previous aproaches, we believe that by expanding our feature set and/or applying a linear/nonlinear feature extraction mechanism, we will achieve much better disambiguation performance.

References

1. Hakkani-Tür, D.Z., Oflazer, K., Tür, G.: Statistical morphological disambiguation for agglutinative languages. Comput. Humanit. **36**(4), 381–410 (2002)
2. Oflazer, K., Hakkani-Tür, D. Z., Tür, G.: Design for a Turkish treebank. In: Proceedings of the Workshop on Linguistically Interpreted Corpora (1999)
3. Sak, H., Güngör, T., Saraçlar, M.: Morphological disambiguation of Turkish text with perceptron algorithm. In: Gelbukh, A. (ed.) CICLING 2007, LNCS 4394, pp. 107–118 (2007)
4. Yüret, D., Türe, F.: Learning morphological disambiguation rules for Turkish. In: Proceedings of HLT-NAACL (2006)
5. Oflazer, K., Kuruöz, I.: Tagging and morphological disambiguation of Turkish text. In: Proceedings of the 4th Applied Natural Language Processing Conference, pp. 144–149 (1994)
6. Oflazer, K.: Two-level description of Turkish morphology. Lit. Linguist. Comput. **9**(2), 137–148 (1994)
7. Sak, H., Güngör, T., Saralar, M.: Turkish language resources: morphological parser, morphological disambiguator and web corpus. In: GoTAL 2008, volume 5221 of LNCS, pp. 417–427, Springer (2008)

A Small Footprint Hybrid Statistical and Unit Selection Text-to-Speech Synthesis System for Turkish

Ekrem Guner and Cenk Demiroglu

Abstract Unit selection based text-to-speech synthesis (TTS) can generate high quality speech. However, The HMM-based text-to-speech (HTS) has also advantages such as the lack of spurious errors that are observed in the unit selection scheme. Another advantage is the small memory footprint requirement. Here, we propose a novel hybrid statistical/unit selection TTS system for agglutinative languages that aims at improving the quality of the baseline HTS system while keeping the memory footprint small. Listeners preferred the hybrid system over a state-of-the-art HTS baseline system in the A/B preference tests.

Keywords Speech synthesis · Hybrid TTS · HMM-based TTS · Turkish TTS · Small memory footprint · Agglutinative languages

1 Introduction

HMM-based and unit selection TTS are currently the two dominant approaches to speech synthesis. In general, unit selection systems can generate higher quality speech than the HTS systems when spurious errors are not present. However, HTS

This Research is supported by TUBITAK. Project no: 109E281

E. Guner (✉) · C. Demiroglu
Ozyegin University, Kusbakisi Cad. No:2,
34662 Istanbul, Turkey
e-mail: ekrem.guner@ozyegin.edu.tr;
ekremgunerozu@gmail.com
URL: http://www.ozyegin.edu.tr

C. Demiroglu
e-mail: cenk.demiroglu@ozyegin.edu.tr
URL: http://www.ozyegin.edu.tr

E. Gelenbe et al. (eds.), *Computer and Information Sciences II*,
DOI: 10.1007/978-1-4471-2155-8_10, © Springer-Verlag London Limited 2012

approach has advantages such as the the small memory footprint of the voice database. As opposed to the large databases needed for the unit selection systems, a couple of megabytes is enough to store the voice database in HTS which is important in where small memory footprint is a key requirement.

Besides HTS and unit selection approaches, there are hybrid systems. In one approach, parameter training for HTS can be improved by minimizing the error of selecting the wrong unit from the database when the HTS parameters are used to calculate the target costs in unit selection [1–4]. In another approach, HTS-generated waveforms are interweaved with the speech units selected from the database [5,6]. The idea is to use smooth HTS-generated waveforms when a unit with a low transition cost cannot be found in the database. In a third approach, synthetic speech generated with unit selection is smoothed at the unit boundaries using an HTS approach [7].

As opposed to most of the existing hybrid methods that are focused on improving the quality of a unit selection system, here, we propose a hybrid HTS/unit selection algorithm to boost the performance of our Turkish HTS system. In the existing hybrid systems, small memory footprint advantage of the HTS system is lost since both a unit selection and an HTS system are used. A key novelty in this work is a hybrid system that keeps the voice database size small while improving the quality of the HTS system. Turkish is an agglutinative language and many different words can be generated from the same root word by using a limited set of suffixes. In the proposed system, a database for the most frequently occurring suffixes in Turkish is created in training. In synthesis, best fitting suffixes are selected using the proposed suffix selection algorithms. Then, the selected suffixes are used in HTS within the proposed parameter generation algorithms. Although, the idea is applied to Turkish TTS here, it can also be used for other morphologically rich languages such as Finnish, Estonian and Czech.

This paper is organized as follows. An overview of the proposed hybrid system is given in Sect. 2. The suffix selection algorithms are presented in Sect. 3, parameter generation algorithms are presented in Sect. 4. Finally, experiment and discussions are given in Sect. 5.

2 Overview of the Hybrid System

An overview of the training and synthesis phases of the proposed system is shown in Fig. 1. In the training phase, HMM models and a decision tree are generated for the target speaker using speaker dependent training with the HTS tools. Then, a morphological analyzer is used to analyze the words in the speech database. To create a suffix database, waveforms that correspond to the suffixes labeled by the morphological analyzer should be extracted from speech. Forced alignment is used with the speaker-dependent HMM models to align text and speech data. Suffix units are then extracted from the speech signal using the alignment information.

Suffix units are parametrized using LPC analysis and only the LSF and pitch parameters are stored. Besides those parameters, each entry in the suffix database

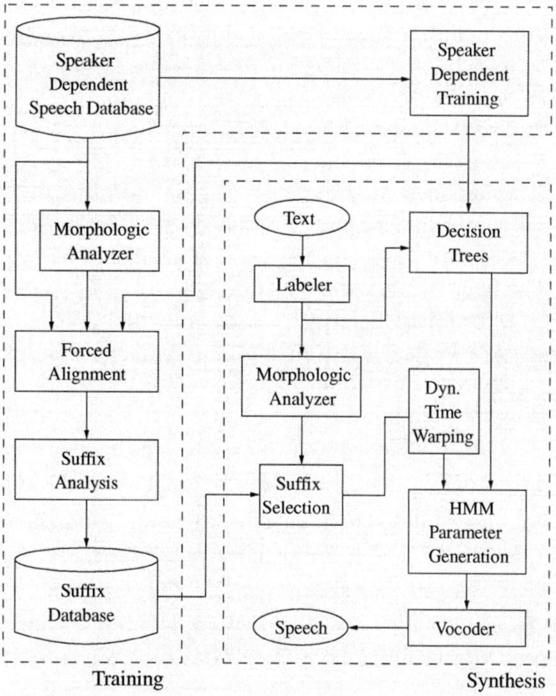

Fig. 1 Overview of the proposed hybrid system

contains a flag that indicates the presence of silence at the right context of the suffix (phrase ending) and another flag that indicates the presence of stress on the suffix. Moreover, beginning and end times of the state-level segments are also stored in the database.

In the synthesis phase, HMM models that correspond to the input text is determined using the decision tree. Input text is analyzed using the morphological analyzer, and, for each suffix in the text, the best fitting suffix is selected using the algorithms described in Sect. 3. The statistics predicted by the decision tree and the parameters of the unit that is selected from the suffix database are combined together and fed into the parameter generation algorithms described in Sect. 4. Finally, the parameter sequences generated are used in an LSF vocoder to synthesize speech. The morphological analyzer described in [8] is used here. The analyzer generates the root word and the morphemes of a given word.

3 Suffix Selection

When synthesizing utterance i, using the morphological analyzer, we determine the set of suffixes $\{S_i^j\}$ in the utterance where $j = 1, 2, \ldots, N_i$ and N_i is the total number of suffixes in the i th utterance. For the j th suffix, the initial set of available units in the database is denoted by $\{U_1^j\}$. In the initial set, only the stress flag of the suffix is used for selection. Two different targets are selected for LSF and pitch parameters. Moreover, the cost calculation for those features are also different.

The proposed unit selection algorithm uses a maximum likelihood (ML) criterion as the target cost. However, we have found in our experiments that the ML criterion does not always return a proper suffix that has a good concatenation cost. To reduce the possibility of an artifact, for the pitch parameter, we have used two heuristics to filter out the set of available units in the database for a given suffix. The heuristics are described below.

During parameter generation, the pitch trajectory of the selected unit is time-warped so that it can fit into the synthetic duration estimated with HTS. In our experiments, expanding the pitch trajectory did not cause any audible artifacts. However, compressing the pitch trajectory occasionally caused sudden pitch changes which are perceived as artifacts by the listener. To avoid that problem, the units in $\{U_1^j\}$ that are R_d percent longer than the synthetic duration of the suffix S_i^j are filtered out. The reduced set of units after filtering is denoted by $\{U_2^j\}$.

Pitch fall at the end of phrases is especially important in perception. Thus, if there is a pause or a short pause at the right context of S_i^j, which indicates the end of a phrase, then the units that do not have a pause or a short pause at their right contexts are filtered out from $\{U_2^j\}$, and the reduced set of units are denoted with $\{U_3^j\}$.

Finally, an ML criterion is used to select the suffix from $\{U_3^j\}$. Log-likelihood for each unit in $\{U_3^j\}$ is computed as follows. For each unit $U_{j,k}$ from $\{U_3^j\}$, the average log-likelihood is computed by

$$L_{j,k} = \frac{1}{N_{j,k}} \sum_{m=1}^{M} \sum_{f=1}^{m_f} log \left[\frac{1}{(2\pi)^{D/2}} \frac{1}{|\Sigma_m|^{1/2}} \right] - \frac{1}{2} (X_m^f - \mu_m)^T \sum_m^{-1} (X_m^f - \mu_m)$$

where $N_{j,k}$ is the total number of frames in the unit, M is the total number of states, m_f is the total number of frames in state m, Σ_m is the covariance matrix in state m, μ_m is the mean vector in state m and X_m^f is the f th observation of state m. X_m^f contains static, delta and delta-delta features.

The heuristics used in calculating the cost function for pitch are not used for the LSF features. Thus, the ML cost is the only criterion in selecting the appropriate suffix for LSFs.

4 Parameter Generation for the Hybrid System

In HTS, each utterance i is composed of a sequence of states and each state s has a set of models $\lambda_{i,s}$ for LSF and pitch features. For LSFs, the probability distribution is defined by a multivariate Gaussian specified by the parameter set

$\lambda_{i,s}^{lsf} = \{\mu_{i,s}^{lsf}, \Sigma_{i,s}^{lsf}\}$. For pitch, the probability distribution is defined by a multivariate Gaussian specified by the parameter set $\lambda_{i,s}^{p} = \{\mu_{i,s}^{p}, \Sigma_{i,s}^{p}\}$. These pdf's are used to generate the feature trajectories using the ML method.

In the hybrid system, once the best matching suffixes are selected for the LSF and pitch parameters, the features extracted from those suffixes are used to modify the parameter generation process of HTS. Similar to suffix selection, different parameter generation algorithms are used for the two features to obtain the best quality speech. The hybrid parameter generation algorithms for the LSF and pitch features are described below.

4.1 Parameter Generation for LSF

If state s in utterance i occurs within a suffix, the LSF model of the state is updated with $\{\mu_{i,s}^{lsf,suf}, \Sigma_{i,s}^{lsf}\}$ where $\mu_{i,s}^{lsf,suf}$ is the feature vector in the suffix that has the smallest distance to the mean parameter $\mu_{i,s}^{lsf}$. The distance is measured with the Itakura-Saito measure which is given by

$$d_{IS} = \int_{-\pi}^{+\pi} \left| \frac{|X_1(w)|^2}{|X_2(w)|^2} + \log \frac{|X_2(w)|}{|X_1(w)|} - 1 \right| \frac{dw}{2\pi} \tag{1}$$

where $X_1(w)$ is the spectrum specified by the LSF parameters of the mean vector $\mu_{i,s}^{lsf}$, and $X_2(w)$ is the spectrum specified by the LSF parameters of a feature vector in the target suffix.

After the pdf's are updated for each state that falls within a suffix, the ML-based HTS parameter generation process is used to create the final LSF trajectories.

4.2 Parameter Generation for Pitch

For the pitch parameters, preserving the intonation pattern of the target unit is important for improving the naturalness of the synthetic speech. Therefore, instead of changing the model parameters of the states, pitch trajectory of the target unit is directly concatenated with the synthetic speech. The trajectory is warped to fit into the HTS estimated suffix duration. Every phoneme in the suffix is warped individually as opposed to warping the whole suffix to compensate for the large phoneme duration variabilities in the suffixes.

Directly concatenating the pitch trajectory of the target unit into the HTS generated trajectory creates discontinuities at the boundaries. To address this problem, HTS parameter generation process is used to smooth out the trajectory around the concatenation points. However, besides smoothing the trajectory at the concatenation point, this approach also distorts the target unit trajectory which is undesirable. That problem is solved by setting the variances of the pitch models on the suffix to ϵ. Because the parameter generation process uses an ML-based measure, the system is effectively enforced to preserve the target trajectory while smoothing out the trajectory at the concatenation points.

5 Experiments and Discussions

500 utterances, 75 minutes of speech, were spoken by a female speaker. Systems in the experiments were trained using HTS 2.1 toolkit with 30 dimensional vectors consisting of 24 LSFs, 1 log F0 coefficient and 5 voicing strength parameters. Voicing strengths are computed using normalized auto-correlation measure for five evenly spaced spectral bands between 0 and 8000 Hz. The duration heuristic parameter R_d for the pitch parameter defined in Sect. 3 is experimentally set to 20 for hybrid system. In the target cost, log-likelihood is computed only for vowels, nasals, liquids and glides. Those long duration sounds have the most impact in the quality. Moreover, consonants can introduce significant amount of noise to the log-likelihood calculation because of their stochastic nature. We have observed that using only vowels, nasals, liquids and glides in the likelihood computation improves the suffix selection process.

A state-of-the-art baseline system is used to assess the quality improvement with the hybrid approach, using A/B preference test. 50 utterances from a Turkish novel were used in the test. Seven listeners took the test. In 34 percent of the utterances, listeners preferred the hybrid system. In 27 percent of the utterances, listeners preferred the baseline system, and in 39 percent of the cases, the listeners thought the quality was the same. Thus, there is a preference for the hybrid system over the baseline system. However, the difference is not large. The test results are further analyzed and it was observed that discontinuities that sometimes occur with the hybrid system had an impact on the listener preference. That same effect was also found to have a significant impact on the Blizzard Challenge tests. In fact, some of the HTS systems outperformed the unit selection systems in those tests due to the discontinuity problem [9].

In Turkish, question sentences typically have special suffixes, such as /mi/, /midir/, at the end of the verbs. In some significant number of cases with the baseline system, we have noticed over smoothed question tags which significantly hurt the listener preference. Most of those issues are resolved with the hybrid system since stress patterns of the question sentences are captured better by the hybrid system. The hybrid system generated more natural prosody for most of the suffixes since their intonation patterns are selected from the natural units in the suffix database.

The hybrid system improved the intonation contours and the clarity of suffixes. The LSF features improved the clarity and reduced the buzzy quality of the long duration sounds such as long vowels. For the shorter sounds, the effect is less noticeable since the trajectory is smoothed by the parameter generation algorithm.

References

1. Lu, H., Ling, Z.H., Lei, M., Wang, C.C., Zhao, H.H., Chen,L.H., Hu,Y. Dai,L.R., Wang, R.H.: The USTC system for Blizzard challenge 2009. In: Blizzard Challenge Workshop (2009)
2. Kawai, H., Toda, T., Ni, J., Tsuzaki, M., Tokuda, K.: XIMERA: a new TTS from ATR based on corpus-based technologies. In: Fifth ISCA Workshop on Speech Synthesis (2004)
3. Rouibia, S., Rosec, O.: Unit selection for speech synthesis based on a new acoustic target cost. In: INTERSPEECH, pp. 2565–2568. (2005)
4. Qian, Y., Yan, Z.J., Wu, Y., Soong, F.K., Zhuang, X., Kong, S.: An HMM trajectory tiling (HTT) approach to high quality TTS. In: INTERSPEECH, pp. 422–425. (2010)
5. Tiomkin, S., Malah, D., Shechtman, S., Kons, Z.: A hybrid text-to-speech system that combines concatenative and statistical synthesis units. In: Audio, Speech, and Language Processing, IEEE Transactions on, vol. pp. 99 (2010)
6. Pollet, V., Breen, A.: Synthesis by generation and concatenation of multiform segments. In: INTERSPEECH, pp. 1825–1828. (2008)
7. Plumpe, M., Acero, A., Hon, H.W., Huang, X.: HMM-based smoothing for concatenative speech synthesis. In: Fifth International Conference on Spoken Language Processing (1998)
8. Oflazer, K., Inkelas, S.: A finite state pronunciation lexicon for Turkish. In: Proceedings of the EACL Workshop on Finite State Methods in NLP, vol. 82, pp. 900–918. Budapest (2003)
9. Black, A.W., Zen, H., Tokuda, K.: Statistical parametric speech synthesis. In: Proceedings of ICASSP, vol. 4, pp. 1229–1232. (2007)

Enhancing Incremental Feature Subset Selection in High-Dimensional Databases by Adding a Backward Step

Pablo Bermejo, Luis de La Ossa, Jose A. Gamez and Jose M. Puerta

Abstract Feature subset selection has become an expensive process due to the relatively recent appearance of high-dimensional databases. Thus, the need has arisen not only for reducing the dimensionality of these datasets, but also for doing it in an efficient way. We propose the design of a new backward search which performs better than other state-of-the-art algorithms in terms of size of the selected subsets and in the number of evaluations, by removing attributes given a smart decremental approach and, besides, it is guided using a heuristic which reduces the needed number of evaluations commonly expected from a backward search.

1 Introduction

In the last 2 decades the evolution in technology has derived in new sources of information which must be stored, automatically classified and retrieved: e.g. microarrays gene expressions or textual databases, which contain records described by thousands or even tens of thousands (high-dimensional databases) variables. This way, lately the task known as *feature subset selection (FSS)* [3, 4]

P. Bermejo (✉) · L. de La Ossa · J. A. Gamez · J. M. Puerta
Castilla-La Mancha University, I3A, Albacete, Spain
e-mail: Pablo.Bermejo@uclm.es; pbermejo@dsi.uclm.es

J. A. Gamez (✉)
e-mail: jose.gamez@uclm.es

J. M. Puerta (✉)
e-mail: jose.puerta@uclm.es

L. de La Ossa
e-mail: luis.delaossa@uclm.es

E. Gelenbe et al. (eds.), *Computer and Information Sciences II*,
DOI: 10.1007/978-1-4471-2155-8_11,

has been an active area of research. This work focuses on hybrid (filter-wrapper) deterministic sequential search. The problem of forward search is that some attributes might be judged as relevant in a moment of the search but, once a new attribute is added to the set of selected features, a former relevant attribute might become non-relevant. This problem is tackled and alleviated in [2]; however, the complexity in practice is increased from $\mathcal{O}(n)$ to $\mathcal{O}(n^{1.2})$ (and having a $\mathcal{O}(n^2)$ worst case complexity). The method we propose in this work tries to detect irrelevant selected attributes by improving the incremental search with a new smart and also sequential backward stage. Backward search is a natural and well-known method to explore the search space, but it might result in a very expensive process. Thus, we provide a new criterion to search and choose features which should be removed from the current selected subset. The obtained algorithm is compared to several state-of-the-art sequential FSS algorithms, resulting that we obtain a better performance in terms of number of selected attributes, while maintaining the linear complexity of the search.

2 Incremental-Backward Wrapper Subset Selection (IWSS$_b$)

IWSS [1, 5] is a sequential hybrid search algorithm which uses a filter measure in order to obtain a marginal ranking of the attributes' relevance with respect to the class. Then, a forward best-first search is run over the ranking by incrementally adding those variables judged as relevant (given an acceptance criteria $\succ$) to the classification process, the relevance of new variables being measured in a wrapper way. The IWSS$_r$ algorithm [2] is an improvement to IWSS in which at each step of the forward search, not only the addition of a new attribute is evaluated but also to swap it with one of the features in the current subset of selected features. The idea is to detect conditional (in)dependencies of attributes respect to the class given the current subset of selected attributes. This improvement increases the worst-case complexity from linear to $\mathcal{O}(n^2)$. We propose the implementation of an heuristic-driven decremental-backward Wrapper Subset Search (IWSS$_b$) which takes the advantages of IWSS (hybrid, linear complexity, smart acceptance criteria), while adding a second step which starts from the reduced subset found by IWSS and then expands the search space by removing attributes given the same acceptance criteria of IWSS but used as removal criteria, which requires that the reduced subset does not perform *significantly* worse. By removing features from the selected subset in a smart way, we aim to discover those attributes which were selected in the first hand but they are not relevant anymore due to some other attribute(s) being selected afterwards. In order to obtain a canonical IWSS$_b$ algorithm, the code shown in Fig. 1 is run over the output $\mathcal{S}$ obtained by IWSS.

By preliminary experimentation we found more suitable for the backward step to analyze the attributes in S in reverse-order of inclusion, that is, backward phase first tries to remove in-S younger attributes. From the study carried out in [1] we propose to use $mf = 2$ as acceptance crietrion both in the forward and backward

backwardSearch method.

In **T** training, $\mathscr{C}$ classifier, $\mathscr{S}$ current subset, *best*$\mathscr{S}$ empty set
Out $\mathscr{S}$ final selected subset

```
BestPerformance=evaluate(𝒞,𝒮,T)
while(best𝒮 != null)
  best𝒮=null
  for i = |𝒮|-1 to 1     //find best&youngest
    𝒮_aux = 𝒮 - 𝒮[i]
    AuxPerformance = evaluate(𝒞,𝒮_aux,T)
    if (AuxPerformance ≻ BestPerformance)
      best𝒮 = 𝒮_aux
      break
  if best𝒮 != null
    BestPerformance = evaluate(𝒞,best𝒮,T)
    𝒮 =best𝒮
```

Fig. 1 Our heuristic-driven backward method $IWSS_b$

phase. That means, that a given attribute is worth to be included (to remain) in *S* only if after an inner 5-cross-validation it gets better mean accuracy (does not get worse mean accuracy) and also better accuracies (not worse accuracies) in at least 2 out of the 5 folds, in the forward and backward stages. Our contribution is then a fully sequential rank-guided forward-backward method, which results in an efficient algorithm with linear complexity and that outperforms compared state-of-the-art algorithms.

3 Experiments

In order to evaluate our proposal we downloaded a corpus composed of 9 datasets from the *ASU Feature Selection Repository*[1], ranging from 2400 to 46151 features. The methodology of our experiments is to find out if our proposed algorithm $IWSS_b$ outperforms other state-of-the-art feature selection algorithms: SFS [3], IWSS and $IWSS_r$. Results are shown in Table 1, where the last two rows show the statistical tests results (paired one-tail Wilcoxon signed-ranks test [6]); with confidence level $\alpha = 0.001$. We first compare by accuracy; then by number of attributes selected and finally those remaining are compared by number of evaluations. At each step, those algorithms found to be statistically worse than $IWSS_b$ are crossed-out

The conclusions shown by the statistical tests are that $IWSS_b$ performs statistically better in both terms of number of selected attributes than SFS, IWSS and

[1] http://featureselection.asu.edu

Table 1 $IWSS_b^{mf-mf}$ against other search algorithms, 10CV and c4.5 classifier

Dataset	#Atts	SFS			$IWSS^{mf}$			$IWSS_r^{mf}$			$IWSS_b^{mf-mf}$		
		Acc.	#Att.	#Evals.	Acc.	#Att.	#Evals.	Acc.	#Att.	#Evals.	Acc.	#Att.	#Evals.
warpPIE10P	2421	77.1	7.4	20295.3	83.3	17.5	2420.0	80.0	9.9	25980.3	78.1	7.8	2487.5
warpAR10P	2400	60.8	7.9	21323.0	70.0	15.2	2400.0	70.0	10.5	26619.7	70.0	7.3	2459.0
pixraw10P	10000	89.0	2.4	33995.8	92.0	4.5	10000.0	94.0	3.9	48832.2	92.0	2.4	10009.0
orlraws10P	10304	70.0	4.3	54599.4	75.0	12.4	10304.0	65.0	7.6	87938.7	76.0	4.8	10346.4
TOX-171	5749	57.3	8.4	53990.8	55.6	20.3	5748.0	59.6	14.0	81281.1	55.6	9.9	5829.9
SMK-CAN-187	19993	53.5	8.2	183897.4	67.9	13.9	19993.0	66.3	10.7	216074.6	70.6	8.5	20028.7
GLI-85	22283	84.7	2.3	73529.9	78.8	3.0	22283.0	82.4	2.8	84579.1	76.5	2.3	22287.6
GLA-BRA-180	46151	48.9	9.0	491464.2	62.2	17.2	49151.0	68.9	11.4	554899.3	64.4	9.4	49217.8
CLL-SUB-111	11340	65.8	4.7	64624.3	58.6	12.7	11340.0	64.0	8.6	104419.6	59.5	6.0	11380.4
pcmac	3289	82.2	15.1	52818.4	83.1	15.8	3289.0	82.8	15.5	44197.5	82.9	13.6	3317.7
basehock	4862	87.6	15.3	79112.1	88.0	17.1	4862.0	88.1	15.8	75321.2	88.3	14.1	4897.4
Mean	**12617.5**	**70.6**	**7.7**	**102695.5**	**74.0**	**13.6**	**12890.0**	**74.6**	**10.1**	**122740.3**	**74.0**	**7.8**	**12932.9**
Tests													
Accuracy				**70.63**			**74.04**			**74.64**			**73.98**
#Attributes				**7.7**			**13.6**			**10.1**			**7.8**
#Evals				**102695.5**									**12932.9**

$IWSS_r$ and number of evaluations. Furthermore, we can observe that our proposal $IWSS_b$ provides a linear complexity, way far from $IWSS_r$ and SFS.

4 Conclusions

Our proposal, $IWSS_b$, presents a design of the backward search which is heuristic-driven and is run over a reduced start set. $IWSS_b$ has been compared to three sequential algorithms known in the literature, using the same acceptance/removal criterion. The obtained results conclude that $IWSS_b$ behaves the same in terms of classification rate than SFS, IWSS and $IWSS_r$, while reducing the cardinality of the final subset and, which is very important for high-dimensional databases, maintaining an in-practice linear complexity for the 11 datasets used in our experiments.

Acknowledgments This work has been partially supported by the JCCM under project PCI08-0048-8577 and CICYT under project TIN2010-20900-C04-03.

References

1. Bermejo, P., Gámez, J., Puerta, J.: On incremental wrapper-based attribute selection: experimental analysis of the relevance criteria. In: IPMU'08: Proceedings of the 12th International Conference on Information Processing and Management of Uncertainty in Knowledge-Based Systems, Málaga (2008)
2. Bermejo, P., Gámez, J.A., Puerta, J.M.: Incremental wrapper-based subset selection with replacement: an advantageous alternative to sequential forward selection. In: Proceedings of the IEEE Symposium on Computational Intelligence and Data Mining (CIDM-2009), Málaga (2009)
3. Guyon, I., Elisseeff, A.: An introduction to variable and feature selection. J. Mach. Learn. Res. **3**, 1157–1182 (2003)
4. Liu, H., Motoda, H.: Feature Extraction Construction and Selection: A Data Mining Perspective. Kluwer Academic Publishers, Boston (1998)
5. Ruiz, R., Riquelme, J.C., Aguilar-Ruiz, J.S.: Incremental wrapper-based gene selection from microarray data for cancer classification. Pattern Recogn. **39**, 2383–2392 (2006)
6. Wilcoxon, F.: Individual comparisons by ranking methods. Biometrics Bulletin **1**, 80–83 (1945)

Memory Resident Parallel Inverted Index Construction

Tayfun Kucukyilmaz, Ata Turk and Cevdet Aykanat

Abstract Advances in cloud computing, 64-bit architectures and huge RAMs enable performing many search related tasks in memory.We argue that term-based partitioned parallel inverted index construction is among such tasks, and provide an efficient parallel framework that achieves this task. We show that by utilizing an efficient bucketing scheme we can eliminate the need for the generation of a global index and reduce the communication overhead without disturbing balancing constraint. We also propose and investigate assignment schemes that can further reduce communication overheads without disturbing balancing constraints. The conducted experiments indicate promising results.

1 Introduction

Inverted index is the de-facto data structure used in state-of-the-art text retrieval systems. Even though it is quite simple as a data structure, Web-scale generation of a global inverted index is very costly [2]. Since the data to be indexed is crawled and stored by distributed or parallel systems (due to performance and scalability reasons), parallel index construction techniques are essential. Despite the popularity of document-based partitioned inverted indices, term-based partitioning has advantages that can be exploited for better query processing [4].

The following studies on index construction [3, 5, 6] extend disk-based techniques for parallel systems. In [5, 6], authors propose a parallel disk-based algorithm with a centralized approach to generate the global vocabulary. In [5] authors analyze the merging phase of the inverted lists and present three algorithms. In [3], authors start from a document partitioned collection and proposed a software-pipelined inversion architecture.

T. Kucukyilmaz · A. Turk · C. Aykanat (✉)
Computer Engineering Department, Bilkent University, 06800 Ankara, Turkey
e-mail: aykanat@cs.bilkent.edu.tr

E. Gelenbe et al. (eds.), *Computer and Information Sciences II*,
DOI: 10.1007/978-1-4471-2155-8_12,

With the advent of 64-bit architectures, huge memory spaces are available to single machines and even very large inverted indices can fit into the total distributed memory of a cluster of such systems, enabling memory-based index construction. Given the current advances in network technologies and cloud computing and the high availability of low cost memory, in-memory solutions for parallel index construction should be considered seriously.

In this work, we propose an efficient, memory-based, parallel index construction framework for generating term-based partitioned inverted indices starting from a document-based partitioned collection (possibly due to parallel crawling). In this framework, we propose to mask the communication costs associated with global vocabulary construction and communication with a term-to-bucket assignment schema. Furthermore, we investigate several assignment heuristics for improving both the final storage balance and the communication costs of inverted index construction. Here, storage balance is important since it relates to query processing loads of processors, whereas the communication cost is important since it determines the running time of parallel inversion. Our contributions in this work are prior to optimizations such as compression [7].

2 Parallel Inversion

Our overall parallel inversion scheme has the following phases: local inverted index construction, term-to-processor assignment, and inverted list exchange and merge. In this section we describe these three phases in detail.

The first task in our framework is generating local inverted indices for all local document collections on each processor. As each processor contains a non-overlapping portion of the whole document collection, this operation can be achieved concurrently without communication.

After local inverted index construction phase, a term-to-processor assignment phase has to follow. In order to achieve a term-to-processor assignment, normally a global vocabulary has to be generated. This could be done by sending each term string, in its word form, to a host processor, where a global vocabulary is constructed. However in such a scheme, a particular term would be sent to the host machine by all processors if all processors contain that specific term. Thus we propose to group terms into a fixed number of buckets prior to the term-to-processor assignment. Using hashing, each word in a local vocabulary is assigned to a bucket. Afterwards, a host processor computes a bucket-to-processor assignment and broadcasts this information to the processors.

At the end of local inversion phase, each processor has a local vocabulary and a set of inverted lists for its terms. However, different processors may contain different portions of an inverted list for each term. For the final term-based partitioned inverted index to be created, the inverted list portions of each term should be accumulated to a single processor. To this end, each term in the global vocabulary should be assigned to a particular processor. This term-to-processor assignment depicts an inverted index partitioning problem. Many different criteria can be considered when finding a suitable index partitioning, but we focus on balancing the storage loads of processors and minimizing the communication overhead of the inversion process. The storage balance guarantees an evenly distribution of the final inverted index. As the memory is assumed to be limited throughout this work, with an even distribution of the storage loads, larger indices can fit in the same set of processors. Storage balancing is also expected to infer balancing on the query processing loads of the processors. Since inversion is a highly communication-bound process, the minimization of the communication overhead ensures that the inverted list exchange phase of the parallel inversion process takes less time. In this work, we model the minimization of the communication overhead as the minimization of the total communication volume while maintaining balance on communication loads of processors.

At the end of the bucket-to-processor assignment phase, all assignments are broadcast to processors and processors exchange their partial inverted lists in an all-to-all fashion. When sending the local inverted lists to their assigned processor, the vocabulary should also be sent since processors do not necessarily contain all vocabulary terms and do not know which terms will be retrieved from other processors. Although such a communication incur additional costs, since the processor-to-host bottleneck due to global vocabulary construction is already avoided, this additional communication cost is easily compensated.

The inverted-list exchange between processors is achieved in two steps. In the first step, the terms (in word form) and their posting sizes are communicated. At the end of this step, all processors obtain their final local vocabularies and can reserve space for their final local inverted index structures. Then the inverted lists are exchanged. At the end of inverted list exchange, posting lists for each term are merged and written into their reserved spaces in local inverted indices.

3 Bucket-to-Processor Assignment Schemes

In the forthcoming discussions we use the following notations: The vocabulary of terms is indicated with $\mathcal{T}$. The posting list of each term $t_j \in \mathcal{T}$ is distributed among the K processors. $w_k(t_j)$ denotes the size of the posting list portion of term t_j that resides in processor p_k at the beginning of the inversion, whereas $w_{tot}(t_j) = \sum_{k=1}^{K} w_k(t_j)$ denotes the total posting list size of term t_j.

We assume that prior to bucket-to-processor assignment, each processor has built its local inverted index $\mathcal{I}_k$ and partitioned the vocabulary $\mathcal{T} = \{t_1, t_2, \ldots, t_n\}$ containing n terms, into a predetermined number m of buckets. The number of buckets m is selected such that $m \ll n$ and $m \gg K$. Let $\mathcal{B} = \Pi(\mathcal{T}) = \{\mathcal{T}_1 = b_1, \mathcal{T}_2 = b_2, \ldots, \mathcal{T}_m = b_m\}$ denote a random term-to-bucket partition, where $\mathcal{T}_i$ denotes the set of terms that are assigned to bucket b_i. In $\mathcal{B}$, $w_{tot}(b_i)$ denotes the total size of the posting lists of terms that belong to b_i.

In an m-bucket and K-processor system, the bucket-to-processor assignment can be represented via a K-way partition $\Pi(\mathcal{B})=\{\mathcal{B}_1,\mathcal{B}_2,\ldots,\mathcal{B}_k\}$ of the buckets among the processors. In $\Pi(\mathcal{B})$, $w_k(b_i)$ denotes the total size of the posting list portions of terms that belong to b_i and reside in processor p_k at the beginning of the inversion. The performance of a bucket-to-processor assignment is measured in terms of two metrics. The storage load balance and the communication cost. The storage load $S(p_k)$ of a processor p_k induced by the assignment $\Pi(B)$ is defined as follows:

$$S(p_k) = \sum_{b_i \in \mathcal{B}_k} \sum_{t_j \in b_i} w_{tot}(t_j). \tag{1}$$

The communication cost of a processor p_k induced by assignment $\Pi(B)$ has two components. Each processor must receive all portions of the buckets assigned to itself from other processors. Thus total receive cost/volume of a processor p_k is:

$$Recv(p_k) = \sum_{b_i \in \mathcal{B}_k} \sum_{t_j \in b_i} (w_{tot}(t_j) - w_k(t_j)). \tag{2}$$

Each processor also sends all postings that are not assigned to it to some other processor. The total send cost of p_k is:

$$Send(p_k) = \sum_{b_i \notin \mathcal{B}_k} \sum_{t_j \in b_i} w_k(t_j) \tag{3}$$

Total communication cost of a processor is defined as the sum of its send and receive costs.

The MCA scheme is based on the following simple observation [1]. If we assign each bucket $b_i \in \mathcal{B}$ to processor p_k that has the largest $w_k(b_i)$ value, we will achieve an assignment with globally minimum total communication volume.

The BLMCA scheme incorporates a storage balance heuristic to MCA [1]. This scheme works in an iterative manner assigning one bucket to a processor at a time. For each bucket, first the processor that will cause the minimum total communication is determined using MCA scheme. If this processor is not the bottleneck processor (in terms of storage load) at that iteration, the bucket is assigned to that processor. Otherwise, the bucket is assigned to the minimally loaded processor.

In BLMCA, two cost metrics, storage load balance and communication cost are calculated and at each iteration an assignment decision that optimizes only one of these metrics is made. The decisions of MCA and BLMCA regarding the communication cost minimization only optimizes the total communication cost and ignores the maximum communication cost of a processor. In order to minimize maximum communication cost, we should consider both the receive cost of the assigned processor and the send costs of all other processors.

To this end we define the energy E of an assignment $\Pi(B)$ based on the storage loads and communication costs of processors. We define two different energy functions for a given term-to-processor assignment $\Pi(B)$:

$$E^1(\Pi(B)) = Max\{Max_{1\leq k\leq K}\{Comm(p_k)\}, Max_{1\leq k\leq K}\{S(p_k)\}\} \quad (4)$$

$$E^2(\Pi(B)) = \sum_{1}^{K}(Comm(p_k))^2 + \sum_{1}^{K}(S(p_k))^2 \quad (5)$$

Utilizing these energy functions, we propose a constructive algorithm that assigns buckets to processors one-by-one. The energy increase in the system by K possible assignments of each bucket are considered, and the assignment that incurs the minimum energy increase is performed. That is, for the assignment of a bucket b_i in the given order, we select the assignment that minimizes $E(\Pi(B_{i-1} \cup \{b_i\})) - E(\Pi(B_{i-1})$, where B_{i-1} denotes the set of already assigned buckets. We call E^1-based and E^2-based assignment schemes as E^1A and E^2A respectively in our experiments.

4 Experiments

In order to test the performance of the proposed assignment schemes for parallel inversion, we conducted two types of experiments. The first set of experiments are simulations to report on the storage imbalance and communication volume performances of the assignment schemes. The second set of experiments are actual parallel inversion runs provided in order to show how improvements in performance metrics relate to parallel running times. These experiments are conducted on a PC-cluster with $K = 32$ nodes, where each node is an Intel Pentium IV 3.0 GHz processors with 1 GB RAM connected via an interconnection network of 100 Mb/sec fast ethernet.

We conducted our experiments on a realistic dataset obtained by crawling educational sites across America. The raw size of the dataset is 30 GB and contains 1,883,037 pages from 18,997 different sites. The biggest site contains 10,352 pages while average number of pages per site is 99.1. The vocabulary of the dataset consists of 3,325,075 distinct terms. There are 787,221,668 words in the dataset. The size of the inverted index generated from the dataset is 2.8 GB.

Tables 1 and 2 compare the storage load balancing and communication performances of the assignment schemes for $K = \{4, 8, 16, 32, 64, 128\}$. We also implemented a random assignment (RA) algorithm, which assigns buckets to processors randomly, as a baseline assignment scheme. As seen in Table 1, MCA achieves the worst final storage imbalance. This is expected since MCA considers only minimization of the total communication cost, disregarding storage balance and as seen in Table 2 MCA achieves lowest average communication cost. BLMCA algorithm on the other hand, achieves best final storage imbalance. This is also expected since the primary objective of BLMCA is to balance the processor loads during the assignments instead of minimizing the communication costs. As seen in Table 2, this storage balancing performance is achieved at the expense of higher average communication values per processor. Experiments indicate that

Table 1 Percent load imbalance values

	Initial	Final				
K		RA	MCA	BLMCA	E^1A	E^2A
4	4.4	12.1	38.3	0.0	6.1	5.5
8	11.7	09.9	60.0	0.1	18.2	14.4
16	18.2	27.4	66.2	1.7	27.2	20.0
32	44.1	29.6	83.0	5.4	35.2	31.1
40	32.2	37.0	77.4	6.2	38.4	31.4
64	44.7	56.6	92.2	11.5	46.9	33.7
128	65.3	94.7	95.6	15.7	64.1	40.4

Table 2 Message volume (send + receive) per processor (in terms of $\times 10^6$ postings)

	RA		MCA		BLMCA		E^1A		E^2A	
K	Avg	Max	Avg	Max	Avg	Max	Avg	Max	Avg	Max
4	131.19	145.71	122.09	150.26	127.45	128.62	124.76	125.29	131.24	131.36
8	76.55	90.58	71.45	119.75	73.40	75.97	72.02	74.53	76.67	76.79
16	41.01	49.25	38.32	77.11	39.22	43.44	38.52	42.33	41.60	41.66
32	21.19	28.75	19.82	71.13	20.28	26.03	19.94	25.08	20.54	21.61
40	17.05	23.96	15.99	44.79	16.32	20.01	16.03	19.22	17.04	17.71
64	10.76	17.77	10.09	74.27	10.34	15.42	10.15	14.75	10.86	11.89
128	5.42	11.97	5.09	65.59	5.22	10.98	5.12	10.02	6.81	7.95

Table 3 Parallel inversion times (in seconds) including assignment and inverted list exchange times for different assignment schemes

K	RA	MCA	BLMCA	E^1A	E^2A
4	69.19	81.34	68.63	68.67	68.49
8	51.42	66.76	46.45	46.59	45.74
16	35.89	60.82	33.04	32.90	32.48
32	19.31	49.45	18.20	17.91	17.20

E^1A and E^2A algorithms both achieve reasonable storage load balance that are either close or better than the performance of RA scheme. Also as seen in Table 2, for K values higher than 8, E^2A achieves the lowest maximum communication volumes. Table 2 also indicates that the average and maximum communication costs induced by E^2A are very close, which means that E^2A manages to distribute the communication loads among processors evenly.

Table 3 shows the running times of our parallel memory-based index inversion algorithm under different assignment schemes. In this table, it is assumed that the local inverted indices are already created and the time for this operation is neglected. As expected from the results presented in Table 1 and Table 2, MCA induces the highest inversion time, RA, BLMCA, E^1A, and E^2A induce similar inversion times and the E^2A scheme achieves the lowest inversion times.

5 Conclusions

In this paper, a memory-based parallel inverted index construction framework was examined. An extensive step-by-step experimentation of our model was presented and further insight were provided using theoretical results and simulations. Also, several problems involving the creation of this framework were identified.

References

1. Aykanat, C., Cambazoglu, B.B., Findik, F., Kurc, T.: Adaptive decomposition and remapping algorithms for object-space-parallel direct volume rendering of unstructured grids. J. Parallel Distrib. Comput. **67**, 77–99 (2007)
2. Cho, J., Garcia-Molina, H.: The evolution of the web and implications for an incremental crawler. In: Proceedings of the 26th International Conference on VLDB (2000)
3. Melink, S., Raghavan, S., Yang, B., Garcia-Molina, H.: Building a distributed full-text index for the web. ACM Trans. Inf. Syst. **19**, 217–241 (2001)
4. Moffat, A., Webber, W., Zobel, J.: Load balancing for term-distributed parallel retrieval. In: Proceedings of the 29th Annual International ACM SIGIR Conference on Research and Development in Information Retrieval, pp. 348–355 (2006)
5. Ribeiro-Neto, B., Moura, E.S., Neubert, M.S., Ziviani, N.: Efficient distributed algorithms to build inverted files. In: Proceedings of the 22nd Annual International ACM SIGIR Conference on Research and Development in IR, pp. 105–112 (1999)
6. Ribeiro-Neto, B.A., Kitajima, J.P., Navarro, G., Sant'Ana, C.R.G., Ziviani, N.: Parallel generation of inverted files for distributed text collections. In: Proceedings of the 18th International Conference of the Chilean Computer Science Society (1998)
7. Zobel, J., Moffat, A., Ramamohanarao, K.: Inverted files versus signature files for text indexing. ACM Trans. Database Syst. **23**, 453–490 (1998)

Dynamic Programming with Ant Colony Optimization Metaheuristic for Optimization of Distributed Database Queries

Tansel Dökeroğlu and Ahmet Coşar

Abstract In this paper, we introduce and evaluate a new query optimization algorithm based on Dynamic Programming (*DP*) and Ant Colony Optimization (*ACO*) metaheuristic for distributed database queries. *DP* algorithm is widely used for relational query optimization, however its memory, and time requirements are very large for the query optimization problem in a distributed database environment which is an NP-hard combinatorial problem. Our aim is to combine the power of *DP* with heuristic approaches so that we can have a polynomial time approximation algorithm based on a constructive method. *DP* and *ACO* algorithms together provide execution plans that are very close to the best performing solutions, and achieve this in polynomial time. This makes our algorithm viable for large multi-way join queries.

Keywords Query optimization · Dynamic programming · Ant colony · Metaheuristic

1 Introduction

Research on distributed database management systems (DDBMSs) has been going on to meet the information processing demands of geographically separated organizations since 1970s. DDBMSs help reduce costs, increase performance by providing parallelism and improve accessibility and reliability [1]. A major

T. Dökeroğlu (✉) · A. Coşar
Computer Engineering Department, Middle East Technical University, Ankara, Turkey
e-mail: tansel@ceng.metu.edu.tr

A. Coşar
e-mail: cosar@ceng.metu.edu.tr

E. Gelenbe et al. (eds.), *Computer and Information Sciences II*,
DOI: 10.1007/978-1-4471-2155-8_13,

problem with DDBMSs is the querying of distributed data. The success of the querying tools mostly depends on their query optimization technologies. Query optimizers enumerate alternative plans, estimate costs, manage cost models and choose the best plan for given queries [2]. The main component of an optimizer is the algorithm it uses to explore the search space of a query. *DP* is one of the most widely used exhaustive enumeration algorithms. IBM's *System R* first used this technique in its query optimizer [3]. Query optimization on a DDB is an NP-hard problem [4]. Every optimizer must take into account and have a tradeoff between CPU time and system memory spent in query optimization and the quality of the generated plans, especially if a query plan will not be saved and thus used only once.

We present and evaluate *DP-ACO* algorithm for DDB queries, where the *ACO* [5] metaheuristic is combined with *DP* and its performance is evaluated. A *metaheuristic* is a general-purpose heuristic method to guide a problem specific heuristic towards more promising regions of the search space containing better solutions. *DP-ACO* also can produce good plans, as problem size grows, whereas *DP* starts suffering from the long execution times and very large memory requirements. For small queries with up to four relations, *DP-ACO* can produce plans very close to the best plans of *DP*. For larger queries involving more than five relations, *DP-ACO* can still produce efficient plans while *DP* takes prohibitively long execution times to come up with an optimal solution for six relations and fails due to insufficient memory for seven relations. *DP-ACO* also has the advantage that it can be easily adapted to existing query optimizers that commonly use *DP*-based algorithms [2]. In Sect. 2, an overview of the related studies is presented. Section 3 describes *DP*, *Processing Trees* (*PT*), and our cost model. Section 4 describes our new algorithm *DP-ACO*. Section 5 presents the results of our experiments on *DP-ACO, DP*, and a genetic algorithm (*SGA1*). Section 6 gives our conclusions and points at future research directions.

2 Related Work

There are many algorithms proposed for query optimization in DDBMSs [6, 7]. Distributed INGRES, R*, and SSD-1 are the basic algorithms [8]. Query optimization algorithms can be classified as exhaustive (EA), approximation, and randomized (RA) algorithms. EAs always guarantee to find the best execution plan. They search the entire problem space and find an optimal solution. The most famous one of them is *DP* [3], which is still widely used in contemporary DBMSs. A* [9] is another representative of this category. Approximation algorithms have polynomial time and space complexity [2]. They produce sub-optimal, but provably within a range of the optimum value, execution plans in shorter times. RAs are non-deterministic algorithms and their results cannot be predicted in advance. The search space and time requirements of RAs can be controlled to achieve almost constant running times and minimal memory requirements.

A constructive algorithm called ACO-MJQA is proposed in [10] where a set of artificial ants build feasible query execution plans. In [11], data allocation problem is solved with *ACO-DAP*. More than one hundred NP-hard problems have been solved by using *ACO* algorithms so far [12]. Machine learning and bioinformatics are some of the other areas where *ACO* has been applied successfully.

3 Dynamic Programming

DP is the most famous query optimization algorithm in DBMSs [13]. *DP* works in a bottom-up manner and constructs all possible sub-plans. Optimal solution for any given sub-plan is calculated only once and is not computed again. The algorithm first builds access plans for every relation in the query and enumerates all two-way join plans in the second phase. All multi-way join plans are built using access-plans and multi-way join plans as building blocks. *DP* continues until it has enumerated all n-way join plans. An efficient plan can prune another plan with the same output. *DP* would enumerate (A $\bowtie$ B) and (B $\bowtie$ A) as two alternative plans, but only the better of the two plans would be kept after pruning. Execution plans are abstracted in terms of PT. The input of the optimization problem is given as a query graph. The query graph consists of all the relations that are to be joined as its leaf nodes. The PT is a simple binary tree. Its leaves correspond to base relations and inner nodes correspond to join operations. Edges indicate the flow of partial results from leaves to the root of the tree. Each plan has an execution cost. The aim of the optimization is to find the evaluation plan with the lowest possible cost [14]. If the inner relation of every join is a base relation, this type of PT is called a *left-deep tree*. There are *n!* ways to allocate *n* base relations to the leaves for this type of PTs. If there is no restriction about the PT, it is called a *bushy tree*. In our work we preferred to use only *left-deep trees*. *Left-deep trees* are used in most query optimizers like the optimizer of IBM System R [3].

A total-processing-time (CPU time + I/O time) based cost model has been used in this work. The cost of a plan is calculated in a bottom-up manner. The first level is comprised of 2-way joins. Here, all possible pairs of *n* relations are evaluated for all sites. If two relations are at the same site, the cost is only the I/O + CPU time for performing the equijoin operation using the nested-loop join algorithm. If only one of the relations, say A, is at the join site, total cost is accessing B from its site, shipping it to the join site over the communication network, matching and joining tuples of A with tuples of B, and writing the resulting joined (A,B) tuples to the local disk. If none of the relations are stored at the join-site, total cost is scanning both A and B, shipping them over the network, performing the join operation and writing the result to the disk. For the upper levels of PT, intermediate join results are treated as base tables.

4 Dynamic Programming with Ant Colony Optimization

Dorigo and his colleagues proposed *ACO* as a method for solving hard combinatorial problems [12]. *ACO* is a metaheuristic inspired by the behavior of real ants that cooperate through self-organization. A substance called *pheromone* is deposited on the ground while ants are foraging. *Pheromone* trails are formed on the ground by this way which also reduces stochastic fluctuations at the initial phases of search. The shorter trails will be used more frequently by ants and they gain more *pheromone*. By modifying appropriate *pheromone* values associated with edges of the graph we can simulate *pheromone* laying by artificial ants. *Evaporation* is another mechanism, and it is used so that artificial ants may forget the history of previously discovered solutions and search for new directions. We simulate the actions of ants on the graph of PTs. Each process trying to calculate the running time of execution plans is considered to be an ant in our algorithm. Solutions represent the food. The earlier the ants reach some food, the more *pheromone* they secrete on the way of solution. The more time it takes for an ant to travel down the path, the more time it has for *evaporation*. Foraging ants continuously look for better execution plans. The paths are the combinational alternatives for the plans of *DP* at each level. If we have five sites for the join operation of (A ⋈ B ⋈ C), there are five different possible paths. (A ⋈ B) is one of the sub-queries of this plan. For five sites, we have to check the response times of each. If we can find an optimal solution at site 2, we increase its *pheromone* whereas the other solutions of (A ⋈ B) have their pheromones reduced with evaporation. We also use a so called *Search Space Limit (SSL)* number to help us control the time and space complexity of our algorithm. The search space is pruned using the *pheromones* of each sub-plan. Without any pruning, *DP-ACO* acts like *DP*. In our experiments we used ($SSL = 1$) to be able to find solutions for large numbers of join operations. Increasing the *SSL* value causes an exponential rise in time and space requirements of the algorithm. *Pheromone* of each path is evaluated as in (1), where Best_Time(i) is the ith best plan.

$$Pheromone\_of\,(k) = \left(\sum_{i}^{SSL} Best\_Time(i) \; \times \; (1/Response\ time(k)) \right) \tag{1}$$

Path decisions can be formulated as in (2). $p^k ij$ is the transition probability of kth ant moving from ith sub-solution to jth sub-solution. l is an element of the sub-solutions. These sub-solutions are ordered as in SQL statement. α and β control the relative importance of pheromone (τij) versus the heuristic information (ηij).

$$p^k ij = \frac{\tau^\alpha ij \cdot \eta^\beta ij}{\sum_{l \in subsolutions} \tau^\alpha il \cdot \eta^\beta il} \tag{2}$$

Site 2	Site 1	Site 6	Site 3	Site 5	Site 4

Fig. 1 *Chromosome* structure for a distributed 7-way chain join

Sub-solutions for each multi-way join are calculated and *pheromones* are updated depending on the cardinality of relations. The amount of pruning is controlled by the *SSL* parameter.

5 Experimental Setup and Results

DP-ACO is experimentally compared with *DP* and a simple GA based algorithm *SGA1*. Quality of plans, running times, and space complexities of optimization algorithms are analyzed. We were able to use *DP* only up to 5-way join queries, while with *DP-ACO* and *SGA1* we were able to find solutions of up to 15-way join queries.

In our DDB environment, each site contains exactly one relation. Network is simulated with a complete graph topology, with no multi-hop transmissions and no store-and-forward delays. The relations have almost the same cardinality and referential integrity is guaranteed to be satisfied by all relations. SQL statements are multi-way chain joins where relations are ordered and unique. We limited our SQL statements to chain equijoin queries with a single selection predicate. An example SQL statement that we used with 7-way join can be given as:

SELECT A.Name
FROM A, B, C, D, E, F, G
WHERE (A ⋈ B ⋈ C ⋈ D ⋈ E ⋈ F ⋈ G) AND G.Income > 1.500

In *SGA1,* we have implemented *truncation* method to simply eliminate individuals with the lowest fitness values in the population. *Chromosomes* are built by *genes* that represent the sites where each join operation will be performed, as can be seen in Fig. 1. Assuming that the execution order of joins is same as given in the above SQL statement, A and B are joined at Site-2 to build (A ⋈ B), resulting join and C are shipped to Site-1 to build ((A ⋈ B) ⋈ C), and so on [15]. In *SGA1,* initial population size is fixed at 100. Five randomized runs have been performed and best plans have been reported in the experimental results. To measure the statistical confidence of *SGA1,* we calculated standard deviations of each plan and validated that quality variation of all plans are within acceptable limits.

DP, *SGA1,* and *DP-ACO* all have similar results up to 3-way joins. We were not able to run *DP* with more than 5-way joins as it exceeds the running time limit when it reaches 6-way joins. Running time requirements of *SGA1* algorithm increases linearly with the increasing number of joins, however it generates lower quality plans than *DP-ACO*. *SGA1's* generated plan quality degrades as search space gets larger. In Fig. 2, the quality of plans produced by each algorithm is given.

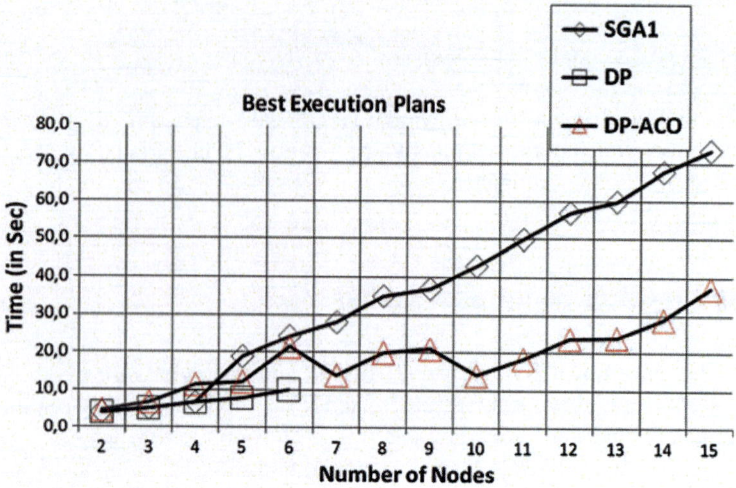

Fig. 2 Quality of execution plans

When comparing the quality of plans we use the generated plans' execution times but we must also consider the optimization times for finding those plans. *DP-ACO* can produce better execution plans than *SGA1* in less time. With the increase in the number of sites, the quality of plans generated by *SGA1* decreases compared to *DP-ACO*. Experiments show that *DP* is not a feasible algorithm with more than 5-way joins whereas *DP-ACO* and *SGA1* can easily generate up to 15-way join execution plans.

6 Conclusions

We have presented a new algorithm for optimization of distributed database queries, *DP-ACO*. This algorithm is based on *DP* and its capabilities have been extended by making use of *ACO metaheuristic*. When there is limited time and memory for coming up with a query execution plan, *DP-ACO* produces good execution plans, quickly and using very little memory. *DP-ACO* is also compatible with and can be easily adopted into the existing *DP*-based query optimizers. Time and space complexity of the system can be adjusted by using *SSL* parameter. As future work, we plan to implement *DP-ACO* algorithms to decide fragmentation and replication choices to design DDBMSs. Comparing *DP-ACO* with Iterative *DP* and analyzing how other exhaustive optimization algorithms behave when implemented together with *ACO* are other areas of interest. Designing a DDB on a parallel cluster machine by extending our model to include updates, replication and fragmentation, and assigning tasks of an optimized query execution plan to nodes of a parallel machine for parallel execution are also promising areas of research [16].

References

1. Ceri, S., Pelagatti, G.: Distributed Databases: Principles and Systems. McGraw-Hill Inc., New York (1984)
2. Kossmann, D., Stocker, K.: Iterative dynamic programming: a new class of query optimization algorithms. ACM Trans. Database Syst. (TODS) **25**(1), 43–82 (2000)
3. Selinger, P.G., Astrahan, M.M., Lorie, R.A., Price, T.G.: Access path selection in a relational database management system. In: Proceedings of the ACM International Conference on Management of Data (SIGMOD), pp. 23–34 (1979)
4. Ibaraki, T., Kameda, T.: On the optimal nesting order for computing n-relational joins. ACM Trans. Database Syst. **9**(3), 82–502 (1984)
5. Dorigo, M.: Optimization, learning and natural algorithms. Ph.D. thesis, Dipartimento di Elettronica, Politecnico di Milano, Milan (1992)
6. Swami, A.: Optimization of large join queries: combining heuristics and combinational techniques. In: Proceedings of the ACM Conference on Management of Data (SIGMOD), pp. 367–376 (1989)
7. Vance, B.: Join-order optimization with cartesian products. Ph.D. Dissertation. Oregon Graduate Institute of Science and Technology, Beaverton (1998)
8. Ozsu, M.T., Valduriez, P.: Principles of Distributed Database Systems, 3rd edn, pp. 245–293. Springer, Berlin (2011)
9. Kemper, A., Moerkotte, G., Peithner, K.: A blackboard architecture for query optimization in object bases. In: Proceedings of the Conference on Very Large Data Bases (VLDB), pp. 543–554 (1993)
10. Nana, L., Yujuan, L., Yongfeng, D., Junhua, G.: Application of ant colony optimization algorithm to multi-join query optimization. Lect. Notes Comput. Sci. **5370**, 189–197 (2008)
11. Adl, R.K., Rankoohi, S.M.T.R.: A new ant colony optimization based algorithm for data allocation problem in distributed databases, pp. 349–373. Springer-Verlag, New York (2009)
12. Dorigo, M., Birattari, M., Stützle, T.: Ant Colony Optimization. Technical Report 2006–023. Université Libre de Bruxelles, Belgium (2006)
13. Kossmann, D.: The state of art in distributed query optimization. ACM Comput. Surv. **32**, 422–469 (2000)
14. Swami, A., Iyer, B.: A polynomial time algorithm for optimizing join queries. In: Proceedings of the 9th International Conference on Data Engineering, pp. 345–354 (1993)
15. Sevinç, E., Coşar, A.: An evolutionary genetic algorithm for optimization of distributed database queries. Comput. J. **54**(5), 717–725 (2011)
16. Aguilar, J., Gelenbe, E.: Task assignment and transaction clustering heuristics for distributed systems. Inf. Sci. **97**(1–2), 199–219 (1997)

Part III
Green ICT

Energy Cost Model for Frequent Item Set Discovery in Unstructured P2P Networks

Emrah Cem, Ender Demirkaya, Ertem Esiner, Burak Ozaydin and Oznur Ozkasap

Abstract For large scale distributed systems, designing energy efficient protocols and services has become as significant as considering conventional performance criteria like scalability, reliability, fault-tolerance and security. We consider frequent item set discovery problem in this context. Although it has attracted attention due to its extensive applicability in diverse areas, there is no prior work on energy cost model for such distributed protocols. In this paper, we develop an energy cost model for frequent item set discovery in unstructured P2P networks. To the best of our knowledge, this is the first study that proposes an energy cost model for a generic peer using gossip-based communication. As a case study protocol, we use our gossip-based approach ProFID for frequent item set discovery. After developing the energy cost model, we examine the effect of protocol parameters on energy consumption using our simulation model on PeerSim and compare push–pull method of ProFID with the well-known push-based gossiping approach. Based on the analysis results, we reformulate the upper bound for the peer's energy cost.

Keywords Energy cost model · Energy efficiency · Peer-to-peer · Gossip-based · Epidemic · Frequent items

This work was partially supported by the COST (European Cooperation in Science and Technology) framework, under Action IC0804, and by TUBITAK (The Scientific and Technical Research Council of Turkey) under Grant 109M761.

E. Cem · E. Esiner · B. Ozaydin · O. Ozkasap (✉)
Department of Computer Engineering, Koc University, Istanbul, Turkey
e-mail: oozkasap@ku.edu.tr

E. Demirkaya
Department of Computer Engineering, Bilkent University, Ankara, Turkey

E. Gelenbe et al. (eds.), *Computer and Information Sciences II*,
DOI: 10.1007/978-1-4471-2155-8_14, © Springer-Verlag London Limited 2012

1 Introduction

Frequent items in a distributed environment can be defined as items with global frequency above a threshold value, where global frequency of an item refers to the sum of its local values on all peers. Frequent Item Set Discovery (FID) problem has attracted significant attention due its extensive applicability in diverse areas such as P2P networks, database applications, data streams, wireless sensor networks, and security applications.

In this study, we propose and develop an energy cost model for a generic peer using gossip-based communication for FID. Gossip-based or epidemic mechanisms are preferred in several distributed protocols [1, 2] for their ease of deployment, simplicity, robustness against failures, and limited resource usage. In terms of their power usage, the efficiency of three models of epidemic protocols, namely basic epidemics, neighborhood epidemics and hierarchical epidemics, has been examined in [3]. Basic epidemics that requires full membership knowledge of peers was found to be inefficient in its power usage. It has been shown that; in neighborhood epidemics, peer's power consumption amount is independent of population size. For hierarchical epidemics, power usage increases with population size. In fact, [3] is the only study that considers power awareness features of epidemic protocols. However, it evaluates different epidemics through simulations only and provides results on latency and power (proportional to the gossip rate). Moreover, effects of gossip parameters such as fan-out and maximum gossip message size were not investigated. In contrast, our study is the first one that proposes an energy cost model for a generic peer using gossip-based communication like in ProFID protocol, and examines the effect of protocol parameters to characterize energy consumption. As a case study protocol, we use our gossip-based approach ProFID for frequent item set discovery [4]. It uses a novel atomic pairwise averaging for computing average global frequencies of items and network size, and employs a convergence rule and threshold mechanism. Due to the page limitation, we refer interested reader to [4] for details of the protocol.

This paper is organized as follows. Section 2 develops energy cost model for a gossip-based peer used in our protocol. Section 3 analyzes the effect of protocol parameters, compares push–pull method of ProFID with the well-known push-based gossiping that we adapted to frequent item set discovery, and reformulates the peer's energy cost. Finally, Sect. 4 states conclusions and future directions.

2 Energy Cost Model

ProFID protocol depends on three main components of operations performed by each peer: energy consumed while (1) computing new state, (2) sending messages and (3) receiving messages. Inspired by studies [5, 6], we propose an energy cost model for a generic peer using gossip-based communication in ProFID. In study [6], energy cost models for client–server and publish–subscribe styles were developed. Then, application and platform specific model parameters were also taken into consideration and energy prediction model was developed. Work of [5] introduces a quorum-based model to compute energy costs of read and write operations in replication protocols, and proposes an approach to reduce the energy cost of tree replication protocol. Different than these prior works, we develop energy cost model for a peer using gossip-based communication and consider the effects of gossip parameters on the cost representation.

We start with the analysis of the energy consumption during an atomic pairwise averaging operation between peers P_i and P_j Different operations consuming energy are explained in Table 1. During an atomic pairwise averaging, energy cost of a peer that initiates a gossip (*gossip starter*) is represented by:

$$E_{gossipStarter} = E_{send} + E_{receive} + E_{compStarter} \tag{1}$$

On the other hand, energy cost of the gossip target can be formulated as follows:

$$E_{gossipTarget} = E_{receive} + E_{send} + E_{compTarget} \tag{2}$$

Note that $E_{compTarget}$ and $E_{compStarter}$ are both proportional to the gossip message size, and they can simply be represented as E_{comp}. Hence, $E_{i,j}$ (the energy consumption of a peer P_i during an atomic pairwise averaging with P_j) can be written as:

$$E_{i,j} = E_{send,j} + E_{receive,j} + E_{comp} + C \tag{3}$$

where $E_{send,j}$ is the energy consumed while sending a gossip message to P_j, $E_{receive,j}$ is the energy consumed while receiving a gossip message from P_j, and E_{comp} is the local computation of the peer. Note that this is the energy cost of a peer that performs an atomic pairwise averaging operation. In real network scenarios, energy consumption may include extra factors such as CPU's energy consumption during I/O. Hence, a constant C is added to the equation. To represent the energy cost of a gossip-based peer during an atomic pairwise averaging operation, the formula was given with respect to the basic conditions (gossip to one neighbor, one round, one item). Step by step, we now extend this cost model of a peer for the ProFID protocol. A peer may initiate multiple gossip operations during a single round depending on the *fanout* value as well as it may become gossip target multiple times. The energy cost of P_i that gossips a single item tuple in a round can be formulated as:

$$E_{P_i}(\text{single round, single item}) = \sum_{j \in V \cup W} E_{i,j} \tag{4}$$

where V is the set of neighbors chosen by P_i as gossip targets, and W is the set of neighbors that initiates an atomic pairwise averaging with P_i. Note that the number of elements in V corresponds to the *fanout* value.

Table 1 Different operations that consume energy

Value	Description
E_{send}	Energy required to send the item tuple
E_{recv}	Energy required to receive the item tuple
$E_{compStarter}$	Energy required to choose tuple to send and update the state
$E_{compTarget}$	Energy required to compute the average and prepare the tuple to send

In general, a gossip message comprises multiple item tuples whose number is upper-bounded by *maximum message size* (*mms*) parameter. Since $E_{send,j}$ and $E_{receive,j}$ are the energies consumed while sending and receiving a single tuple respectively, total energy consumed during a gossip round would linearly increase with the *mms*. Hence, energy cost of P_i in a round can be expressed as:

$$E_{P_i}(\text{single round}) \leq mms \cdot \sum_{j \in V \cup W} E_{i,j} \tag{5}$$

Since a peer repeats those operations in every round, number of rounds R would increase the energy cost of a peer proportionally. Hence, the overall energy cost of P_i can be written as:

$$E_{P_i} \leq R \cdot mms \cdot \sum_{j \in V \cup W} E_{i,j} \tag{6}$$

3 Analysis and Results

We have developed a simulation model for ProFID protocol [7] on PeerSim simulator [8] and analyzed the effects of protocol parameters on the energy consumption. As presented in Eq. 6, energy cost of a peer is proportional to the convergence time, that is the number of rounds R. In this section, we analyze the effects of protocol parameters on R, compare push–pull based method of ProFID with the well-known push-based gossiping, evaluate the effects of convergence parameters on frequency error (i.e. the percentage of items which were identified as frequent though they are actually not) and reformulate the upper bound of the overall energy cost of a peer in terms of protocol parameters.

We performed our evaluations through extensive large-scale distributed scenarios (up to 30,000 peers) on PeerSim. We tested different topologies such as random topology and scale-free Barabasi–Albert topology with average degree 10. All the data points presented in graphs are the average of 50 experiments. The default values of parameters used in the experiments are given in Table 2.

Convergence Parameters (convLimit, ε**):** Convergence parameters are used for self-termination of peers and they have direct effects on R. Figure 1a shows that R is inversely proportional to $\log \varepsilon$. This is because *convCounter* will be

Table 2 Default parameter values

Parameter	Value	Parameter	Value	Parameter	Value
N	1000	M (number of items)	100	*convLimit*	10
ε	10	*mms*	100	*fanout*	1

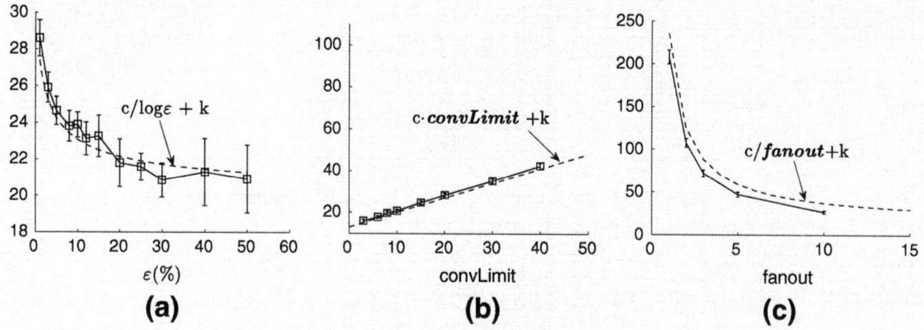

Fig. 1 Effects of **a** ε on R, **b** *convLimit* on R, **c** *fanout* on R

incremented with less chance and it will take longer time to reach *convLimit*. However, R is directly proportional to *convLimit* as depicted in Fig. 1b, and this is because *convCounter* needs to be incremented more to take convergence decision.

Fanout: Intuitively, increasing *fanout* will cause to consume more energy in a single round. On the other hand, algorithm will converge faster since a peer exchanges its state with more peers in a single round. Figure 1c depicts that *fanout* has an inverse proportion with R. Note also that *fanout* has a direct proportion with the upper bound given in Eq. 6 since *fanout* is the cardinality of set V.

Gossip message size: Parameter *mms* is the upper bound for a gossip message size in terms of number of ⟨item,frequency⟩ tuples. Large *mms* means more state information is sent in a single gossip message. On one hand, this causes faster convergence, but on the other hand, the energy consumption of sending a single gossip message increases. Results in Fig. 2a verify that *mms* is inversely proportional to R. Note also that *mms* is directly related with the energy cost of a peer in a single round, and these cancel each other in our cost formulation. Recall that ProFID assumes each peer knows about its neighboring peers only and gossips with them, and hence it is based on neighborhood epidemics. In this respect, our results are also consistent with [3] that reports the efficiency of neighborhood epidemics in its power usage.

Comparison with Adaptive Push-sum: We have compared ProFID with the Push-sum approach [9] to observe different gossip-based approaches as a solution to the FID problem. In order to compute aggregates of items, Push-sum protocol assumes that all peers are aware of all items in the network which is not practical for the case of FID. For this reason, we have developed an Adaptive Push-sum protocol on PeerSim by modifying the Push-sum algorithm and included the convergence rule in order to adapt it to the FID problem. As depicted in Fig. 2b,

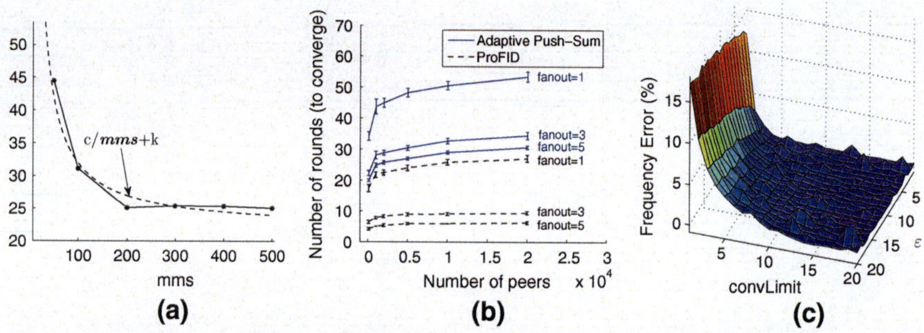

Fig. 2 **a** Effect of *mms* on frequency error, **b** ProFID versus Adaptive Push-Sum, **c** Effect of convergence parameters on frequency error

ProFID converges faster than Adaptive Push-Sum algorithm in all different fanout values. We also observed that ProFID outperforms Adaptive Push-Sum in terms of message complexity in these simulations.

Energy Cost and Frequency Error in Terms of Protocol Parameters: Combining the experimental analysis results, effects of protocol parameters on convergence time R can be represented as:

$$R \approx (1/log\,\varepsilon) \cdot log\,N \cdot convLimit \cdot (1/fanout) \cdot (1/mms) \quad (7)$$

Based on these findings above, we can reformulate the energy cost of P_i (in Eq. 6) as follows:

$$E_{P_i} \leq (1/log\,\varepsilon) \cdot log\,N \cdot convLimit \cdot (1/fanout) \cdot \left(\sum_{j \in V \cup W} E_{i,j} \right) \quad (8)$$

We should also consider the frequency error while minimizing the energy cost since obtaining unreasonable results with low energy cost would not be meaningful. Frequency error can be written in terms of protocol parameters by combining experimental result shown in Fig. 2c as follows:

$$FrequencyError \approx \varepsilon / convLimit \quad (9)$$

4 Conclusions and Future Work

Frequent item set discovery problem in P2P networks is relevant for several distributed services such as cache management, data replication, sensor networks and security. Our study is the first one that introduces and develops an energy cost model for a generic peer using gossip-based communication. Different than the prior works, we also studied the effect of protocol parameters through extensive large-scale simulations, compared push–pull and push-based gossiping methods.

As future work, we plan to deploy our protocol on PlanetLab and analyze its energy cost on this network testbed. We also aim to extend our energy cost model to hierarchical gossip approaches.

References

1. Ozkasap, O., Caglar, M., Yazici, E., Kucukcifci, S.: An analytical framework for self-organizing peer-to-peer anti-entropy algorithms. Performance Evaluation Journal, Elsevier Science **67**(3), 141–159 (2010)
2. Ozkasap, O., Genc, Z., Atsan, E.: Epidemic-based reliable and adaptive multicast for mobile ad hoc networks. Comput. Netw. **53**, 1409–1430 (2009)
3. van Renesse, R.: Power-aware epidemics. In: Proceedings of IEEE Symposium on Reliable Distributed Systems (2002)
4. Cem, E., Ozkasap, O.: Profid: practical frequent item set discovery in peer-to-peer networks. In: Proceedings of ISCIS, pp. 199–202 (2010)
5. Basmadjian, R., de Meer, H.: An approach to reduce the energy cost of the arbitrary tree replication protocol. In: Proceedings of e-Energy, pp. 151–158 (2010)
6. Seo, C., Edwards, G., et al.: A framework for estimating the energy consumption induced by a distributed system's architectural style. In: Proceedings of SAVCBS (2009)
7. The ProFID implementation. https://sites.google.com/a/ku.edu.tr/emrah-cem/projects/profid
8. The Peersim simulator. http://peersim.sf.net
9. Kempe, D., Dobra, A., Gehrke, J.: Gossip-based computation of aggregate information. In: Proceedings of FOCS, pp. 482–491 (2003)

Towards a Fairer Energy Consumption in Wireless Ad hoc Networks

Maurizio D'Arienzo

Abstract A well known and still open issue of wireless ad hoc networks is the unfair energy consumption among the nodes. The specific position of certain nodes composing an ad hoc network makes them more involved in network operations than others, causing a faster drain of their energy. To better distribute the energy level and increase the lifetime of the whole network, we propose to periodically force the cooperation of less operative nodes while overwhelmed nodes deliberately stop their service. A dedicated ad hoc network routing protocol is introduced to discover alternative paths without losses in the overall network performance.

1 Introduction

Ad hoc networks are composed of several wireless nodes with limited power resources usually provided by accumulators. In these networks each node is an end system and a router at the same time. The limited energy is then not only used to deliver own packets to the destinations but also to serve other nodes as message relayer. All of the nodes accept to cooperate in these operations thanks to a blind trust agreement among them that allows any node to send its traffic. In general the node position in a given topology raises some node inequalities. For instance, inner nodes may be requested to relay packets more often than outer ones, thus leading to a different consumption of energy. Outer nodes can save the energy for their own transmissions and have the chance to last longer [1, 2].

Due to these issues, certain nodes can then decide to refuse to cooperate in relay operations acting as defecting nodes. This behaviour can be malicious, because the

M. D'Arienzo (✉)
Dipartimento di Studi Europei e Mediterranei,
Seconda Università di Napoli, Napoli, Italy
e-mail: maudarie@unina.it

E. Gelenbe et al. (eds.), *Computer and Information Sciences II*,
DOI: 10.1007/978-1-4471-2155-8_15,

nodes try to maximise their lifetime and the delivery ratio at the expenses of other nodes, or it can be simply needed to avoid a premature shut down. In a previous work [3] we studied the effect of nodes defection and proposed an algorithm nested in an ad hoc routing protocol to identify and isolate the malicious nodes. The *legacy* routing protocols are in fact not able to detect the behaviour of the nodes composing the network, and we gave some first proofs on how the presence of parasites nodes can starve the performance of good behaving ones, especially in terms of the final delivery ratio.

In this paper we consider another possibility, that is, certain nodes heavily involved in relay operations decide to defect to save power for themselves and to increase their life. We aim at minimising the residual energy variance, that means to improve the fairness, and at increasing the minimum residual energy, that means to extend the entire network lifetime. To this purpose, a subset of nodes, for instance the inner nodes, can periodically switch off to save their energy and then balance the energy among the nodes. By making use of the routing protocol for the detection of defecting nodes, the transmitting nodes can soon realise that some paths are temporally unavailable and have the chance to search for alternative paths. This way, less working nodes are pushed in packet relay operations and the overall network performance, like for instance the delivery ratio, are kept at good level.

Rather than introducing energy aware features into routing protocols, we rely on a decentralised algorithm to only discover the nodes' behaviour in order to quickly discovery alternative paths, and we demonstrate that this can balance the energy among the nodes. The algorithm exploits and enhances an existing ad hoc routing protocol. It has been implemented and tested in a simulated environment. Next sections introduce the algorithm, the modifications brought to the routing protocol, the experimental results, and finally last section provides related work and conclusive remarks.

2 Tracking Nodes Behaviour

In an ad hoc network, the number of nodes and links can change during time, so we consider the number of nodes $N(t)$ as a function of time t. We also define a dynamic array $C(t)$ of $N(t)$ elements for each node of the network. The generic element $c_i(t)$ of $C(t)$ assumes the values (UNKNOWN, COOPERATE, DEFECT) meaning that the behaviour of node i at time t is respectively unknown, cooperative or non cooperative. At time $t = 0$ all the values are set to UNKNOWN, since at the beginning each node is not aware of the behaviour of the other nodes.

Suppose the generic node s of the network needs to send some traffic to the destination d. The first task is to discover an available path, if it exists, to reach the destination. To this purpose, we consider a source based routing protocol capable of discovering a list $A(t)_{(s,d)i}$ $\forall i : 0<i<P$ of P multiple paths. All the

nodes in the list $A(t)_{(s,d)i}$ are considered under observation and marked as probably defecting in the array $C(t)$ unless a positive feedback is received before a timeout expires. The sender s starts sending his traffic along all the discovered paths. If the destination node generates D acknowledgement messages containing the list of all the nodes $L_{(s,d)i}$ $0<i<D$ traversed, as it happens in some source based routing protocols, the sender s is informed about the behaviour of intermediate nodes. For each acknowledgement message received, the sender s can make a final update of the array $C(t)$ by setting the matching elements $c_i(t)$ contained in the list $L_{(s,d)i}$ as cooperative. Notice that the last update overwrites the previous stored values and represents the most recent information concerning the behaviour of a node.

Given this algorithm, each node is aware of the behaviour of other nodes and can react in the most appropriate way. For example, a node can refuse to relay packets of defecting nodes, or operate a selective operation like queuing their packets and serving them only if idle and not busy with the service requested by cooperative nodes. The array $C(t)$ is not static over time and its values are continuously updated. In fact, due to the dynamic situation of ad hoc networks, the search of available paths is frequently repeated, and the list $A_{(s,d)}$ consequently updated. Hence, if a defecting node decides to cooperate, its identification address will be included in one of the acknowledgement messages $L_{(s,d)i}$ sent to the sender s and its aim to cooperate will be stored in the array $C(t)$.

The situation described here for the pair (s,d) is replicated for all the possible pairs of nodes that try to interact, but each node stores only one array $C(t)$ that is updated upon reception of any acknowledgement message, wherever it comes from. Furthermore, not all the packets relayed are checked in order to verify the nodes' behaviours, but only a sample of them, thus keeping the total overhead under control.

The algorithm introduced in the previous section has been implemented in AH-CPN (Ad Hoc Cognitive Packet Network) [4–7]. We first modified this protocol to support the search of multiple paths, and then included the new algorithm for the selection of cooperative paths.

3 Testbed and Experiments

To trace the energy consumption of the ad hoc network nodes under different working conditions we tested the proposed routing protocol on a simulated testbed in ns-2 simulator. The testbed is composed of 25 nodes arranged in a 5x5 grid topology, depicted in Fig. 1. The distance between each pair of nodes is enough to make a transmitting node in condition to make use of adjacent nodes rather than non adjacent ones as the next hop. Four sessions of constant bit rate traffic are generated between the pairs 0–24, 4–20, and backwards on a round robin cycle every 1 s. All the nodes start from an initial energy of 1,000 J. The total duration of each experiment is 720 s.

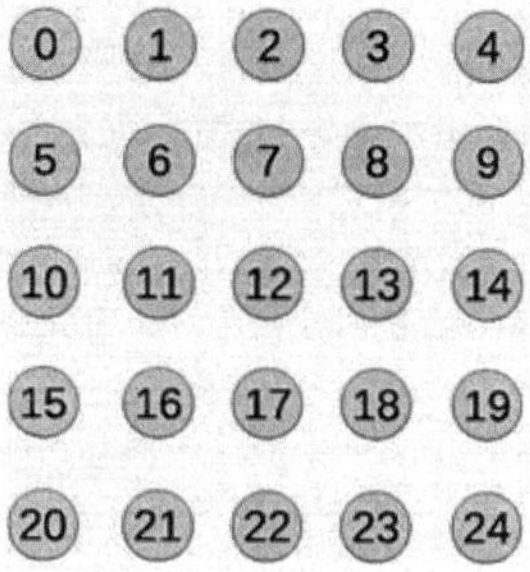

Fig. 1 The simulated testbed

During the first series of experiments the percentage of cooperation of the inner nodes 6, 7, 8–11, 12, 13–16, 17, 18 is fixed at 100% (full cooperation), in the second and third series of experiments this percentage is respectively set at 85 and 75% (partial cooperation). We compared the proposed algorithm with the basic AH-CPN version and the multiple path version with the trace detection disabled. They are marked in the following graphs as *basic* (AH-CPN), *mPath*, the multiple path version without detection of defecting nodes, and *mPath-T* the version with the proposed tracing algorithm. At the end of each experiment we measure the delivery ratio dr_i, the average residual energy of all the nodes μ, their variance v, and the energy e_i spent by node i to successfully deliver one single byte to the destination, which is calculated as:

$$e_i = \frac{Ec_i}{(s_i + rl_i)} * \frac{s_i}{r_i}$$

being Ec_i the energy consumed by node i, s_i the total number of bytes sent to the destination, rl_i the number of bytes relayed from node i, and r_i the bytes correctly received at destination. e_i has a dimension of $[Joule/bytes]$.

All the final values are collected in the Table 1. To better illustrate a comparative result, the measures related to the single session 0–24 are packed in four histograms, one for each of the measure considered. Each graph reports the final values of a single parameter under the three inner nodes' cooperation cases described above.

4 Results

In the experiments with full cooperation of the inner nodes, the delivery ratio is the best value for all the three routing protocols, showed in Fig. 2a. The average remaining energy graphed in Fig. 2c is in favour of the *basic* protocol as well as the variance value. The variance (Fig. 2b) is significantly reduced when the percentage of cooperation of inner nodes decreases at 85 and 75%, which means a fairer distribution of energy consumed by each node. This is achieved by all the three protocols, independently from the activation of tracing functionality. Notice

Table 1 Experimental results

	100%			85%			75%		
	Delivery ratio								
Session	basic	mPath	mPath-T	basic	mPath	mPath-T	basic	mPath	mPath-T
0-24	0.931	0.928	0.940	0.853	0.847	0.936	0.818	0.804	0.938
24-0	0.711	0.685	0.940	0.693	0.644	0.937	0.650	0.653	0.935
4-20	0.934	0.926	0.938	0.857	0.858	0.937	0.809	0.814	0.934
20-4	0.663	0.553	0.941	0.666	0.516	0.935	0.599	0.492	0.937
	Average residual energy (J)								
	864.5	852.1	838.8	874.3	865.0	849.3	881.7	871.6	855.3
	Variance								
	207.2	264.82	330.2	59.7	78.9	88.37	68.0	79.9	132.0
	Energy per byte μJ								
0-24	2.46	2.67	2.88	2.56	2.74	2.80	2.53	2.80	2.75
24-0	3.26	3.61	2.89	3.18	3.60	2.79	3.22	3.46	2.79
4-20	2.53	2.78	2.90	2.67	2.85	2.81	2.75	2.95	2.73
20-4	3.47	4.54	2.89	3.35	4.61	2.79	3.69	4.74	2.76

however how the delivery ratio of *basic* and the *mPath* versions decreases as the percentage of cooperation of inner nodes decreases, while the delivery ratio of *mPath-T* protocol is kept at its highest value. The first two protocols are in fact unable to realise that certain nodes, and then certain paths, are no longer available, and do not try to discover alternative paths. The *mPath-T* is instead suddenly aware of nodes and paths unavailability, and react by discovering alternative paths to still reach the destination. These results are consistent with all the other sessions of traffic. The split among the values is even wider for the traffic session 4–20 and the backward session 20–4 as showed in the Table 1, where the delivery ratio is not high for the *basic* and the *mPath* versions even in full cooperation.

As far as the energy spent per byte delivered e_i presented in Fig. 2d, the *mPath-T* value is slightly higher than the other two protocol versions. This result was expected and is in line with the higher average energy spent registered at the end of the experiments, and is due to the overhead introduced by the multiple path feature and the detection of non cooperative nodes. In any case this value is decreasing as the percentage of cooperation decreases and the difference with the other protocols is quite negligible. Furthermore, the other values e_4, e_{20} and e_{24} reported in Table 1 are instead in favour of *mPath-T* protocol with a more important difference with the values of *basic* and *mPath* protocols.

5 Related Work and Conclusions

To address the fair energy consumption issue, in [8] it is proposed a fair cooperative protocol (FAP) to improve the overall performance of the whole network. Each node calculates a *power reward* to evaluate the power contributed to and by

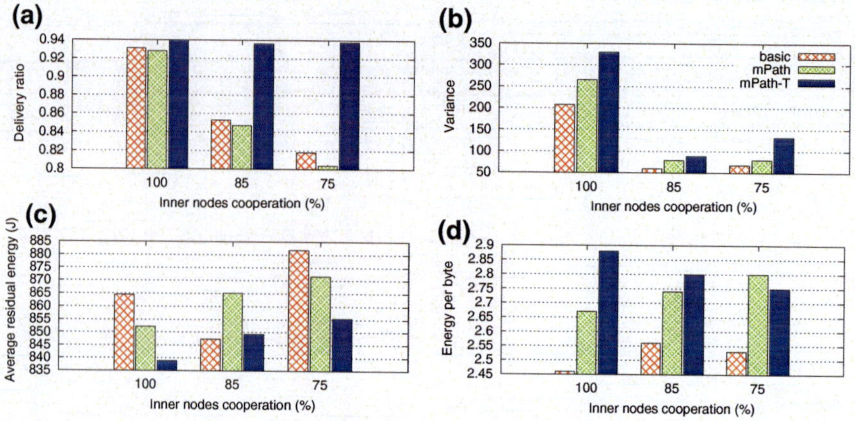

Fig. 2 Results for session "0–24". **a** Delivery ratio. **b** Energy variance. **c** Average energy. **d** Energy per byte

the others. The adoption of such a fair scheme can also lead to substantial throughput gains over the direct transmission and the full cooperation case. There exists however a trade off between the fairness and the throughput achieved. An analytical study helps in the selection of the working point to achieve the best trade off between the fairness and the throughput. In a similar way, [9] takes into account the ratio between the residual node energy and its drain rate. The routing algorithm then selects the path that maximise this ratio.

The work done in [10] still relies on cooperation among nodes. However, it makes the assumption that only a subset of nodes belonging to the same group can be interested in a mutual cooperation instead of a full cooperation involving all nodes. If nodes belonging to a group form a *coalition* and route packets without intervention of nodes outside the coalition, then the routing shows some benefits with respect to the full selfish case.

In this paper we propose to introduce some modifications in ad hoc routing protocols to support the identification of nodes behaviour. The knowledge of nodes behaviour gives the chance to the busiest nodes to temporarily stop the service for the others while the routing protocol helps discover alternative paths to keep the residual energy and network performance fair. A metric to evaluate the ratio between the job done and the job requested is under evaluation and will be included in this modified version of ad hoc routing protocol.

References

1. Ephremides, A.: Energy concerns in wireless networks. Wirel. Commun. IEEE [IEEE Pers. Commun.] **9**(4), 48–59 (2002)
2. Feeney, L.M., Nilsson, M.: Investigating the energy consumption of a wireless network interface in an ad hoc networking environment. In: IEEE INFOCOM 2001, vol. 3, pp. 1548–1557 (2001)

3. D'Arienzo, M., Oliviero, F., Romano, S.P.: Smoothing selfishness by isolating non-cooperative nodes in ad hoc wireless networks. In: Advances in Future Internet, AFIN '10, pp. 11–16. IEEE Computer Society, Washington (2010)
4. Gelenbe, E., Lent, R.: Power-aware ad hoc cognitive packet networks. Ad Hoc Netw. **2**(3), 205–216 (2004) (ISSN: 1570-8705)
5. Gelenbe, E., Lent, R., Xu, Z.: Design and performance of cognitive packet networks. Perform. Eval. **46**(2–3), 155–176 (2001)
6. Gelenbe, E.: Steps toward self-aware networks. Commun ACM **52**(7), 66–75 (2009)
7. Gelenbe, E.: Search in unknown random environments. Phys. Rev. E **82**, 061112 (2010) (Published December 7, 2010)
8. Dai, L., Chen, W., Cimini, L.J., Letaief, K.B.: Fairness improves throughput in energy-constrained cooperative ad-hoc networks. Wirel. Commun. IEEE Trans. **8**(7), 3679–3691 (2009)
9. Kim, D., Garia Luna Aceves, J.J., Obraczka, K.J., Cano Manzoni, P.: Power-aware routing based on the energy drain rate for mobile ad hoc networks. In: Proceedings of IEEE 11th International Conference on Computer Communications and Networks, pp. 562–569 (2002)
10. Guha, R.K., Gunter, C.A., Sarkar, S.: Fair coalitions for power-aware routing in wireless networks. Mobile Comput. IEEE Trans. **6**(2), 206–220 (2007)

Energy Efficient Resource Allocation Strategy for Cloud Data Centres

Dang Minh Quan, Robert Basmadjian, Hermann De Meer, Ricardo Lent, Toktam Mahmoodi, Domenico Sannelli, Federico Mezza, Luigi Telesca and Corenten Dupont

Abstract Cloud computing data centres are emerging as new candidates for replacing traditional data centres. Cloud data centres are growing rapidly in both number and capacity to meet the increasing demands for highly-responsive computing and massive storage. Making the data centre more energy efficient is a necessary task. In this paper, we focus on the organisation's internal Infrastructure as a Service (IaaS) data centre type. An internal IaaS cloud data centre has many distinguished features with heterogeneous hardware, single application,

D. M. Quan (✉) · L. Telesca · C. Dupont
CreateNet, Trento, Italy
e-mail: quan.dang@create-net.org; quandm@upb.de

L. Telesca
e-mail: luigi.telesca@create-net.org

C. Dupont
e-mail: cdupont@create-net.org

R. Basmadjian · H. De Meer
University of Passau, Passau, Germany
e-mail: Robert.Basmadjian@uni-passau.de

H. De Meer
e-mail: hermann.demeer@uni-passau.de

R. Lent · T. Mahmoodi
Imperial College, London, UK
e-mail: r.lent@imperial.ac.uk

T. Mahmoodi
e-mail: t.mahmoodi@imperial.ac.uk

D. Sannelli · F. Mezza
ENI, Milan, Italy
e-mail: domenico.sannelli@eni.com

F. Mezza
e-mail: Federico.Mezza@eni.com

E. Gelenbe et al. (eds.), *Computer and Information Sciences II*,
DOI: 10.1007/978-1-4471-2155-8_16,

stable load distribution, lived load migration and highly automated administration. This paper will propose a way of saving energy for IaaS cloud data centre considering all stated constraints. The basic idea is rearranging the allocation in a way that saving energy. The simulation results show the efficiency of the method.

Keywords Energy efficient · Resource allocation algorithm · Cloud data centre

1 Introduction

An IaaS cloud data centre is used to host computer systems and associated components. Cloud computing data centres are emerging as new candidates for replacing traditional data centres that are growing rapidly in both number and capacity to meet the increasing demands for computing resources and storages. Clearly this expansion results in a significant increase in the energy consumption of those data centres. According to [1], data centres in USA consumed about 61 BkWh and accounted for 1.5% of total US electricity consumption in 2006.

The reported work in this paper is shaped based on the research collaboration within the Fit4Green project on energy efficiency in data centres, where the ENI's IaaS data centre is used for the investigation purposes. An internal IaaS data centre like the ENI's data centre has the following characteristics: Mixed hardware environment with different technologies, Single application with stable load, Live load migration capability, Highly automated administration.

To this end, we propose an energy efficient resource allocation approach that utilises the capability of VMs live migration to reallocate resources dynamically. The basic idea is to use a heuristic that is consolidating and rearranging the allocation in an energy efficient manner.

This paper is organised as follows. Section 2 describes the related works. After analysing the energy consumption model of different entities in the data centre in Sect. 3, the energy efficient resource allocation and performance investigation are discussed in Sect. 4. Finally, Sect. 5 concludes the paper with a short summary.

2 Related Work

Currently, resource allocation mechanisms which are widely used in data centres include load balancing, round robin and greedy algorithms. The load balancing module tries to distribute the workload evenly over the computing nodes of the system. In the round robin algorithm, the servers are in a circular list. There is a pointer pointing to the last server that received the previous request. The system sends a new resource allocation to the next node with enough physical resources to handle the request. Having the same set up as the round robin algorithm, the greedy algorithm continues to send new resource allocation request to the same node until it is no longer has enough physical resources, then the system goes to the next one.

The work in [2] studies general energy saving approaches. Saving energy in ICT divides into two main fields, saving energy for network [3, 4] and saving energy for computing nodes [5, 6]. Related to saving energy in cloud data centre, the work in [7] adjusts the working mode of the server according to load to save energy. In [8], the authors focus on saving energy for PaaS (Platform as a Service) cloud data centre. Our work is different from previous in two main ways. We focus on IaaS scenario and use moving workload as the main method. [9] is the most closely work with our work. However, in [9] the authors use the predefined MIPS to present the capacity and the load which is quite difficult to the user. We use the measured load value instead.

3 Energy Consumption Model

3.1 Server Power Consumption

3.1.1 Processor

Based on our observations of a custom benchmark that puts on each core variations of load ranging from 10 to 100%, we noticed that the power consumption of individual core of a processor increases linearly with its utilisation, and is given by:

$$P_{Core} = P_{\max} * \frac{L_{Core}}{100}, \tag{1}$$

where L_{Core} is the utilization, the workload, of the corresponding core and $P_{\max}$ denotes the maximum power of the processor. Then, the power consumption of multi-core processors is:

$$P_{CPU} = P_{idle} + \sum_{i=1}^{n} P_{Core_i}, \tag{2}$$

where P_{idle} is constant and denotes the power of the processor in the idle state.

3.1.2 Memory

Given an unbuffered SDRAM of type DDR_3, then its power consumption is:

$$P_{RAM} = P_{RAM\_idle} + \delta * \beta, \tag{3}$$

where $\delta = 1.3$, $\beta = 7.347$, and P_{RAM_idle} is the idle power consumption given by:

$$P_{RAM\_idle} = \sum_{i=1}^{n} s_i * p, \tag{4}$$

such that i denotes the number of installed memory modules of size s, whereas

$$p = \frac{f}{1000} + \alpha\sqrt{f}(f_c - f),$$

where $fc = 1600$, f denotes the input frequency, and $\alpha = 0.000026$.

3.1.3 Hard Disk

We noticed that the startup and accessing mode power consumptions are in average respectively 3.7 and 1.4 times more than that of the idle (P_{idle}):

$$P_{HDD} = a * 1.4 * P_{idle} + b * P_{idle} + c * 3.7 * P_{idle}, \tag{5}$$

where $a, b, c \in [0, 1]$ denote respectively the probability of accessing, idle and startup.

3.1.4 Mainboard

The power consumption of the mainboard is given by the following equation:

$$P_{Mainboard} = \sum_{i=1}^{l} P_{CPU} + P_{RAM} + \sum_{j=1}^{m} P_{NIC} + \sum_{k=1}^{n} P_{HDD} + t, \tag{6}$$

where t is a constant having a value of 55.

3.1.5 Server Power

Given a server composed of several components, then its power consumption is:

$$P_{Server} = \sum_{i=1}^{l} P_{Mainboard} + \sum_{j=1}^{m} P_{Fan} + \sum_{k=1}^{n} P_{PSU},$$

such that j and k denote respectively the number of fans and power supply units whose power consumption models can be found in [10].

3.2 Network Power Consumption

Given that a data centre consists of N stations (servers) that connected through S network switches, which all have the same number of ports n. Switches are interconnected forming a mesh topology, and each computing device i can generate and consume a total traffic of λ_i packets per second. Let us define T as the total network traffic (in packets) generated or consumed by all stations. The power consumption of all network switches, P_S can be estimated from the expression,

$$P_S = \sum_{j=0}^{S-1} \Phi_j \left(\alpha T + (1-\alpha) \sum_{i=jn}^{jn+n+1} \lambda_i + \Gamma_j \right), \tag{7}$$

where α is the fraction of inter-switch traffic. Γj represents external traffic demand entering and leaving the switch. A linear approximation of Φ can be utilised basing on the maximum (nominal) power consumption, ρ, and the switch bandwidth, Λ that can be obtained from the specification sheets of the device.

$$\Phi^*(\lambda) = \rho \left(\alpha + \frac{\beta - \alpha}{\Lambda} \lambda \right) \tag{8}$$

3.3 Data Centre Power Consumption

Let $E(i)$ be the function calculating energy consumption for node i, and w_i denotes the number of edges at node i. Then the overall energy consumption of a data centre is given by the following equation:

$$E_{tot} = \sum_{i=1}^{n} \left(\sum_{j=0}^{wi} E(c_{ij}) \right) + E(i), \tag{9}$$

4 Optimisation Algorithms

4.1 Problem Statement

Assume that we have a set of servers S. Each server $s_i \in S$ is characterised with number of cores, amount of memory and amount of storage ($s_i.nr\_Core$, $s_i.nr\_RAM$, $s_i.nr\_Stor$).

Each server s_i has a set of running virtual machine V_i including k_i virtual machines. Each virtual machine $v_j \in V_i$ is characterised with required number of virtual CPU, amount of memory, amount of storage (v_j*.r_vCPU*, v_j*.r_RAM*, v_j*.r_Stor*) and the average CPU usage rate computed in %, amount of memory, amount of storage (v_j*.a_Urate*, v_j*. a_RAM*, v_j*. a_Stor*). Because the load in each VM is quite stable, we assume that those values do not change through time.

When there is a new resource allocation request, the new VM must be allocated to a server in a way that total usage of VMs does not exceed the capacity of the server and the added energy usage is minimum.

For the global optimisation request, the VMs must be arranged in a way that total usage of VMs does not exceed the capacity of the server and the energy usage is minimum.

4.2 F4G-CS (Traditional Single Algorithm)

When a new VM comes, the F4G-CS algorithm will check all computing nodes to find the suitable node using the least amount of energy.

Step 0: Determine all servers meet the constraint and store in array A
Step 1: If the array A is empty, stop the algorithm and inform no solution
Step 2: Set index i at the beginning of the array A
Step 3: Calculate energy consumption E_i of the data centre if the VM is deployed on the server at array index i
Step 4: Store (i, E_i) in a list L
Step 5: Increase i to the next index of A
Step 6: Repeat from Step 3 to Step 5 until i goes out of the scope of A
Step 7: Determine min E_i in the list L
Step 8: Assign the VM to the server at index i_{min}
Step 9: If server at index i_{min} is OFF, put action turn ON to action list

It is noted that E_i is calculated for servers having workload. The server without workload will be shutdown.

4.3 F4G-CG (Cloud Global Optimisation) Algorithm

The F4G-CG algorithm has two main phases. In the first phase, we will move the VMs from low load servers to higher load servers if possible in order to free the low load server. The free low load server can be turned off. As the increasing energy by increasing the workload is much smaller than the energy consumed by the low load servers, we can save energy. The algorithm for phase 1 is as following:

Step 0: Forming the list LS of running servers
Step 1: Find the server having the smallest load rate.The load rate of a server is defined as the maximum (CPU load rate, Memory load rate, Storage load rate)

Step 2: Sort the VMs in the low load server according to the load level in descending order list LW
Step 3: Remove the low load server out of running server list
Step 4: Use F4G-CS algorithm to find the suitable server in LS for the first VM in the list LW
Step 5: If F4G-CS finds out the suitable server, update the load of that server, remove the VM out of LW
Step 6: Repeat from Step 4 to Step 5 until LW is empty of F4G-CS cannot find out the suitable server
Step 7: If F4G-CS cannot find out the suitable server, reset the state of found servers and mark Stop = true
Step 8: Repeat from Step 1 to Step 7 until Stop = true

In the second phase, we will move the VMs from the old servers to the modern servers. The free old servers can be turned off. As one modern server can handle the workload of many old servers, the energy consumed by the modern server is smaller than the total energy consumed by those many old servers. The algorithm is:

Step 0: Sort free servers into a descending order list LFS according to resource level.The resource level of a server is defined as the minimum(s_i*.nr_Core/max_Core*, s_i*.nr_RAM/max_RAM*, s_i*.nr_Stor/max_Stor*) with *max_Core*, *max_RAM*, *max_Stor* are the maximum number of Cores, memories, storages of the server in the pool.
Step 1: Sort running servers into an ascending order list LRW according to load rate.The load rate of a server is defined as the maximum(CPU load rate, Memory load rate, Storage load rate)
Step 2: Set m_count = 0
Step 3: Remove the first server *s* out of LFS
Step 4: Remove the first server *w* out of LRW
Step 5: If we can move workload from *w* to *s* m_count + = 1
Step 6: Repeat from Step 4 to Step 5 until we cannot move workload from *w* to *s*
Step 7: If m_count < = 1, reset the state of moved *w*,*s* and mark Stop = true
Step 8: Repeat from Step 2 to Step 7 until Stop = true

4.4 Simulation Result

The simulation is done to study the saving rate of the resource allocation mechanism in different resource configuration scenarios. To do the simulation, we use 4 server classes correlated to single core, duel cores, quad cores and six cores. We generated 3 resource scenarios: modern data centre, normal data centre and old data centre. In the modern data centre, the percentage of many cores server is dominated. In the old data centre, the percentage of small number of cores server is dominated. With each resource scenario, we generated a raw set of jobs.

Table 1 F4G-CS/F4G-CG simulation result

Scenario	1	2	3
Round robin(MWt)	61.23	48.45	41.89
Round robin + F4G-CG (MWt)	47.64	40.83	37.21
Greedy (MWt)	50.74	40.35	35.78
Greedy + F4G-CG (MWt)	45.67	37.78	33.47
Load balance (MWt)	60.46	47.87	41.15
Load balance + F4G-CG (MWt)	46.56	39.67	36.85
F4G-CS (MWt)	46.58	36.13	32.18
F4G-CS + F4G-CG (MWt)	44.76	34.47	30.7

Those jobs randomly come to the system within the period of 1,000 time slots. We select the runtime for each job from 1 to 100 time slots. To perform simulation for global optimisation request, with each resource and workload scenario, we still use linear resource allocation algorithms to allocate the new coming job to the resource. However, every 5 time slots, we execute the F4G-CG to perform global optimisation. The simulation result is in Table 1.

The simulation result shows the efficiency of the energy aware global optimisation algorithm. It can significantly reduce the energy consumption of a data centre.

5 Conclusion

This paper has presented a method that potentially reduces the energy consumption of the internal IaaS data centre. To save energy, we rearrange the resources allocation by the workload consolidation and frequency adjustment. In the reallocation algorithm, we take advantage of the fact that new generation computer components have higher performance and consume less energy than the old generation. Thus, we use the heuristic that move the heavy load applications to the new servers with larger number of cores while moving light load applications to the old servers with smaller number of cores, and thus switch off the old servers as many number as possible. The investigations show that our presented algorithm can enhance the performance further when data centre consists of larger number of old server, and also in the case when many old servers are working with heavy load rate and many modern servers are working with light load rate.

References

1. U.S. Environmental Protection Agency: Report to congress on server and data center energy efficiency. Technical report, ENERGY STAR Program, Aug. 2007
2. Berl, A., Gelenbe, E., di Girolamo, M., Giuliani, G., de Meer, H., Dang, M.-Q., Pentikousis, K.: Energy-efficient cloud computing. Comput. J. **53**(7), 1045–1051 (2010). doi:10.1093/comjnl/bxp080

3. Gelenbe, E., Morfopoulou, C.: A framework for energy aware routing in packet networks. Comput. J. **54**(6), 850–859 (2011)
4. Gelenbe, E., Mahmoodi, T.: Energy-aware routing in the cognitive packet network. In: International Conference on Smart Grids, Proceeding of Energy 2011 Conference (2011)
5. Heath, T., Diniz, B., Carrera, E. V., Jr, W. M., Bianchini, R.: Energy conservation in heterogeneous server clusters. In: Proceedings of the Tenth ACM SIGPLAN Symposium on Principles and Practice of Parallel Programming-PPoPP'05, pp. 186–195 (2005)
6. Xian, C., Lu, Y.H.: Dynamic voltage scaling for multitasking real-time systems with uncertain execution time. In: Proceedings of the 16th ACM Great Lakes Symposium on VLSI (2006)
7. Do, T.V.: Comparison of allocation schemes for virtual machines in energy-aware server farms. Comput. J. 54(4), 433–441 (2011)
8. Leverich, J., Kozyrakis, C.: On the energy (in)effciency of hadoop clusters. SIGOPS 44(1): 61–65(2010)
9. Beloglazov, A., Buyya, R.: Energy efficient resource management in virtualized cloud data centres. In: Proceeding of the 10th IEEE/ACM International Conference on Cluster, Cloud and Grid Computing (2010)
10. Basmadjian, R., Ali, N., Niedermeier, F., Meer, H. d., Giuliani,G.: A methodology to predict the power consumption for data centres. In: Proceedings of e-Energy 2011 (2011)

A Survey of Recent Work on Energy Harvesting Networks

H. Erkal, F. M. Ozcelik, M. A. Antepli, B. T. Bacinoglu and E. Uysal-Biyikoglu

Abstract Harvesting ambient energy from the environment has become an attractive possibility for nodes in wireless networks. In addition to being environmentally friendly, energy harvesting networks promise almost perpetual operation which has been a holy grail for autonomous networks such as wireless sensor networks, and more recently M2M. However, achieving these benefits requires rethinking communication and network algorithms in the presence of energy harvesting. In particular, energy consumption schedules are restricted by arbitrary *energy causality* constraints. This paper attempts to make a concise survey of the recent body of literature which have defined and solved scheduling and optimization problems, and derived information-theoretic bounds for communication systems with energy harvesting transmitters.

H. Erkal (✉) · F. M. Ozcelik · M. A. Antepli · B. T. Bacinoglu · E. Uysal-Biyikoglu
Department of Electrical and Electronics Engineering, METU, 06531, Ankara, Turkey
e-mail: hakanerkal@gmail.com

F. M. Ozcelik
e-mail: fatihmehmetozcelik@gmail.com

M. A. Antepli
e-mail: akifantepli@gmail.com

B. T. Bacinoglu
e-mail: tbacinoglu@gmail.com

E. Uysal-Biyikoglu
e-mail: elif@eee.metu.edu.tr

E. Gelenbe et al. (eds.), *Computer and Information Sciences II*,
DOI: 10.1007/978-1-4471-2155-8_17,

1 Introduction

Energy efficiency has been considered as an important issue in the design of network architectures since the emergence in the last decade of wireless sensor networks composed of small energy-limited autonomous units. The *lifetime* of such a network is seriously dependent on the initial energies stored at nodes as well as the energy efficiency of network protocols. There is a rich literature [1] of transmission schemes and network protocols that strive to maximize network lifetime or related objective functions. Recent developments in ambient energy harvesting technologies allow battery-limited devices to bear their own energy cost [2–6], so that the network can sustain itself. As network lifetime is extended with each energy harvest, it can in principle be unbounded. However, the amount harvested during each finite time interval may be erratic, which can make it challenging to maintain a continuous communication rate or quality, respecting delay constraints. Different harvesting technologies lead to entirely different *energy harvesting profiles*, which is a major factor in determining the energy consumption schedules. Also, battery inefficiencies and capacity limits are among significant parameters in practical systems. This new set of constraints which introduce a new twist on communication as well as network problems have sparked active research effort in recent years. This paper will try to make a short review of this very recent body of literature. We shall start by reviewing information theoretic studies on energy harvesting systems, then we will turn our attention offline and online transmission *scheduling* formulations with different objectives, and the solutions of these problems. We shall conclude with an outline of future directions.

2 Information Theoretic Bounds

An average energy harvest rate $\bar{P}$, introduces an energy consumption constraint on the communication system. It follows that the information-theoretic AWGN channel capacity of a point-to-point (p2p) communication system with energy harvesting transmitter having a perfect battery equals the AWGN channel capacity with an average power constraint $\bar{P}$ [7, 8]. While many studies focused solely on transmission energy, in practical systems other types of energy consumption, such as that occurs in the processor, sensor unit etc., can be significant. Information-theoretic analysis of an energy harvesting device with energy consuming components has been considered in [8] and the AWGN channel capacity of the system is found to be equal to the AWGN channel capacity of the ideal system with average energy harvest rate reduced by the average processor power consumption. A *randomized sleep policy* is found to be useful under processor energy consumption assumption. Also in [8], the AWGN channel capacity of an energy harvesting transmitter with an imperfect battery is shown to be equal to that of a system with mean energy harvest rate $\bar{P}$, that is the maximum average transmission power the battery can supply.

Note that the capacity results mentioned above require coding over long blocks of data and harvests, and the transmission time required to approach these in practice may be excessive. In the following sections we will review studies that consider finite-horizon performance. We start by considering studies that assume deterministically known energy harvest and data arrival times.

3 Offline Algorithms

Transmission time minimization of a given finite number of bits on a p2p energy harvesting link with AWGN noise is considered in [9, 10]. Energy is assumed to be harvested in discrete units at known arbitrary points in time. Assuming an infinite capacity battery at the transmitter, the transmission power is monotone non-decreasing in time, over the duration of the schedule [9], yet it may decrease with a finite capacity battery assumption since any schedule causing overflow of the battery is suboptimal [10].

The problem in [9] has been extended to AWGN *broadcast channel* (BC) in the concurrent works [11, 12] with the assumption that whole data to be transmitted is ready at the beginning of transmission. It is shown that both the total transmit power as well as individual power levels allocated to each of the two users exhibit the same properties as in the single-user case. In [12], the problem is shown to reduce to a problem of finding a cut-off power level for the strong user that splits power schedules of users for optimal power allocation and an iterative algorithm to attain optimal schedule has been proposed. In [11] the problem has been solved with a polynomial-time algorithm, a modified version of the *FlowRight* algorithm [13] that iteratively reaches the globally optimal schedule through local optimization on two consecutive energy harvest intervals (epochs) at a time, and passes through all consecutive epoch pairs in each iteration. The problem in [11, 12] is reformulated by considering the data arrivals during transmission in [14] and it is shown that an optimal schedule can never decrease total transmission power, yet individual powers may decrease due to *data causality* constraints. A special case of the problem where all of the weak user's bits are available at the beginning is shown to have a unique solution. Finally, an algorithm called *DuOpt*, that is shown to approach the optimal schedule iteratively, is proposed.

While the offline assumption may not be practical in all scenarios, offline solutions provide benchmarks and bounds on performance of online algorithms. Also, it may be possible to extend these to obtain online algorithms through the use of *look-ahead* buffers [15]. In general, arrivals of data and harvests may be modeled as a stochastic processes. The analysis of such models are known to be less tractable, yet several different approaches have appeared to date. In the next section, we review those approaches.

4 Online Algorithms

In many practical scenarios, times and amounts of energy harvests as well as data arrivals are not known in advance, which makes the online approach to the formulation of the problem more relevant. Online approaches range from optimizing sleep cycles to dynamically maximizing throughput or a utility function under delay constraints [16–23]. Online algorithms need to adapt to the dynamics of the energy harvesting and data arrival processes, to satisfy criteria such as stability of data and energy buffers, delay requirements and fairness among nodes in the network. An important performance criterion is *energy-neutral operation*, which means always keeping a positive amount of available energy. Finite battery capacity makes the problem more interesting as a full battery cannot take advantage of incoming harvests.

5 Conclusion

The ambient energy harvesting ability of wireless devices can enable self-sustaining systems and diminish the importance of network lifetime considerations in networks such as wireless sensor networks. On the other hand, depending on the energy harvesting technology and the resulting harvest profile, satisfying communication requirements with the available energy profile is a nontrivial problem. This paper has reviewed a number of recent approaches to optimizing communication systems and networks with energy harvesting transmitters. Among these are schedules that optimize communication rate (or minimize delay) assuming an arbitrary but known sequence of harvests, as well as online schedules that maximize utility under a proportional fairness setup. At present, online scheduling to optimize delay, or throughput are less well understood and remain open. Yet, as online solutions would make the most impact for most real-time practical applications, developments in that area are much needed.

Acknowledgments This work was funded by TUBITAK Grant 110E252

References

1. Tekbiyik, N., Uysal-Biyikoglu, E.: Energy efficient wireless unicast routing alternatives for machine-to-machine networks. In: J. Netw. Comput. Appl. (to appear 2011)
2. Chalasani, S., Conrad, J.M.: A survey of energy harvesting sources for embedded systems. In: Southeastcon, 2008. IEEE In Southeastcon, 2008. IEEE (2008), pp. 442–447
3. Seah, W.K.G., Eu, Z.A., Tan, H.P.: Wireless sensor networks powered by ambient energy harvesting—survey and challenges. In: Proceedings of the 1st International Conference on Wireless Communications, Vehicular Technology, Information Theory and Aerospace and Electronic Systems Technology (Wireless VITAE), Aalborg, Denmark, May 17–20 (2009)

4. Seah, W.K., Euy, Z.A., Tan, H.-P.: Wireless sensor networks powered by ambient energy harvesting (WSN-HEAP)—survey and challenges. In: 1st International Conference on Wireless Communication, Vehicular Technology, Information Theory, and, Aerospace and Electronic Systems Technology (Wireless VITAE), pp. 1–5 (2009)
5. Lin, K., Hsu, J., Zahedi, S., Lee, D., Friedman, J., Kansal, A., Raghunathan, V., Srivastava, M.: Heliomete: enabling long-lived sensor networks through solar energy harvesting. In: Proceedings of the 3rd International Conference on Embedded Networked Sensor Systems (SenSys), San Diego, CA, USA, p. 309 (2005)
6. Torah, R., Glynne-Jones, P., Tudor, M., O'Donnell, T., Roy, S., Beeby, S.: Self-powered autonomous wireless sensor node using vibration energy harvesting. In: Measurement Science and Technology, vol. **19**(12), pp. 1–8 (2008)
7. Ozel, O., Ulukus, S.: Information-theoretic analysis of an energy harvesting communication system. In: IEEE PIMRC Workshops (2010)
8. Rajesh, R., Sharma, V., Viswanath, P.: Information capacity of energy harvesting sensor nodes. 2011 IEEE international symposium on information theory (2011, submitted for publication)
9. Yang, J., Ulukus, S.: Optimal packet scheduling in an energy harvesting communication system. IEEE Trans. Commun. (2010, submitted for publication)
10. Tutuncuoglu, K., Yener, A.: Optimum transmission policies for battery limited energy harvesting nodes. IEEE Trans. Wirel. Commun. (2010, submitted for publication)
11. Antepli, M.A., Uysal-Biyikoglu, E., Erkal, H.: Optimal packet scheduling on an energy harvesting broadcast link. IEEE J. Sel. Areas Commun (to appear), Special Issue on Energy-Efficient Wireless Communications (2011)
12. Yang, J., Ozel, O., Ulukus, S.: Broadcasting with an energy harvesting rechargeable transmitter. IEEE Trans. Wirel. Commun. (2010, submitted for publication)
13. Uysal-Biyikoglu, E., El Gamal, A.: On adaptive transmission for energy efficiency in wireless data networks. IEEE Trans. Inf. Theory **50**, 3081–3094 (2004)
14. Ozcelik, F.M., Erkal, H., Uysal-Biyikoglu, E.: Optimal offline packet scheduling on an energy harvesting broadcast link. 2011 IEEE International symposium on information theory (2011, submitted for publication)
15. Uysal-Biyikoglu, E., Prabhakar, B., El Gamal, A.: Energy-efficient Packet Transmission over a Wireless Link. IEEE/ACM Trans. Netw. **10**, 487–499 (2002)
16. Kansal, A., Hsu, J., Zahedi, S., Srivastava, M.B.: Power management in energy harvesting sensor networks. In: ACM. Transaction Embedded Computing Systems (2006)
17. Hsu, J., Zahedi, S., Kansal, A., Srivastava, M., Raghunathan, V.: Adaptive duty cycling for energy harvesting systems. In: Proceedings 2006 International Symposium Low Power Electronics and Design (ISLPED '06), pp. 180–185 (2006)
18. Vigorito, C.M., Ganesan, D., Barto, A.G.: Adaptive control of duty cycling in energy-harvesting wireless sensor networks. In: IEEE SECON07, pp. 2130, June (2007)
19. Sharma, V., Mukherji, U., Joseph, V., Gupta, S.: Optimal energy management policies for energy harvesting sensor networks. IEEE Trans. Wirel. Commun. **9**, 1326–1336 (2010)
20. Huang, L., Neel, M.J.: Utility optimal scheduling in energy harvesting networks. In: 2011 Proceedings of MobicHoc, May 2011 (2010, submitted for publication)
21. Liu, R.S., Sinha, P., Koksal, C.E.: Towards achieving perpetual operation in rechargeable sensor networks. IEEE/ACM Trans. Netw. (TON) (2010, submitted for publication)
22. Gatzianas, M., Georgiadis, L., Tassiulas, L.: Control of Wireless Networks with Rechargeable Batteries. Trans. Wirel. Commun. **9**, 581–593 (2010)
23. Levron, Y., Shmilovitz, D., Martinez-Salamero, L.: A power management strategy for minimization of energy storage reservoirs in wireless systems with energy harvesting. IEEE Trans. Circuits Sys. I **58**(3), 633–643 (2011, Regular Papers)

Distributed Energy-Aware Routing Protocol

Erol Gelenbe and Toktam Mahmoodi

Abstract This paper presents an energy aware routing protocol (EARP) whose purpose is to minimise a metric that combines the total consumed power in the network and the QoS bounds requested by the incoming flows. The algorithm performs in a fully distributed manner thanks to the functionalities provided by the Cognitive Packet Network (CPN) which runs at each node of the network. Measurements on an experimental test-bed are presented showing a reduction in power consumption, as compared to a purely QoS driven approach. We also observe that the requested QoS level is also respected.

1 Introduction

Energy efficient protocols have been extensively studied for wireless networks, because energy savings for battery powered nodes is crucial [8], while research on energy consumption is relatively new in wired networks even though the amount consumed is a significant fraction of the energy used for ICT systems [1]. Energy can be saved in the Internet by modifying routing policies, e.g. aggregating traffic alonga few routes and putting some network nodes to sleep [12]. The scheme in [2] examines various configurations of line cards so that switching on/off some components can minimise power consumption. A heuristic approach is presented in [4] that switches off some nodes in the network to minimise power. The work in [15] combines rate adaptation for active nodes with putting idle nodes to sleep

E. Gelenbe · T. Mahmoodi (✉)
Intelligent Systems and Networks Group, Department of Electrical
and Electronic Engineering, Imperial College, London SW7 2BT, UK
e-mail: t.mahmoodi@imperial.ac.uk

E. Gelenbe
e-mail: e.gelenbe@imperial.ac.uk

E. Gelenbe et al. (eds.), *Computer and Information Sciences II*,
DOI: 10.1007/978-1-4471-2155-8_18,

to save energy in the network. An energy-aware congestion control technique is presented in [16], which controls the capacity each network node can offer to meet the actual traffic demand. Finally, [3] provides an estimate of potential energy savings that may be obtained in the Internet.

In this paper we present an energy aware routing protocol (EARP) that attempts to minimise the total consumed power in the network, and also respects the QoS requested by each incoming flow. EARP relies on the underlying Cognitive Packet Network (CPN) [6] for the information it requires, and uses it to minimise power consumption. In other words, the smart packets in CPN [9] collect information with regard to the power usage at the nodes so that the CPN source routing scheme can include power consumption as a decision criterion. In previous work CPN has been proposed as a means to optimise energy consumption [11], and we continue this previous research.

The remainder of this paper is organised as follows. Section 2 elaborates the energy-aware routing protocol. After presenting the configuration of our network testbed, Sect. 3 illustrates the experimental results. Conclusions are presented in Sect. 4.

2 Energy and QoS Aware Routing

At each node i, let us denote by T_i, the traffic this node carries in packets per seconds (pps). Assuming that a flow l carries a traffic of rate t_l pps, then T_i can be computed as:

$$T_i = \sum_{l \in F(i)} t_l \tag{1}$$

where $F(i)$ denotes the set of flows that use node i. Let $p_i(T)$ and $Q_i(T)$ be the power consumption and QoS requirements of node i when the traffic it carries is T. Adding a new flow k to node i will result in a change in power consumption and QoS at that node. Let $p_i(x)$ be the instantaneous power consumption in watts at node i when it carries x packets per seconds, including all aspects of packet processing: storing, routing, and forwarding them through appropriate link drivers.

Define the *Power Cost* associated with the kth flow at node i by $m_i^k(t_k, T_i)$, as a combination of the flow's own power consumption, and of the impact it has on other flows which are using the node:

$$m_i^k(t_k, T_i) = a p_i(T_i + t_k) + b[p_i(T_i + t_k) - p_i(T_i)] \tag{2}$$

where $a,\ b \geq 0$. Here the first term is the total power (watts) due to adding the kth flow, multiplied by some constant a. The second term represents the increase in wattage for the other flows, multiplied by some constant b. The power related cost functions for the kth traffic flow of rate t_k on a path $\pi(i)$ originating at node i is written as:

$$m^k_{\pi(i)}(t_k, \overline{T_{\pi(i)}}) = \sum_{n\in\pi(i)} m^k_n(t_k, T_n), \tag{3}$$

Similarly, we would have the QoS criterion, such as loss, delay or some other metric:

$$Q^k_{\pi(i)}(t_k, \overline{T_{\pi(i)}}) = \sum_{n\in\pi(i)} Q^k_n(t_k, T_n) \tag{4}$$

where $\overline{T_{\pi(i)}} = (T_{n_1}, \ldots, T_{n_{|\pi(i)|}})$ where $n_1 = i$, and the n_j, with $1 \leq j \leq |\pi(i)|$ are the successive nodes of path $\pi(i)$.

The power related information are gathered from across the network using the approach presented by CPN, which runs autonomously at each node using Reinforcement Learning with a recurrent Random Neural Network [5]. The measurement results for this protocol are summarised in [7]. Since EARP is expected to minimise the overall cost of power while satisfying the requested QoS, the goal G_i to be optimised will combine the power consumption with the QoS constraint. All quantities of interest for some flow k will relate to the forward path from any node i to the destination node of that flow. Thus the goal will take the form:

$$G_i = m^k_{\pi(i)}(t_k, \overline{T_{\pi(i)}}) + \gamma 1[Q^k_{\pi(i)}(t_k, \overline{T_{\pi(i)}}) - Q^k_o > 0](Q^k_{\pi(i)}(t_k, \overline{T_{\pi(i)}}))^{\nu} \tag{5}$$

where $\nu \geq 1$, and $\gamma > 0$ are constants meant to match the delay units with respect to power, while Q^k_o is the QoS value that *should not be exceeded* for flow k. Moreover, $1[X]$ is the function that takes the value 0 if X is true, and takes the value 1 if X is false. The CPN reward function, R, is defined based on the function in (5) i.e. $R = G^{-1}$. This reward function is utilised by the RL algorithm in the similar way as [9] to seek the most optimal path.

3 Experiments

Our experimental testbed and the parameters that are used in the experiments are the same as those described in [10, 14]. Each experiment ran for 600 s, and measurements were collected from each node every five seconds. The performance of our proposed energy-aware routing scheme was compared with the CPN routing protocol that is only QoS driven, i.e., its aim is to minimise end-to-end delay, and also with the routing algorithm that seeks shortest path. The shortest path here is selected by the CPN protocol and based on the available information in the smart packets.

We use the presented model in [13] for the power consumption model of the routers that is defined based on the offline measurements from our testbed's nodes. It has been shown that a polynomial function can represent the relation between power consumption and traffic of a router, where the polynomial coefficients depend on the router's operating mode i.e. either the router receives, transmits or forward data.

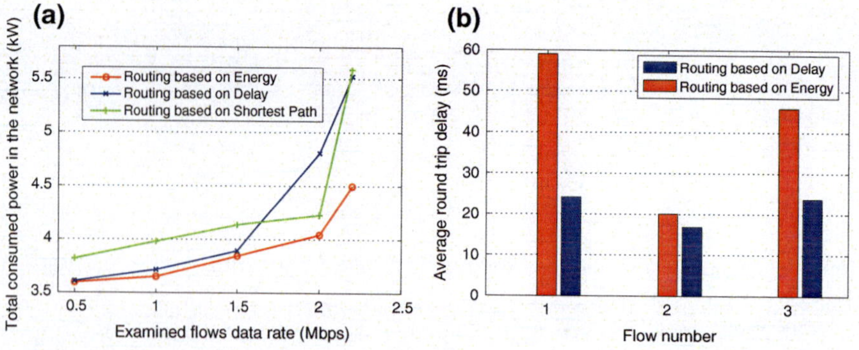

Fig. 1 Scenario one: power consumption and round trip delay of the three examined flows. **a** Total power consumption in the network versus traffic rate. **b** Average round trip delay

In the first scenario, we choose three pairs of source-destination nodes, and have set up three flows (23-12, 30-14, and 33-2). All three flows were first run at a data rate of 0.5 Mbps, which was then increased to 1, 1.5, 2, and 2.2 Mbps in successive rounds of the experiment simultaneously for all three flows. Figure 1 shows an interesting comparison of the measured total power consumption versus the five different data rates for the three flows. On the other hand, as would be expected, EARP results in higher end to end delays, shown in Fig. 1. It can be seen that the delay achieved by the delay minimisation scheme is approximately 45% smaller in comparison to EARP. A similar experiment was s conducted with CPN, seeking shortest paths without consideration for the QoS. Although shortest path routing can potentially save energy by engaging fewer routers, our experimental results summarised in Fig. 1a seem to contradict this, probably due to the non-linear relation between the routers' power consumption and traffic.

In the second scenario there are nine pairs of source-destination nodes. Three flows are initially active: (23-12, 30-14, and 33-2). Another three flows are initiated 100 s later: (20-26, 28-10, and 35-7), and the remaining three flows are set up 200 s after the experiment begins: (11-8, 13-24, and 29-6). Each flow's lifetime is 400 s, and the experiment lasts 600 s. All nine flows were first run at a data rate of 0.5 Mbps, which was then increased to 1, 1.5, 2, and 2.2 Mbps in successive rounds, simultaneously for all the flows.

The power consumption in the network for data rates per flow of 1 and 2 Mbps is shown in Fig. 2 for the first 400 s of the experiment (while the first three flows are yet active), versus the elapsed time. This figure shows a step increase in power consumption when the experiment's elapsed time is 100 s, and another step increase when the elapsed time is 200 s, which refers to the newly initiated flows in the network at those two time instances. Further observation from Fig. 2 reveals the savings in power consumption with EARP when the data rates for both flows is 1 and 2 Mbps.

We report the end-to-end delay observed with EARP in this scenario and compare it with the delay minimisation scheme in Fig. 3a, where the average delay

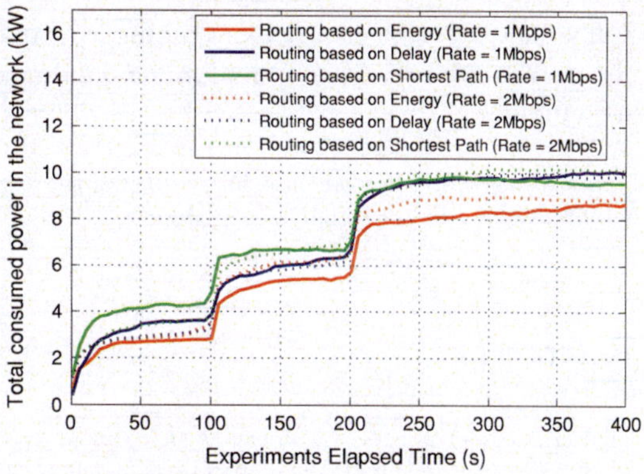

Fig. 2 Scenario two: Total power consumption in the network versus the experiment's elapsed time

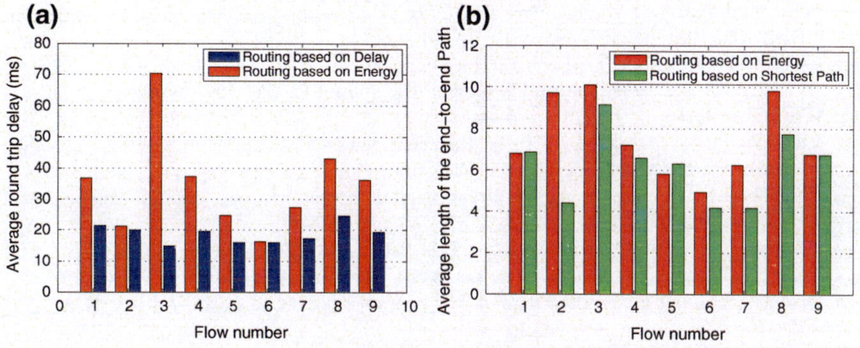

Fig. 3 Scenario two: round trip delay and the route length of the active flows. **a** Average round trip delay. **b** Average length of the end-to-end path

experienced by each flow is presented (the average is computed over all five data rates). It is seen that the QoS-only driven scheme experiences approximately 45% less latency as compared to EARP. Investigations on the path used by EARP are summarised in Fig. 3b; the average route length of each flow using EARP appears to be (0.6) hops longer that of routes chosen by the shortest path algorithm.

4 Conclusions

This paper introduces the energy-aware routing protocol (EARP) that attempts to minimise the total power consumption of each flow in a packet network, while keeping the principal QoS metric below an acceptable upper bound. EARP is fully distributed and uses the functionalities of CPN. We have experimented with EARP to compare its performance to a version of CPN that only attempts to minimise delay. Further experiments are carried out to compare EARP's performance with CPN when it seeks the shortest path. Although shortest path routing protocol can potentially save energy by engaging less routers, the non-linear dependence of the routers' power consumption on the data rates seem to contradict this intuitive assumption.

Acknowledgments We are grateful for the support and technical motivation that has been provided by the EU FP7 Fit4Green Project.

References

1. Berl, A., Gelenbe, E., Girolamo, M.D., Giuliani, G., de Meer, H., Dang, M.Q., Pentikousis, K.: Energy-efficient cloud computing. Comput. J. **53**(7), 1045–1051 (2010)
2. Chabarek, J., Sommers, J., Barford, P., Estan, C., Tsiang, D., Wright, S.: Power awareness in network design and routing. In: The 27th Conference on Computer Communications (INFOCOM '08). pp. 457–465. IEEE (April 2008)
3. Chiaraviglio, L., Ciullo, D., Leonardi, E., Mellia, M.: How much can the internet be greened? In: IEEE Global Communications Conference (GLOBECOM '09) Workshops. pp. 1–6 (December 2009)
4. Chiaraviglio, L., Mellia, M., Neri, F.: Reducing power consumption in backbone networks. In: IEEE International Conference on Communications (ICC '09). pp. 1–6 (June 2009)
5. Gelenbe, E.: Learning in the recurrent random neural network. Neural Comput. **5**(1), 154–164 (1993)
6. Gelenbe, E.: Cognitive packet network. U.S. Patent 6,804,201, 11 Oct 2004
7. Gelenbe, E.: Steps toward self-aware networks. Commun. ACM **52**(7), 66–75 (2009)
8. Gelenbe, E., Lent, R.: Power-aware ad hoc cognitive packet networks. Ad Hoc Netw. **2**(3), 205–216 (2004)
9. Gelenbe, E., Lent, R., Nunez, A.: Self-aware networks and qos. Proc. IEEE **92**(9), 1478–1489 (2004)
10. Gelenbe, E., Mahmoodi, T.: Energy-aware routing in the cognitive packet network. In: International Conference on Smart Grids, Green Communications, and IT Energy-Aware Technologies (Energy '11) (May 2011)
11. Gelenbe, E., Silvestri, S.: Reducing power consumption in wired networks. In: The 24th International Symposium on Computer and Information Sciences (ISCIS '09). pp. 292–297. IEEE (September 2009)
12. Gupta, M., Singh, S.: Greening of the Internet. Comput. Commun. Rev. **33**(4), 19–26 (2003)
13. Lent, R.: Simulating the power consumption of computer networks. In: The 15th IEEE International Workshop on Computer Aided Modeling, Analysis and Design of Communication Links and Networks (CAMAD '10). pp. 96–100 (December 2010)
14. Mahmoodi, T.: Energy-aware routing in the cognitive packet network. Perform. Eval. **63**(4), 338–346 (2011)
15. Nedevschi, S., Popa, L., Iannaccone, G., Ratnasamy, S., Wetherall, D.: Reducing network energy consumption via sleeping and rate-adaptation. In: The 5th USENIX Symposium on Networked Systems Design and Implementation (NSDI '08). pp. 323–336 (April 2008)
16. Panarello, C., Lombardo, A., Schembra, G., Chiaraviglio, L., Mellia, M.: Energy saving and network performance: a trade-off approach. In: The 1st International Conference on Energy-Efficient Computing and Networking (e-Energy '10). pp. 41–50. ACM (April 2010)

Part IV
Wireless Networks

A Distributed Scheme to Detect Wormhole Attacks in Mobile Wireless Sensor Networks

Oya Simsek and Albert Levi

Abstract Due to mostly being unattended, sensor nodes become open to physical attacks such as wormhole attack, which is our focus in this paper. Various solutions are proposed for wormhole attacks in sensor networks, but only a few of them take mobility of sensor nodes into account. We propose a distributed wormhole detection scheme for mobile wireless sensor networks in which mobility of sensor nodes is utilized to estimate two network features (i.e. network node density, standard deviation in network node density) through using neighboring information in a local manner. Wormhole attack is detected via observing anomalies in the neighbor nodes' behaviors based on the estimated network features and the neighboring information. We analyze the performance of proposed scheme via simulations. The results show that our scheme achieves a detection rate up to 100% with very small false positive rate (at most 1.5%) if the system parameters are chosen accordingly. Moreover, our solution requires neither additional hardware nor tight clock synchronization which are both costly for sensor networks.

Keywords Mobile wireless sensor networks · Security · Wormhole attacks

This work was supported by the Scientific and Technological Research Council of Turkey (TUBITAK) under grant 110E180.

O. Simsek (✉) · A. Levi
Sabanci University, Orhanli, Tuzla, 34956 Istanbul, Turkey
e-mail: oyasimsek@su.sabanciuniv.edu

A. Levi
e-mail: levi@sabanciuniv.edu

E. Gelenbe et al. (eds.), *Computer and Information Sciences II*,
DOI: 10.1007/978-1-4471-2155-8_19,

1 Introduction

As a result of significant advances in hardware manufacturing and wireless communication technology along with efficient software algorithms, wireless sensor networks [1] emerged as a promising network infrastructure for various applications. Due to being mostly unattended and the open nature of wireless communication channels, sensor nodes become open to physical attacks which may lead to various attacks including wormhole attack. In wormhole attack, an attacker tunnels messages received in one part of the network over a wormhole link and replays them in a different part of the network. This low-latency tunnel attracts network traffic on the wormhole link which can empower the attacker to perform traffic analysis, denial of service attacks; collect data to compromise cryptographic material; or just selectively drop data packets through controlling these routes using the wormhole link. Moreover, an attacker can perform this attack without compromising any legitimate nodes, or knowing any cryptographic materials since the attacker neither creates new packets nor alters existing packets.

There are several approaches for wormhole detection in wireless sensor networks which mostly focus on static networks. These solutions are mainly based on detecting the maximum distance any message can travel, or the maximum time of travel of any message [2], discovering one-hop neighbors in a secure way [3], or monitoring the data traffic of neighbor nodes [4]. Also, most of these approaches require additional hardware (e.g. directional antennas in [5], GPS in [2], a specialized hardware for one-bit challenge request-response [3] protocol), special trusted nodes such as guards in [6], highly accurate time or location measurements [3], or tight clock synchronization [2], which seems infeasible for large scale wireless sensor networks because of its resource limitations and economic costs. In this paper, we propose a distributed wormhole detection scheme for mobile wireless sensor networks. Our scheme aims to utilize the mobility feature of the sensor nodes to examine the environment and network properties, and derive new features which help understanding the network better.

2 Proposed Scheme

Our scheme includes two main phases: (1) stabilization, and (2) detection phases. Stabilization phase is for sensor nodes to collect information from the network through using neighboring information to estimate the node density of the network locally, d_i^r, for node i at rth round, and to compute the standard deviation of the change in the estimated node density, σ_i^r. This phase runs once right after the uniform random deployment of the sensor nodes. In detection phase, based on the pre-computed statistical values, the detection mechanism is activated to check for anomalies in the network, and detected nodes are revoked from the network. Without a wormhole attack being performed, the difference between the number of neighbors of a node and its estimated network density does not exceed the standard deviation of its network density. However, under wormhole attack, this difference can be higher due to fake neighboring connections, especially when a node is close to the wormhole ends.

2.1 Network Assumptions and Threat Model

The network is composed of mobile nodes having same communication range as well as same physical properties. The sensor nodes are deployed randomly using uniform distribution in the sensing area. None of the nodes know their location information. Nodes can obtain the neighbor count information of their neighbors as well as their own neighboring information. Secure neighbor discovery is out of the scope of the paper. There are proposed solutions for neighbor discovery, [7, 8], addressing node mobility as well as energy efficiency in the literature. Necessary link level security requirements (i.e. confidentiality, authentication, and integrity) are assumed to be fulfilled by the lower layers. It is sufficient for an attacker to capture two legitimate nodes and create a low-latency tunnel between them. We assume that the wormhole link is bidirectional. In other words, both ends of wormhole link overhear the packets; tunnel these packets to other node via this low-latency tunnel so that the receiving node can replay these packets at that end of the wormhole. The attacker may drop the packets selectively in a random way. However, by doing so, the wormhole link becomes less attractive and this is not a desired situation for the attacker. Thus, we assume that the attacker does not drop any packets.

2.2 Details of the Proposed Scheme

Stabilization Phase. Stabilization phase starts right after the uniform random deployment of N sensor nodes, and runs S rounds. In a round, each node discovers their neighbors securely, broadcasts its neighbor count, and locally computes statistical features of the network (i.e. d_i^r and σ_i^r) after receiving all neighbor counts of its neighbors.

Share Neighboring Information. When a node learns its neighbors, it broadcasts an information packet including its own identity, i, and the number of its neighbors, Ψ_i. This information is critical while estimating the network features.

Calculate and Update Statistical Metrics. After all nodes share the number of their neighbors, each node i has the following information: its own neighbors, N_i, the number of its own neighbor number, Ψ_i, and neighbor count information of its neighbors, $\Psi_j, \forall j \in N_i$. Then, node i computes the network density, d_i^r, and standard deviation in d_i^r, σ_i^r, in a local way using equations (initial conditions are $d_i^0 = 0$ and $\sigma_i^0 = 0$):

$$d_i^r = \frac{\Psi_i + \sum_{j \in N_i} \Psi_j}{\Psi_i + 1} \times (1 - \alpha) + d_i^{r-1} \times \alpha \tag{1}$$

$$\sigma_i^r = \sqrt{\frac{1}{\Psi_i + 1} \times \left[\left(\sum_{j \in N_i} \left((\Psi_j - d_i^{r-1})^2 \right) \right) + (\Psi_i - d_i^{r-1})^2 \right]} \times (1 - \alpha) + \alpha_i^{r-1} \times \alpha \tag{2}$$

We use exponential averaging, which we are inspired by its usage in TCP round trip time estimation, to give more importance to the latest data retrieved from neighbors without losing the previous calculated values. α and $(1 - \alpha)$ are the weights which are used to estimate standard deviation and local network density of a node. At each round, each node estimates a candidate density value which is calculated by averaging the neighbor counts received from neighbors along with its own neighbor count (Eq. 1). After that, the node updates its density via using the exponential average of the previous value and the new estimated value. The procedure is same for the calculation of standard deviation in the node density (Eq. 2).

Detection Phase. In detection phase, pre-computed network features along with round threshold, T_{round}, (i.e. the maximum number of rounds in which a node a needs to witness an anomaly about a node b to keep node b in its local suspected nodes list), alarm threshold, T_{alarm}, (i.e. the minimum number of alarm to broadcast a node as globally *suspected*), and revocation threshold, T_{revoc}, (i.e. the number of nodes required to revoke a node), are used to detect the anomaly created by the wormhole link. A round in detection phase is composed of neighbor discovery, sharing the number of neighbors, testing detection criteria along with broadcasting specific messages when necessary, and finally revocation of detected nodes.

Check for Suspicious Nodes based on Statistical Metrics. After obtaining the neighborhood information, each node i has locally-estimated network density, d_i^s, and locally-estimated standard deviation in d_i^s, σ_i^s, and the neighboring information $\Psi_j, \forall j \in N_i$. To detect an anomaly, node i first checks whether the number of its own neighbors exceeds d_i^s more than σ_i^s. If the difference exceeds σ_i^s, it accuses its neighbors and adds them to its list which is for tracking locally suspicious nodes. Otherwise, node i checks its neighbors one by one with the same method to detect a suspicious behavior and updates its list accordingly. If the alarm counter for a locally suspected node j exceeds T_{alarm}, then node i broadcasts a message deeming j is a globally *suspected* node. If any node in the list of locally suspected nodes does not show an anomaly during T_{round}, then node i deletes that node from its list. When a node i receives an alarm saying node j is a potential malicious node, it runs the following check: If j is already in its globally suspected nodes list, it updates the alarm counter of j; otherwise, it adds j to the list. To revoke node j, the number of nodes deeming node j as suspected must exceed T_{revoc} which is basically a preset percentage of the total number of nodes in the network.

Revoke Detected Node. A globally *suspected* node can be revoked from network through node self-destruction mechanisms proposed in [9] and [10]. When a node i receives a message saying node j is a malicious node, it sends a message to the base station for revocation of j and updates its list which is for keeping track of revoked nodes accordingly.

3 Performance Evaluation

We analyzed the performance of our scheme via simulations, and present our results in a comparative way. We analyzed the effects of the change in the system parameters on the detection rate under the simulation setup defined below. Due to limitation of space, only a small subset of the simulation results is included in the paper.

The results presented in the graphs are average of 20 simulations. $N = 200$ nodes are distributed over a field of $A = 100$ m × 100 m. We use random movement model in which each node chooses a random direction; and moves towards it with a uniformly distributed random speed in the range of (5, 15 m/s). Nodes have a communication range of 15 m. We simulated various values of T_{alarm} and T_{revoc}. The results show that the more optimal and stable value for α is 0.5. Therefore, we choose α as 0.5 in our simulations. We assume that 5% of all nodes are static all the time. Also, we assume that wormhole attack is not performed during stabilization phase. Stabilization phase runs once and lasts $S = 1{,}000$ rounds. Detection phase runs during the lifetime of a sensor node due to the possibility of wormhole attack being performed at any time. However, we limit this value to 2,000 rounds in our simulations.

Detection and false positive rates are our main metrics while evaluating the success of the simulations. Detection rate is the ratio of the number of simulation runs where the wormhole is detected successfully, *D#*, over total number of simulation runs, *S#*, which is computed as *D#*/*S#*. False positive rate per simulation run is computed as the ratio of falsely detected nodes, *F#*, over total node number, *N*. False positive rate is the average of this ratio of all simulation runs, hence, it is computed as $\frac{\sum_{1}^{S\#}\left(\frac{F\#}{N}\right)}{S\#}$.

If T_{alarm} increases, node i needs to witness more suspicious behavior of node j to broadcast it as globally *suspected*, and detection probability of wormhole decreases considering the mobility of the nodes. Since nodes are mobile, they may not be under the effect of wormhole for such long time to exceed that high T_{alarm} value for a suspected node. Hence, there may not be enough nodes to broadcast that suspected node as globally *suspected*. The number of revoked nodes decreases, thus, detection and false positive rate decreases. Similarly, T_{revoc} is inversely proportional to the number of revoked nodes since high T_{revoc} means more nodes are required to agree on revoking a node. The simulation results which are presented in Figs. 1 and 2 verify those observations. Increasing T_{alarm} or T_{revoc}

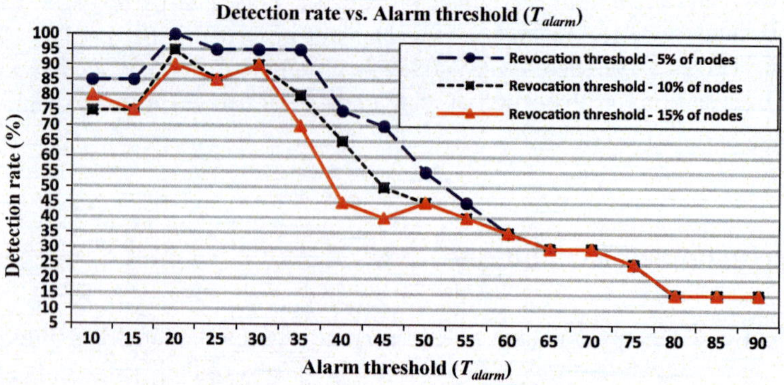

Fig. 1 Detection rate versus Alarm threshold (T_{alarm}) for $T_{\mathrm{revoc}} = 10$, $T_{\mathrm{revoc}} = 20$, and $T_{\mathrm{revoc}} = 30$; $T_{\mathrm{round}} = 20$; wormhole ends are chosen randomly

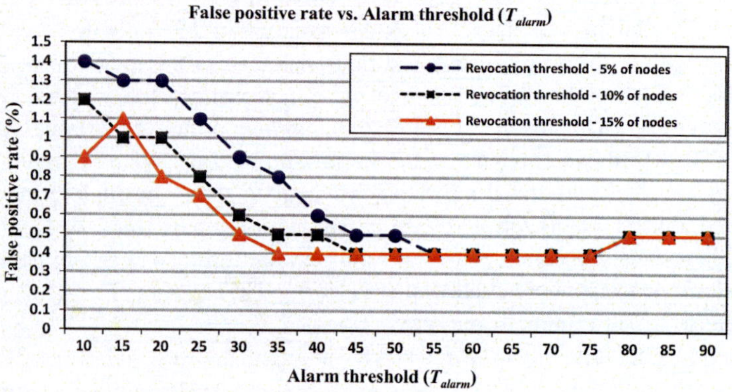

Fig. 2 False positive rate versus Alarm threshold (T_{alarm}) for $T_{\mathrm{revoc}} = 10$, $T_{\mathrm{revoc}} = 20$, and $T_{\mathrm{revoc}} = 30$; $T_{\mathrm{round}} = 20$; wormhole ends are chosen randomly

decreases the detection and false positive rates, but they do not change much after a high enough T_{alarm} value.

4 Conclusions

In this paper, we propose a distributed wormhole detection scheme for mobile wireless sensor networks which utilizes mobility of sensor nodes to detect wormhole attack to estimate new features in a local way which helps understanding the network better. Wormhole attack is detected via observing anomalies in the neighbor nodes' behaviors based on these estimated network features and the neighboring information. We analyzed the performance of proposed scheme via simulations using different system parameters. The results show that our scheme achieves a detection rate up to 100% with very small false positive rate (at most 1.5%) if the system parameters are chosen accordingly. Moreover, our solution requires neither additional hardware nor tight clock synchronization which are both costly for sensor networks.

References

1. Akyildiz, I.F., Su, W., Sankarasubramaniam, Y., Cayirci, E.: Wireless sensor networks: a survey. Comput. Netw. **38**(4), 393–422 (2002)
2. Hu, Y.C., Perrig, A., Johnson, D.B.: Packet leashes: a defense against wormhole attacks in wireless ad hoc networks. IEEE INFOCOM **3**, 1976–1986 (2003)
3. Capkun, S., Buttyan, L., Hubaux, J.: SECTOR: secure tracking of node encounters in multi-hop wireless networks. SASN, pp. 21–32 (2003)
4. Khalil, I., Bagchi, S., Shroff, N.B.: LITEWORP: a lightweight countermeasure for the wormhole attack in multihop wireless networks. DSN, pp. 612–621 (2005)
5. Hu, L., Evans, D.: Using directional antennas to prevent wormhole attacks. NDSS, pp. 22–32 (2004)
6. Lazos, L., Poovendran, R., Meadows, C., Syverson, P., Chang, L.W.: SeRLoc: secure range-independent localization for wireless sensor networks. Wise, pp. 21–30 (2005)
7. Kohvakka, M., Suhonen, J., Kuorilehto, M., Kaseva, V., Hannikainen, M., Hamalainen, T.D.: Energy-efficient neighbor discovery protocol for mobile wireless sensor networks. Ad hoc Netw. **7**(1), 24–41 (2009)
8. Bagchi, S., Hariharan, S., Shroff, N.: Secure neighbor discovery in wireless sensor networks. ECE Technical Reports. Paper 360 (2007)
9. Curiac, D.-I., Plastoi, M., Banias, O., Volosencu, C., Tudoroiu, R., Doboli, A.: Combined malicious node discovery and self-destruction technique for wireless sensor networks. SENSORCOMM, pp. 436–441 (2009)
10. Plastoi, M., Curiac, D.-I.: Energy-driven methodology for node self-destruction in wireless sensor networks. SACI, pp. 319–322 (2009)

Cross-Layer Optimization with Two-Group Loading for Ad hoc Networks

Hadhrami Ab Ghani, Mustafa Gurcan and Zhenfeng He

Abstract A two-group resource allocation scheme enhanced with a capacity-approaching modulation and coding scheme is proposed to improve the effective transmission bit rate over wireless ad hoc networks. Improvement in the transmission bit rate reduces the end-to-end delay of the selected routing paths. With computationally efficient loading scheme and reduced gap values, the energy is efficiently used to improve the total bit rate.

Keywords Two-group allocation · Ad hoc networks · Reduced gap values

1 Introduction

The end-to-end delay of a routing path in wireless ad hoc networks limits the total realizable transmission bit rate over the path and causes the rise in the energy consumption of the energy-limited nodes [1]. As it depends on the transmission bit rates over the transmission links between the adjacent nodes in the selected paths, the end-to-end delay can be minimized by increasing the bit rates whilst keeping the required energy low for transmitting a given number of data packets. High-speed multi-code transmission systems [2] are often used to enhance the transmission bit rate over a wireless link. Although a large number of bit rates, with

H. A. Ghani · M. Gurcan (✉) · Z. He
Department of Electrical and Electronic Engineering, Imperial College, London, UK
e-mail: m.gurcan@imperial.ac.uk

H. A. Ghani
e-mail: hadh.ghani05@imperial.ac.uk

Z. He
e-mail: zhenfeng.he06@imperial.ac.uk

E. Gelenbe et al. (eds.), *Computer and Information Sciences II*,
DOI: 10.1007/978-1-4471-2155-8_20,

relatively small bit granularities or differences between the adjacent bit rates, are offered and the ISI components can be removed in [2], this equal energy loading results in a wasted SNR problem due to the varying SNRs which differ from the minimum required SNR to realize the desired bit rate. Alternatively, the SNR can be made equal per channel by iteratively adjusting energy using margin adaptive optimization. This method however results in a relatively large residual energy.

To utilize the residual energy and maximize the total bit rate, rate adaptive optimization is used to test different combinations of bit allocation, each of which require iterative energy calculation. To avoid this high computational load, a modified two-group loading approach, which was formerly presented in [3], in this paper which considers only two adjacent bit rates with reduced gap values [4] to improve the total bit rate. With reduced gap values, the energy consumption of an energy-limited node in wireless ad hoc network is minimized. The next section formulates and describes these addressed problems.

2 Problem Description

Consider a block of data with m_z bits, to be transmitted using a multi-code transmission model over path z in packets of size L_z bits with K orthonormal signature sequences having a spreading factor of N and a chip period of T_c. The end-to-end delay T_D^z [1] over path z can be written as

$$T_D^z = \left(\left\lfloor \frac{m_z}{L_z} \right\rfloor - 1 \right) \max_{i,l \in J_z} \overline{T}_{i,l} + \sum_{i,l \in J_z} \overline{T}_{i,l} \tag{1}$$

where the average service time $\overline{T}_{i,l}$ between node i and l is defined as [1]

$$\overline{T}_{i,l} = \overline{T}^s_{i,l} + \left(\overline{T}^s_{i,l} p^i_s + \overline{T}^i p^i_i + \overline{T}^c p^i_c \right) B + C \tag{2}$$

with [5] $\overline{T}^s_{i,l} = \frac{L_z N T_c}{R_T^{i,l}} + A$ as the average time the channel is busy because of the successful transmission with a probability of p^i_s, $\overline{T}^i$ as the average time the channel is in the idle state with a probability of p^i_i, $\overline{T}^c$ as the average time the channel falls into a collision state, with a probability of p^i_c and A, B, C are derived from the MAC layer parameters as given in [5]. Therefore, the aim of this paper is to maximize the total bit rate $R_T^{i,l}$ in bits per symbol using a near-optimal two-group resource allocation scheme incorporated with a capacity-approaching coding scheme. Assuming that multi-code transmission with K code channels which are realizable with a set of bit rate $\{b_p\}_{p=1}^P$ bits per symbol is run between node i and l, this paper addresses the following points

1. Maximize $R_T^{i,l} = \sum_{k=1}^K b_{p_k}$ for a given total energy $R_T^{i,l}$ with a relatively low computational complexity for $p_k \in \{1, 2, \ldots, P\}$.

Fig. 1 An example of 5-node wireless ad hoc networks

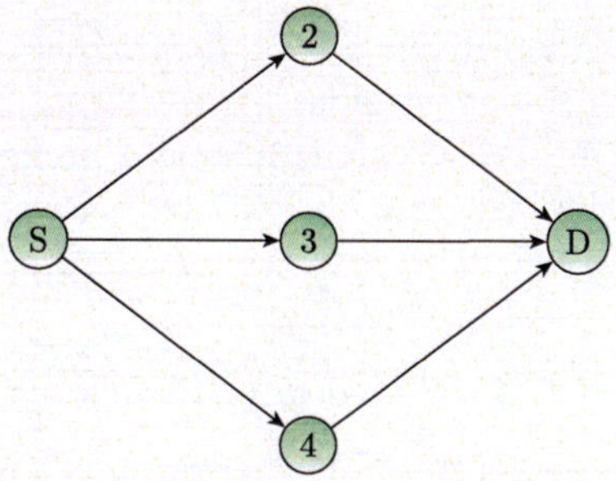

2. Minimize the required SNR, hence the energy, by minimizing the gap value $\Gamma^{(r)}$ [4] for realizing b_{p_k}, in order to improve $R_T^{i,l}$ for a given E_T.

The next subsection briefly describes the existing resource allocation techniques before presenting the proposed modified two-group loading scheme Fig. 1.

3 Bit Rate Maximization Schemes

For a given set of rates $\{b_p\}_{p=1}^{P}$ in bits per symbol, a total energy of E_T and K code channels, a general bit maximization model for a multi-code transmission system is formulated as

$$\begin{aligned} \max R_T^{i,l} &= \max_{k=1,2,,K} \sum_{k=1}^{K} b_{p_k} \\ s.t.\ E_{sum} &= \sum_{k=1}^{K} E_k\left(b_{p_k}\right) \le E_T \\ &and\ p_k \in \{1,,P\} \end{aligned} \tag{3}$$

where $E_k\left(b_{p_k}\right)$ is the symbol energy required to realize the bit rate b_{p_k} over channel k given by

$$E_k\left(b_{p_k}\right) = \frac{\Gamma\left(2^{b_{p_k}}-1\right)}{1+\Gamma\left(2^{b_{p_k}}-1\right)}\left(\boldsymbol{q}_k^H \mathbf{C}^{-1} \boldsymbol{q}_k\right)^{-1}, \tag{4}$$

when minimum mean square error despreading filters are used where $\boldsymbol{q}_k$ is the receiver signature sequence [3]. Both $E_k(b_{p_k})$ and $\mathbf{C}$ are the converged energy and covariance matrix, which are iteratively computed since they are a function of each other to achieve a target minimum SNR, $\gamma_k^*(b_{p_k}) = \Gamma\left(2^{b_{p_k}}-1\right)$

As there are P^K possible combinations of bit rates b_{p_k} to be allocated over K employed channels, the above optimization method is very computationally inefficient. The incurred computational complexity will further increase when the iterative energy calculation process is run for each of these P^K combinations.

Alternatively a two-group resource allocation, which tests at most $P+K-2$ bit rate combinations, is developed in this paper by extending the margin adaptive loading technique described in the previous section to utilize the residual energy for enhancing the total bit rate. An account of this scheme is presented next.

3.1 Two-Group Allocation Over Channels with ISI

The enhanced total bit rate aimed by the two-group resource allocation scheme, which improves the existing margin adaptive equal rate loading, can be written as follows

$$R_T^{i,l} = (K-m)b_p + mb_{p+1}. \tag{5}$$

with the total required energy $\sum_{k=1}^{K} E_k^{(m)}(y_k)$ to realize $R_T^{i,l}$ bits per symbol must satisfy

$$\max_m \left(\sum_{k=1}^{K} E_k^{(m)}(y_k) \right) \leq E_T, \tag{6}$$

where $E_k^{(m)}(y_k)$ is the energy required to load y_k bits per symbol at channel k such that the target minimum SNR $\gamma^*(y_k)$ for y_k is met when m channels are loaded with b_{p+1} bits per symbol and $K-m$ channels are loaded with b_p bits per symbol for $y_k \in \{b_p, b_{p+1}\}$.

If the target minimum SNR $\gamma_k^*(y_k) = \Gamma(2^{y_k}-1)$ as given in Eq. (4) is reduced, the allocated energy $E_k(y_k)$ in Eq. (6) can be reduced to further enhance the total bit rate improvement. The only way to reduce the target minimum SNR is by reducing the gap value Γ, as will be explained in the next subsection.

3.2 Gap Value Reduction with Coded Parity Packet Approach

A modulation and coding scheme called coded parity packet (CPP) coding scheme is developed in [4] is proposed to be incorporated with the two-group resource allocation scheme in order to reduce the gap value Γ. It controls the bit rate

$$b_p = \frac{V_p}{E_p}\frac{N_p}{B_p}\log_2 M \tag{7}$$

by increasing V_p and E_p whilst keeping b_p the same to provide additional external extrinsic information values to the decoding unit to reduced the gap value, where V_p and $E_p - V_p$ are the number of source and parity packets respectively, $r_{in,p} = \frac{N_p}{B_p}$

is the inner code rate and M is the constellation size. It will be demonstrated in the next section that the gap value reduction minimizes the consumed amount of energy in wireless ad hoc networks.

4 Numerical Results

The proposed two-group resource allocation scheme is implemented with the CPP coding technique which reduces the gap value of the realized bit rate. This gap value reduction results in an improved transmission bit rate whilst keeping the energy consumption per bit relatively low. Three bit rates and the corresponding initial gap values, which are relatively smaller than those of the existing methods, produced by the CPP coding scheme are tested, which are $b_p \in \{1.33, 1.50, 2.25\}$ bits per symbol and $\Gamma_p \in \{3.43, 2.88, 2.85\}$ dB respectively. The same bit rates are also provided using a Turbo coding scheme with gap values of 5.8,5.13 and 5.85 dB.

Figure 2a and b show the new effective transmission bit rates in bits per symbol when up to $N_{re} = 10$ retransmissions are permissible if some of the data packets originally sent at $b_p = \frac{V_p}{E_p}\frac{N_p}{B_p}\log_2 M$ bits per symbol fail, where V_p is the number of data packets, E_p is the number of packets transmitted including the parity packets, $r_{in} = \frac{N_p}{B_p}$ is the inner Turbo code rate and M is the constellation size for the CPP coding scheme, which is 16 in this case. If the packets fail at a rate of $PER(\frac{E_s}{N_0})$ at an input SNR of $\frac{E_s}{N_0}$ dB, the new effective rate when retransmissions are permissible is written as

$$b_p^{(r)} = \frac{V_p}{E_p + V_p \sum_{i=1}^{N_{re}} \left(PER(\frac{E_s}{N_0})\right)^i} \frac{N_p}{B_p} \log_2 M, \tag{8}$$

The adjustment parameter $\alpha = \frac{V_p}{V_{p,o}} = \frac{E_p}{E_{p,o}}$, as can be seen in Fig. 2a and b, increases the actual number $V_{p,o}$ of data packets to provide more additional soft information values to the decoding process to reduce the gap value. The reduced gap value, $\Gamma^{(r)}$ which is the difference between the required input SNR and the lower bound in dB, for the case of $\alpha = 3$ is about $\Gamma^{(r)} = 1.5$ dB in Fig. 2a and about $\Gamma^{(r)} = 2$ dB in Fig. 2b. These gap values are almost 2 dB less than those of the Turbo-QAM schemes.

Figure 3 shows the achievable system, $C_T = \frac{R_T^{i,l}}{NT_c}$, when the two-group research allocation scheme is implemented with the new effective bit rates after retransmissions, as described previously, where $N = 16$ and $1/T_c = 3.84$ Mchips/s. The target SNRs to realize these bit rates are taken from the previous two figures, where these target minimum SNRs are observed to be the input SNRs when the target packet error rate is roughly 1×10^{-2}. It can be seen that the system

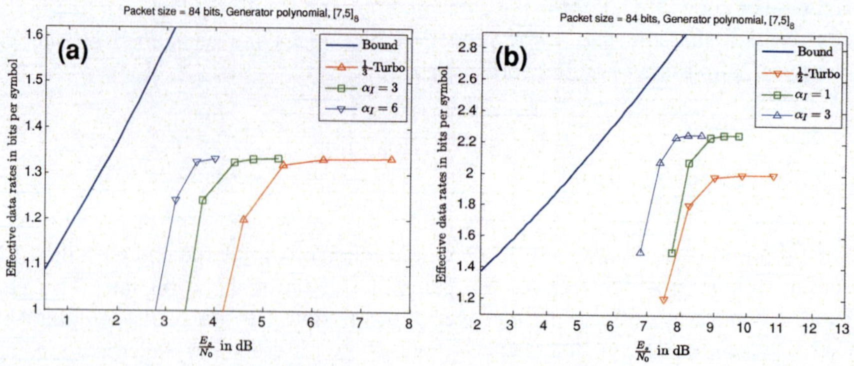

Fig. 2 Effective rates after retransmission **a** $b_p =$ bits per symbol, **b** $b_p = 2.25$ bits per symbol

Fig. 3 Total bit rate after retransmissions and gap value reduction

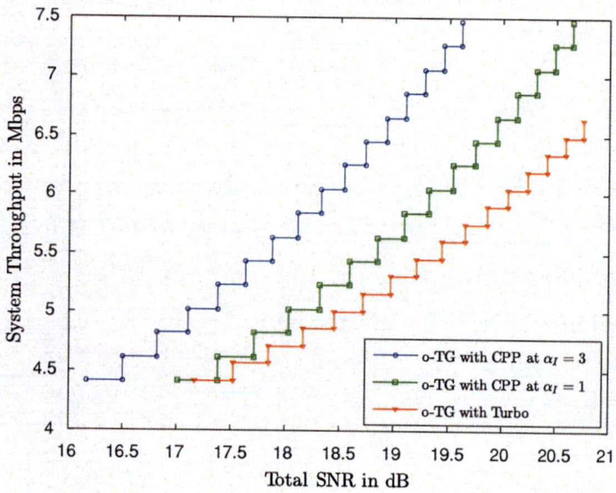

throughput achieved by the two-group scheme improves significantly when the gap values are reduced using the employed CPP coding scheme. It can be seen in this figure that the total bit rate increased by more than 1 Mbps for a given total SNR.

4.1 Practicality

With these results, the proposed scheme is observed to offer a practical multi-code transmission model over the links in wireless ad hoc networks for a few reasons. Firstly, the computational complexity is lowered as only two bit rates are considered. Secondly, the energy required is minimized since the gap value is reduced, which is suitable for energy-limited nodes in wireless ad hoc networks. Therefore, this model is practically deployable in the networks. As the computational complexity is independent of the network size, this proposed model is also applicable for any scales or sizes of the networks.

5 Conclusions

A modified two-group resource allocation scheme, which is incorporated with a capacity-approaching coding scheme, is demonstrated to reduce the effective transmission bit rate over the links in wireless ad hoc networks, where retransmissions are considered. The improved transmission bit rate for a given SNR is essential to reduce the end-to-end delay over the paths used in the networks.

References

1. Gurcan, M., Ab Ghani, H., Zhou, J., Chungtragarn, A.: Bit energy consumption minimization for multi-path routing in ad hoc networks. Comput. J. (2011)
2. GPP TS 25.214: Physical layer procedure (FDD), V10.1.0 ed., 3GPP, December (2010)
3. He, Z., Gurcan, M., Ghani, H.: Time-efficient resource allocation algorithm over HSDPA in femtocell networks. In: IEEE 21st International Symposium On Personal, Indoor and Mobile Radio communications Workshops (PIMRC Workshops), 2010. IEEE, pp. 197–202 (2010)
4. Gurcan, M., Ghani, H.A.: Coded parity packet approach for small-sized packet transmission in wireless ad-hoc networks, submitted to european transactions on telecommunications in December (2010)
5. Carvalho, M.M., Garcia-Luna-Aceves, J.J.: A scalable model for channel access protocols in multihop ad hoc networks. In: ACM Mobicom 2004, pp. 330–344. ACM, New York, USA (2004)

Algorithms for Sink Mobility in Wireless Sensor Networks to Improve Network Lifetime

Metin Koç and Ibrahim Körpeoglu

Abstract Sink mobility is an effective solution in the literature for wireless sensor network lifetime improvement. In this paper, we propose a set of algorithms for sink site determination (SSD) and movement strategy problems of sink mobility. We also present experiment results that compare the performance of our algorithms with other approaches in the literature.

1 Introduction

Several schemes are proposed in the literature to minimize the total energy consumption in the WSN and thus improve the wireless sensor network lifetime: power adjusting when transmitting messages, developing energy-efficient MAC or routing protocols, etc. Making the data collection node (or sink) mobile appears to be another approach for improving the lifetime of WSNs. In such a network, the packet traffic flows from the base station to the leaf nodes, that is, all packets of the network are delivered to the sink node via its first-hop neighbors. This situation causes these nodes to deplete their energy faster than the other nodes in the network. Therefore, sink changes its position periodically by fairly delegating the sink's neighbor role among the sensor nodes to balance the remaining energy levels of the nodes to improve the wireless sensor network lifetime.

In this paper, we propose two sink site determination (SSD) algorithms and give a movement strategy for the sink used after the sojourn time expires. Simulation results show that proposed algorithms perform better than its counterparts and improve the wireless sensor network lifetime effectively.

M. Koç (✉) · I. Körpeoglu
Department of Computer Engineering, Bilkent University, 06800 Ankara, Turkey
e-mail: mkoc@cs.bilkent.edu.tr

E. Gelenbe et al. (eds.), *Computer and Information Sciences II*,
DOI: 10.1007/978-1-4471-2155-8_21,

The rest of the paper is organized as follows: Sect. 2 describes the proposed algorithms. Results of the experiments are presented in Sect. 3. Finally, Sect. 4 concludes the paper.

2 Proposed Algorithms

The main motivation behind sink site determination (SSD) algorithms is to decrease candidate migration points in the deployment area to minimize the time needed to determine which sink site to next visit after the sojourn time at the current sink site expires.

2.1 Neighborhood-Based SSD Algorithm

Sometimes it can be difficult to know the exact boundaries of the deployment area and the coordinates of each sensor node in the region. In such cases, neighborhood information of the nodes can be used for determining candidate sink positions.

We present a greedy heuristic algorithm for dealing with dominating set problem (choose q nodes from out of n nodes such that the union of the neighbors of these nodes covers all the nodes in the area.). In the beginning, after determining the neighborhood information of each node, the sink node sorts these nodes in descending order according to their number of neighbors. Then the heuristic algorithm takes the coordinate of the node (a *contributed* node) with the most number of neighbors in the beginning and put those neighbors to the current neighbor list. After first contributed node is chosen (the node with the most number of neighbors), its neighbors are saved in *coveredNodes* list. The *uncoveredNodes* list is simply calculated via taking set difference of universal set (i.e., all nodes) and *coveredNode* list. After initialization of those lists, node that has the maximum number of common elements (neighbors) with *uncoveredNodes* is chosen as the next contributed node. Then its neighbors are added to *coveredNodes* list and *uncoveredNodes* list is updated. This iteration continues until *uncoveredNodes* list becomes empty (*coveredNodes* equal to universal list).

2.2 Coordinate-Based Sink Site Determination Algorithm

It is possible to group nodes using their coordinate values (if they are known) on the sink side. In the coordinate-based sink-site determination algorithm, we divide the deployment area into squares such that each one's length is equal to the transmission range. That enables us to group (cluster) nodes that can be a sink's neighbors in any round and compare their energy levels and decide which sub-area to move to in the next round. The number of areas dynamically changes according to the transmission range values.

The distance between any two neighbor sink sites is R, where R is the maximum transmission range. Each sink site is ideally placed at the center of the allocated area. After determining the centers of each sub-square, sparse areas are eliminated if their *density* is below the threshold, where the threshold is determined by dividing the number of nodes by the number of sub-squares.

A dynamic sink site selection algorithm (either neighborhood- or coordinate-based) provides us to eliminate the areas that are on inaccessible terrains which prevents the sink to move and stay at that point.

2.3 Movement Strategy

After candidate sink sites are determined, the sink node moves to the densest point of the area (first migration point). If the sojourn time expires (either exceeds t_{min}—the minimum time that a sink should stay on the current site—or a change in energy level occurs), the sink examines the minimum remaining energy value in each candidate migration point, which means the minimum energy value among the nodes' energy values that fall into the squares, using the information in the last received packets. Then it moves to the point where the minimum remaining energy level is maximum among the sites that have not been visited yet [visited max–min (VMM)]. When we say 'have not been visited', we mean that a site cannot be visited until the sink has moved to all of the candidate migration points once. After all visits have been completed, then the *visited* flag will be set to zero for all of the sink sites and they all become available to visit again.

3 Simulation Results

3.1 System Model and Main Parameters of the Simulation

Sensor networks in the simulation have N static sensor nodes and a mobile base station. Those nodes are deployed to a region of interest randomly. Square areas are used in the simulations, which are generally either $300 \times 300\,\text{m}^2$ or $400 \times 400\,\text{m}^2$. The energy model and the radio characteristics used in the simulations comes from [2]. In this work, we define the network lifetime as the period of time until the first node dies, which is a commonly used definition in the literature.

Three existing SSD approaches in the literature are summarized in Fig. 1. P1 and P2 are given as sink sites in [3]. In Fig. 1a, center points of four grids are chosen as possible sink sites, whereas the second one takes four corner points and the center of the big square (coordinates are given for a $100 \times 100\,\text{m}$ square).

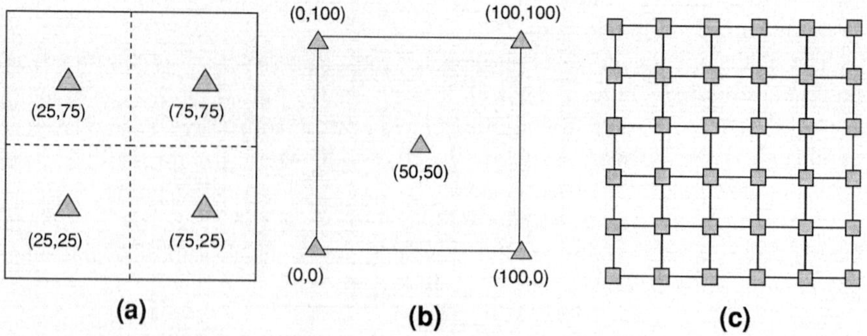

Fig. 1 Different SSD Approaches in the literature. **a** Sink Sites—Approach 1 (P1). **b** Sink Sites—Approach 2 (P2). **c** Sink Sites—Approach 3 (B1)

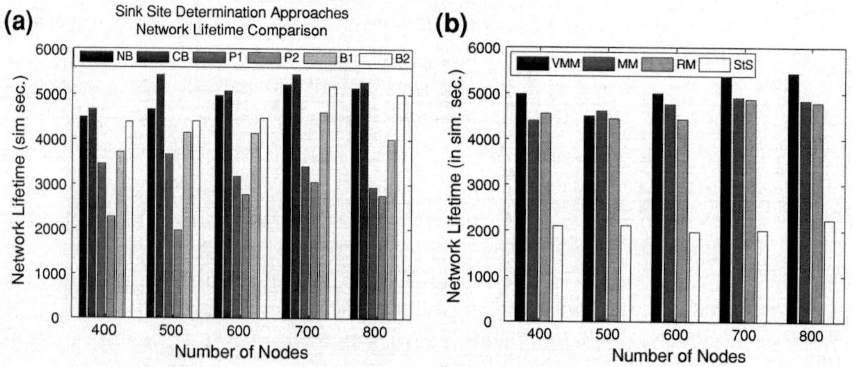

Fig. 2 Network Lifetime Comparison of SSD and Movement Algorithms. **a** SSD Approaches: Network Lifetime Comparison. **b** Network lifetime for 400 nodes (SSD = NB)

In the third approach, which comes from Basagni et al. [1], the area is divided into 3×3 (5×5) grids, totally 16 (36) corner points of sub-squares are taken as candidate migration points (B1 and B2 respectively). We evaluate the performance of our approaches (neighborhood-based set covering heuristic (NB), and coordinate-based (CB)) with these four methods.

As can be seen in Fig. 2a, both neighbor- and coordinate-based approaches perform better than other four in terms of network lifetime. The CB approach is three times better than P2 for 500 nodes as well. P1 has the best data latency (average hop count) because four different sites have been optimally placed in the center of the four grids (not shown here due to page limitations). Although the NB and CB approaches have a 25% worse data latency than P1, they have up to 60% better network lifetimes and better data latency than the other three in all cases as well. Figure 2b shows that VMM performs better than max–min (MM), random movement (RM) and static sink (STS) approaches in almost all cases (SSD is fixed to NB and transmission range is 35 m).

4 Conclusion

In this paper, we investigated sink site determination and movement strategy parts of the sink mobility problem. We propose two sink site determination algorithms and a movement strategy. We compare the performance of our algorithms with different approaches via simulation experiments. Our sink site determination algorithms perform better than the other four approaches (also lower data latency values than three of them) in the literature. Our movement strategy's (VMM) results are also better than the other three strategies almost in all cases.

References

1. Basagni, S., Carosi, A., Melachrinoudis, E., Petrioli, C., Wang, M.: Controlled sink mobility for prolonging wireless sensor networks lifetime. ACM J Wirel. Netw. (WINET) **14**, 831–858 (2007)
2. Heinzelman, W., Chandrakasan, A., Balakrishnan, H.,: Energy-efficient communication protocol for wireless microsensor networks. In: Proceedings of the 33rd International Conference on System Sciences (HICSS '00) (2000)
3. Papadimitriou, I., Georgiadis, L.: Energy-aware routing to maximize lifetime in wireless sensor networks with mobile sink. J. Commun. Softw. Syst. **2**, 141–151 (2006)

A Hybrid Interference-Aware Multi-Path Routing Protocol for Mobile Ad hoc Network

Phu Hung Le and Guy Pujolle

Abstract In this paper, we present a formula of interference and a novel Hybrid Interference-Aware Multi-Path Routing protocol (HIA-MPOLSR) that was based on the Optimized Link State Routing protocol (OLSR) (Clausen T, Jacquet P IETF request for comments: 3626, optimized link state routing protocol OLSR, October 2003) for mobile ad hoc network. HIA-MPOLSR minimizes the influence of interference to increase the stability and reliability of the network. HIA-MPOLSR calculates interference by considering the geographic distance between nodes instead of hop-by-hop. From our simulation results, we show that the HIA-MPOLSR outperforms IA-OLSR, the original OLSR and OLSR-Feedback (UM-OLSR, http://masimum.dif.um.es/?Software:UM-OLSR) in terms of packet delivery fraction, routing overhead and normalized routing load.

1 Introduction

In recent years, many multi-path routing protocols have been proposed such as SR-MPOLSR [1], and MPOLSR [2] etc. However, only a few multi-path routing protocols deal with interference in the network. In MANET, interference is a key factors that has the greatest impact to network performance because interference causes data loss, conflict, retransmission etc. In this paper, we introduce our formula of interference and build a novel hybrid interference-aware multi-path routing protocol (HIA-MPOLSR) to enhance the network performance.

P. H. Le (✉) · G. Pujolle
LIP6, University of Pierre and Marie Curie, 4 Place Jussieu, 75005 Paris, France
e-mail: Phu-Hung.Le@lip6.fr; hung@rp.lip6.fr

G. Pujolle (✉)
e-mail: Guy.Pujolle@lip6.fr

E. Gelenbe et al. (eds.), *Computer and Information Sciences II*,
DOI: 10.1007/978-1-4471-2155-8_22, © Springer-Verlag London Limited 2012

2 Interference and Measurement of Interference

The interference of a node, a link and a path is defined in [5].

In this paper, we consider the whole interference of a node as a circle with a radius of Rcs(carrier sensing range of a node), then we divide the interference region into four regions that are determined by R_1, R_2, R_3 and R_4 as follows.

$$\begin{aligned}&\text{zone1: } 0 < d \le R_1, R_1 = 1/4Rcs;\\&\text{zone2: } R_1 < d \le R_2, R_2 = 2/4Rcs;\\&\text{zone3: } R_2 < d \le R_3, R_3 = 3/4Rcs;\\&\text{zone4: } R_3 < d \le R_4, R_4 = Rcs\end{aligned}$$

where d is the transmitter- receiver distance.

This choice is a compromise between the precision and the calculation complexity.

For each zone, we assign an interference weight which represents the interference level. If the weight of interference of zone1 is 1, the weight of interference of zone2, zone3 and zone4 are α, β and γ respectively ($\gamma < \beta < \alpha < 1$). We can calculate the interference of a node u in MANET as follows:

$$I(u) = n1 + \alpha \cdot n2 + \beta \cdot n3 + \gamma \cdot n4. \tag{1}$$

where n1, n2, n3 and n4 are the number of nodes in zone1, zone2, zone3 and zone4 respectively. Parameters α, β and γ are determined by Eq. (2) (Two-Ray Ground path loss model detailed in [6]).

$$P_r = P_t G_t G_r h_t^2 h_r^2 / d^k \tag{2}$$

Here, we assume that MANET is homogeneous, that is all the radio parameters are identical at each node and the common path loss model has k as 2.

$$\begin{aligned}\alpha &= (P_t G_t G_r h_t h_r / R_2^k) / (P_t G_t G_r h_t h_r / R_1^k) = R_1^k / R_2^k = 0.5^k\\ \beta &= (P_t G_t G_r h_t h_r / R_3^k) / (P_t G_t G_r h_t h_r / R_1^k) = R_1^k / R_3^k = 0.33^k\\ \gamma &= (P_t G_t G_r h_t h_r / R_4^k) / (P_t G_t G_r h_t h_r / R_1^k) = R_1^k / R_4^k = 0.25^k\end{aligned}$$

Therefore, α=0.25, $\beta = 0.11$, $\gamma = 0.06$ and

$$I(u) = n1 + 0.25n2 + 0.11n3 + 0.06n4. \tag{3}$$

A Link interconnecting two nodes u and v, e = (u,v); I(u) and I(v) are the interference of u, v respectively.

$$I(e) = (I(u) + I(v))/2 \tag{4}$$

Based on the formula (4), we can calculate interference of a path P that consists of links e_1 $e_2, \ldots, e_n$

$$I(P) = I(e_1) + I(e_2) + \cdots + I(e_n)$$

3 Modelling MANET as a Weighted Graph and Algorithm of Hybrid Multi-Path

A MANET can be considered as a weighted graph where nodes of MANET are vertices of the graph and the edges of the graph are the links between any two neighbor nodes. We calculate the interference of each node based on the formula (3). To determine n1, n2, n3, and n4 for a node, we calculate the distances from the considered node to the other nodes of the network. Then the interference of each link is determined by the formula (4). The weight of each edge is the interference of the corresponding link.

To build the algorithm of hybrid multi-path(the paths may be some common links and nodes.), we perform as following steps :

- *Step* 1: Using the Dijsktra' algorithm, we can get the minimum interference path (called IA-OLSR) from a source to a destination.
- *Step* 2: Dijkstra's algorithm is repeated for a number of times k (k = 2, …, n) and while avoiding at least one node or one link between the source and the destination along the paths found in the previous steps to find k-minimum interference path.

4 Topology Information, Route Recovery and Forwarding

Like OLSR, our protocol HIA-MPOLSR also uses "HELLO" message, "Topology Control" (TC), Multipoint Relays (MPRs). Moreover, HIA-MPOLSR updates the position of all nodes, the interference level of all nodes and links.

Before transmitting packets to next node in the path that found, the sent node checks status of the received node. The packet will be transmitted without any problem on the path. Otherwise, a different path will be used immediately. When there have been no available paths, the paths will be recomputed. HIA-MPOLSR is also able to detect the failed link as in OLSR-FB [4].

5 Environment and Performance Evaluation

The protocol is implemented in NS2 with 10 Mbps channel. The distributed coordination function (DCF) of IEEE 802.11 for wireless LANs is used as the MAC layer. The Two-Ray Ground and the Random Waypoint models have been used. Each node has a transmission range of 160 m and a carrier sensing range of 400 m and moves from 4 to 10 m/s, the packet size of 512 bytes and Constant Bit Rate (CBR) changes from 320 to 1024 Kbps.

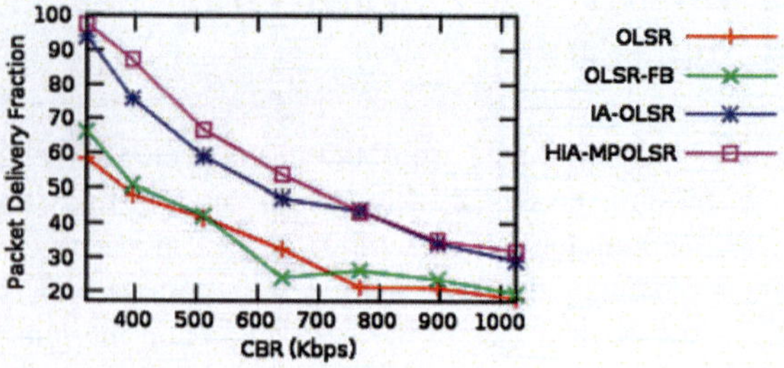

Fig. 1 Packet delivery fraction

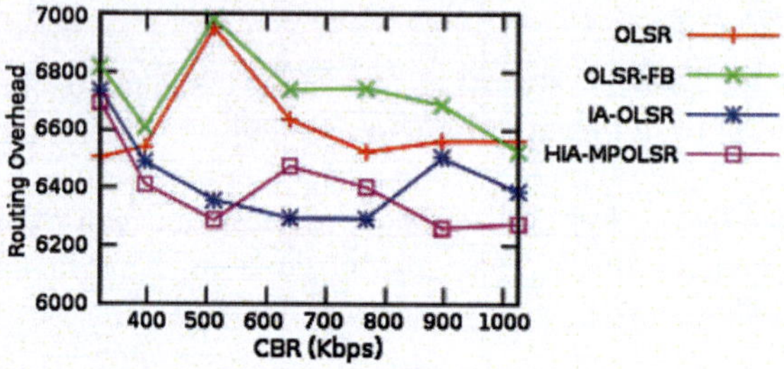

Fig. 2 Routing overhead

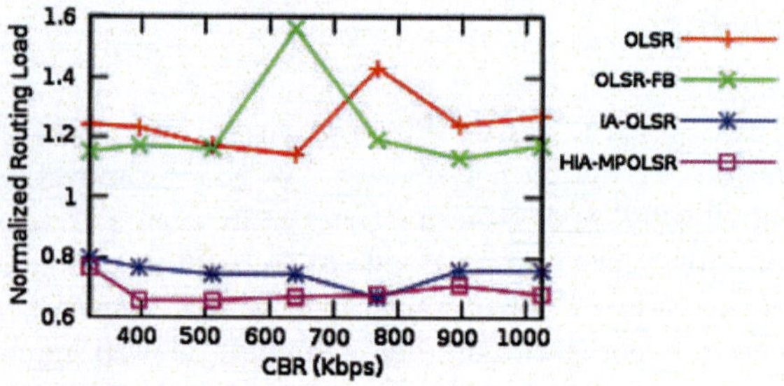

Fig. 3 Normalized routing load

As shown in Fig. 1, the PDF of HIA-MPOLSR can be approximately 11% higher than that of IA-OLSR, 38% than that of the original OLSR and OLSR-FB.

Routing overhead of HIA-MPOLSR is possibly 5% lower than that of IA-OLSR, 11% than that of the original OLSR and OLSR-FB as shown in Fig. 2.

Figure 3 shows that the NRL of HIA-MPOLSR decreases possibly 12% compared to that of IA-OLSR, 49% to that of the original OLSR and OLSR-FB.

References

1. Zhou, X., Lu, Y., Xi, B.: A novel routing protocol for ad hoc sensor networks using multiple disjoint paths. In: 2nd International Conference on Broadband Networks, Boston, MA, USA, 2005

2. Jiazi, Yi., Eddy, C., Salima, H., Benoît, P., Pascal, L.: Implementation of Multipath and Multiple Description Coding in OLSR, 4thIntrop/Workshop, Ottawa, Canada
3. Clausen, T., Jacquet, P.: IETF request for comments: 3626, optimized link state routing protocol OLSR, October 2003
4. UM-OLSR, http://masimum.dif.um.es/?Software:UM-OLSR
5. Xinming, Z., Qiong, L., Dong, S., Yongzhen, L., Xiang, Y.: An Average Link Interference-aware Routing Protocol for Mobile Ad hoc Networks, Conference on Wireless and Mobile Communications (ICWMC'07)
6. Xu, K., Gerla, M., Bae, S.: Effectiveness of RTS/CTS handshake in IEEE 802.11 based ad hoc networks. J. Ad hoc Netw. **1**(1), 107–123 (2003)

Providing Automated Actions in Wireless Multimedia Sensor Networks via Active Rules

Hakan Öztarak, Kemal Akkaya and Adnan Yazici

Abstract Manual processing of multimedia data in Wireless Multimedia Sensor Networks (WMSNs) may not always be possible. This necessitates autonomous operation of data processing at the sink for taking actions as well as communicating with the appropriate personnel whenever needed. In this paper, we propose a framework for fusing and automated processing of incomplete/imprecise WMSN data using active rules. First, data fusion is performed via fuzzy logic to handle the uncertainty in the received data at the sink. We then extend Event-Condition-Action (ECA) rules to process the data and infer predefined actions. The proposed approach has been shown to provide automated actions with higher accuracies.

1 Introduction

Wireless Sensor Networks (WSNs) can collect huge amount of data from the environment with inexpensive deployment and labor costs. After the collection of data, in a lot of applications, autonomous operation at the sink is crucial for taking right on-time actions according to the events occurring in the area [1]. These type of networks are referred to as sensor-actuator networks [2] and are gaining popularity through cyber-physical systems as well [3]. In such automated

H. Öztarak (✉) · A. Yazici
Department of Computer Engineering, Middle East Technical University, 06531 Ankara, Turkey
e-mail: hoztarak@aselsan.com.tr; hakan.oztarak@ceng.metu.edu.tr

A. Yazici
e-mail: yazici@ceng.metu.edu.tr

K. Akkaya
Department of Computer Science, Southern Illinois University Carbondale, Carbondale, IL 62901, USA
e-mail: kemal@cs.siu.edu

E. Gelenbe et al. (eds.), *Computer and Information Sciences II*,
DOI: 10.1007/978-1-4471-2155-8_23, © Springer-Verlag London Limited 2012

applications of WSNs, typically, after the collection of data, reactive solutions are employed at the sink node. Reactive solutions are asynchronous processes that are widely used in the past [4–7]. In these studies, Event-Condition-Action (ECA) Rules are used for taking the necessary actions in response to these asynchronous processes. The event may occur at any time in the network and thus the action is performed against that event asynchronously. While there has been a number of previous studies in WSNs to provide such a reactive framework [4, 5], this problem has not been studied for WMSNs which additionally deploy wireless camera sensors that can collect image/video data.

We propose to extend ECA rules to take action against asynchronous events in WMSNs to provide autonomous operation. We target border surveillance applications and assume limited information coming from camera sensors in order to extend their lifetime. Such limited information is processed at the sink node using fuzzy logic. The goal is to be able to handle the uncertainty and identify the detected objects as well as the nature of the events. To this end ECA rules are extended to WECA rules where "W" refers to "when". Data fusion is performed as a part of "when" on the received objects. The actions are decided based on the predefined rules which are stored in a Data/RuleBase. These rules are retrieved from the Data/RuleBase in reference to existing events. An inference mechanism is then employed in order to decide the desired actions.

We have implemented the rules in a sample border surveillance application and conducted experiments on the accuracy metric when performing object fusion with uncertain data. The results show that the rules can be processed with acceptable object identification accuracies without putting a heavy burden on camera sensors.

2 Related Work

In the literature, there exist two different working models for sensor-actuator applications [4]:

- Demand-Driven Model: The system initiates the action when an external request/query is injected to the system. For instance, in order to do an environmental monitoring, a query may be injected to the system periodically [8]. Each sensor in this model is considered as a data source to the database virtual table stored in sink [9]. An example query for this may be as follows: "*Tell me the Average Temperature in the region once every five seconds*". These type of applications are not very prevalent and thus we will not follow this approach.
- Event-Driven Model: An individual sensor node produces an event and propagates through network and then the system initiates an action against the produced event. An example application may be detecting the forest fires with temperature sensors. It is used for reactive applications [8] which is our focus in this work. Typically, Active Database (ECA Rules) approach [4] is followed in these works. The aim of ECA systems is to respond against the events occurring in the area under the control of the active rules asynchronously [6–8, 10].

While these approaches can respond to data collected from a WSN, they cannot be applied to WMSNs due to the nature of the data produced. The data contains both redundancy and imprecision and thus ECA rules need to be extended to provide data fusion before the rules can be processed. None of the above studies considers these issues. While [5] adds fuzziness support to ECA rules, it does not provide any fusion. However, it complements our approach in providing fuzzy querying support.

3 Problem Definition

Our problem can be formally defined as follows: "*Assume that we have N wireless camera sensors distributed randomly in a critical environment for surveillance purpose. At a specific time T, let us assume that there exist* $M > 0$ *different moving objects (human intruders, vehicles or animals) in the area. Our goal is to define a rule-based reactive model at the sink which will process the received uncertain and limited data regarding these M moving objects and decide the appropriate actions to be taken if any in an autonomous manner*".

Eventually, the ultimate goal is to minimize the human intervention on the application. We would like to note that M objects can be captured by K camera sensors where $K < N$ and thus redundancy at the sink node is expected. The sink should be able to identify redundant objects. However, limited data will be available at the sink due to resource constraints on the camera sensors. Specifically, the camera sensors will apply data elimination/reduction, etc. to reduce the size of the sent data to the sink. Therefore, uncertainty is expected to be produced from multiple camera sensors sending similar data about the objects.

4 Solution Overview

The overall framework depends on the processing of data at both the camera sensors and the sink node. The extraction of the objects, localization and determination of speed are not the focus of this work. This type of processing has been studied in the past in a variety of works such as [11]. In this paper, we explain the four phases after receiving events from different camera sensors as explained below briefly. Note that fuzzy logic is utilized to capture the uncertainty regarding events/objects. Given the limited data from camera sensors and incomplete information about the events, we cannot come up with crisp event definitions and conditions at the sink node.

1. *Fuzzy Object/Event Fusion*: The first phase where events coming from different camera sensors are processed and fused if applicable (i.e., deciding whether two different moving object events sent by two different camera sensors belong to the same intruder).

2. *Fuzzy Event Identification*: The semantic event performed by moving objects at the area under surveillance is extracted
3. *Fuzzy Condition Matching*: The conditions are processed to check whether the detected object/event is in particular state to fire the rules
4. *Fuzzy Inference and Combination*: The phase where the actions are inferred when all the preliminaries are suitable

5 Extended ECA for WMSNs

With the purpose of handling four steps given above, we propose an extended ECA rule syntax (given in an example below) which is also able perform data fusion before the ECA rules can be applied. Consider two different camera sensors, S_1 and S_2, detecting the same intruder trying to enter into the monitored region. Both cameras send different events (e.g., MOE1 and MOE2) to the sink. Hence the rule defined at the sink should appear as below. The rule fires when the detected intruder is sensed by at least two different camera sensors (Ph.1) and it approaches to the water supply (Ph.2). Then, when the intruder is close enough to a powerful high resolution camera (Ph.3), the action to be taken is to request the original video frames from the high resolution camera (Ph.4).

```
WHEN Camera Sensor_1 sends MOE1
     Camera Sensor_2 sends MOE2
     MOE1 and MOE2 contains human object and
     they are very similar and
     they are fused to the same human object O
ON O is approaching to water supply
IF O is close to high resolution camera
THEN Request original frame from high resolution camera
```

6 Proof of Concept

In order to test the feasibility of the rules that we have proposed, we use JESS, the rule engine for JAVA platform [12]. Because of space considerations, we give examples only about sub-rule for fusion part and the final rule for the whole process in JESS syntax. The final rule uses sub rules' results and ensures that if an object is sensed by two different camera sensors and if that object is human and running, when that object is near to one of the camera sensors, then we request original video data from the camera sensor.

Table 1 AVERAGE FUSION ACCURACIES

Object Count	%25 Coverage	%50 Coverage	%75 Coverage	%100 Coverage
3 Objects	0.38	0.71	0.81	1.0
6 Objects	0.38	0.74	0.81	1.0
9 Objects	0.39	0.73	0.83	1.0
12 Objects	0.39	0.75	0.83	1.0

```
(defrule TWO_SENSED
  (MOE (object_name ?O)(sensor ?x)(time ?t1))
  (MOE (object_name ?O)(sensor ?y&~?x)(time ?t1))
=>
  (printout t ''TWO SENSOR SENSE THE OBJECT '' ?O '' AT '' ?t1 crlf)
  (assert (two_sensed ?O)) )
```

The final rule is shown below:

```
(defrule REQUEST_ORIGINAL_FRAME
  (two_sensed ?O)(human_running ?O)(near_sensor ?O ?s)
=>
  (printout t ''REQUEST ORIGINAL FRAME FROM '' ?s crlf))
  (request_original_frame ?s)
```

7 Experimental Evaluation

The metric that we use in our experiment is Fusion Accuracy. This metric indicates how successful the algorithm is when identifying the objects. It is defined as follows: At each time unit, the real object count under surveillance and the determined object count by the sink are compared. If these values are not equal to each other, the ratio of the difference of these values to the real object count is noted as *false ratio*. $(1 - falseratio)$ is defined as the fusion accuracy. Our goal is to maximize the accuracy.

Overall accuracies for an area of size $1000 \times 600\,\text{m}^2$, 100 camera sensors placed randomly and objects generated randomly are presented in Table 1. The results indicate that when the coverage is 100%, our approach guarantees accurate tracking of any number of objects by maximum fusion accuracy. In all other coverage cases, the fusion accuracy is always more than the coverage percentage.

8 Conclusion

In this paper, we propose an active rule processing framework for surveillance applications of WMSNs. Instead of the whole frame or the detected object, the sink gathers MOEs (incomplete and imprecise information about the objects) from different camera sensors and processes them using predefined active rules. In order to fuse data flowing from different camera sensors, we have extended ECA rules according to the needs of WMSNs. Sink performs four phases by using rule-based processing after getting those MOEs from different camera sensors: Fuzzy Object/Event Fusion, Fuzzy Event Identification, Fuzzy Condition Matching and Fuzzy Inference. The feasibility of the rule-based framework has been shown via a JESS implementation. In addition, we have conducted experiments for assessing the fusion accuracy of the proposed framework. The experiment results have shown that acceptable accuracy values can be obtained via the proper processing of incomplete and uncertain information using fuzzy logic.

References

1. Akyildiz, I.F., Su, W., Sankarasubramaniam, Y., Cayirci, E.: Wireless sensor networks: a survey. Comput. Netw. **38**, 393–422 (2002)
2. Akyildiz, I.F., Kasimoglu, I.H.: Wireless sensor and actor networks: Research challenges. Elsevier Ad hoc Netw. J. **2**, 351–367 (2004)
3. Noble, B.D., Flinn, J.: Wireless, self-organizing cyber-physical systems. In: NSF Workshop on Cyber-physical Systems (2009)
4. Zoumboulakis, G.R.M., Poulovassilis, A.: Active rules for sensor databases, DMSN 2004 International Workshop on Data Management for Sensor Networks (2004)
5. Bostan-Korpeoglu, B., Yazici, A., Korpeoglu, I., George, R.: A new approach for information processing in wireless sensor network database application. Proceedings of NETDB, p. 1 (2006)
6. Ceri, S., Widom, J.: Deriving incremental production rules for deductive data, information systems. Inform. Syst. **16**(6) (1994)
7. Ghanem, N., DeMenthol, D., Doermann, D., Davis, L.: Representation and recognition of events in surveillance video using petri nets. Conference on Computer Vision and Pattern Recognition Workshop vol. 7, pp. 23–39 (1993)
8. Zaniolo, C.: On the unification of active and deductive databases. Advances in 11th British National Conference on Database pp. 23–39 (1993)
9. Gehrke, J., Madden, S.: Query processing in sensor networks. IEEE Pervasive Comput. **3**(1), 46–55 (2004)
10. Aiken, J.H.A., Widom, J.: Static analysis techniques for predicting the behavior of active database rules. ACM TODS **20**(1), 3–41 (1995)
11. Oztarak, K.A.H., Yazici, A.: Lightweight object localization with a single camera in wireless multimedia sensor networks. Proceedings of the GLOBECOM, pp. 12–13 (2009)
12. JESS.: http://herzberg.ca.sandia.gov/jess

File Transfer Application For Sharing Femto Access

Mariem Krichen, Johanne Cohen and Dominique Barth

Abstract In wireless access network optimization, today's main challenges reside in traffic offload and in the improvement of both capacity and coverage networks. The operators are interested in solving their capacity problems in areas where the macro network signal is not able to serve the demand for mobile data. Thus, the major issue for operators is to find the best solution at reasonable expanses. The femto cell seems to be the answer to this problematic. In this work, we focus on sharing femto access between a same mobile operator's customers. This problem is modeled as a game where service requesters customers (SRCs) and service providers customers (SPCs) are the players. This article considers one SPC. SRCs are static and have some similar and regular connection behavior. Moreover, the SPC and each SRC have a software embedded respectively on its femto access, user equipment. After each SRC's connection request, its software will learn the strategy increasing its gain using local information. We present a distributed learning algorithm with incomplete information running in SRCs software. This work answers these questions: Does the distributed algorithm converge to a stable state? If yes, does this state a Nash Equilibrium and how many iterations we need to reach it?

Keywords Component: game theory · Sharing femto access · Nash Equilibrium · Distributed learning algorithm

---

M. Krichen (✉) · J. Cohen · D. Barth
PRiSM, University of Versailles, 45 avenue des Etats-Unis,
78035 Versailles, France
e-mail: mariem.krichen@prism.uvsq.fr

J. Cohen
e-mail: johanne.cohen@prism.uvsq.fr

D. Barth
e-mail: dominique.barth@prism.uvsq.fr

E. Gelenbe et al. (eds.), *Computer and Information Sciences II*,
DOI: 10.1007/978-1-4471-2155-8_24,

1 Introduction

Today, one of the biggest issues for Mobile Operators is to provide acceptable indoor coverage for wireless networks. Among the several in-building solutions, the femto cell is the one which is gaining significant interest. A *femto cell* is a small cellular base station characterized by low transmission power, limited access to a closed user group designed for residential or small business use. This solution could help operators solve localized coverage problems and extend their network. This would be feasible if access points owners accept to be part of a Club where each member is willing to open up its access point to the other members. A femto club member could share its 3G/LTE signal securely with other club members. Sharing femto access is a service proposed by the Mobile Operator to its clients divided into *service providers customers (SPCs)* and *service requesters customers (SRCs)*: *SPCs* are the owners of femto cells accesses for which they have contracted with a *Mobile operator* denoted by *MO*. *SRCs* are customers using a mobile terminal in an area covered by some SPCs access points and requesting to use them. Since the interests of all the actors could be antagonist, especially between many SRCs requesting a same SPC, we model our system as a game to determine equilibria of such situations.

1.1 Related Works

The potential games introduced by Rosenthal [1] are classical games having at least one pure Nash equilibrium. These games have a potential function such that each of its local optimums corresponds to a pure Nash equilibrium. This property has been used for the congestion games (see [2]), with Resource Reuse in a wireless context (see [3]) and for a real-time spectrum sharing problem with QoS provisioning [4].

A decentralized learning algorithm of Nash equilibria in multi-person stochastic games with incomplete information has been presented by M.A.L. Thathachar et all. In the considered game, the distribution of the random payoff is unknown to the players and no player knows the strategies or the actual moves of other players. It is proved that all stable stationary points of the algorithm are Nash equilibrium for the game [5].

1.2 Our Contribution

Section 2 presents the model for sharing femto access and the actors involved in this model are described. The game considering only SRCs competition is presented in Sect. 3. Section 4 details the principle of a distributed algorithm used to learn the game NE if any exists and some simulations results are given in Sect. 4.2. Finally, Sect. 5 draws a general conclusion and gives some perspectives.

2 Model for Sharing Femto Access

Now, we present the model for sharing femto access. Our work considers a unique SPC denoted by X and several SRCs. The SPC shares some amount of bandwidth with SRCs. The SPC's resource reserved for sharing is an amount of bandwidth denoted by $B_S(X)$ and is divided into two parts:

- $B_{S_G}(X)$ is the part of bandwidth in which SRCs communications can never be preempted.
- $B_{S_Y}(X)$ is the part of bandwidth in which SRCs communications can be preempted due to the fact that the SPC has priority on this part of bandwidth.

Let's consider a SRC Y who needs an amount of bandwidth equal to bw. bw will be allocated to Y only if it is free. If X will need bw, he will not be able to use it if the connection he allocated to Y is green. However, he will be able to preempt the connection allocated to Y if it is a yellow one. Note that yellow connections are cheaper than green ones due to the risk of preemption and that failed yellow connections are free.

2.1 Actors Description

The following section describes the actors involved in the model we have just presented.

2.1.1 SPC Actor

A SPC is a customer of the MO. SPC proposes to share an amount of its bandwidth with SRCs for a price per bandwidth unit which depends on the type of connection (Green,Yellow). The bandwidth split into Green and Yellow parts is determined following its *sensitivity to Gain* denoted by $\mu \in [0,1]$ and its *sensitivity to its own connection QoS* denoted by $\Gamma \in [0,1]$. These two parameters are dual: $\mu + \Gamma = 1$.

The *Gain sensitivity* parameter indicates its sensitivity degree to the price of the connection shared while the *QoS sensitivity* parameter indicates the SPC's tolerance degree towards preemption risk. When $\mu > 1/2$, we say that the SPC is sensitive to gain. Otherwise, the SPC is considered as sensitive to its access QoS.

2.1.2 SRC Actor

A SRC is also a customer of the MO who wants to use the SPC's resources. SRC is characterized by a*QoS sensitivity* parameter α and a*price sensitivity* parameter β.

The *QoS sensitivity* parameter indicates the SRC's tolerance degree towards the QoS degradation while the *price sensitivity* parameter indicates the SRC's tolerance degree towards the cost of the connection. These two parameters are dual: $\alpha + \beta = 1$. When $\alpha > 1/2$, the SRC is sensitive to QoS. Otherwise, the SRC is sensitive to price.

3 Game Presentation

This work focuses on the competition between SRCs when the SPC's bandwidth split is fixed. So, sharing femto access problem denoted by the SRCs game is thus modeled as a game where N SRCs are the players. We assume that SRCs are static and have some regular and similar connection behavior: each SRC requests the SPC's connection in nearly same time slots with almost invariant QoS needs.

SRCs game is defined as follows: given a fixed SPC's bandwidth split, what would be the best strategy to be played by SRCs in order to have a stable situation where the strategy of each SRC is optimal for him considering the other SRCs strategies. In game theory, this situation is called *pure Nash equilibrium*: for each player, there is no unilateral strategy deviation that increases its utility [6].

3.1 SRCs QoS Needs

SRCs QoS needs are relative to the type of application requested. Our work considers the File Transfer Application. QoS is defined as the time transfer file denoted by t. The SRC's QoS satisfaction is related to t and is defined as follows:

- Case $t = T_1$: BW_{Max} corresponds to the required bandwidth to download a file in $t = T_1$. If the SPC provides an amount of bandwidth equal to BW_{Max}, then the SRC's QoS satisfaction is at the top.
- Case $t = T_2$: BW_{Min} corresponds to the required bandwidth to download a file in $t = T_2$. If the SPC provides an amount of bandwidth equal to BW_{Min}, then the SRC's QoS satisfaction is minimal.

Since the SPC's bandwidth is limited and several SRCs could request the SPC's connection at the same time, one possible response that a SRC could receive is a deny one. To decrease the chances to receive such a response, each SRC requests a Minimum and a Maximum amount of bandwidth following their profiles.

Note that the parameters characterizing a SRC's profile are real in [0, 1]. We aim at translating these parameters into intervals of bandwidth requests which are actually integers. So, we introduce a parameter ε representing the discretization of the bandwidth resquested. Let S_{SRC_i} be the set of possible strategies of SRC_i. In the following, we focus on the request of a SRC denoted by SRC_i considering only the case where SRC_i has its QoS sensitivity parameter α_i greater than $1/2$. SRC_i fixes a revenue threshold, denoted by Rev_Th_i, under which he denies any proposed connection. This parameter depends on the QoS sensitivity α_i of the SRC. $Rev\_Th_i = \alpha_i$.

$$S_{SRC_i} = \{Rev\_Th_i, Rev\_Th_i + \varepsilon, Rev\_Th_i + 2\varepsilon, \ldots, 1\}.$$

Each element s_i of S_{SRC_i} permits to define an interval of bandwidth to be requested in Green and Yellow connection. This represents a couple $(g_i = [m_i^G, M_i^G], y_i = [m_i^Y, M_i^Y])$ of couples of integers.

3.2 SPC's bandwidth allocation

Let B_S be SPC's shared bandwidth and Ψ_S be the proportion of B_{S_G} regarding B_S. Each SRC sends a request to the SPC. At reception, the SPC decides the way its bandwidth is allocated to SRCs. The request of each SRC_i is represented by one element in S_{SRC_i}. According to a set Π of SRCs requets $\Pi = <s_1, s_2, \ldots, s_N>$ where s_i corresponds to the request of SRC_i, for any i, $1 \leq i \leq N$, SPC gives an answer to each SRC_i represented by a triple (G_i, Y_i, bw_i) defined as follows:

- bw_i represents the amount of bandwidth given by the SPC to SRC_i.
- $G_i = 1$ (resp. $Y_i = 1$) means that the SPC proposes a Green (resp. Yellow) connection to SRC_i. Note that the case where $G_i = 1$ and $Y_i = 1$ is not possible.
- If $G_i = 0$ and if $Y_i = 0$, then the SPC proposes no connection to SRC_i.

Let $config(\Pi)$ be the set of all the SPC's answers to Π (one answer per SRC). So,

$$config(\Pi) = <(G_1, Y_1, bw_1), (G_2, Y_2, bw_2), \ldots, (G_N, Y_N, bw_N)>$$

The answers respect the two following properties:

1. The SPC gives SRC_i an amount of bandwidth equal to bw_i where bw_i is in the interval requested: if $(G_i = 1)$ then $bw_i \in g_i$ or if $(Y_i = 1)$ then $bw_i \in y_i$.
2. The SPC provides bandwidth to SRCs in the limits of its bandwidth availability:

$$\sum_{i=0}^{N} G_i \times bw_i < \Psi_S B_S \text{ and } \sum_{i=0}^{N} Y_i \times bw_i < (1 - \Psi_S) B_S.$$

The SPC allocates to each SRC a connection in Green or Yellow such as this outcome function is maximized:

$$U_{SPC}(config(\Pi)) = \sum_i Prop(bw_i)(\mu - \Gamma)(G_i + Y_i(1-\delta)) \tag{1}$$

3.3 SRC Game Definition

The formulation of the SRC game $G = \langle \mathcal{N}, \mathcal{S}, \mathcal{U}_{\|} \rangle$ can be described as follows:

- The set of players is $\mathcal{N}$. Each player is a SRC. There are N SRCs.
- The space of pure strategies S formed by the Cartesian product of each set of pure strategies

$$S = S_{SRC_1} \times S_{SRC_i} \times \ldots \times S_{SRC_N}$$

- A set of utility functions $\{U_1, U_2, ..., U_N\}$ that quantifies the players' outcomes.

According to a set Π of SRCs requets $\Pi = <s_1, s_2, \ldots, s_N>$ where s_i corresponds to the strategy of SRC_i, for any i, $1 \le i \le N$, the SRCs utilities are determined through the SPC's allocation. Since several allocation decisions could maximize the SPC's outcome function given by Equation 1 the SRC's utility corresponds to a mean of all these allocation decisions. Let $sol(\Pi)$ be the set of $config(\Pi)$ that maximizes the SPC's outcome function with $M = |sol(\Pi)|$.

The utility $U_i(\Pi)$ of SRC_i from $sol(\Pi)$ is expressed as follows:

$$U_i(\Pi) = \sum_{c \in sol(\Pi)} \frac{gain_i(c)}{M}. \tag{24.2}$$

The gain of SRC_i from the SPC's allocation decision c (c represents a triple (G_i, Y_i, bw_i) to SRC_i) in $sol(\Pi)$ is expressed as follows:

$$\begin{aligned} gain_i(c) = {} & [Rev(bw_i)G_i\alpha_i + Rev(bw_i)Y_i(1-\delta)\alpha_i] - [Cost(bw_i)G_i\beta_i \\ & + Cost(bw_i)Y_i(1-\delta)\beta_i] \end{aligned}$$

where $Rev(bw_i) \in [0, 1]$ and $Cost(bw_i) \in [0, 1]$.

3.4 SRC Game Equilibrium

Since each SRC_i has a finite set of strategies, this game has a mixed Nash equilibrium [6]. In the following, we study the existence of a pure Nash equilibrium in the SRC game using the properties of potential games defined in [1].

Theorem *Each instance of the SRC game admits at least one pure Nash equilibrium.*

4 Learning the SRCs Game Nash Equilibrium

In this section, we denote by A_{dist} a decentralized learning algorithm of Nash equilibrium in multi-person stochastic games with incomplete information.

4.1 Algorithm Principle

First, each SRC sends a request using A_{dist}. Then, the SPC defines its decision $sol(\Pi)$ and send it to all the SRCs. Finally, each SRC computes its utility following $sol(\Pi)$. In [2], it has been proved that if a game has at least one pure Nash equilibrium and if there exists a sufficiently small value of the learning speed parameter for which A_{dist} converges, then the convergence point is a pure Nash equilibrium. A_{dist} is described as follows:

- Each SRC_i will update its strategy at a step $t(s_i^t)$ following its local mixed strategy and then computes its new probability vector s_i^{t+1} using a set of local information (s_i^t, $margin_i^t$, $gain_i^t$) where $gain_i^t$ is the gain from the SPC's allocation decision and $margin_i^t$ is the pure strategy in S_{SRC_i}.
- At each step t, each SRC chooses randomly one strategy ($margin_i^t$) and increases its probability (s_i^t) according to its gain ($gain_i^t$) and a learning parameter $b \in [0,1]$.

The learning technique is based on the following update rule:

$$\text{If } j \neq margin_i^t \text{ then } s_{i,j}^{t+1} = s_{i,j}^t - b.u_i^t.s_{i,j}^t$$
$$\text{else } s_{i,j}^{t+1} = s_{i,j}^t - b.u_i^t. \sum_{k \neq margin_i^t} s_{i,k}^t$$

Such that: $u_i^t = \frac{gain_i^t}{U_i^t}$ is the normalized utility. The variables ($U_i^t = max_{k \leq t} gain_i^k$) correspond respectively to the maximum utility of SRC_i at iteration t.

4.2 Simulation Results

For a number of SRCs varying from 2 to 8, we have checked by simulations that A_{dist} converges in a finite number of iterations to a pure Nash equilibrium. We propose two solutions to reduce the number of requested connections denoted by *Conn* to reach convergence. In the first one, we consider that the system is stable if all the SRCs have a strategy with a probability equal to p ($0 < p < 1$). In the second one, for a connection of duration D, the SRC's learning process is triggered each $d = q * D$ slot time ($q \in [0,1]$).

5 Conclusions

The article investigates the problem of sharing femto access for a file transfer application. The competition between SRCs is modeled as a game considering a fixed SPC's bandwidth split. Simulations have shown that the Distributed Learning Algorithm always converges to a pure Nash equilibrium using only local information.

Acknowledgments This work is supported by the COMET project AWARE. http://www.ftw.at/news/project-start-for-aware-ftw

References

1. Rosenthal, R.W.: A class of games possessing pure-strategy Nash equilibria. Int. J. Game Theory **2**(1), 65–67 (1973)
2. Nisan, N., Roughgarden, T., Tardos, E., Vazirani, V.V.: Algorithmic game theory. Cambridge University Press, (2007)
3. Ahmad, S.H.A., Liu, M., Wu, Y.: Congestion games with resource reuse and applications in spectrum sharing. *CoRR*, abs/0910.4214, (2009)
4. Xing, Y., Mathur, C.N., Haleem, M.A., Chandramouli, R., Subbalakshmi, K.P.: Real-time secondary spectrum sharing with qos provisioning. In: Consum. Commun. Netw. Conf. (CCNC), 630–634 (2006)
5. Thathachar, M.A.L., Sastry, P.S., Phansalkar, V.V.: Decentralized learning of Nash equilibria in multi-person stochastic games with incomplete information. IEEE Trans. Sys. Man Cybern. **24**(5), 769–777 (1994)
6. Nash, J.F.: Equilibrium points in *n*-person games. Proc. Natl. Acad. Sci. USA **36**, 48–49 (1950)

Part V
Computer Networks

Time Parallel Simulation and hv-Monotonicity

J. M. Fourneau, I. Kadi and F. Quessette

Abstract We show how we can make more efficient the time parallel simulation of monotone systems adapting Nicol's approach. We use the monotonicity of a model related to the initial state of the simulation and we prove an algorithm with fix-up computations which minimises the number of runs before we get a consistent sample-path. We obtain proved upper or lower bounds of the sample-path of the simulation and bounds of some estimates as well.

1 Introduction

Time Parallel Simulation (TPS in the following) is a powerful technique to conduct parallel simulations but it can only be applied for some types of stochastic model which exhibit strong properties (see [4] Chap. 6 and references therein). This technique partitions the simulation time horizon $[0, T)$ into K consecutive time intervals which are simulated independently. Each processor is assigned to build one sample-path of the system for one time interval. But the ending instant of one time segment is the first instant of the next segment and the states of the sample-path at this particular instants cannot be computed in general without building the sample-path. The efficiency of TPS depends on our ability to guess the state of the system at the beginning of the simulation intervals or to efficiently correct the guessed states to finally obtain a consistent sample-path after a small number of trials. Several solutions have already

J. M. Fourneau (✉) · I. Kadi · F. Quessette
PRiSM, Université de Versailles-Saint-Quentin, CNRS UMR 8144 Versailles, France
e-mail: jean-michel.fourneau@prism.uvsq.fr

I. Kadi
e-mail: imene_yamina@yahoo.fr

F. Quessette
e-mail: qst@prism.uvsq.fr

E. Gelenbe et al. (eds.), *Computer and Information Sciences II*,
DOI: 10.1007/978-1-4471-2155-8_25,

been published: regeneration and efficient forecasting of some points of the sample-path [5], parallel prefix computation [6], a guessed initial state followed up by some fix-up computations when the first state has a weak influence on the sample-path [7]. Another approach consists in relaxing these assumptions to obtain an approximation of the result [8]. We have proposed in [2] a new property of the model which can be used to improve the efficiency of TPS: the input sequence-monotonicity. The simulation model is seen as a black box with one input sequence and an initial state, which computes one output sequence. We define some orderings on the sequences. A model is input sequence-monotone when it preserves the ordering: i.e. if two input sequences are ordered, then the output sequences produced by the simulation with the same initial state are also ordered. If this property holds, we have proposed a new approach to compute a bound of the sample path with an improved version of TPS. In performance evaluation or reliability studies, one must prove that the systems satisfy some quality of service requirements. Thus it is sufficient to prove that the upper bound (or lower bound, depending on the measure of interest) of the output sequence satisfies the constraint. Here we present a new property related to monotonicity of models (hv-monotonicity or monotonicity related to hidden variables or initial state) and we apply it to TPS. Like in [2] we obtain a bound on the sample-path and on the rewards. We show how we can build a sample-path for a bound of the exact simulation using some simulated parts and ordering relations without some of the fix-up phases proposed in [7]. Both monotonicity concepts are related to the stochastic monotonicity [3]. Hv-monotonicity may also be seen as a formal version of the non crossing property in [1]. The rest of the paper is as follows. In Sect. 2, we give a detailed presentation of Nicol's approach of TPS with fix-up computation phases. Then in Sect. 3, we define the comparison of sequences and the hv-monotonicity and we present the main theorem, the algorithm and some numerical results to illustrate the effectiveness of the approach.

2 Time Parallel Simulation with Fix-Up Phases

Let us now detail Nicol's approach [7]. In a first step, the interval $[0, T)$ is divided into K equal intervals $[t_i, t_{i+1})$. A point in the sample-path is a state and a time instant. Let $X(t)$ be the state at time t during the exact simulation obtained with a sequential algorithm. The aim is to build $X(t)$ from t in $[0, T)$ through an iterative distributed algorithm. For the sake of simplicity, we assume that for all i, logical process LP_i simulates the i-th time interval. We denote as $Y_j^i(t)$ the state of the system at time t obtained during the j-th iteration of logical process LP_i. We only know the initial state of part 1 (i.e. the initial state of the simulation). During the first run, the other initial states (say $Y_1^i(t_i)$ for the initial state of simulation at LP_i) are chosen at random. Simulations are ran in parallel. The ending states of each simulation (i.e. $Y_1^i(t_{i+1})$) are computed at the end of the time intervals. They are compared to the initial state we have previously used (i.e. $Y_1^{i+1}(t_{i+1})$). The equality of both states is a necessary condition for consistency of the path. Otherwise one must conduct a new set of

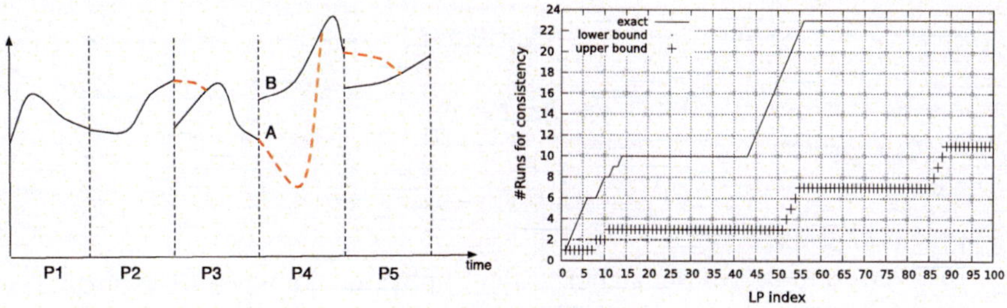

Fig. 1 TPS, coupling and fixing the sample-path (*left*), Time for LP to be consistent TPS, 100 processes, tandem finite buffer network of 5 queues at heavy load (*right*)

parallel simulations for the inconsistent parts using $Y_1^i(t_{i+1})$ as starting point (i.e. $Y_2^{i+1}(t_{i+1})$) of the next run on logical process LP_{i+1}. These new runs are performed with the same sequence of random inputs. The simulations are ran until all the parts are consistent. The number of rounds before the whole simulation is consistent is smaller than the number of *LP*. It is clear that at the end of the first run, the simulation performed at LP_1 is consistent. Similarly by induction on i, at the end of round i, LP_i is consistent. It is the worst case we may obtain, and in that case the time to perform the simulation in parallel is equivalent to the time in sequential. Note that performing the simulation with the same input sequence may speed up the simulation due to coupling. Suppose that we have stored the previous sample-paths computed by LP_i. Suppose now that for some t, we find that the new point $Y_k^i(t)$ is equal to a formerly computed point $Y_m^i(t)$. As the input sequence is the same for both runs, both sample-paths have now merged and $Y_m^i(u) = Y_k^i(u)$ for all $u \geq t$. Thus, it is not necessary to build the new sample-path. Such a phenomenon is defined as the coupling of sample-paths. Note that it is not proved that the sample-paths couple and this is not necessary for the proof of the TPS that it happens. Clearly, coupling allows to reduce the number of rounds before the whole simulation becomes consistent. [hbtp] Consider an example of TPS with one fix-up step in Fig. 1. During the first step, five parts of the simulation are ran in parallel. Part 1 and Part 2 are consistent as the initial point of LP2 is the final point of LP1 due to a correct guess. The other parts are not consistent and the simulations are ran again with a speculative initialisation (*A* instead of *B* for the second run of part 4 for instance) and the same random inputs to obtain new sample-paths. The former sample-paths are compared to the new ones and as soon that the simulations couple, they are stopped as simulations after coupling will follow the same paths. If the simulations do not couple, we only keep the new sample-path. At each step, the number of consistent sample-paths will increase. The efficiency of the approach is defined as the number of runs to obtain a globally consistent sample-path. In Fig. 1 the new runs are represented in dotted red lines and we see that all the simulations couple during the first fix-up step. We assume that the logical processes $LP_i (i = 1..K)$ perform simulations and exchange information with a Master process which must check the consistency of the partial sample-paths. Thus the simulation effort is distributed among $K+1$ processes.

Algorithm LP_i

1. $k \leftarrow 0$. Read in shared memory *I(t)* for all t in $[t_i, t_{i+1}[$.
2. $Y_1^i(t_i) \leftarrow Random.$
3. Loop
   (a) k++.
   (b) Perform run k of Simulation with Coupling on the time interval $[t_i, t_{i+1}[$ to build $Y_k^i(t)$ using input sequence *I(t)*.
   (c) Send to the Master: the state $Y_k^i(t_{i+1})$.
   (d) Receive from the Master: Consistent(i) and a new initial state U.
   (e) If not Consistent(i) then $Y_{k+1}^i(t_i) \leftarrow U$.
4. Until Consistent(i).
5. $\forall t \in [t_i, t_{i+1}[, X(t) \leftarrow Y_k^i(t)$ and write *X(t)* in shared memory.

Algorithm Master

1. For all i Consistent(i) ←False.
2. Consistent(0) ← True ; LastConsistant ← 0 ; $k \leftarrow 0$.
3. Loop
   (a) k++.
   (b) For all i, if not Consistent(i) Receive from LP_i the state $Y_k^i(t_{i+1})$.
   (c) $Y_1^0(t_1) \leftarrow Y_1^1(t_1); i \leftarrow$ LastConsistent.
   (d) Loop
      i. i++.
      ii. If $(Y_k^i(t_i) = Y_k^{i-1}(t_i))$ and Consistent(i-1) then Consistent(i) ← True.
   (e) Until (not Consistent(i)) or (i>K).
   (f) LastConsistent $\leftarrow i - 1$.
   (g) For all i send to LP_i, Consistent(i) and the state $Y_k^{i-1}(t_i)$.
4. Until LastConsistent $= K$.

The first logical process LP_1 slightly differs for this general pseudo-code: $Y_1^1(t_1)$ is equal to $X(t_1)$ the known initial state of the simulation.

3 Fast Parallel Computation of Bounds and Monotonicity

A simulation model [2] is defined as an operator on a sequence of parameters (typically the initial state) and an input sequence (typically arrivals and service) and produces an output sequence. Let $\mathcal{M}$ be a simulation model. We denote by $\mathcal{M}(a, I)$ the output of model $\mathcal{M}$ when the parameter sequence is a and the input sequence I. As usual an operator is monotone iff its application on two comparable sequences provides two output sequences which are also comparable.

Definition 1 *Let I_1 and I_2 be two sequences with length $n.I(m)$ is the* mth *element of sequence $I.I_1 \preceq_p I_2$ if and only if $I_1(t) \leq I_2(t)$ for all index $t \leq n$.*

This order is interesting because it implies a comparison on the non negative rewards computed on the output sequence. Many rewards such as moments and tails of distribution are non negative. But various orders are possible.

Definition 2 [hv-monotone Model] *Let $\mathcal{M}$ be a simulation model, $\mathcal{M}$ is hv-monotone with respect to orderings $\preceq_\alpha$ and $\preceq_\beta$ if and only if for all parameter sets a and b such that $a \preceq_\alpha b$ then $\mathcal{M}(a,I) \preceq_\beta \mathcal{M}(b,I)$ for all input sequence I.*

We now assume that the model is hv-monotone. We show how to modify the TPS to converge faster on a bound of the sample-path. To speed up the computation we change the rules about the consistency of the simulations performed by the logical processes. We accept to compute a proved point-wise bound of the sample-path. We now say that two parts are consistent if they are a proved bound of the sample-path that we do not have computed. To illustrate the approach let us suppose that we want to compute faster a lower bound of the sample-path. We modify the TPS as follows: the simulation processes $LP_i(i = 1..K)$ are not modified but the Master process performs a slightly different computation. The first assumption is, as before, that all parts except the first one are marked as not consistent while the first one is consistent. Suppose that at the end of round k, the points $Y_k^i(t_{i+1})$ have been computed using $Y_k^i(t_i)$ as starting points. If $Y_k^i(t_{i+1}) = Y_k^{i+1}(t_{i+1})$ and part i is consistent, mark part $i+1$ as consistent. Furthermore if $Y_k^i(t_{i+1}) > Y_k^{i+1}(t_{i+1})$ and part i is consistent, the concatenation of the parts provide a sample-path of a lower bounding sequence. Thus, we mark part $i+1$ as consistent. Two consecutive parts are consistent if the first one is consistent and if the second one is a proved lower bound. Let us more precisely describe the new version of the Master Process, we only report the inner loop.

———————Lower Bounding sample-path———————

3.d Loop

i) i++;
ii) if $(Y_k^i(t_i) = Y_k^{i-1}(t_i))$ and Consistent(i-1) then Consistent(i) $\leftarrow$ True;
iii) elsif $(Y_k^i(t_i) < Y_k^{i-1}(t_i))$ and Consistent(i-1) then Consistent(i) $\leftarrow$ True;

3.e Until (not Consistent(i)) or (i>K);

In the original method [7], instruction 3.d..iii) of our approach is not included.

Theorem 1 *Assume that the simulation model is hv-monotone, the new version of the Master for the TPS algorithm makes the logical simulation processes build a*

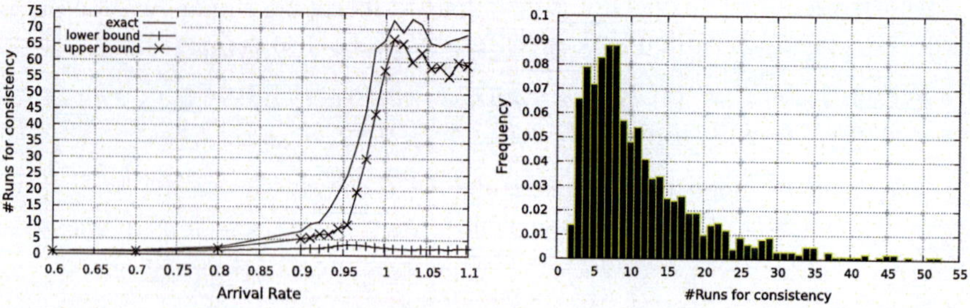

Fig. 2 Tandem network, 5 nodes, Time to be consistent vs load (*left*), Distribution of the time to be consistent on 100 processes, 1000 simulations, high load (*right*)

point-wise lower bound of the sample-path faster than the original approach. Thus the number of runs is smaller than K.

Proof By induction on the part number we prove that part i is a lower bound of the exact sample-path. At the end of the first run the Master process computes the values of Consistent(i) for all i. The first part is found consistent because it is exact. Now suppose that at the end of run $k-1$, we have found that part $i-1$ is consistent and part i is not. We perform a new run for parts i to K. Now part i begins with the ending point of part $i-1$ due to instruction 4.e.ii). Thus part i is found to be consistent because it is the exact following of part $i-1$. Note however that we only know that it is a lower bound of the exact solution due to the induction. Now consider part $i+1$, if condition 4.e.ii) holds, part $i+1$ is consistent due to the same argument. Furthermore if condition 3.d..iii) holds, part $i+1$ starts in a state smaller than the ending state of part i. Due to the hv-monotonicity, part $i+1$ will be point-wise smaller than a simulation starting with the ending state of part i. As part i is exact or a lower bound of the exact simulation, part $i+1$ is also a lower bound of the exact simulation due to the transitivity of the ordering. □

For instance, Fig. 1 shows that after step 2, we have coupled and we have obtain an exact solution. But at the end of step 1 we have obtained a proved lower bound if the system is hv-monotone. Thus we can immediately stop the computation and consider the sample-path after the first run. This sample-path is not an exact one but it is a point-wise lower bound of the exact sample-path. The expectations of non negative rewards on this sample-path are lower bounds of the exact expected rewards. Note that the bound is proved while the approximations such as [8] are not proved. Finally it must be remark that changing the initial state when the model is not hv-monotone does not always result in a bound of the sample-path. Of course we have also designed a method to compute upper bound (i.e. replace "$>$" by "$<$" in instruction 4.e.iii) of the method. It is worthy tp remark that we do not make any assumptions on the ordering on the states and typical ordering used in the examples are partial orderings. Finally we present in the right part of Fig. 1 and in both parts of Fig. 2 some experiments. We reports the number of runs for a consistent simulation on 100 processes for a network of 5 finite queues in tandem. Clearly, the bounds are computed much faster than the exact path.

Acknowledgments We thank UVSQ for a partial support (grant BQR 2010).

References

1. Andradottir, S., Hosseini-Nasab, M.: Parallel simulation of transfer lines by time segmentation. Eur. J. Oper. Res. **159**(2), 449–469 (2004)
2. Fourneau, J.M., Kadi, I., Pekergin, N.: Improving time parallel simulation for monotone systems. In: Turner, S.J., Roberts, D., Cai, W., El-Saddik, A. (eds.) DS-RT, pp. 231–234. IEEE Computer Society, Singapore (2009)
3. Fourneau, J.M., Pekergin, N.: An algorithmic approach to stochastic bounds. In: Calzarossa, M., Tucci, S. (eds.) Performance Evaluation of Complex Systems: Techniques and Tools, Performance 2002, Tutorial Lectures, Rome, Italy. LNCS, vol. 2459, pp. 64–88. Springer, London (2002)
4. Fujimoto, R.M.: Parallel and Distributed Simulation Systems. Wiley Series on Parallel and Distributed Computing. John Wiley & Sons, New York (2000)
5. Fujimoto, R.M., Cooper, C.A., Nikolaidis, I.: Parallel simulation of statistical multiplexers. J. Discret. Event Dyn. Syst. **5**, 115–140 (1994)
6. Greenberg, A.G., Lubachevsky, B.D., Mitrani, I.: Algorithms for unboundedly parallel simulations. ACM Trans. Comput. Syst. **9**(3), 201–221 (1991)
7. Nicol, D., Greenberg, A., Lubachevsky, B.: Massively parallel algorithms for trace-driven cache simulations. IEEE Trans. Parallel Distrib. Syst. **5**(8), 849–859 (1994)
8. Turgut, D., Wang, G., Boloni, L., Marinescu, D.C.: Speedup-precision tradeoffs in time-parallel simulation of wireless ad hoc networks. In: DS-RT '06: Proceedings of the 10th IEEE International Symposium on Distributed Simulation and Real-Time Applications, pp. 265–268. IEEE Computer Society, Malaga (2006)

A Prediction Based Mobility Extension for eHIP Protocol

Zeynep Gurkas Aydin, A. Halim Zaim, Hakima Chaouchi and Tulin Atmaca

Abstract By revealing of heterogeneous networks, importance of cooperation of different technologies and mobility management became an important issue of wireless and mobile networks. The idea of separation the role of IP addresses for location and identification of a mobile node in the network has also increased its importance. In this study, our motivation is to extend our previously proposed handover management method (early update for host identity protocol (eHIP). We propose to introduce the idea of simply employing movement prediction idea to regular eHIP and so enhancing the handover latency.

Keywords Host identity protocol · Early update for HIP · Mobility management · Handoff decision · Movement prediction

---

Z. G. Aydin (✉)
Computer Engineering Department, Istanbul University, Istanbul, Turkey
e-mail: zeynepg@istanbul.edu.tr

A. H. Zaim
Institute of Science and Engineering, Commerce University, Istanbul, Turkey
e-mail: azaim@iticu.edu.tr

H. Chaouchi
Telecom and Management SudParis, Dept LOR, CNRS SAMAVOR, UMR, 5157 Paris, France
e-mail: hakima.chaouchi@it-sudparis.eu

T. Atmaca
Telecom and Management SudParis, Dept RST, CNRS SAMAVOR, UMR, 5157 Paris, France
e-mail: tulin.atmaca@it-sudparis.eu

E. Gelenbe et al. (eds.), *Computer and Information Sciences II*,
DOI: 10.1007/978-1-4471-2155-8_26,

1 Introduction

Over last decades, there have been great advance in wireless and mobile communications. The revealing of heterogeneous networks also increased the importance of cooperation of different technologies. Mobility managements also became an important issue of wireless and mobile networks. There are also new contributions about mobility management by IETF [1] and IRTF [2] such as Host Identity Protocol (HIP) to replace Mobile IP in terms of managing mobility [3]. The basic idea of HIP is based on the idea of separation the role of IP addresses for location and identification of a mobile node in the network.

In this study, our motivation is to extend our previously proposed handover management method that we named "eHIP [4]". We propose to introduce the idea of simply employing movement prediction idea to regular eHIP and so enhancing the handover latency.

2 Prediction Based Extension for Early Update (p-eHIP)

In order to run p-eHIP mode in an eHIP architecture, we assume that rendezvous servers (RVS) have the network topology information considering the all other rendezvous servers and access points in the network. According to this new design, actually two modes of eHIP operates together according to the success of our decision and update. Note that the operating responsibility of this new mode is on rendezvous servers. If a RVS match the conditions for prediction then it completes a decision based early update (p-EU) for the mobile node. When a mobile node continues its movement and requests the regular early update initiation from its current RVS, then RVS replies it with the information that it has been successfully updated before its request base on movement prediction. If there is no prediction based update or if the prediction is not successful, then RVS switches on to regular early update mode and continues early update for the mobile node. Figure 1 represents the general process flowchart of p-eHIP extension.

We also propose that mobile nodes are responsible for sending their location information in the network to their current RVS periodically. RVS keep these location information tables for each mobile node to track the changes in its movement. In order to apply this method, a decision method is needed. Whenever a movement is detected due to the records that RVS keep for mobile nodes' location, our prediction calculation method is invoked. The movement of mobile node is simply detected by any change on the location information of the mobile node. Figure 2 shows the prediction method for p-EU decision in p-eHIP. L1, …, L5 show the location information of mobile node and Prediction 1, …, Prediction 3 show the prediction made for the mobile node's next AP which is calculated according to network topology information. A prediction for next AP and RVS can be made after each location information. However, the decision of triggering an early update for

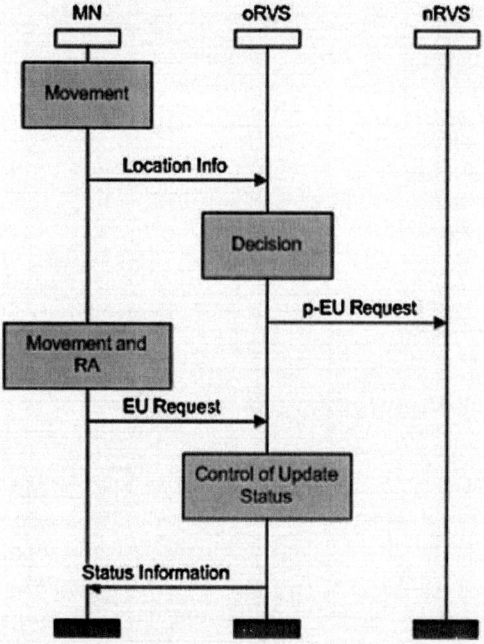

Fig. 1 p-eHIP general process flowchart

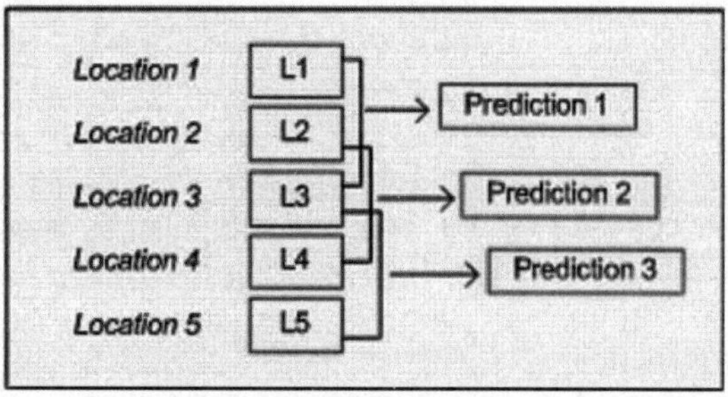

Fig. 2 Prediction method according to location

this mobile node is made after iterative three identical results in order to avoid unnecessary false updates for the mobile node. These early updates are made for each prediction that satisfies the conditions and can be cancelled due to false predictions or any timeout condition.

In order to improve the prediction based p-EU decisions, a method for considering the velocity of a mobile node has been also figured on for p-eHIP. The aim of this variation is to minimize the false or too early update decisions for a MN. The proposed control mechanism in this variation is based on calculating the average velocity of a mobile node from the beginning of its movement (path) records and taking into account this average velocity and its distance to nearest predicted AP.

Table 1 Numericals results for Scenario 1

Mode	Period	Total p-EU decisions	False	True	Number of handoffs	Number of handoffs without p-EU decision
Classic	1	50	7	43	6	0
Classic	2	11	0	11	6	2
Classic	3	1	0	1	6	5
Velocity	1	9	3	6	6	0
Velocity	2	5	0	5	6	2
Velocity	3	1	0	1	6	5

3 Simulation and Results

The methods for p-eHIP are developed on MATLAB [5] environment and numerical results are obtained and analyzed. The results presented in this section are related to the mobile node's path and the networks topology chosen. The prediction and p-EU decisions are done while considering that MN is sending its location information on its each step/movement during its path as period (p). In this section, one of the sample scenario is presented and analyzed for p-eHIP method.

In the following network topology and scenario, five different RVS_2 and three APs connected to each are located. As the period increases, the number of p-EU decisions is reduced due to the process of p-eHIP. The number of true or false predictions is determined by comparing the decisions to the requested AP by mobile in EU message. Some numerical results of this scenario for both methods of p-eHIP (classic and velocity based) are presented in Table 1.

The numerical results depend of the mobile node's path and its average velocity during this path. When the path of mobile node is drawn manually, the velocity is not constant. It changes according to the mobile node's step interval, so mobile node can get over different distances for each unit time in the simulation.

According to the results in Table 1, for $p = 1$, p-EU decisions are made before all EU requests in both modes. An 86% true decision rate is obtained for this period. As expected, number of total decisions reduced in $p = 2$ and $p = 3$. The increase in the period means, the prediction for mobile node cannot be done for each location information, so p-EU decisions' interval is increased and RVS cannot satisfy the conditions for p-EU decision before EU request. Nevertheless, true decision rate is 100%, which means all predictions and decisions are true. Especially for velocity mode, number of unnecessary updates and false decisions significantly reduced as intended. Table 1 shows that 80% enhancement for total number of decisions (updates) and 70% enhancement in number of false decisions are obtained.

4 Conclusion

In this study, a prediction extension has been introduced for eHIP method. This extension, aims to trigger the early update of a mobile node by investigating the path during its mobility earlier than eHIP. The success of this method has been examined with integration of this extension to eHIP method and successful decisions made both with and without taking into account the mobile node's speed. Mostly about 60–80% enhancements on early update decisions have been observed with this extension while considering the consistency of mobile node's movement in the network. As future work, we are studying on extending this method by using other location positioning methods and integrating to eHIP protocol to enhance the accuracy more.

Acknowledgments This study is a part of PhD thesis entitled "Design of a New Mobility Management System for Next Generation Wireless Networks" at Istanbul University, Institute of Science. Telecom SudParis also supported this study as co-supervising of this thesis.

References

1. IETF: Internet Engineering Task Force. http://www.ietf.org/ (2011)
2. IRTF: Internet Research Task Force. http://www.irtf.org/ (2011)
3. Moskowitz, R., Nikander, P., Henderson, T.: RFC 5201, Host Identity Protocol. http://tools.ietf.org/html/rfc5201 (2008)
4. Gurkas Aydin, Z., Chaouchi, H., Zaim, A.H.: eHIP: Early update for host identity protocol. In: ACM Mobility Conference 2009, Article No.: 55, September, Nice, France, 2009
5. MATLAB: The language of technical computing. http://www.mathworks.com/products/matlab/

Performance Evaluation of a Multiuser Interactive Networking System: A Comparison of Modelling Methods

Tadeusz Czachórski, Krzysztof Grochla, Adam Jozefiok, Tomasz Nycz and Ferhan Pekergin

Abstract The article presents a queueing model for performance evaluation of a large database system at an assurance company. The system includes a server with a database and a local area network with a number of terminals where the company employees run applications that introduce documents or retrieve them from the database. We discuss a model of clients activities. Measurements were collected inside the working system: the phases at each users application performance were identified and their duration was measured. The collected data are used to construct a synthetic model of applications activities which is then applied to predict the system behaviour in case of the growth of the number of users. We apply simulation, Markov and diffusion models—their comparison, based on real data, may better verify the utility of particular methods than usual academic examples.

T. Czachórski (✉) · K. Grochla
Institute of Theoretical and Applied Informatics, Polish Academy of Sciences, Baltycka 5, 44-100 Gliwice, Poland
e-mail: tadek@iitis.gliwice.pl

K. Grochla
e-mail: kil@iitis.gliwice.pl

A. Jozefiok · T. Nycz
Institute of Informatics, Silesian University of Technology ul. Akademicka, 16, 44-100 Gliwice, Poland
e-mail: jozefiok@gmail.com

T. Nycz
e-mail: tomasz.nycz@polsl.pl

F. Pekergin
LIPN, Université Paris-Nord, 93430 Villetaneuse, France
e-mail: pekergin@lipn.univ-paris13.fr

E. Gelenbe et al. (eds.), *Computer and Information Sciences II*,
DOI: 10.1007/978-1-4471-2155-8_27,

1 Introduction

The performance of large computer systems, serving thousands of users on daily basis is crucial to many company activities. Network speed limitation and computing power of the servers influence the response time, slowing down the work of many employees and introducing large costs. The evaluation of such a system should determine the performance bottleneck, localize and show to the administrators the parts of the system (e.g. networking devices or servers) which should be replaced or improved. In this work we try to analyze a large networking system providing access to database servers in one of the largest Polish insurance companies. The goal of the work is to identify the time needed to entirely serve one client during a session (this time is composed of multiple operations and multiple data accesses) as a function of the number of active users.

The investigated computer network is built on the Microsoft Windows software and called the Complex IT System (CITS). The whole system performs about 100 million of accountancy operations annually and about 40 billion calculations in special data centres but we investigate only one its branch.

The system includes multiple database servers and a local area network with a number of terminals running applications that introduce documents or retrieve them from the database. Measurements were collected inside the working system: the phases at each users application were identified and their duration was measured. The collected data are used to construct a synthetic model of applications activities which is then applied to predict the system behaviour in case of the growth of the number of users.

Typical activities of an interactive application are shown in Fig. 1. A user defines data concerning a taxpayer to be found, sends a request for a defined document, and processes it for a certain time. Next, after a break, demands the access to another document concerning the same taxpayer—and this pattern repeats several times. Then the user starts the same application concerning another taxpayer. Hence four essential times are to be defined: time to download a payer, time to download a document, time between downloading the documents and time between the end of the service of one payer and beginning the treatment of the next one. The times and the number of retrieved documents are random and their distribution depends on the type of the application. Five types of applications, say $A_1, \ldots, A_5$, were distinguished. The total load of the system is a mixture of these five kinds of applications. Some histograms of measured data are displayed in Figs. 2 and 3.

2 The System Model

The measurements indicated that the overhead given by the network and transmissions is negligible when compared with the times of retrieving the documents in the database. Hence, the queueing model represents only the database server.

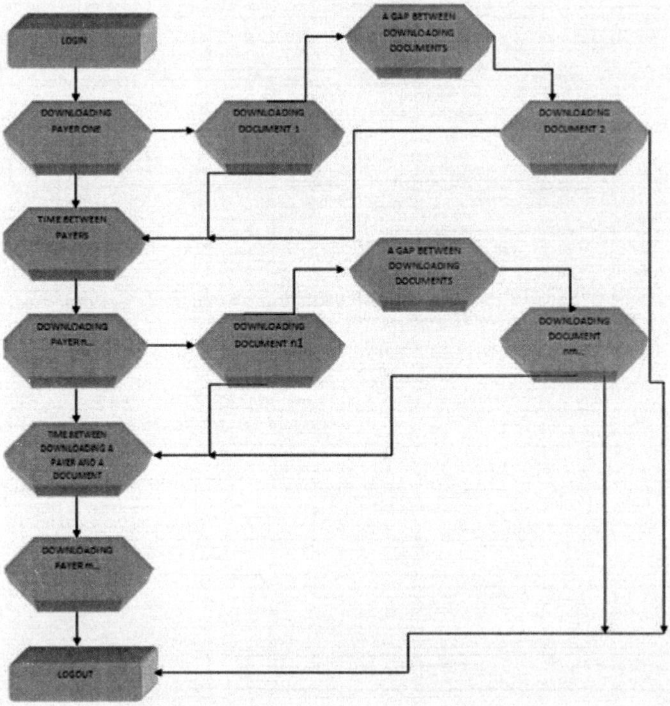

Fig. 1 Diagram of an application activities

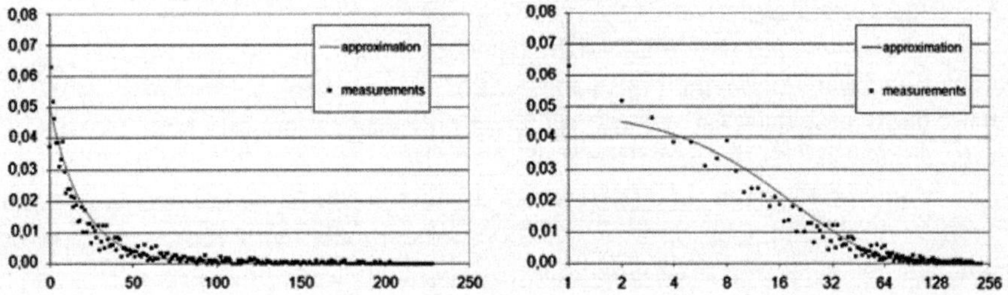

Fig. 2 Application A_1, distribution of the download time of a single payer—linear and logarithmic scale

As in the investigated configuration are 6 servers, each able to treat one demand at a moment, we assumed 6 parallel service channels. This way the model of the system was reduced to a kind of machine-repairman model dealing with a pool of clients, see Fig. 4. Each application gives two types of clients: one representing the retrieval of the main document on the taxpayer and the second type concerning the downloads of consecutive documents related to it. In the whole set of applications, we have ten types of customers. This model was investigated with the use of three

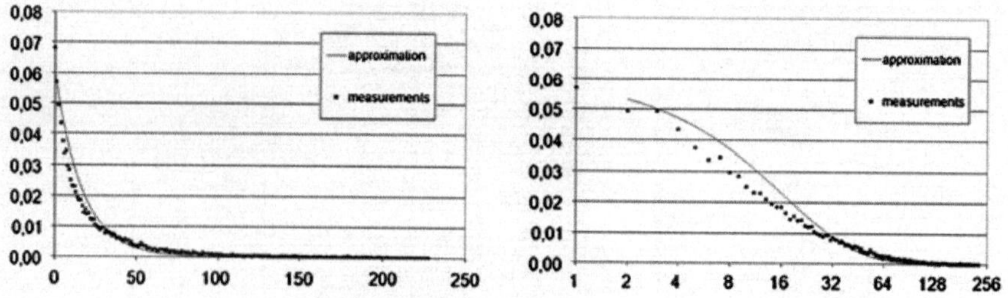

Fig. 3 Application A_1, distribution of download time of documents—linear and logarithmic scale

Fig. 4 Queueing model of the system

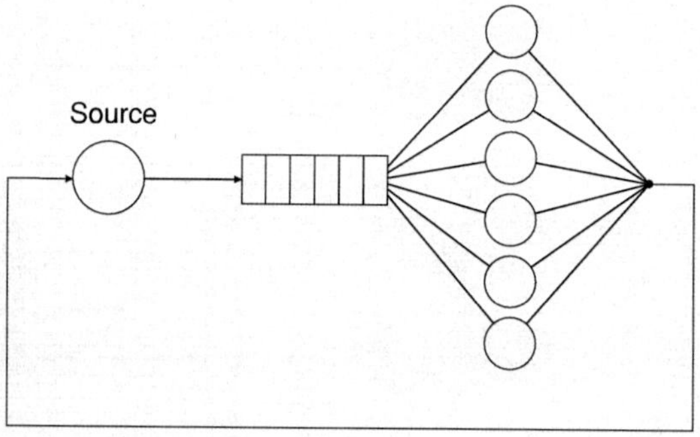

methods: simulation, Markov chains and a mixture of diffusion approximation and mean value analysis.

Simulation model was made with the use of OMNET++ software [1]. The simulations were performed with histograms gathered for each of five types of applications and then for the mix of them following their observed frequency. Some simulation results are given below. Figure 5 presents the distribution densities of the total session time (retrieval of all documents related to one payer) as a function of the number of active stations (the model with only one type of applications). The time of simulations was about 1.5 h for one curve.

Markov model. Measured distributions were approximated by exponential, 2nd and 3rd order Erlang distributions as well as 2nd and 3rd order Cox distributions, the parameters of which were found by the method of least squares. The minimum of the sum of squared residuals, a residual being the difference between an observed value and the fitted value provided by a density function, Mathlab LSQNONLIN, FMINCON functions were used. To assure global minimum, for each minimization 1600 starting points were randomly chosen following uniform distribution at ranges [0, 1], [0,10], [0,100], [0.1000], [0,10000], and [0,1E150]. In Figs. 2 and 3 we see exemplary results of this approximation (continuous line). Then the hypothesis on the type of distributions was tested using χ square and λ-Kolmogorov tests, e.g. [2]. In majority of cases (but not all) the positive results were achieved, i.e. the hypothesis was not rejected with confidence level 95 %.

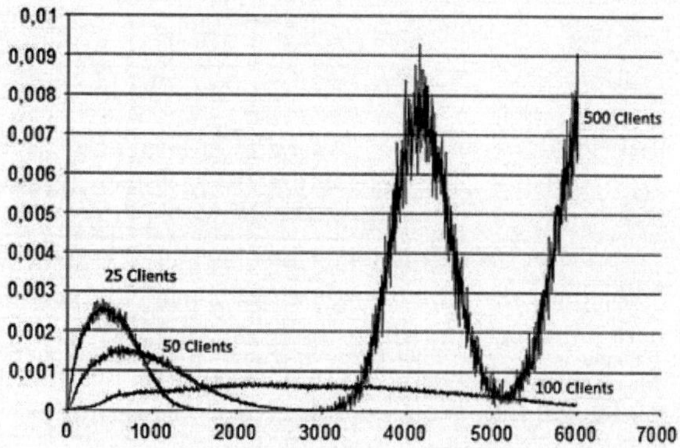

Fig. 5 Density of total time for retrieval of all documents related to one payer as a function of the number of active stations, application A_1, simulation

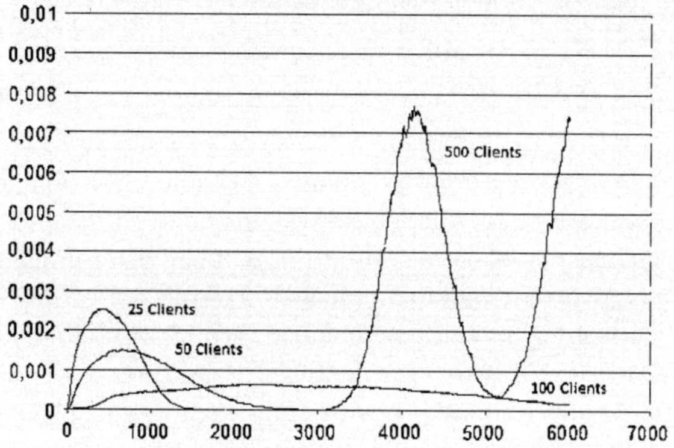

Fig. 6 Density of total time for retrieval of all documents related to one payer as a function of the number of active stations, application A_1, Markov model

This gave us a basis to construct the Markov chain. The equations were generated and solved with the use of a special tool OLYMP [3] designed in IITiS PAN enabling the solution of very large (up to a hundred of million states) Markov chains. Because of states explosion we were obliged to limit the model to one type of applications and to assume the random service (with probability proportional to the number of customers of specified class) instead of FIFO queue. Even then, for 500 active terminals and for exponentially distributed service and sojourn times, the number of states was 125 millions and the computations performed on a typical workstation lasted about 12 h (one curve). We used a projection method based on Krylov subspaces and Arnoldi procedure, e.g. [4]. Figure 6 presents the output of the Markov model to be compared with simulations results in Fig. 5.

Diffusion model. Diffusion approximation is a classical method, proposed in the present form in [5, 6] to model single *GI/GI/1*, *GI/GI/1/N* stations and their networks and used frequently since then, e.g. [7, 8]. Here, we developed a multichannel model with finite multiclass population of customers to be conformed with the model in Fig. 4. The method replaces the process $N(t)$ (the number of customers inside a station) by a continuous diffusion process $X(t)$. The diffusion equation

$$\frac{\partial f(x,t;x_0)}{\partial t} = \frac{\alpha}{2}\frac{\partial^2 f(x,t;x_0)}{\partial x^2} - \beta\frac{\partial f(x,t;x_0)}{\partial x}$$

defines the conditional pdf $f(x,t;x_0)dx = P[x \le X(t) < x + dx|X(0) = x_0]$ of $X(t)$ approximating the distribution of customers present in the queue, $f(n,t;n_0) \approx p(n,t;n_0)$.

In the diffusion model of the considered system we should take into account the finite dimension of the source of customers and multiple service channels—the both influence the diffusion parameters α, β and make them state-dependent. To determine the input flows we use an iterative model based on the diffusion model and the mean value analysis. The results are similar to those obtained with the use of precedent models.

3 Conclusions

The article reports our attempts to capture the characteristics of client activities in a large computer data base system and to construct a queueing model to predict the system performance while the number of customers is growing. The results are based on a simulation model, Markov and diffusion approximation models and show a significant resemblance. We observe a small influence of approximations made in Markov (exponential, Erlang or Cox distributions, random order of service) and diffusion models (information on only two moments of time distributions, omission of the exact sequence of events) on the results. Diffusion approximation, as it operates only on mean values and variances of measured time distributions and is merging states seen by a Markov chain, demands considerably less programming and computation effort than other methods. The proportions of typical computation time for simulation, Markov and diffusion models were: 1.5 h/12 h/few seconds.

Acknowledgments This research was partially financed by a grant no. 4796/B/T02/ 2011/40 of Polish National Council of Science (NCN).

References

1. OMNET ++ site. http://www.omnetpp.org
2. DeGroot, M.H., Schervish, M.J.: Probability and Statistics, 3rd edn. Addison-Wesley, Boston (2002)
3. Pecka, P.: An object-oriented software system for numerical solution of Markovian queueing models. PhD Thesis, IITiS PAN Gliwice, Poland (2002)
4. Stewart, W.J.: An Introduction to the Numerical Solution of Markov Chains. Princeton University Press, Princeton (1994)
5. Gelenbe, E.: On approximate computer systems models. J. ACM **22**(2), 261–269 (1975)

6. Gelenbe, E., Pujolle, G.: The behaviour of a single queue in a general queueing network. Acta Informatica **7**, 123–136 (1976) Fasc. 2
7. Gelenbe, E.: A diffusion model for packet travel time in a random multi-Hop medium. ACM Trans. Sens. Netw. **3**(2),1–19 (2007)
8. Gelenbe, E.: Search in unknown random environments. Phys. Rev. E **82**, 061112 (2010)

Auction-Based Admission Control for Self-Aware Networks

Georgia Sakellari, Timothy Leung and Erol Gelenbe

Abstract In the future, networks will be increasingly Quality-of-Service (QoS) oriented, and admission control will play a vital role in ensuring that relevant QoS requirements of flows are satisfactorily met. Recent research viewing the network as a chargeable resource, associating a pricing mechanism for different types of data, is also becoming increasingly prevalent. In this paper, we introduce the idea of using auctions to coordinate the distributed admission control system for self-aware networks we have developed in the past. The proposed system employs the use of quota at each of the nodes in order to regulate congestion in the network. The current level of quota at a node indicates the amount of additional burden the node is permitted to introduce into the network through the acceptance of flows. If a node does not have enough quota to admit an incoming flow, it can request additional top-up quota whose allocation is controlled by an auctioning mechanism.

G. Sakellari (✉)
School of Computing, Information Technology and Engineering,
University of East London, London E16 2RD, UK
e-mail: g.sakellari@imperial.ac.uk; g.sakellari@uel.ac.uk

T. Leung
AESOP Group, Department of Computing, Imperial College London,
London SW7 2AZ, UK
e-mail: timothy.leung08@imperial.ac.uk

E. Gelenbe
ISN Group, Electrical & Electronic Engineering Department,
Imperial College London, London SW7 2BT, UK
e-mail: e.gelenbe@imperial.ac.uk

E. Gelenbe et al. (eds.), *Computer and Information Sciences II*,
DOI: 10.1007/978-1-4471-2155-8_28,

1 Introduction

Data transmission requests by QoS-oriented services, or flows, can be parameterised by a set of QoS requirements and a source-destination pair. A network will only be able to admit and satisfy a certain number of such flows before resource and topology limitations of the network mean that flow QoS requirements cannot be met. An admission control (AC) system governs the admission of each flow given the current congestive state of the network and the flow's QoS requirements, such as required bandwidth, delay, jitter and packet loss. AC systems can either be centralised or distributed; limitations of having a centralised system are such that it acts as a single point of failure and it bottlenecks the system by limiting flow admission requests to the rate at which it can process them. Distributed systems, on the other hand, base their decisions on limited knowledge and might not be very accurate. Here, we propose a novel auction-based, measurement-based admission control system that tries to overcome the limitations of both a centralised and a completely distributed system. It uses a quota system, derived from the burden of a flow on the network, which input nodes use to restrict flow admission. Our scheme is implemented atop the Cognitive Packet Network (CPN) [1–3], since it already employs a measurement-based infrastructure, and atop the distributed AC algorithm presented in [4, 5], which is also measurement-based and allows different users to have different and multiple QoS criteria when entering the network.

This paper's purpose is to provide a proof of concept of how auctions could help improve the performance of the distributed AC presented in [4, 5]. We do not elaborate on how to dynamically calculate quota in each machine. This is something that we are currently working on.

2 Our Proposed Algorithm

Our auction-based distributed admission control system employs the use of quota at each of the nodes in order to regulate congestion in the network. The current level of quota at a node indicates the amount of additional burden it is permitted to introduce into the network through the acceptance of flows. If the distributed admission control (DAC) algorithm described in [4, 5] decides that a new flow requesting to enter the network can be accommodated without affecting the existing flows of that node and the node has enough quota, then the flow will be accepted. If the DAC decides a new flow can be accepted but the node does not have enough quota, it can request additional top-up quota whose allocation is controlled by an auctioning mechanism. An advantage of auction-based DAC is that it is able to exercise a degree of coordination and control over the admission of flows into the network only when the network's available capacity becomes scarce.

In order to be admitted, an incoming flow will follow these steps:

1. Given that q is a measure of network burden, the flow is assigned a value of Δq, which can be interpreted as the extra burden the additional flow will place on the network. It can be computed through the probing process used by the DAC. By also applying the DAC we can have an estimation on whether the network can accommodate the new flow or not. If the flow cannot be accommodated in the network, it will be rejected outright.
2. Each input node has pre-allocated quota and is permitted to grant immediate entry to any flow for which it currently has sufficient available quota. If the incoming flow is admitted, the amount of available quota at the input node is decremented accordingly, and only when the flow frees up the connection is the used quota restored. If, however, the input node has insufficient quota, the admission request from the incoming flow is turned over to the auctioning mechanism.
3. The auctioning mechanism allocates top-up quota, denoted by δq, to input nodes in order to cover the difference between the amount required for the input node to admit a flow and the amount that the input node has available. The auctioning mechanism may be viewed as a reverse auction that will, over time, receive decreasing bids for top-up quota from input nodes. Depending on the auction algorithm, the auction will terminate with an appropriately low bid being accepted and the amount of quota corresponding to that bid will be lent to the input node that submitted the bid in order to accept a flow; the top-up quota is to be returned when *any* connections terminate from that input node.

Many auction mechanisms can be used. The simplest one would be to have a constant amount of time that the auctioneer waits for bids before awarding the top-up quota to the lower bidder. The auction mechanism described in [6] can also be used. The goal of the AC is to admit as many flows as possible as fast as possible into the network, while satisfying all QoS requirements for admitted flows. Quota at each input node governs how much autonomy the input node has over flow admission and increases the speed at which the AC can make admission decisions. Too low a level of quota at an input node will tend to increase the likelihood of invoking the auctioning mechanism which acts as a centralised admission control system by serialising admission requests, while a high level of quota will reduce the rate of traffic directed towards the auctioning mechanism and allow input nodes to admit flows that require a disproportionate amount of quota by constituting a larger burden to the network than other flows vying for quota awaiting admission at other input nodes.

The goal of the auctioning mechanism includes (i) to minimise top-up quota expenditure and conserve its resources so it can continue to run auctions, and (ii) to try to accept as many bids as possible and as soon as possible since each new bid means rejecting or stalling the flow that submitted the previous bid. The balance between these two goals can be pressurised by the amount of top-up quota the auctioning mechanism has at its disposal. We are currently investigating algorithms to allocate quota to both the input nodes and the auctioneer. In the experiments conducted for this paper though, each input node has pre-allocated quota and is permitted to grant immediate entry to any flow for which it currently has sufficient available quota. The quota values we have chosen are explained in Sect. 3.

3 Configuration of the Experiments

In order to evaluate our mechanisms we conducted our experiments on the real testbed located at Imperial College London (same as in [4]). There are three sets of experiments, with different amount of users constantly making requests to send traffic; in the first set we had 5 users corresponding to 5 Source-Destination (S-D) pairs, in the second we had 9 users, and finally in the third and more demanding set of experiments, we had 17 users independently making requests to send traffic into the network. We set a random waiting time W between requests, chosen to be uniformly distributed in the range of [0,15] seconds. We set the probing rate at 40% of the user's rate and the probing duration at 2 s. In our experiments all links have 10 Mbits/s capacity and all users have the same QoS requirements: round-trip $delay \leq 50$ ms and $jitter \leq 2.5$ ms. When a call is accepted, the source generates UDP traffic of 300 kbps constant bit rate that lasts for a duration which is uniformly distributed in the range of values [10,200 s]. Also, we have introduced constant background traffic of 3.2 Mbps. Each experiment lasts for 5 min (300 s) and is conducted 5 times. The results presented here are the average values of those runs. At a specific input node, in order to avoid having more than one users requesting to enter the network at the same time each user request enters a queue ("request queue"). In the centralised version there is one such queue and it is located in the data collection point, so that all users from all input nodes will queue there in order to enter the network.

Each set of experiments covers six cases: (i) The centralised AC, proposed in [7] (CAC), (ii) The AC in each input node simply allows everybody to enter the network (DAC-All), (iii) The DAC proposed in [4, 5] (DAC), (iv) The DAC is enabled and the nodes are fully coordinated (DAC-Full) as explained in [4, 5], (v) The DAC is enabled and the nodes are randomly coordinated (DAC-Rand) as in [4, 5], and (vi) The DAC is enabled and our proposed auction mechanism controls the admittance when a node does not have enough quota (DAC-Auction).

Each input node has pre-allocated quota and is permitted to grant immediate entry to any flow for which it currently has sufficient available quota. Because all users in our experiments have the same QoS criteria, the quota each user needs in order to be accepted into the network depends on (equals) the duration they will occupy the network for. Based on experimental trials we have decided that the total amount of quota to be distributed among the source (input) nodes of the network is 5000. So, the auction quota at each input node when we have 2 users is 2500, when we have 5 users is 1000, and for 13 users making requests into the network the value drops to 385 per input node. Also, the auction quota when we have 2 users is 7500, when we have 5 users is 3000, and for 13 users making requests into the network the auctioneer's quota is 1150. The auctioneer's total quota also depends on the number of source nodes in the networks so that a source node will not be allocated more quota that it can handle. Another way of assuring this will not happen is by having a maximum amount of quota a node can have. As the auction mechanism is concerned, we have chosen a basic auction where the auctioneer waits for a constant amount of time, in this case 1 s, before deciding the winner.

4 Experimental Results

In order to evaluate the performance of our algorithm in each set of experiments, four areas are covered. First, we measure the satisfaction of each user (see Fig. 1). We consider the satisfaction of a user as the percentage of time, over the whole lifetime of the user, where all the QoS requirements of that user are fulfilled. The results presented here are the average values for all users in the network. Secondly we measure the average time a user has to queue before being given a decision on whether to enter the network or not (see Fig. 2). Finally we count the total number of requests that were served during the duration of the experiments (see Fig. 3) and how many of them were actually accepted (see Fig. 4).

Figure 2 reveals that the shortest average time the users have to wait until being admitted is when no AC is used. This waiting time is close to zero but not zero, since in the cases where multiple users request to enter the network at the same input node there is still a queue at the input node. The longest waiting time is when the CAC is used since all users queue at the central point. When we apply the DAC, the time a user waits until being either admitted or rejected, is the shortest between all the distributed versions. The co-ordination slightly increases the waiting time due to the overhead caused by message exchanges. Finally, with the auction-based scheme, the waiting time is comparable to the DAC, case except when having 9 users in the network. In that scenario, the waiting time is quite high, which is also reflected in the total number of requests in Fig. 3. This could be due to the sub-optimality of the quota value.

Waiting for service affects the number of users that are actually accepted into the network. Figure 3 shows that centrally controlled admission of users results in small number of users served and accepted into the network, which was expected based on the waiting times shown prior, while in all other cases the number of requests served is more or less the same. The fact that in some cases less requests were made when all users were accepted (DAC-all) is due to the randomness in the generation of the requests.

In practice, the centralised algorithm offers greater user satisfaction but fewer users (see Fig. 1 and 4). The satisfaction of the DAC is lower probably, because multiple probes are present in the network and each input node has limited information, making the estimation of the algorithm less accurate. The use of coordination does not improve the DAC algorithm since it introduces the exchanging of messages which make the estimations less accurate. By allowing the DAC mechanism to work while the network traffic is low and serialising

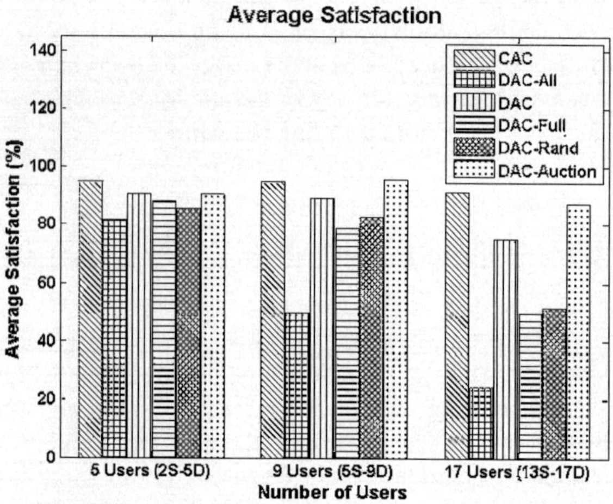

Fig. 1 Average satisfaction of the users

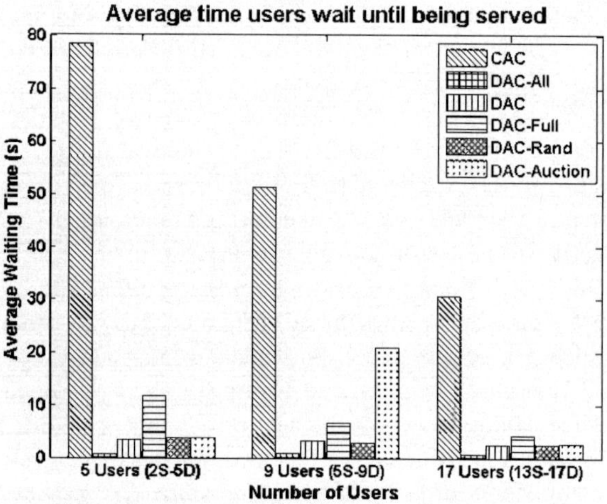

Fig. 2 Average time a user waits in the "request queue" before being served

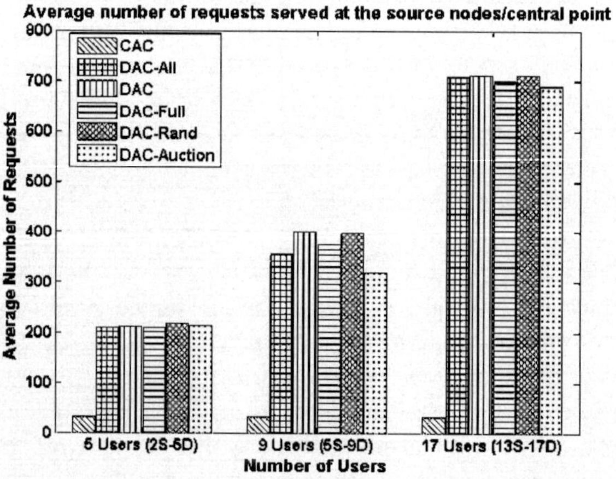

Fig. 3 Average number of requests

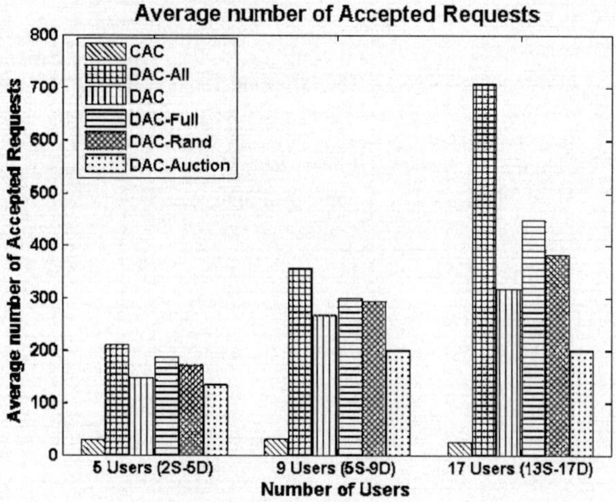

Fig. 4 Average number of accepted requests

the admission mechanism when the network is close to congestion through the use of an auction process, we are able to achieve satisfaction rates close to the ones of the centralised AC but with considerably more users accepted into the network.

5 Conclusions

This paper constitutes a proof-of-concept that combines the advantages of both the distribution and centralised admission control systems, by way of an admission control system that implements an auction mechanism. Our future work will concentrate on how to calculate the quota and how to allocate these quota to the various input nodes and to the auction mechanism, in a way that reflects the congestive state of the network and the additional burden that nodes are permitted to inflict on the network.

References

1. Gelenbe, E., Xu, Z., Seref, E.: Cognitive packet networks. In: Proceedings of the 11th International Conference On Tools With Artificial Intelligence (ICTAI '99), Chicago, IL, USA, IEEE Computer Society Press (Nov. 1999) 47–54
2. Gelenbe, E., Lent, R., Xu, Z.: Design and performance of cognitive packet networks. Perfor. Evaluation **46**(2–3), 155–176 (2001)
3. Gelenbe, E., Lent, R., Montuori, A., Xu, Z.: Cognitive Packet Networks: QoS and Performance. In: Proceedings of the 10th IEEE International Symposium On Modeling, Analysis, and Simulation Of computer and Telecommunications Systems (MASCOTS' 02), Fort Worth, Texas, USA, Opening Keynote Paper, 3–9 (Oct. 2002)
4. Sakellari, G., Gelenbe, E.: A distributed admission control mechanism for multi- criteria QoS. In: Proceedings of the IEEE GLOBECOM 2010 workshop on advances in communications and networks, Miami, FL, USA (Dec. 2010) 1995–1999

5. Sakellari, G.: A Decentralised, measurement-based admission control mechanism for self-aware networks. In: Proceedings of the 25th International Symposium on Computer and Information Sciences (ISCIS 2010), London, UK, 177–182 (Sep. 2010)
6. Gelenbe, E.: Analysis of single and networked auctions. ACM Trans. Internet Technol. (TOIT) **9**(2), 1–24 (2009)
7. Gelenbe, E., Sakellari, G., D' Arienzo, M.: Admission of QoS aware users in a smart network. ACM Trans. Auton. Adapt. Sys. **3**(1), 4:1–4:28 (2008)

Complexity Reduction for Multi-hop Network End-to-End Delay Minimization

Mustafa Gurcan, Irina Ma, Hadhrami Ab Ghani and Zhenfeng He

Abstract A new receiver scheme is developed to reduce the end-to-end delay in the ad hoc networks without increasing the energy required. The proposed Pulse Amplitude Modulation–Minimum Mean Square Error (PAM–MMSE) receiver reduces the granularity of the discrete bit values compared to the current multi-code Code-Division Multiple Access (CDMA) network, increasing transmission data rate in an ad hoc network without reducing the network lifetime

Keywords Complexity · Two-group allocation · Iterative energy calculation

1 Introduction

Minimizing the end-to-end delay [1] over a selected routing path in any wireless ad hoc network is the key to increase the overall transmission rates. A multi-code transmission model is considered for this purpose in this paper. The current multi-code transmission technique employs quadrature amplitude modulation (QAM) [2] and uses equal rate and equal energy (EREE) allocation calculated from the worst signal to interference plus noise ratio (SINR) channel. An equal rate iterative energy algorithm is proposed in [3] to improve the performance and reduce noise and interference but suffers residue energy problem.

M. Gurcan (✉) · I. Ma · H. A. Ghani · Z. He
Department of Electrical and Electronic Engineering,
Imperial College, London, UK
e-mail: m.gurcan@imperial.ac.uk

I. Ma
e-mail: irina.ma06@imperial.ac.uk

H. A. Ghani
e-mail: hadh.ghani05@imperial.ac.uk

Z. He
e-mail: zhenfeng.he06@imperial.ac.uk

E. Gelenbe et al. (eds.), *Computer and Information Sciences II*,
DOI: 10.1007/978-1-4471-2155-8_29,

A two-group allocation scheme has been proposed to improve system throughput by using the remaining residue energy to load some channels with the next adjacent discrete bit rate with QAM [1]. It has been proved in [4] that PAM may achieve the same capacity as QAM if the total number of channels $K_{PAM} = 2K_{QAM}$. However, a PAM incorporated two-group allocation design may provide more discrete values and thus, a higher transmission rate. Unlike real valued WBE codes [5], the complex Grassmannian codes [6] is used as it eliminates real-valued cross correlations between codes for PAM. Due to the use of complex-valued signature sequence, the two-group allocation used with QAM cannot be applied directly, and the PAM–CCC method is proposed in [4]. This method suffers from a high computational complexity which decreases the network lifetime. Therefore, a PAM–MMSE receiver is proposed to achieve complexity reduction. A general two-group loading model with QAM is first briefly described in the next section.

2 QAM System with Two-Group Allocation

Iterative energy adjustment for achieving target SNRs [3] reduces the wasted SINR but produces residue energy. This residue energy may be minimized by loading two groups of channels at two adjacent discrete bit rates, b_p and b_{p+1}, to improve the total bit rate as $R_T = (K - m)b_p + mb_{p+1}$ bits per symbol for K code channels. The lower bit rate b_p is computed by satisfying $\max_p\left(\sum_{k=1}^{K} E_{k,I_{max}}\right) \leq E_T$ for an equal bit rate b_p per channel. The number m is found by minimizing the residue energy E_{res} such that $\min_m\left(E_{res} = E_T - \sum_{k=1}^{K} E_{k,I_{max}}\right) \geq 0$ for a bit rate b_{p+1} targetted in channel $k = 1, \ldots, m$ and a bit rate b_p elsewhere, where $m = 0, \ldots, K-1$. The energy is updated with $E_{k,i+1}(r_k) = \frac{\gamma^*}{1+\gamma^*}\left(\overline{q}_k^H \mathbf{C}_i^{-1} \overline{q}_k\right)^{-1}$ from $i = 1, \ldots, I_{max}$ for $k = 1, \ldots, K$, where $[\overline{q}_1, , \overline{q}_{3K}] = \mathbf{Q}_e$ are the receiver signature sequences, $\left[\overline{q}_{K+1}, \ldots, \overline{q}_{3K}\right]$ are the receiver signature sequences for the ISI components, $\widetilde{\mathbf{A}}$ is the diagonal amplitude matrix, σ^2 is the noise variance and $\mathbf{C} = \mathbf{Q}_e \widetilde{\mathbf{A}} \mathbf{Q}_e^H + 2\sigma^2 \mathbf{I}_{N+2\alpha}$ is the covariance matrix. The MMSE despreading coefficient is $\overline{w}_k = \frac{\mathbf{C}^{-1}\overline{q}_k}{\overline{q}_k^H \mathbf{C}_i^{-1} \overline{q}_k}$. The next section describes the PAM–CCC scheme.

3 PAM–CCC

For PAM, the number of channels is doubled, $K = K_{PAM} = 2K_{QAM}$, to have the same capacity as that of QAM, with the discrete bit values of $b_p = b_{p,PAM} = \frac{1}{2} b_{p,QAM}$. With CCC, complex-valued signature sequences are employed, where the complex-valued concatenated received matrix is $\mathbf{R}_{cc} = \frac{1}{\sqrt{2}}\left[\mathbf{R}^T \quad (\mathbf{R}^*)^T\right]^T$ is a $2(N + 2\alpha) \times N^{(x)}$ dimensional matrix. The corresponding despreading filter sequence $\overline{w}_{k,cc}$ is extended with length $2(N+2\alpha)$. Similarly, the signature sequence matrix and noise matrix is concatenated such that:

$$\mathbf{R}_{cc} = \frac{1}{\sqrt{2}}\begin{bmatrix}\mathbf{R}\\ \mathbf{R}^*\end{bmatrix} \approx \frac{1}{\sqrt{2}}\left(\begin{bmatrix}\mathbf{Q}_e\\ \mathbf{Q}_e^*\end{bmatrix}\widetilde{\mathbf{A}}\widetilde{\mathbf{X}} + \begin{bmatrix}\mathbf{N}\\ \mathbf{N}^*\end{bmatrix}\right) = \mathbf{Q}_{cc}\widetilde{\mathbf{A}}\widetilde{\mathbf{X}} + \mathbf{N}_{cc} \tag{1}$$

Symbols can be estimated by using $\widehat{\overline{x}}_k = \mathrm{Re}\left(\overline{w}_{k,cc}^H \overline{r}_{cc}\right)$ and the output SINR of the coded channel k is:

$$\gamma_k = \frac{E_k\left(\mathrm{Re}\left(\overline{\mathrm{w}}_{\mathrm{k,cc}}^{\mathrm{H}}\overline{\mathrm{q}}_{\mathrm{k,cc}}\right)\right)^2}{\sum_{j=1,j\neq k}^{3K} E_j\left(\mathrm{Re}\left(\overline{w}_{k,cc}^H \overline{q}_{j,cc}\right)\right)^2 + \sigma^2 \mathrm{Re}(\overline{w}_{k,cc}^H \overline{w}_{k,cc})} \tag{2}$$

where $E_j = E_{j-K}$ for $j > K$ and $\gamma^*(b_p) = 2^{2b_p} - 1$ with b_p as the set of discrete bit values having index of p. The energy is iteratively calculated as:

$$E_{k,i+1} = \frac{\gamma_k^*\left(\sum_{j=1,j\neq k}^{3K} E_j(\mathrm{Re}\left(\overline{w}_{k,cc}^H \overline{q}_{j,cc}\right))^2 + \sigma^2 \mathrm{Re}(\overline{w}_{k,cc}^H \overline{w}_{k,cc})\right)}{\left(\mathrm{Re}\left(\overline{w}_{k,cc}^H \overline{q}_{k,cc}\right)\right)^2} \tag{3}$$

with the despreading filter coefficient vector $\overline{w}_{k,cc}$ defined as:

$$\overline{w}_{k,cc} = \frac{\mathbf{C}^{-1}\overline{q}_{k,cc}}{\overline{q}_{k,cc}^H \mathbf{C}^{-1}\overline{q}_{k,cc}} \qquad k = 1, 2, \ldots, K \tag{4}$$

where $\mathbf{C} = \mathbf{Q}_{cc}\mathbf{A}^2\mathbf{Q}_{cc}^H + 2\sigma^2\mathbf{I}_{2(N+2\alpha)}$ is the covariance matrix.

To achieve $R_T = (K-m)b_p + mb_{p+1}$ bits per symbol, the two-group loading method described in Sect. 2 is run with the energy equation given in (3).

3.1 Complexity of PAM–CCC

An $n \times n$ matrix inversion using Guass-Jordan elimination has a complexity of $O(n^3)$. Therefore solving a $2N \times 2N$- dimensional complex-valued inverse matrix $\mathbf{C}^{-1}$ has a complexity of $4O((2N)^3)$ in each iteration. Taking the average number of p and m iterations, the overall complexity will be $I_{max} \times [(P+K)/2] \times 4O((2N)^3)$. Alternatively, a computationally lower PAM–MMSE scheme is proposed and presented in the next section.

4 PAM–MMSE

Due to the high complexity in the PAM–CCC scheme, a new PAM–MMSE receiver is introduced which does not require received matrix to be concatenated. With only real value operations, the complex despreading filter coefficient $\overline{w}_k$ is calculated using a real-valued stacked despreading coefficient vector $\overline{w}_{e,k}$ such that $\overline{w}_k$ can be reconstructed using:

$$\overline{w}_k = \overline{w}_{k,r} + j\overline{w}_{k,i} = \left[w_{e,k,1}, \ldots, w_{e,k,N}\right]^T + \left[w_{e,k,N+1}, \ldots, w_{e,k,2N}\right]^T \tag{5}$$

The stacked despreading coefficient vector can be calculated by:

$$\overline{w}_{e,k} = \left(\chi^T\right)^{-1} \overline{p}_k \tag{6}$$

where $\chi = \begin{bmatrix} \chi_{11} & \chi_{12} \\ \chi_{21} & \chi_{22} \end{bmatrix}$ is a $2(N+2\alpha) \times 2(N+2\alpha)$ dimensional matrix containing 4 real-valued $(N+2\alpha) \times (N+2\alpha)$ dimensional submatrices defined as follows:

$$\begin{aligned} \chi_{11} &= \mathrm{Re}\left(\mathbf{Q}_e\mathbf{A}^2\mathbf{Q}_e^H\right) + \mathrm{Re}\left(\mathbf{Q}_e\mathbf{A}^2\mathbf{Q}_e^T\right) + 2\sigma^2\mathbf{I} \\ \chi_{12} &= \mathrm{Im}\left(\mathbf{Q}_e\mathbf{A}^2\mathbf{Q}_e^H\right) + \mathrm{Im}\left(\mathbf{Q}_e\mathbf{A}^2\mathbf{Q}_e^T\right) \\ \chi_{21} &= -\mathrm{Im}\left(\mathbf{Q}_e\mathbf{A}^2\mathbf{Q}_e^H\right) + \mathrm{Im}\left(\mathbf{Q}_e\mathbf{A}^2\mathbf{Q}_e^T\right) \\ \chi_{22} &= \mathrm{Re}\left(\mathbf{Q}_e\mathbf{A}^2\mathbf{Q}_e^H\right) - \mathrm{Re}\left(\mathbf{Q}_e\mathbf{A}^2\mathbf{Q}_e^T\right) + 2\sigma^2\mathbf{I} \end{aligned} \tag{7}$$

and the $2N$-length real-valued vector $\overline{p}_k = [2\sqrt{E_k}\mathrm{Re}(\overline{q}_k)^T \quad 2\sqrt{E_k}Im(\overline{q}_k)^T]^T$.

With the despreading coefficient vector $\overline{w}_k$, the theoretical output SINR of the kth coded channel can be calculated:

$$\gamma_k = \frac{E_k\left(\mathrm{Re}\left(\overline{\mathrm{w}}_\mathrm{k}^\mathrm{H}\overline{\mathrm{q}}_\mathrm{k}\right)\right)^2}{\sum_{j=1, j\neq k}^{3K} E_j\left(\mathrm{Re}\left(\overline{w}_k^H\overline{q}_j\right)\right)^2 + \sigma^2\mathrm{Re}\left(\overline{w}_k^H\overline{w}_k\right)} \tag{8}$$

and the energy can be iteratively updated with:

$$E_{k,i+1} = \frac{\gamma_k^*\left(\sum_{j=1, j\neq k}^{3K} E_j(\mathrm{Re}\left(\overline{\mathrm{w}}_\mathrm{k}^\mathrm{H}\overline{\mathrm{q}}_\mathrm{j}\right))^2 + \sigma^2\mathrm{Re}(\overline{\mathrm{w}}_\mathrm{k}^\mathrm{H}\overline{\mathrm{w}}_\mathrm{k})\right)}{\left(\mathrm{Re}\left(\overline{w}_k^H\overline{q}_k\right)\right)^2} \tag{9}$$

The changes in the PAM–MMSE algorithms are based on the original algorithms described for the QAM scheme. The input parameters for $\mathbf{Q}_e$ remains unchanged while the set of b_p is halved. With the above energy equation, the two-group loading scheme to realize $R_T = (K-m)b_p + mb_{p+1}$ bits per symbol is run using the method described in Sect. 2.

4.1 Complexity

With only real value operations, the complexity of the PAM–MMSE is lower than one from PAM–CCC. Without considering ISI effects, the χ matrix has a dimension of $2N \times 2N$. Using Gauss-Jordan elimination, the number of

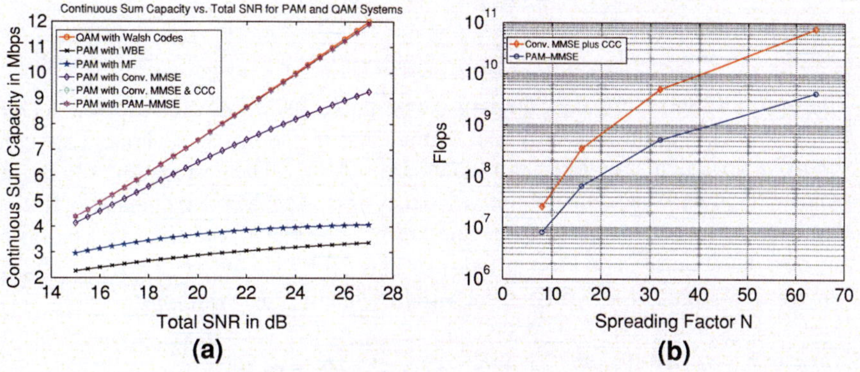

Fig. 1 System performance. **a** System throughput versus total SNRs. **b** FLOPS versus spreading factor

multiplications is $O((2N)^3)$, four times less than the complexity achieved in the PAM–CCC scheme.

5 Performance Evaluation

This section compares the proposed PAM–MMSE receiver scheme against other QAM and PAM receiver designs. Using the vehicular A model model [2], and assuming channel impulse response is known in the transmitter, the simulations are performed with a spreading factor $N = 16$, chip rate $1/T_c = 3.84Mcps$, noise variance per dimension $\sigma^2 = N_0/2$ where $N_0 = -174dBm/Hz$. The number of channels and discrete bit value for QAM is $K_{QAM} = 15$ and $\{b_{p,QAM}\}_{p=1}^{6} = \{0.5, 1, 1.5, 2, 3, 4\}$; while PAM has $K = 2K_{QAM} = 30$ and $b_{p,PAM} = \frac{1}{2} b_{p,QAM} = \{0.25, 0.5, 0.75, 1, 1.5, 2\}$.

Although the theoretical capacity are the same for both the QAM and PAM with K_{QAM} and K_{PAM} channels, the PAM incorporating the two-group allocation have more discrete values, giving a 3–5% increase in system performance for both PAM–MMSE and PAM–CCC receivers compared to the equivalent QAM system(Fig. 1a). The FLOPs (Floating Point Operations) in Fig. 1b further confirms the complexity of PAM–MMSE receiver scheme is four times less than PAM–CCC as discussed in this section.

6 Conclusion

This paper confirms that PAM may achieve higher capactiy than the equivalent QAM modulation scheme. It further verifies that the proposed PAM–MMSE receiver scheme have four time less computational complexity compared to the existing PAM–CCC receiver scheme, yet achieving the same capacity.

7 Appendix

This appendix shows the derivation of the proposed PAM–MMSE receiver. As real PAM symbols transmitted, only real components of the decision variable are considered when calculating the mean square error (MSE). The despreading filter coefficient vector, $\boldsymbol{w}_k$ is designed to minimize the MSE in each coded channel k. Noting that the noise vector and its conjugate are not correlated and the expectation of any correlated noise is σ^2.Further consider that $E(x_k^* x_k) = 1$ and $E(x_k^* \mathbf{R}) = \sqrt{E_k}\boldsymbol{q}_k$. The Mean Square Error (MSE) in each coded channel k is written as:

$$
\begin{aligned}
MSE_k &= E\left(\left|x_k - \mathrm{Re}\left(\boldsymbol{w}_k^H \mathbf{R}\right)\right|^2\right) = E\left(\left|x_k - \frac{1}{2}\left(\boldsymbol{w}_k^H \mathbf{R} + \boldsymbol{w}_k^T \mathbf{R}^*\right)\right|^2\right) \\
&= E\left(x_k^* x_k + \frac{1}{4}\left(\boldsymbol{w}_k^H \mathbf{R} + \boldsymbol{w}_k^T \mathbf{R}^*\right)\left(\mathbf{R}^H \boldsymbol{w}_k + \mathbf{R}^T \boldsymbol{w}_k^*\right)\right. \\
&\quad \left. - \frac{1}{2} x_k\left(\boldsymbol{w}_k \mathbf{R}^H + \boldsymbol{w}_k^* \mathbf{R}^T\right) + \frac{1}{2} x_k^*\left(\mathbf{R}\boldsymbol{w}_k^H + \mathbf{R}^* \boldsymbol{w}_k^T\right)\right) \\
&= 1 + \frac{1}{4}\left[\boldsymbol{w}_k^H\left(\mathbf{Q}\mathbf{A}^2\mathbf{Q}^H + N_0\mathbf{I}\right)\boldsymbol{w}_k + \boldsymbol{w}_k^T\left(\mathbf{Q}^*\mathbf{A}^2\mathbf{Q}^H\right)\boldsymbol{w}_k + \boldsymbol{w}_k^H\left(\mathbf{Q}\mathbf{A}^2\mathbf{Q}^T\right)\boldsymbol{w}_k^*\right. \\
&\quad \left. + \boldsymbol{w}_k^T\left(\mathbf{Q}^*\mathbf{A}^2\mathbf{Q}^T + N_0\mathbf{I}\right)\boldsymbol{w}_k^*\right] - \sqrt{E_k}\left(\boldsymbol{w}_k^H \boldsymbol{q}_k + \boldsymbol{w}_k^T \boldsymbol{q}_k^*\right)
\end{aligned} \tag{10}
$$

By breaking down the despreading filter coefficient vector into its real and imaginary components:$\boldsymbol{w}_k = \boldsymbol{w}_{k,r} + j\boldsymbol{w}_{k,i}$, $\partial MSE_k/\partial \boldsymbol{w}_{k,r}$ and $\partial MSE_k/\partial \boldsymbol{w}_{k,i}$ can be acquired:

$$
\begin{aligned}
\frac{\partial MSE_k}{\partial \boldsymbol{w}_{k,r}} &= \frac{1}{2}\boldsymbol{w}_k^H\left(\mathbf{Q}\mathbf{A}^2\mathbf{Q}^H + N_0\mathbf{I}\right) + \boldsymbol{w}_k^T\left(\mathbf{Q}^*\mathbf{A}^2\mathbf{Q}^H\right) + \boldsymbol{w}_k^H\left(\mathbf{Q}\mathbf{A}^2\mathbf{Q}^T\right) \\
&\quad + \boldsymbol{w}_k^T\left(\mathbf{Q}^*\mathbf{A}^2\mathbf{Q}^T + N_0\mathbf{I}\right) - 2\sqrt{E_k}\mathrm{Re}(\boldsymbol{q}_k)^T
\end{aligned} \tag{11}
$$

$$
\begin{aligned}
\frac{\partial MSE_k}{\partial \boldsymbol{w}_{k,i}} &= \frac{1}{2}\boldsymbol{w}_k^H\left(\mathbf{Q}\mathbf{A}^2\mathbf{Q}^H + N_0\mathbf{I}\right) + \boldsymbol{w}_k^T\left(\mathbf{Q}^*\mathbf{A}^2\mathbf{Q}^H\right) - \boldsymbol{w}_k^H\left(\mathbf{Q}\mathbf{A}^2\mathbf{Q}^T\right) \\
&\quad - \boldsymbol{w}_k^T\left(\mathbf{Q}^*\mathbf{A}^2\mathbf{Q}^T + N_0\mathbf{I}\right) - 2\sqrt{E_k}\mathrm{Im}(\boldsymbol{q}_k)^T
\end{aligned} \tag{12}
$$

Setting $\partial MSE_k/\partial \boldsymbol{w}_{k,r}$ and $\partial MSE_k/\partial \boldsymbol{w}_{k,i}$ to be equal to 0, they can be simplified as:

$$
\begin{aligned}
2\sqrt{E_k}\mathrm{Re}(\boldsymbol{q}_k)^T &= \left(\boldsymbol{w}_{k,r}\right)^T\left[\mathrm{Re}\left(\mathbf{Q}\mathbf{A}^2\mathbf{Q}^H\right) + \mathrm{Re}\left(\mathbf{Q}\mathbf{A}^2\mathbf{Q}^T\right) + 2\sigma^2\mathbf{I}\right] \\
&\quad + \left(\boldsymbol{w}_{k,i}\right)^T\left[\mathrm{Im}\left(\mathbf{Q}\mathbf{A}^2\mathbf{Q}^H\right) + \mathrm{Im}\left(\mathbf{Q}\mathbf{A}^2\mathbf{Q}^T\right)\right]
\end{aligned} \tag{13}
$$

$$
\begin{aligned}
2\sqrt{E_k}\mathrm{Im}(\boldsymbol{q}_k)^T &= \left(\boldsymbol{w}_{k,r}\right)^T\left[-\mathrm{Im}\left(\mathbf{Q}\mathbf{A}^2\mathbf{Q}^H\right) + \mathrm{Im}\left(\mathbf{Q}\mathbf{A}^2\mathbf{Q}^T\right)\right] \\
&\quad + \left(\boldsymbol{w}_{k,i}\right)^T\left[\mathrm{Re}\left(\mathbf{Q}\mathbf{A}^2\mathbf{Q}^H\right) - \mathrm{Re}\left(\mathbf{Q}\mathbf{A}^2\mathbf{Q}^T\right) + 2\sigma^2\mathbf{I}\right]
\end{aligned} \tag{14}
$$

The solutions of the despreading coefficient vector, $\boldsymbol{w}_k$ is solved by rewriting (13) and (14) into matrix and vector components as shown in Sect. 5.

References

1. Gurcan, M.K., Ab Ghani, H., Zhou, J., Chungtragarn, A.: Bit energy consumption minimization for multi-path routing in ad hoc networks. Comput. J. **54**(6), 944–959 (2011)
2. Holma, H., Toskala, A.: WCDMA for UMTS-HSPA evolution and LTE, 4th edn. John Wiley, Chichester (2007)
3. Sayadi, B., Ataman, S., Fijalkow, I.: Joint downlink power control and multicode receivers for downlink transmissions in high speed umts. EURASIP J. Wirel. Commun. Netw. **2006**(2), 60 (2006)
4. He, Z., Gurcan, M.K.: The rate adaptive throughput maximization in pam-modulated overloaded system. In:Wireless Communications and Networking Conference, 2009. WCNC 2009, pp. 1–6. IEEE, April 2009
5. Ulukus, S., Yates, R.: Iterative construction of optimum signature sequence sets in synchronous cdma system. Inf. Theory IEEE Trans. **47**(5), 1989–1998 (2001)
6. Strohmer, T., Heath, R.W.: Grassmannian frames with applications to coding and communication . Appl. Comput. Harmon. Anal. **14**(3), 257–275 (2003)

Performance Improvement of an Optical Network Providing Services Based on Multicast

Vincent Reinhard, Johanne Cohen, Joanna Tomasik, Dominique Barth and Marc-Antoine Weisser

Abstract We focus on circuit switching optical networks and on repetitive multicast demands whose source and destinations are à priori known by an operator. He may have corresponding trees "ready to be allocated" and adapt his network infrastructure according to these transmissions. This adjustment consists in setting branching routers in the selected nodes of a predefined tree. The branching nodes are opto-electronic nodes which are able to duplicate data and retransmit it in several directions. We are interested in the choice of nodes of a multicast tree where the limited number of branching routers should be located to minimize the amount of required bandwidth. After formally stating the problem we solve it by proposing a polynomial algorithm whose optimality we prove. We perform computations for different methods of the tree construction and conclude by giving dimensioning guidelines.

keywords Circuit switching optical networks · Multicast · Complexity · Optimisation

---

V. Reinhard · J. Tomasik (✉) · M.-A. Weisser
SUPELEC Systems Sciences, Computer Science Department,
91192 Gif sur Yvette, France
e-mail: Joanna.Tomasik@supelec.fr

V. Reinhard
e-mail: Vincent.Reinhard@supelec.fr

M.-A. Weisser
e-mail: Marc-Antoine.Weisser@supelec.fr

V. Reinhard · J. Cohen · D. Barth
PRiSM, University of Versailles, 45 avenue des Etats-Unis,
78035 Versailles, France
e-mail: Johanne.Cohen@prism.uvsq.fr

D. Barth
e-mail: Dominique.Barth@prism.uvsq.fr

E. Gelenbe et al. (eds.), *Computer and Information Sciences II*,
DOI: 10.1007/978-1-4471-2155-8_30,

1 Introduction

Optical networks have become a dominant technology in modern networks covering large areas. They face a growing demand on the part of service providers. New offered services are more complex than the simple connectivity service assured traditionally by network operators. Providers sell services like storage and computation together with connectivity service. To meet the demands of their customers, they have to purchase a guaranteed connectivity service at network operators. At the same time, operators can deal with numerous service providers. They are interested in using their network resources the most efficiently and in this way minimize the cost of a prospective extension of their infrastructure.

We studied the mechanisms to execute distributed applications in an optical mesh network in the context of the CARRIOCAS project [1, 8]. Unlike a customary approach applied in Grids, this project went into the study of the coexistence of distributed applications in a network whose operator should make financial profit. Routers which are able to duplicate data and send it in several directions allow an operator to lower the bandwidth amount necessary to construct a multicast tree. On the other hand, these branching nodes (BrNs) are more expensive and more energy consuming than the transparent ones. The realistic assumption is thus that only a subset of routers is equipped with the duplicating functionality. In [4] we studied the tree construction to any multicast with minimization of the used bandwidth and a limited number of BrNs.

Our study mentioned above inspired us to explore certain special cases of multicasts. A network operator can know in advance recurrent multicast transmissions. Being aware of frequent demands for identical multicasts an operator may have corresponding trees "ready to be allocated" and adapt his network infrastructure according to these recurrent transmissions. This adjustment may consist in setting available BrNs in the selected nodes of the predefined tree. We are interested in the choice of nodes of a tree where the BrNs should be located in order to minimize the amount of required bandwidth. This approach allows an operator to make his network more efficient without any additional cost.

In Sect. 2 our problem and its solution properties are stated. Next, we propose an algorithm to solve our problem, compute its complexity, and prove that it is optimal. Our problem is evidenced to be polynomial. Section 4 presents the results. Finally, we give the conclusions and outline further work.

2 Optimization Problem and Properties of its Solution

An optical network is modelled by a directed connected symmetrical graph $G = (V, E)$. A multicast request is a pair $\epsilon = (e, R)$, where $e \in V$ is a source and $R \subset V$ is a set of destinations. We suppose that all requests which we deal with can be transmitted in the network as a set of unicasts [2]. Therefore, we do not have to make precise the amount of data to transfer. For a given ϵ we first determine its tree, $A_\epsilon = (V_{A_\epsilon}, E_{A_\epsilon})$. This tree is a subgraph of G rooted in e, whose leaves are in R and whose arcs are directed from the root towards the leaves. We note D_{A_ϵ} the branching nodes in $A_\epsilon, D_{A_\epsilon} \subseteq V_{A_\epsilon}, |D_{A_\epsilon}| \leq k$.

We determine the choice of BrNs to minimize the bandwidth consumption. We adopt as a metric of the bandwidth used by a multicast a total number of arcs which construct its tree taking into account the fact that an arc may transport the same data more than once. We start by determining the situations in which ϵ is satisfied by a set of paths in the multicast tree, $\mathcal{S}(D_{A_\epsilon})$:

- every node of R is the final extremity of exactly one path in $\mathcal{S}(D_{A_\epsilon})$,
- every node of D_{A_ϵ} is the final extremity of at most one path in $\mathcal{S}(D_{A_\epsilon})$,
- the origin of a path in $\mathcal{S}(D_{A_\epsilon})$ is either e or a node of D_{A_ϵ}; in the latter case, the node of D_{A_ϵ} is also the final extremity of a path in $\mathcal{S}(D_{A_\epsilon})$,
- any node of $a \in D_{A_\epsilon}$ is in a path $p \in \mathcal{S}(D_{A_\epsilon})$ only if it is the final extremity or the origin of p.

The metric load_{A_ϵ} is thus defined as a sum of lengths of all paths in $\mathcal{S}(D_{A_\epsilon})$.

Definition 1 *Diffusing Nodes in Multicast Tree Problem (DNMTP) Data*: a directed connected symmetrical graph $G = (V, E)$, a multicast $\epsilon = (e, R)$, a rooted tree corresponding to this multicast A_ϵ, and a natural number k. *Goal:* Find $D_{A_\epsilon}, |D_{A_\epsilon}| \leq k$ so that load_{A_ϵ} is minimal.

We now focus on a given multicast ϵ and we omit the subscript ϵ in the formulæ for their clarity. We study properties of the solution induced by a set D of branching nodes, $\mathcal{S}(D)$. For any u, $u \in V_A$ in A we define A^u as a sub-tree of A rooted in u. A set D^u is a set of BrNs in tree $A^u(D^u \subseteq D)$. A set R^u is a set of destinations in tree $A^u(R^u \subseteq R) \cdot a^u$ is the arc connecting A^u from the remainder of A. We propose:

Definition 2 Let D be a set of vertices in A and u be a vertex in A. The path number $pn(u)$ is a number of paths in a solution $\mathcal{S}(D)$ spanned on A which pass through u or which terminate in u. The window of the solution $\mathcal{S}(D)$ on arc a^u is an triplet (β, d, load) where β is its path number $pn(u), d = |D^u|$ and load represents the load of $\mathcal{S}(D)$ in A^u.

The proofs of the following lemmas et properties can be found in [7].

Lemma 1 *Let D be a set of vertices in A. Let u be a node having one child u_1 of A. The window on arc a^u is equal to*

$$\begin{cases} (1, d+1, load+1) & \text{if } u \in D \\ (b+1, d, load+b+1) & \text{if } u \notin D \text{ and } u \in R \\ (b, d, load+b) & \text{if } u \notin D \text{ and } u \notin R \end{cases} \tag{1}$$

where the window on arc a^{u_1} is $(b, d, load)$.

Lemma 2 *Let u be a node of A having ℓ children $u_1, , u_\ell$. The window on arc a^u is equal to*

$$\begin{cases} (1, & 1+\sum_i^\ell d_i, & \sum_i^\ell load_i + 1) & \textit{if } u \in D \\ (1+\sum_i^\ell b_i, & \sum_i^\ell d_i, & \sum_i^\ell load_i + b + 1) & \textit{if } u \notin D \quad \textit{and } u \in R \\ \left(\sum_i^\ell b_i, \right. & \sum_i^\ell d_i, & \left. \sum_i^\ell load_i + b\right) & \textit{if } u \notin D \quad \textit{and } u \notin R \end{cases} \tag{2}$$

where the window on arc a^{u_i} is $(b_i, d_i, load_i)$ *for any* $i, 1 \leq i \leq \ell$.

Definition 3 Let D and D' be two subsets of vertices in A. Let v be a vertex. $\mathcal{S}(D) \preceq_v \mathcal{S}(D')$ if and only if the following three conditions are simultaneously satisfied: (i) $b \leq b'$; (*ii*) $d \leq d'$; (*iii*) load $\leq$ load$'$ where the window of the solution $\mathcal{S}(D)$(resp. $\mathcal{S}(D')$) on arc a^v is (b, d, load) (resp. (b', d', load')).

Property 1 *Let* u *be a vertex in* A. *Let* D *and* D' *be two subsets of vertices in* A *such that* $\mathcal{S}(D) \preceq_u \mathcal{S}(D')$. *Then the solution induced by* D'' *where* $D'' = (D' \setminus D'^u) \cup D^u$ *satisfies:* $\mathcal{S}(D'') \preceq_v \mathcal{S}(D')$ *for all nodes* v *not in* A^u.

Definition 4 Let u be a vertex in A. Let D be a subset of vertices in A. D is sub-optimal for A^u if and only if for any D' which is a subset of vertices in A such that $d = d'$ and $b = b'$, we have load $\leq$ load$'$ where the window of the solution $\mathcal{S}(D)$(resp. $\mathcal{S}(D')$) on arc a^u is (b, d, load) (resp. (b', d', load')).

Property 2 *Let* u *be a vertex of* A *having* ℓ *children* $u_1, \ldots, u_\ell$. *Let* D *be a subset of vertices in* A. *If* D *is sub-optimal for node* u, *then* D *is also sub-optimal for node* u_i, *for any integer* $i, 1 \leq i \leq \ell$.

3 Algorithm, its Complexity and Optimality

Our algorithm is based on the dynamic approach. For any $u, u \in V_A$ in A we define A^u as a sub-tree of A rooted in u. We also define two parameters of u in A. The height $h(u)$ is a distance between u and e in A, $h_{\max} = \max_{u \in V} h(u)$. We remind that $\text{pn}(u)$ is a number of paths in $\mathcal{S}(D_A)$ spanned on A which pass through u or which terminate in u (if u is a BrN then $\text{pn}(u) = 1$).

The idea of our algorithm is to compute for any $u, u \in V(A)$, some sub-optimal sets D of BrNs for A^u where the window of $\mathcal{S}(D)$ on arc a^u is (b, d, load). One set D is constructed for any value $b, 1 \leq b \leq |R^u|$, any value $d, 0 \leq d \leq k$. As the reader might already guess, a sub-optimal set D for the root e gives a solution to our problem. We want therefore to find these sets starting from the leaves and ending up in the root of A. As u may be or may not be a BrN, we have to know how to compute the two sets for both the cases.

As u may not be equipped with the branching property, the minimal load of the sub-optimal set for it should be stored in the matrix $M(u)$ whose rows are indexed by $\text{pn}(u)$ and whose columns are indexed by the number of BrNs deployed in A^u. As u may become a BrN, the minimal load of the sub-optimal set can be stored in a line vector $L(u)$ because $\text{pn}(u) = 1$ in this case.

In a nutshell: $M_{i,j}(u) = \alpha \neq 0$ ($L_i(u) = \alpha \neq 0$, resp.) iff a sub-optimal set D exists in A^u having its window on arc a^u equal to (j, i, α) (resp. to $(1, i, \alpha)$).

The destinations u, which are leaves of A, have "unitary" matrix and vector attributed: $M_{1,0}(u) = 1$, $L_1(u) = 1$ and all other elements are zero.

Our algorithm attributes to each u its $M(u)$ and $L(u)$ starting from the leaves whose height is $H = h_{\max}$ and performing the bottom-up scheme with $H = H - 1$ until the root is reached. The attribution of $M(u)$ and $L(u)$ to u is realised by the procedure Mat_Vec_Filling. It takes u and its corresponding sub-tree as data. Intuitively, this is a modified breadth-first search in which one arbitrarily chosen successor, treated first, computes its matrix and vector (the First_Succ_Mat_Vec procedure) in a different way from its brothers (the Others_Succ_Mat_Vec procedure).

First_Succ_Mat_Vec operates on a node u and one of its successors v for which $M(v)$ and $L(v)$ are already known as Mat_Vec_Filling follows a bottom-up approach. The main idea is based on the observation that the weight of the multicast tree in $A^v \cup \{u\}$ is equal to the multicast weight in A^v incremented by the weight of reaching u which is itself equal to $\text{pn}(u)$. This procedure computes the formulæ deduced from Lemma 1 and Property 1 (see [7] for details).

Others_Succ_Mat_Vec operates on a node u and its successors w different from v which has already been examined in First_Succ_Mat_Vec. It leads from the fact that this time the weight of the multicast tree in $A^v \cup A^v \cup \{u\}$ is equal to the sum of the multicast weights in $A^v \cup \{u\}$ and in A^w with the BrNs deployed in both $A^v \cup \{u\}$ and A^w. This procedure computes the formulæ deduced from Lemma 2 and Property 2 (see [7] for details).

Theorem 1 *The optimal set of BrNs is obtained by the configuration associated to* $min_{i:1\leq i\leq k} L_i(e)$. *Its complexity is* $\mathcal{O}(k^2|R|^2|V_A|)$. (see [7] *for details*).

4 Numerical Results

Our algorithm determines the optimal localizations for k BrNs in a multicast tree which has already been created for a multicast $\epsilon = (e, R)$. We selected two heuristic methods of construction of trees. The first one establishes a shortest path (ShP) between e and each $r \in R$. The corresponding multicast tree A_ϵ^{ShP} is a union of these shortest paths. The second method, which is based on the solution of the Steiner tree problem (StT) proposed in [6], gives A_ϵ^{StT} tree. To generate a 200-node graph of we apply the Waxman model [9] of BRITE [3]. We estimate with the 5% precision at $\alpha = 0.05$ the average weight of tree.

To perceive the impact of BrNs on the tree weight we perform the computations with four nodes placed by our algorithm, and without them. In Fig. 1 we observe that the weight reduction obtained for ShP with the BrNs is significant. The improvement obtained by the introduction of BrNs into the trees built with StT (Fig. 2) is even more substantial. These two figures exhibit that ShP generates trees whose weight is less than those generated by StT.

We now fix the number of destinations to 20 and we estimate the weights of trees obtained with ShP and StT algorithms in function of the number of BrNs.

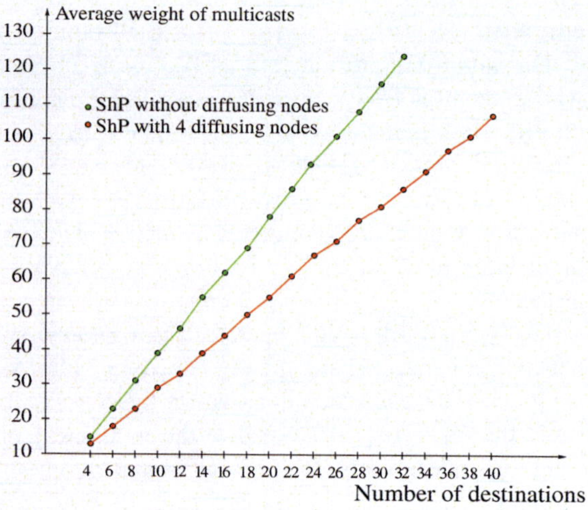

Fig. 1 Average A_{ϵ}^{ShP} weight as a function of the destination number

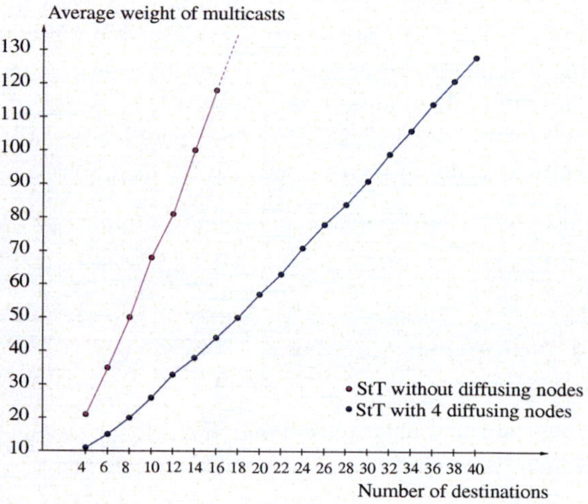

Fig. 2 Average A_{ϵ}^{StT} weight as a function of the destinations number

Figures 3 and 4 also show the average weight of ShP and StT trees estimated with the absence of BrNs. In accordance with the comment made above in the context of the absence of BrNs, ShP trees are almost twice as good as StT ones. The introduction of three BrNs reduces the weight of A_{ϵ}^{ShP} by about 20% (Fig. 3). The influence of the BrNs on the reduction of the tree weight in the StT case is striking (Fig. 4): an improvement of almost 60% until almost 75%.

5 Conclusions and Further Work

We studied an infrastructural design of an optical meshed network with a circuit switching routing. This problem was stated within the context of services based on

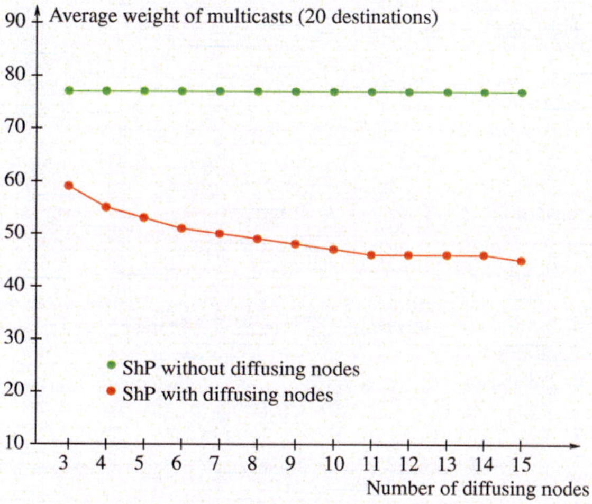

Fig. 3 Average $A_{\epsilon}^{\mathrm{ShP}}$ weight as a function of the number of BrNs

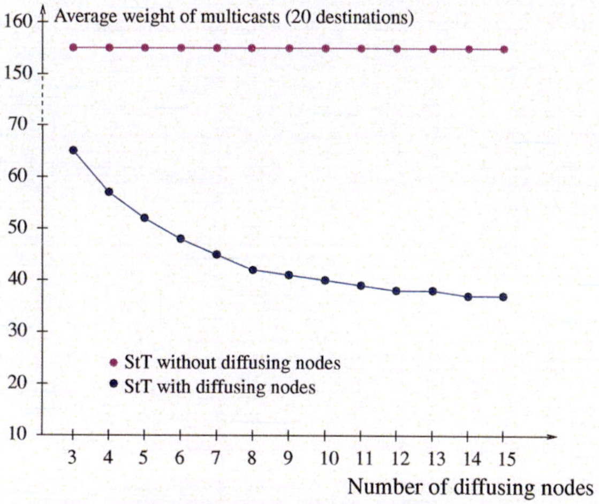

Fig. 4 Average $A_{\epsilon}^{\mathrm{StT}}$ weight as a function of the number of BrNs

multicast transmission. It concerns multicasts whose source and destinations are à priori known and its solution determines the localizations of BrNs.

After formally stating the problem we proposed an algorithm to solve it. We proved its optimality and found its complexity which is polynomial. We computed a gain in terms of the used bandwidth compared with multicast trees without branching nodes. Generally speaking, we found ShP more efficient in finding a multicast tree than StT. It is not excluded, however, that a more precise StT algorithm [5] may give better results.

We plan to continue this work to determine a specific solution in particular graphs (e.g. bounded treewidth). We also consider pursuing our work on optimal multicast deployment by studying the Steiner problem in certain oriented graphs.

References

1. Audouin, O.: CARRIOCAS description and how it will require changes in the network to support Grids. In: 20th Open Grid Forum (2007)
2. Malli, R., Zhang, X., Qiao, C.: Benefit of multicasting in all-optical networks. In: SPIE All Optical Networking, pp. 209–220 (1998)
3. Medina, A., Lakhina, A., Matta, I., Byers, J.: BRITE: an approach to universal topology generation. In: MASCOTS, Cincinnati, OH, USA (2001)
4. Reinhard, V., Tomasik, J., Barth, D., Weisser, M.-A.: Bandwidth optimisation for multicast transmissions in virtual circuits networks. In: IFIP Networking (2009)
5. Robins, G., Zelikovsky, A.: Improved Steiner tree approximation in graphs. In: ACM-SIAM (2000)
6. Takahashi, H., Matsuyama, A.: An approximate solution for the Steiner problem in graphs. Math. Jap. **24**(6), 573–577 (1980)
7. Tomasik, J., Reinhard, V., Cohen, J., Barth, D., Weisser, M.-A.: Performance improvement of an optical network providing services based on multicast. Technical Report 105.0438, CoRR (2011)
8. Verchère, D. et al.: Automatic network services aligned with grid application requirements in CARRIOCAS project. In: GridNets, pp. 196–205 (2008)
9. Waxman, B.M.: Routing of multipoint connections. J-SAC **6**(9), 1617–1622 (1988)

Part VI
Discovery Science

FLIP-CPM: A Parallel Community Detection Method

Enrico Gregori, Luciano Lenzini, Simone Mainardi and Chiara Orsini

Abstract Uncovering the underlying community structure of networks modelling real-world complex systems is essential way to gain insight both into their structure and their functional organization. Of all the definitions of community proposed by researchers, we focused on the *k*-clique community definition as we believe it best catches the characteristics of many real networks. Currently, extracting *k*-clique communities using the methods available in the literature requires a formidable amount of computational load and memory resources. In this paper we propose a new parallel method that has proved its capability in extracting *k*-clique communities efficiently and effectively from some real-world complex networks for which these communities had never been detected before. This innovative method is much less resource intensive than Clique Percolation Method and experimental results show it is always at least an order of magnitude faster. In addition, tests run on parallel architectures show a noticeable speedup factor, in some cases linear with the number of cores.

1 Introduction

The automatic discovery of network communities provides an insight into the mesoscale structure of real-world complex networks, which are far too large to be made sense of manually, even with the help of visualization techniques. There

E. Gregori (✉) · S. Mainardi · C. Orsini
Italian National Research Council, Institute of Informatics and Telematics, Pisa, Italy
e-mail: enrico.gregori@iit.cnr.it

S. Mainardi
e-mail: simone.mainardi@iet.unipi.it

C. Orsini
e-mail: chiara.orsini@iet.unipi.it

L. Lenzini · S. Mainardi · C. Orsini
Department of Information Engineering, University of Pisa, Pisa, Italy
e-mail: l.lenzini@iet.unipi.it

E. Gelenbe et al. (eds.), *Computer and Information Sciences II*,
DOI: 10.1007/978-1-4471-2155-8_31,

have been numerous definitions of community in the literature including k-clique communities [6], which are fine-grained and tightly connected. Unfortunately, their extraction requires a substantial amount of memory and computational load for large-scale complex networks. The first technique for extracting k-clique communities, called Clique Percolation Method (CPM), was proposed in 2005 by Palla et al. [6]. CPM can be broadly partitioned into the following three phases. The first consists in extracting all the maximal cliques of the input graph. Since the number of maximal cliques is exponential with the number of nodes in the graph, it is unlikely to find an algorithm with provably good execution times with reference to the number of nodes. An exhaustive review of these algorithms can be found for example in [1, Sect. 5]. The second phase of CPM consists in building up a clique-clique overlap matrix, which was proposed by Everett in 1998 [3] as a tool for describing and analysing the amount of overlap between cliques. In the third phase CPM extracts k-clique communities by carrying out a component analysis of binary matrices obtained from the clique-clique overlap matrix. Unfortunately CPM does not scale in terms of space required and computational time. The first step toward the enhancement of CPM was made in 2008 by Kumpula et al. [5] with the Sequential Clique Percolation method (SCP). SCP was designed for discovering communities with a given k. However, as also highlighted by the authors, since it generates and processes cliques rather than maximal cliques, it is extremely slow on graphs with medium-large cliques (e.g. with more than 10 nodes). To the best of our knowledge, no software tool has so far been able to extract k-clique communities from the global Internet topology at the Autonomous System[1] (AS) level of abstraction, which is the main topic of our research activities. This encouraged us to design and develop an innovative parallel method capable of alleviating CPM drawbacks (i.e. space requirements and execution time). In Sect. 2 we illustrate this method, which processes the clique-clique overlap matrix in chunks of configurable size and in parallel analyses the overlap by exploiting a multi-processor computing architecture, which nowadays is available almost everywhere (e.g. laptops and standard desktops). We present experimental results in Sect. 3 and discuss conclusions and future work in Sect. 4.

2 Fast Lightweight Parallel CPM

To improve CPM performance, we developed an optimized parallel version of the method, named Fast LIghtweight Parallel Clique Percolation Method (FLIP-CPM), which relies upon: (a) a new technique, called incremental G_k connected components; and (b) a *sliding window* mechanism which enables parallelization.

[1] A connected group of one or more IP prefixes run by one or more network operators that has a single and clearly defined routing policy.

While (a) is designed to strongly reduce memory requirements associated with the size of the clique-clique overlap matrix, (b) is designed to reduce the execution time. At first, FLIP-CPM extracts all the ϱ maximal cliques from the input graph G. Then, for each κ between 3 and κ_{max}, it creates an initial set consisting of n_k singletons, one for each maximal clique with size greater than or equal to κ (κ_{max} is the size of a maximum clique). At this point FLIP-CPM slides the window, starting from the beginning down to the end of the clique-clique matrix and for each new window, it performs parallel operations described in paragraph (b) hereafter.

(a) *Incremental G_k Connected Components*. Assuming that, for each k, the binary matrix is an adjacency matrix of a graph G_k, k-clique communities are equivalent to the connected components of G_k, which in turn correspond to the k-clique connected components of G. The graph G_k is made up of $\sum_{i=k}^{k_{max}} l_i = n_k$ nodes, where l_k is the number of maximal cliques with size k in G. However, our goal is not to have a complete topology of G_k. In fact, although such a topology can tell a lot more about the interactions between cliques, in order to extract k-clique communities, the connected components of G_k are sufficient. Maintaining these (runtime growing) connected components can be easily classified as *partially* dynamic problems on undirected graphs [2, Chap. 8]. This problem may be efficiently addressed with the use of the so called *set union* data structures [2, Chap. 5]. These data structures enable us to maintain a collection of disjoint sets under an intermixed sequence of *union* and *find* operations, starting from a collection of singleton sets *Union(h,i)* combines the two sets h and i into a new set. *Find(j)* returns the set containing element j. A detailed analysis of set union algorithms together with their worst-case execution time can be found in [7]. The technique we have designed, called *incremental G_k connected components*, is outlined below. For each of the $\binom{n_k}{2}$ possible combinations of maximal cliques, overlap is computed. Whenever at least $(k-1)$ nodes are found to be shared between a pair of maximal cliques, the corresponding nodes in G_k, let them be u and v, are checked via two find operations: $U \leftarrow$ *Find(u)* and $V \leftarrow$ *Find(v)*. If $U = V$, nodes are already connected and hence the overlap for another pair can be immediately processed. If $U \neq V$, nodes are in two separate components that have to be merged with a *Union(U,V)* before analysing the next pair of maximal cliques. With the technique described above, space complexity has a linear dependence, i.e. $O(l)$, on the number of maximal cliques, rather than quadratic as in CPM where the clique-clique overlap matrix is used.

(b) *Sliding Window and Parallelization*. The basic idea behind the new method we designed is to process the clique-clique overlap matrix (here in after referred to simply as matrix) in parallel, in *chunks* of configurable size, through a *sliding window* on the matrix itself. The sliding window enables multiple threads to process the matrix as if it were in memory in its entirely, while actually only a *chunk* physically resides in memory. Let W be the size, in bytes, of the window. If w bytes are used for each element in the matrix, then the maximum number of elements that can be placed in the window is $\eta = \lfloor W/w \rfloor$, constant and known a priori. Furthermore, if s is the index of the first row in the window, there is a way to also compute, a priori, the index e s.t. the maximum number of consecutive rows can fit in the window; i.e. the rows with indices i s.t. $i \in [s, e]$. In fact, the range $[s,e]$ varies according to the position of the sliding window on the matrix. Assuming that the window is slid across the upper triangular part, excluding the diagonal, of the $l \times l$ matrix; if the rows have indices between 0 and $(l-1)$, we can solve the following equation:

$$\sum_{j=s}^{j=x} (l-(j+1)) = \eta. \tag{1}$$

After some straightforward algebra, taking into account that $\sum_{i=1}^{i=r} i = r(r+1)2^{-1}$ for each positive integer r, we can rewrite the (1) as $-\frac{x^2}{2} + x\left(l - \frac{3}{2}\right) + (s-1)\left(\frac{s-2l+2}{2}\right) = \eta$, finding it has two solutions for x: $x_1 = \frac{2l-3}{2} - \frac{1}{2}\sqrt{\Delta}$ and $x_2 = \frac{2l-3}{2} + \frac{1}{2}\sqrt{\Delta}$, where $\Delta = (2s - 2l + 1)^2 - 8\eta$. It can be verified that (1) always has at least one real solution (i.e. $\Delta \geq 0$) if the following constraints hold: (a) $l \geq 2$ because the problem only makes sense only if the number of maximal cliques is greater than 1; (b) $\eta \geq l-1$ because the window must be sized to contain, at least, the largest row, i.e. the one with index 0; (c) $0 \leq s \leq l-1$ since the starting index must be one of the possible indices of the matrix. So, assuming that the first row in the window has index s, the index e can be computed by solving (1) and considering, in order, the following cases: (i) if $x_1 = l-2$ and $x_2 = l-1, e = x_2 = l-1$; (ii) else if $x_1 \geq 0, e = \lfloor x_1 \rfloor$; (iii) else, i.e. $x_1 < 0, e = l-1$ because all the rest of the matrix can be placed in the window. That said, it is easy to design an API that implements the following functions: (i) *Slide(s)* that, given s as input, computes and returns e; and (ii) *Read(i,j)* and *Write(i,j,value)* which provide read/write access to the elements with indices i,j s.t. $i \in [s, e]$ and $i < j \leq l-1$. Such elements can easily be located in memory by adding the offset $w\left(\sum_{h=s}^{h=i} [l-(h+1)] + j - i - 1\right)$ to the first address of the window.

Chunks are processed by two pools of threads. First, a pool $T_B = \{\tau_0, \ldots, \tau_{b-1}\}$ of b threads computes overlap values and writes them through the API discussed in the previous paragraph. Each thread is assigned a subset of rows in $[s,e]$ to process. That is, for each thread $\tau_t \in T_B$, the subset consists of all the rows i in $[s,e]$ s.t. $t = i (\mathrm{mod}\, b)$. Since the subsets are disjoint, there is no need for any mechanism to ensure that write operations are carried out in mutual exclusion. Furthermore, the previous round-robin like row assignment, although very simple, ensures that most likely multiple threads perform an almost equal number of operations if maximal cliques are ordered by decreasing (increasing) size. When each thread in T_B has finished processing its own rows, another pool $T_C = \{\tau_0, \ldots, \tau_{c-1}\}$ of $k_{max} - 2 = c$ threads reads every row with index in $[s, e]$ and keeps the connected components of c graphs $G_3, G_4, \ldots, G_{k_{max}}$ updated accordingly, using the incremental G_k connected components technique previously described. That is, each thread $\tau_t \in T_C$

is responsible for the incremental G_{t+3} connected components. Once again, mutual exclusion mechanisms have been avoided.

3 Experimental Results

In this section we show the experimental results obtained by running a C implementation of FLIP-CPM and CFinder [6], a free closed-source implementation of CPM. They have been run both on a standard personal computer, the iMac[2] and on a 4 CPUs (24 cores) server machine, the Dell.[3]

In Table 1 we summarize some characteristics of the input graphs. Specifically, *n*, *m* and *l* are the number of nodes, the number of edges and the number of maximal cliques respectively. $\mu = l^{-1} \sum_k k \cdot l_k$ is the average maximal clique size and $\sigma^2 = l^{-1} \sum_k l_k (k - \mu)^2$ the variance. Finally, ρ is the density of the clique-clique overlap matrix, i.e. the ratio between the number of elements with a value greater than zero and the total number of elements. Figure 1a shows the computation time experienced by running both FLIP-CPM and CPM on the iMac. Several small Internet Exchange Point (IXP) induced subgraphs were used as inputs. In fact they are the only IXP-induced subgraphs on which CFinder has enough memory to complete its execution. It was not possible to run it on the other graphs as the square of their number of maximal cliques is much larger than the amount of memory available on the iMac. The new method was always at least one order of magnitude faster.

Figure 1b shows the computation time experienced by running FLIP-CPM on the Dell, with an exponentially growing number of threads. Two large graphs were used as inputs. The dashed line in the figures represents the ideal case where doubling the number of threads implies halving the execution time (i.e. a speedup linear with the number of threads). With a number of threads less than or equal to eight, the new method achieves the best performance: in this range, the execution time decreases exponentially with the number of threads. When the number of threads becomes greater than or equal to sixteen, the speedup becomes less significant. FLIP-CPM seems to work well also on graphs much larger than the ones used in the previous paragraphs when running time analyses and comparisons. For example, an analysis of the *k*-clique communities of the Internet at the AS level, recently published in [4], has been carried out using FLIP-CPM. Moreover, we have been able to execute it on the Enron email network and the Condense Matter collaboration network in a few minutes.

---

[2] (2x) Intel E7600 CPU @ 3.06 GHz; (4x) 2 GB RAM modules @ 1067 MHz; Mac OS X 10.6.4 operating system, Darwin 10.4.0 kernel

[3] (4x) Intel E7540 CPU @ 2 GHz; (16x) 4 GB RAM modules @ 1067 MHz; GNU/Linux operating system; Linux 2.6.35.22 kernel

Table 1 Graphs properties

	n	m	l	μ	σ^2	ρ
Internet	35,390	152,233	2,747,484	22.64	20.54	0.976
AMS-IX	322	13,434	752,108	22.20	9.55	1
LINX	345	14,188	384,494	23.01	11.29	1
NL-IX	224	2,619	4,127	11.09	5.89	0.976
MSK-IX	293	4,225	1,593	12.36	9.42	0.994
SwissIX	116	1,110	669	8.62	2.46	0.993
MIX-IT	76	861	714	8.32	2.50	0.939
KleyReX	119	932	332	8.08	5.63	0.968

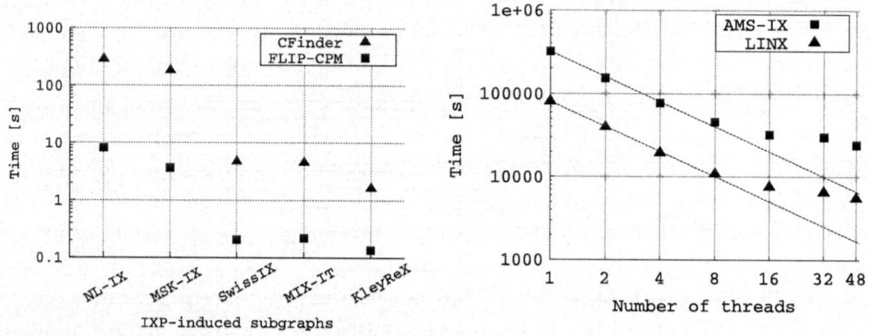

(a) (b)

Fig. 1 Computation times experienced by running FLIP-CPM and CPM on our iMac using some small graphs as inputs (**a**) and computation time versus number of threads on AMS-IX and LINX, respectively (**b**)

4 Conclusions and Future Work

In this paper we tackled the problem of extracting *k*-clique communities from real-world complex networks. We have described Fast LIghtweight Parallel Clique Percolation Method (FLIP-CPM), which greatly reduces the memory required for the extraction of communities, through the use of special data structures that have a *linear* (rather than quadratic) dependence on the number of maximal cliques. The new method also reduces the execution time by exploiting a parallel computing architecture. With far fewer stringent requirements in terms of memory, the new method is shown to be experimentally able to extract, for the first time, *k*-clique communities also from the Internet AS-level topology graph. FLIP-CPM is highly effective and, even when run on standard hardware architectures, it turns out to be at least one order of magnitude faster than CPM. The effectiveness is confirmed by the speedup, in a multi-processor environment, which in several cases, proves to be linear with the number of cores. The efficiency can be boosted by parallelizing the algorithm which keeps the incremental connected components updated at runtime. We plan to carry out this parallelization as part of our future work.

Furthermore the analysis of more and more dense and large complex systems, such as Wikipedia or Facebook networks, is very well under-way in order to test the capability of FLIP-CPM on networks of this scale.

References

1. Bomze, I.M., Budinich, M., Pardalos, P.M., Pelillo, M.: The maximum clique problem. In: Du, D.-Z., Pardalos, P.M. (eds.) Handbook of Combinatorial Optimization. Kluwer Academic Publishers, Boston (1999)
2. Eppstein, D., Galil, Z., Italiano, G.F.: Dynamic graph algorithms. In: Atallah, M.J. (ed.) Algorithms and Theory of Computation Handbook. Purdue University, CRC Press, West Lafayette (1998)
3. Everett, M.G., Borgatti, S.P.: Analyzing clique overlap. Connections **21**(1), 49–61 (1998)
4. Gregori, E., Lenzini, L., Orsini, C.: k-clique communities in the internet AS-level topology graph. In: SIMPLEX 2011 (2011)
5. Kumpula, J.M., Kivelä, M., Kaski, K., Saramäki, J.: Sequential algorithm for fast clique percolation. Phys. Rev. E **78**(2), 026109 (2008)
6. Palla, G., Derenyi, I., Farkas, I., Vicsek, T.: Uncovering the overlapping community structure of complex networks in nature and society. Nature **435**(7043), 814–818 (2005)
7. Tarjan, R.E., van Leeuwen, J.: Worst-case analysis of set union algorithms. J. ACM **31**, 245–281 (1984)

G-Network Modelling Based Abnormal Pathway Detection in Gene Regulatory Networks

Haseong Kim, Rengul Atalay and Erol Gelenbe

Abstract Gene expression centered gene regulatory networks studies can provide insight into the dynamics of pathway activities that depend on changes in their environmental conditions. Thus we propose a new pathway analysis approach to detect differentially behaving pathways in abnormal conditions based on G-network theory. Using this approach gene regulatory network model parameters are estimated from normal and abnormal samples using optimization techniques with corresponding constraints. We show that in a "p53 network" application, the proposed method effectively detects anomalous activated/inactivated pathways related with MDM2, ATM/ATR and RB1 genes, which could not be observed from previous analyses of gene regulatory network normal and abnormal behaviour.

1 Introduction

One of the fundamental problems of biology is to understand complex gene regulatory networks (GRNs), and various mathematical and statistical models have been introduced for inference from GRNs [1]. Based on such networks, over-represented biological processes or pathways of a group of genes are identified by mapping them onto the gene ontology (GO) terms or regulatory structures [2].

H. Kim · E. Gelenbe (✉)
Department of Electrical and Electronic Engineering, Imperial College, London, UK
e-mail: e.gelenbe@imperial.ac.uk

R. Atalay
Department of Molecular Biology and Genetics, Bilkent University, Ankara, Turkey

E. Gelenbe et al. (eds.), *Computer and Information Sciences II*,
DOI: 10.1007/978-1-4471-2155-8_32,

These pathway analyses provide the annotations and functional insight of the group of genes which are usually determined by conventional statistical tests such as the t-test. However, these differentially expressed gene (DEG) derived analyses are limited in detecting defective pathways since they only observe the amount of expression of a gene itself rather than considering the flows of expression signals that communicate with neighboring genes.

Here we aim to detect the abnormal pathways of GRNs by modelling them using G-Networks [3] which is a probabilistic model of a system with special agents such as positive and negative customers, signals and triggers. In contrast to normal queuing networks, the negative customers of G-Networks describe the inhibitory effects of GRNs [4, 5]. G-networks have a product form solution which enables us to handle the dynamics of complex GRNs without heavy computation times. The parameters of the modelled GRN are inferred from normal samples with the assumed transition probabilities of gene expression signals. Then the transition probabilities of abnormal conditioned samples are estimated by minimizing the difference between the observed and predicted steady-state probabilities with constraints. Finally, permutation tests are performed to determine the statistical significance of the estimated transition probabilities.

2 G-Networks for Gene Regulatory Networks

Following [4] consider the notion of a "packet" that contains the gene expression signals, and a network node that represents a gene consisting of a queue where its packets are stored and a server where the packets' fates are determined. Let λ_i^+ and λ_i^- be the positive and negative packet input rates to the ith node, respectively. μ_i is the packet firing rate (service rate) of the ith node. Furthermore we define $\mathbf{x} = \{x_1, ..., x_n\}$ a non-negative integer n-vector with $\mathbf{x}_i^+ = \{x_1, ..., x_i + 1, ..., x_n\}$, $\mathbf{x}_i^- = \{x_1, ..., x_i - 1, ..., x_n\}$, and $\mathbf{x}_{ij}^{+-} = \{x_1, ..., x_i + 1, x_j - 1, ..., x_n\}$. Let p_{ij}^+ and p_{ij}^- be the transition probabilities for packet motion from the ith node to the jth node as a positive and a negative packet, respectively. Note that a negative packet has the effect of disappearing after it destroys one packet of the target node, or it disappears also if it does not find a positive packet to destroy. Lastly, d_i denotes the probability that a packet leaves the system so that $\sum_{j=1}^{n}(p_{ij}^+ + p_{ij}^-) + d_i = 1$.

Consider now a random process $\mathbf{X}(t) = \{X_1(t), ..., X_n(t)\}$ where $X_i(t)$ is an integer-valued random variable representing the number of packets in the ith node at time $t \geq 0$. If $Pr(\mathbf{x},t)$ is the probability that $\mathbf{X}(t)$ takes the value $\mathbf{x}$ at time t, then the G-network equations are:

$$
\begin{aligned}
Pr(\mathbf{x}, t+\Delta t) = \sum_{i=1}^{n} \Bigg[& (\lambda_i^+ \Delta t + o(\Delta t)) Pr(\mathbf{x}_i^-, t) I(\mathbf{x}_i > 0) + (\lambda_i^- \Delta t + o(\Delta t)) Pr(\mathbf{x}_i^+, t) \\
& + \sum_{j=1}^{n} \Big\{ (p_{ij}^+ \mu_i \Delta t + o(\Delta)) Pr(\mathbf{x}_{ij}^{+-}, t) I(\mathbf{x}_j > 0) \\
& + (p_{ij}^- \mu_i \Delta t + o(\Delta)) Pr(\mathbf{x}_{ij}^{++}, t) + (p_{ij}^- \mu_i \Delta t + o(\Delta)) Pr(\mathbf{x}_i^+, t) I(\mathbf{x}_j = 0) \\
& + \sum_{l=1}^{n} \big((p_{ijl} \mu_i \Delta t + o(\Delta t)) Pr(\mathbf{x}_{ijl}^{++-}, t) + (p_{jil} \mu_j \Delta t + o(\Delta t)) Pr(\mathbf{x}_{ijl}^{++-}, t) \big) I(\mathbf{x}_l > 0) \Big\} \\
& + (d_i \mu_i \Delta t + o(\Delta t)) Pr(\mathbf{x}_i^+, t) + (1 - (\lambda_i^+ + \lambda_i^- + \mu_i) \Delta t + o(\Delta t)) Pr(\mathbf{x}, t) \Bigg]
\end{aligned}
\tag{1}
$$

where $I(C)$ is 1 if C is true and 0 otherwise, and $o(\Delta t) \rightarrow 0$ as $\Delta t \rightarrow 0$. The complete equilibrium solution of (1) was given in [4]. Let q_i be the steady-state probability that the ith gene is activated:

$$
q_i = \min\left[1, \frac{\lambda_i^+ + \Lambda_i^+}{\mu_i + \lambda_i^- + \Lambda_i^-}\right] \tag{2}
$$

with

$$
\Lambda_i^+ = \sum_{j=1}^{n} q_j \mu_j p_{ji}^+ + \sum_{j,l=1, l \neq j}^{n} q_j q_l \mu_j p_{jli} \text{ and } \Lambda_i^- = \sum_{j=1}^{n} q_j \mu_j p_{ji}^- + \sum_{j,l=1, l \neq j}^{n} q_l \mu_l p_{lij}
$$

then the steady-state probability that there are x_i packets of ith node in each of the n cells is:

$$
\lim_{t \to \infty} Pr(X_1 = x_1, ..., X_i = x_i, ..., X_n; t) = \Pi_{i=1}^{n} q_i^{x_i} (1 - q_i) \tag{3}
$$

3 Abnormal Edge Detection

The packets in the G-network represent latent objects containing the gene expression signal, and we assume that the number of packets is proportional to the mRNA expression levels which are actually observable data. We also assume that the mRNA levels are observations of the steady-state. Therefore the steady-state probability that there is at least one mRNA of ith gene is $q_i = \frac{a_i}{a_i+1}$ from (3) if we denote by a_i the average mRNA level (average queue length) of ith gene, also given by $a_i = q_i/(1 - q_i)$.

To determine the G-network parameters under normal conditions we use (2) where there are four sets of unknown parameters $p_{ji} = \{p_{ji}^+, p_{ji}^-, q_{jli}, q_{lij}\}$, λ_i^+, λ_i^-, and μ_i. We initially set $p_{ji} = 1/(n_j^{out} + 1)$ where n_j^{out} is the out-degree of gene j. We set the packet output rate μ_i based on the values of λ_i^+ and λ_i^- which are

$\lambda_i^+ = 0.0062\,\text{sec}^{-1}$ and $\lambda_i^- = 0.002\,\text{sec}^{-1}$ [6], with $\mu_i = c \cdot n_i^{out}$ where c is a scaling constant. From (2) we have $q_i = f_i(\lambda_i^+, \lambda_i^-, \mu_i | \mathbf{q}_j, p_{ji})$ where $\mathbf{q} = (q_1, .., q_n)$. Then c can be found by minimizing the following equation given the initial values of λ_i^+ and λ_i^-;

$$\tilde{c} = \arg\min_c \sum_i (q_i - f_i(c | \mathbf{q}, p_{ji}, \lambda_i^+, \lambda_i^-))^2 \tag{4}$$

Once each μ_i is determined, we can find the optimal positive input rate λ_i^+ which minimizes $(q_i - f_i(\lambda_i^+ | \mathbf{q}, p_{ji}, \tilde{\mu}_i, \lambda_i^-))^2$ for each gene with the initial value λ_i^- and a constraint $0 \leq \tilde{\lambda}_i^+ \leq \mu_i + \lambda_i^- + \Lambda_i^- - \Lambda_i^+$. Then we determine $\tilde{\lambda}_i^-$ which produces exactly the same values of q_i.

3.1 Transition Probabilities in Abnormal Conditions

In an abnormal condition, let q_i' be the steady-state probability that ith gene is activated and p_{ji}' be a packet transition probability from the ith gene to jth gene in the same condition. If there are k unknown p_{ji}' for ith gene, then we will denote them by a vector $\mathbf{p}_{ki}$. For the detection of the abnormally behaving pathways, $\mathbf{p}_{ki}$ needs to be estimated given the input ($\tilde{\lambda}_i^+$ and $\tilde{\lambda}_i^-$) and output ($\tilde{\mu}_i$) rates found in normal conditions. $\mathbf{p}_{ki}$ can be determined by minimizing the following sqaured error with two constraints, $0 \leq \tilde{p}_{ji}'$ and $\sum_i \tilde{p}_{ji}' \leq 1$;

$$\tilde{\mathbf{p}}_{ki}' = \arg\min_{\mathbf{p}_{ki}^{(h)}} (q_i' - f_i(\mathbf{p}_{ki}^{(h)} | \mathbf{q}_j', \tilde{\lambda}_i^+, \tilde{\lambda}_i^-, \tilde{\mu}_i))^2 \tag{5}$$

where $\mathbf{p}_k^{(h)}$ is the hth hypothesis in the constrained parameter space. Our algorithm searches for the optimal solution iteratively with different initial starting values to reduce the possibility of remaining in a local minimum.

3.2 Permutation Test for the Estimated Transition Probabilities

When the estimated $\tilde{p}_{ij}'$ differs from its initially assumed value p_{ij}, it is necessary to determine if the difference is statistically significant. The null hypothesis of this test will be $\tilde{p}_{ij}' = p_{ij}$. To proceed with the test the set of samples is shuffled at random and divided into normal and abnormal groups with the same sample size of the original group. Then the proposed method is applied in the same way as the original data. Let M be the number of permutations and $\tilde{p}_{ij}^{(m)}$ be the estimated transition probability of the mth permutation. Then we can compute the emperical p-value of the $\tilde{p}'_{ij}$ as follows,

$$p\text{−value of } \tilde{p}'_{ij} = \begin{cases} \frac{1}{M}\sum_{m=1}^{M} I(p'_{ij} \leq \tilde{p}_{ij}^{(m)}) & \text{if} \quad p'_{ij} > p_{ij} \\ \frac{1}{M}\sum_{m=1}^{M} I(p'_{ij} \geq \tilde{p}_{ij}^{(m)}) & \text{if} \quad p'_{ij} \leq p_{ij} \end{cases}$$

where $I(C)$ is the indicator function. Thus if the p-value is less than α_2 then the null hypothesis is rejected. In our study, $\alpha_2 = 0.1$.

4 The p53 Pathway

In order to evaluate our approach using experimental data, we selected the p53 pathway which is a well studied system in human cells whose most important feature is tumor suppression when DNA is damaged. The regulatory structure of the p53 pathway with 30 genes was constructed on the basis of the KEGG database, and we also downloaded two microarray mRNA expression datasets from GEO. The first dataset (GSE12941) consists of 10 non-tumor liver tissue and 10 hepatocellular carcinoma (HCC) samples. The second (GSE6222) is a dataset for the study of liver cancer progression in HCC. In this dataset, we use 2 normal and 10 HCC samples. Before applying the proposed method, the data was normalized and scaled with mean 3 so that the average number of mRNAs of a gene without its interactions in a single cell are assumed to be approximately 3, while the variance was scaled to 1. The gene input and output rates are assumed to be $0.0062\,\text{sec}^{-1}$ and $0.002\,\text{sec}^{-1}$ from [6] so that $0.0062/0.002 \approx 3$.

Figure 1 shows the average expression levels of genes in each dataset, and the corresponding p-values of t-tests which detect DEGs. The two datasets share nine significant DEGs with 0.05 significance level while GSE12941 has four more DEGs. This similarity can be confirmed by observing their expression patterns in Fig. 1.However interpreting the DEGs even when we know their regulatory structure is yet another challenge. Figure 2 shows the results from our proposed method. Despite the apparent lack of significance of p53 and MDM2 in the t-test, the *p*53-*MDM* feedback loop was clearly activated in cancer samples in both datasets. In [7], the *p*53-*MDM2* feedback loop appears to produce oscillatory expression patterns. Thus in terms of the system dynamics, the activation of two pathways between p53 and MDM2 in our result might be more appropriate than the activation of only one pathway from MDM2 to p53. One of the significant pathways in both datasets is *TP53-IGFBP3-IGF1* [8]. Also our method properly detects two pathways, *ATM-CHEK2-TP53* and *ATR-CHEK1-TP53* as expected from [9] in both datasets, which cannot be detected merely by observing the p-values of the DEG test.

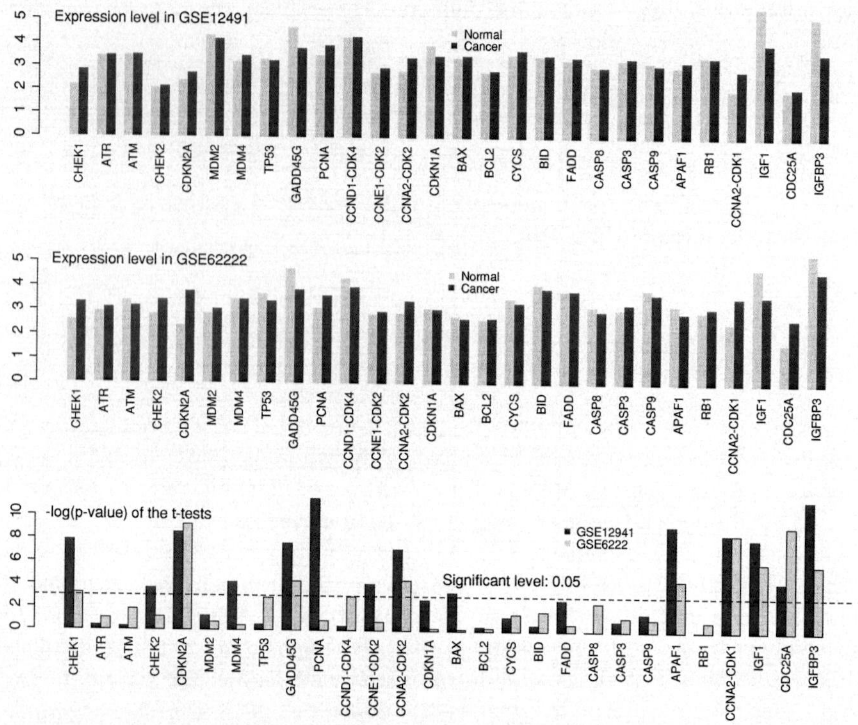

Fig. 1 Average mRNA expression levels of genes in two datasets, GSE12941 (*top*) and GSE6222 (*middle*). The *p*-values of the *t*-tests are shown in the bottom panel where the *y*-axis represents negative natural logarithms of the *p*-values. Note that multiple testing corrections are not applied in these i-tests

5 Discussion

We have proposed a new approach for detecting abnormal pathways in GRNs based on G-network modelling. This method provides an effective way to describe the flows of gene expression signals including negative or inhibitory effects on gene expression. Using some experimental data, we show that one advantage of our approach is that it can detect abnormal information flows in the dynamics of gene pathways. Thanks to existing G-network theory, the model uses a computationally tractable steady-state analysis and therefore does not require a large number of samples from time-dependent data. Moreover the analytic solution provided by G-network theory offers the possibility that our approach may be extended to very large-scale GRN systems.

In order to exploit this analytical tool, our work shows that a successful application of this method requires that the model be started with a reliable prior network structure based on real experimental data, or carefully calibrated GRN information. Though our intial experimental evaluation appears quite positive,

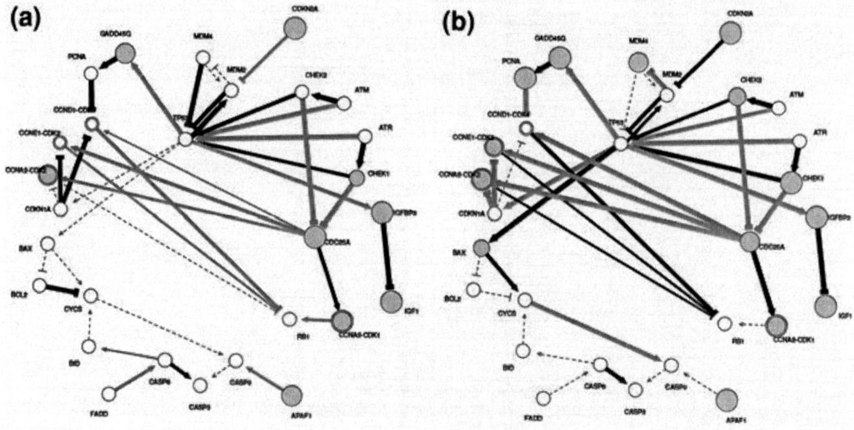

Fig. 2 The p53 network with the results of (**a**) GSE12941 and (**b**) GSE6222 dataset analysis. The solid line represents significantly activated (*black*) or inactivated (*grey*) pathways while the dashed line indicates non-significant pathways. Wider lines represent more significant pathways. The *grey nodes* are the selected DEGs from the *t*-test. The radius of a node is larger if its DEG is more significant with a 0.05 significant level. White nodes indicate non-significant genes

further experimental studies will be needed to validate the proposed approach and apply it to attain biological meaningful and clinically useful results.

Acknowledgments We would like to thank to Omer Abdelrahman and Zerrin Isik for helpful discussions.

References

1. Opgen-Rhein, R., Strimmer, K.: Learning causal networks from systems biology time course data: an effective model selection procedure for the vector autoregressive process. BMC Bioinformatics **8**(suppl 2), S3 (2007)
2. Beissbarth, T., Speed, T.: GOstat: find statistically overrepresented Gene Ontologies within a group of genes. Bioinformatics **20**(9), 1464–1465 (2004)
3. Gelenbe, E.: G-networks with triggered customer movement. J. Appl. Probab. **30**(3), 742–748 (1993)
4. Gelenbe, E.: Steady-state solution of probabilistic gene regulatory networks. J. Theor. Biol. Phys. Rev. E **76**, 031903 (2007)
5. Kim, H., Gelenbe, E.: Anomaly detection in gene expression via stochastic models of gene regulatory networks. BMC Genomics **10**(suppl 3), S26 (2009)
6. Thattai, M., van Oudenaarden, A.: Intrinsic noise in gene regulatory networks. In: Proceedings of the National Academy of Sciences **98**(15), 8614–8619 (2001)
7. Wilkinson, D.J.: Stochastic modelling for quantitative description of heterogeneous biological systems. Nature Rev. Genetics **10**(2), 122–133 (2009)
8. Schedlich, L., Graham, L.: Role of insulin-like growth factor binding protein-3 in breast cancer cell growth. Microsc. Res. Tech. **59**(1), 12–22 (2002)
9. Brown, C., Lain, S., Verma, C., Fersht, A., Lane, D.: Awakening guardian angels: drugging the p53 pathway. Nature Rev. Cancer **9**(12), 862–873 (2009)

DeDALO: A Framework for Distributed Systems Dependencies Discovery and Analysis

Emiliano Casalicchio, Antonello Paoletti and Salvatore Tucci

1 Introduction

The welfare of our daily life depends, even more, on the correct functioning of complex distributed applications. Moreover, new paradigms such as Service oriented computing and Cloud computing encourage the design of application realized coupling services running on different nodes of the same data center or distributed in a geographic fashion. Dependencies discovery and analysis (DDA) is core for the identification of critical and strategical assets an application depends on, and it is valid support to risk and impact analysis [1].

The goal of this research, framed in the context of the MOTIA[1] project, is to define methodologies and metrics to quantitatively and qualitatively evaluate service level dependencies in critical distributed systems.

In literature there is a pletora of network monitoring tools, working at layer 2 and 3, that offer discovery dependencies features and that allow to building a dependency map of the observed system (an updated list could be found here[2]). On the contrary few works concentrate their attention on application level DDA [3–9]. Often, DDA is used as a tool for distributed application management and

[1] This research is partially supported by the MOTIA project, http://www.motia.eu, funded by the European Commission-Directorate-General Justice, Freedom and Security: JLS/2009/CIPS/AG/C1-016.

E. Casalicchio (✉) · S. Tucci
Department of Computer Science, University of Roma "Tor Vergata", Rome, Italy
e-mail: emiliano.casalicchio@uniroma2.it

S. Tucci
e-mail: tucci@uniroma2.it

A. Paoletti
Lab Nazl Frascati, Ist Nazl Fis Nucl, 00044 Frascati, Italy
e-mail: antonello.paoletti@lnf.infn.it

E. Gelenbe et al. (eds.), *Computer and Information Sciences II*,
DOI: 10.1007/978-1-4471-2155-8_33, © Springer-Verlag London Limited 2012

typically gives a qualitative picture of system dependencies. At the best of our knowledge there are no examples of works oriented to application level dependency quantification that is, no indicators has been defined to quantify how much two services are dependent.

This paper briefly describe DeDALO, the *DEpendency Discovery and AnaLisys using Online traffic measurement* framework we have designed and implemented.

2 The DeDALO Framework

The architecture of the DeDALO framework is sketched in Fig. 1. DDA is realized through four main steps (implemented by related software modules): Traffic acquisition, Flow identification, Flow sequencing and Dependency analysis. In the following we give a brief description of each phase and therefore we concentrate our attention on Flow Identification and Dependency Analysis.

Traffic acquisition. In this phase DeDALO accesses a network interface card or a PCAP file to extract an IP packet and to convert it in a manageable data structure. The DeDALO observation system, inspired to existing works, has been implemented using an agent-based architecture, deploying several agents to observe and acquire the system activities. To implement the network traffic collection module we use the `libpcap` library.
Flow identification. As second step, DeDALO works on the extracted packet trying to match it with a suitable group of packets (flow) according to its header. Each flow is an instance of an access, performed by a client node to obtain a service from a server node.
Flow sequencing. For any starting flow, DeDALO matches it with all ended flows, originated by the same client node, saving the corresponding interarrival time. For any couple of flows, DeDALO keeps a frequency distribution which counts the occurrencies of a given time.
Dependancy analysis. In this phase DeDALO analizes all distributions to evaluate a possible dependency between the observed flows, applying an inference engine to all samples. The result of this analysis is a qualitative evaluation of dependencies (let say Yes/No/Maybe). The analysis phase is completed by the evaluation of a dependency metric we propose and that measures the intensity of a dependency. As result DeDALO builds a wheighted graph with the flows represented as vertex and the dependencies as weighted edges.

2.1 Flow Identification and Sequencing

We define *data flow* F_i a continuos sequence of information exchanged by two nodes. A service S_i consists of specific software components, running on a networked node, and providing its functionalities though the network.

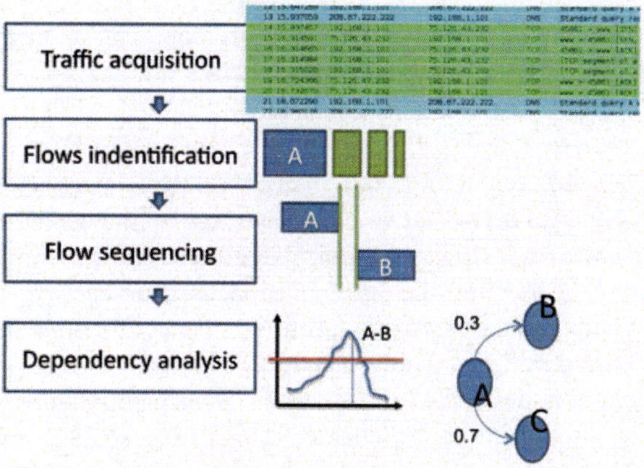

Fig. 1 Phases of the DeDALO framework

Each data flow considers a tuple composed by the two end-point nodes and the time interval in which they exchanged data. The end-points are called *origin* and *destination*, identifying the source of the request and the element offering the service. Two data flows, let say $F_{i,j}$ and $F_{i,k}$, originated by the same node i, towards two different destinations (j and k), may occur in a sequence interleaved by an interarrival time $\lambda_{i,j,k}$. It is possible to study the correlation between the two data flows by analyzing the distribution of $\lambda_{i,j,k}$.

The assumption made by DeDALO is that the relation of dependency between two services S_j and S_k is a condition related to a systematic and sequential use of both services inside a common application context (of course, this assumption impose some limitation we will discuss later on). If a sequence of data flow (representing service accesses) is repeatly spaced by similar interarrival time, than we assume that the two flows (and by consequence, the two services) are bound by a dependency relation. For example, if data exchanged through the first flow $F_{i,j}$, toward S_j, are vital to the origin end-point i to successfully access S_k, access identified by $F_{i,k}$. Therefore S_j and S_k are dependent, as well as the application/service S_i, S_j and S_k.

A common example is the domain name resolution process (very simplified here). The sequence of a DNS flow and a HTTP flow analyzed from the vantage point of a node running a web browser is likely to be found many times, because the node will need to know the HTTP server IP address, querying the DNS. If the DNS is unavailable the HTTP server will be not reachable and, even if it is properly working, the client will perceive a denial of service.

DeDALO's goal is to retrieve couples of IP services bound by a dependency relation. It carries on the analysis by fetching every IP packet exchanged between an origin and two or more destination services. The algorithm considers only packets headers and timings, discarding the payload.

DeDALO identifies a data flow by aggregating packets with same connection information (IP address, TCP ports, protocol) saving the timing of first and last packet of each flow, in order to identify the interarrival time. It's important to point out that all timing information are coherent, as they are referred to packets originated by the same node, and there's no need for synchronization.

As a new data flow $F_{i,j}$ is identified, DeDALO tries to correlate it with all flows $F_{i,*}$ previously originated by the same node. Each of these flows is coupled with $F_{i,j}$, by calculating the interarrival time between it and $F_{i,j}$, meaning that a sequence between two data flows has been detected with a given interarrival time. When two flows are correlated, we're also correlating the nodes which are responsible for the offered services. For each couple of flows (and nodes) is computed the interarrival time frequency distribution, counting the occurencies of timing samples. The distribution represents the frequency with which one or more client nodes access two services subsequently. Any distribution can have different shapes and can be characterized by a specific mean, standard deviation and sparseness.

2.2 Dependency Discovery Model

DeDALO uses all interarrival time frequency distributions to identify a dependency between the corresponding services. It looks for one or more interarrival times with a frequency higher than a given *dependancy threshold d*.

To build our discovery model we considering that only data flows $F_{i,j}$ and $F_{i,k}$ with an interarrival time $\lambda_{i,j,k} \leq 3$ s are considered to contribute to possible dependencies. The distribution of interarrival time is computed binning observed interarrival times with a granularity of 10^{-2} s. Therefore, being μ_f the average value for interarrival time frequency and σ_f its standard deviation, we compute the dependency threshold as $d = \mu_f + \alpha \cdot \sigma_f$, where $1.96 \leq \alpha \leq 3$, for example in [5] the authors use $\alpha = 3$.

If one or more interarrival time show a frequency higher than the dependency threshold, than such interarrival time value are evidence of a systematic behavior that is likely to represent a dependency.

3 Concluding Remarks

The main lessons learned from the design and implementation of DeDALO are the followings. First, there is a set of limitation to draw a map of service level dependencies in a network regardless its extension. Such limitation can be classified as structural factors (e.g. switched networks) and encoding factors (e.g. packet encryption). Second, a dependency model is strictly related to the observation points used, and therefore it is influenced by the policy used to deploy agents.

References

1. Macaulay, T.: Critical Infrastructure: Understanding its Component Parts, Vulnerabilities, Operating Risks, and Interdependencies. CRC Press, Boca Raton (2009)
2. Standfor Linear Accelerator Center, Network Monitoring Tools http://www.slac.stanford.edu/xorg/nmtf/nmtf-tools.html (2001)
3. AggreGate Network Manager. http://aggregate.tibbo.com (2011)
4. Bahl, P., Ranveer Albert, C., Greenberg, A.G., Kandula, S., Maltz, D.A., Zhang, M.: Towards highly reliable enterprise network services via inference of multi-level dependencies, ACM SIGCOMM, Kyoto, August (2007)
5. Chen, X., Ming Zhang, Z., Mao, M., Bahl, P.: Automating network application dependency discovery: experiences, limitations, and new solutions. Microsoft Research, University of Michigan (2008)
6. HP Network management center. https://h10078.www1.hp.com/cda/ (2011)
7. IBM Tivoli. http://www-01.ibm.com/software/tivoli (2011)
8. OpenNMS. http://http://www.opennms.org (2011)
9. ServiceNow. http://www.service-now.com (2011)

A Content-Based Social Network Study of Evliyâ Çelebi's *Seyahatnâme-Bitlis Section*

Ceyhun Karbeyaz, Ethem F. Can, Fazli Can and Mehmet Kalpakli

Abstract Evliyâ Çelebi, an Ottoman writer, scholar and world traveler, visited most of the territories and also some of the neighboring countries of the Ottoman Empire in the seventeenth century. He took notes about his trips and wrote a 10-volume book called *Seyahatnâme (Book of Travels)*. In this paper, we present two methods for constructing social networks by using textual data and apply it to *Seyahatnâme-Bitlis Section* from book IV. The first social network construction method is based on proximity of co-occurence of names. The second method is based on 2-pair associations obtained by association rule mining by using sliding text blocks as transactions. The social networks obtained by these two methods are validated using a Monte Carlo approach by comparing them with the social network created by a scholar-historian.

1 Introduction

Evliyâ Çelebi; a seventeenth century Ottoman writer, scholar, and world traveler (born on 1611, died circa 1682); visited most of the territories and also some of the neighboring countries in Africa, Asia and Europe of the Ottoman Empire over a

C. Karbeyaz · E. F. Can · F. Can (✉)
Department of Computer Engineering, Bilkent University, 06800 Ankara, Turkey
e-mail: canf@cs.bilkent.edu.tr

C. Karbeyaz
e-mail: karbeyaz@cs.bilkent.edu.tr

E. F. Can
e-mail: efcan@cs.bilkent.edu.tr

M. Kalpakli
Department of History, Bilkent University, 06800 Ankara, Turkey
e-mail: kalpakli@bilkent.edu.tr

E. Gelenbe et al. (eds.), *Computer and Information Sciences II*,
DOI: 10.1007/978-1-4471-2155-8_34,

period of 40 years. His work *Seyahatnâme (The Book of Travels)* is known by its distinguished style and detailed descriptions of people and places that he visited during his long journeys [1]. UNESCO has declared 2011 as the anniversary year of Evliyâ Çelebi on the four hundredth anniversary of his birth. This provides an additonal motivation for this study.

One of the methods we present in this study is based on association rules, which are the derived relations between the items of a dataset. Let I = $\{i_1, i_2, ..., i_n\}$ be a set of items with size n and T = $\{t_1, t_2, ..., t_m\}$ is set of transactions (in market data analysis a transaction involves the group of items purchased together) with size m. Then an association rule is shown as $X \Rightarrow Y$ where $X, Y \subseteq I$ and $X \cap Y = ø$. In a related work Raeder and Chawla [2] model a store's product space as a social network using association rules. Their work shows the use of social networks for market basket analysis.

The important aspects of this study can be summarized as follows. We present two different methods for constructing social networks from textual data and apply them to *Seyahatnâme-Bitlis Section* from book IV. For this purpose, we use the text in transcribed form [3]. We employ the social network created by a human expert as the ground truth, and assess the effectiveness of the methods by comparing the generated network structure with that of the ground truth. Finally, we use a Monte Carlo approach and show that the social network structures obtained by our methods are significantly different from random, i.e., are not by chance and hence valid.

2 Methods and Measuring Effectiveness

In this paper, we present two social network construction methods. These are the text proximity-based method (ProxiBM) and the association rule-based method (RuleBM). Both methods are based on co-occurrence of names in close proximity within a text block. For determining text blocks with a meaningful cohesive context we use two approaches. In the first blocking approach we use the paragraph information provided in the transcribed text. We manually identified and tagged 164 paragraphs. Each paragraph is used as a block. In the second method we employ a sliding window-based blocking approach (see Fig. 1).

In ProxiBM, edges for the undirected graph of social network are derived by creating a link between every character that appear in the same paragraph within a close word proximity (using a threshold). The proximity threshold between any two names is varied between 5 words and 500 words in steps. This approach is inspired by the use of term closeness as an indicator of document relevance [4].

In RuleBM we use the sliding text window for blocking and treat each block as a transaction where the existing names correspond to shopping items. We use the 2-pair association rules as relational edges of the social network by using the Apriori algorithm [5]. We employ different support threshold values and repeat

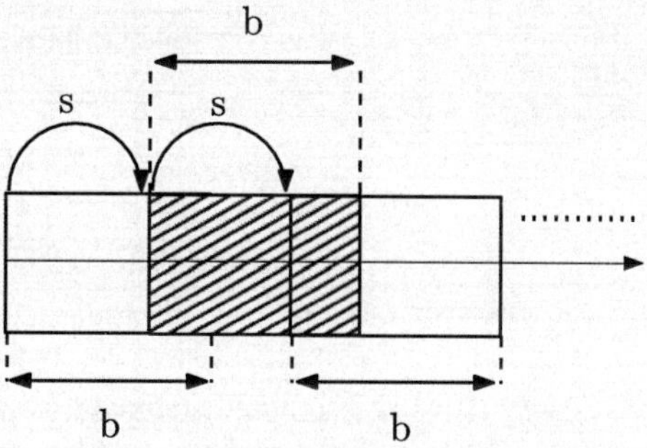

Fig. 1 Sliding window-based blocking:l total text length, b block size $(0 < b \leq l)$, s step size, nb number blocks, $nb = 1 + \lceil\frac{(l-b)}{s}\rceil$ for $0 < s \leq b$, $nb = 1 + \lfloor\frac{(l-b)}{s}\rfloor$ for $s > b$

Table 1 Performance results of ProxiBM over paragraphs for different proximity threshold (Θ) values in terms of no. of words

Measure	$\Theta = 5$	$\Theta = 10$	$\Theta = 25$	$\Theta = 50$	$\Theta = 100$	$\Theta = 250$	$\Theta = 500$
Precision	0.47	0.52	**0.54**	0.49	0.48	0.46	0.45
Recall	0.16	0.39	0.66	0.70	0.70	**0.71**	**0.71**
F-measure	0.24	0.44	**0.59**	0.58	0.57	0.56	0.56

the blocking operation for different block and step sizes in order to find the best performing parameters. The agreement between automatically constructed social networks and the manually constructed (actual) social network is measured by precision, recall, and the F-measure [6, pp. 142–144].

3 Experimental Results

In the experiments, both methods are tested in various conditions in order to find their best matching configuration to the ground truth. For ProxiBM we use various proximity threshold values and for RuleBM different blocksize, step size and support threhold values. Table 1 shows the precision, recall and F-measure results of ProxiBM for different proximity threshold values. The best configuration (the highest F-measure value) for this method is observed when proximity threshold is 25 words.

A similar experiment is done for RuleBM. Association rules correspond to frequencies of these 2-pair items (character names) appearing together in different transactions (blocks). They are derived from the blocks for different support thresholds ranging from 5 to 20%. The precision, recall and F-measure results of the RuleBM for the best configuration which is 500 words, stepsize: 300 words and support threshold: 5% can be seen in Table 2 among with other support threshold values.

Table 2 Performance results of RuleBM for blocksize: 500 words, stepsize: 300 words and for different support threshold (β) values

Blocksize	Stepsize	Measure	$\beta = 5\%$	$\beta = 10\%$	$\beta = 15\%$	$\beta = 20\%$
		Precision	0.24	**0.31**	0.20	*
500	300	Recall	**0.38**	0.07	0.01	*
		F-measure	**0.29**	0.11	0.01	*

* No association rules are found for that configuration.

Automatically generated social networks are further tested to understand if they are significantly different from random. For this purpose Monte Carlo experiments are performed [7]. In all Monte Carlo experiments, we generate a random version of the social network which is being evaluated 1000 times and measure the average F-measure values. In order to achieve this the Erdos-Renyi random network generation algorithm is used [8]. Monte Carlo results show that both methods with proper parameters generate networks which are significantly different from random.

4 Conclusion and Future Work

We present two methods ProxiBM and RuleBM for constructing social networks by using textual data and apply it to *Seyahatnâme-Bitlis Section* from book IV. The experimental results show that the networks created by ProxiBM show a higher similarity to the manually created social network than those of RuleBM. However, the disadvantage of ProxiBM is that it requires more focused (cohesive) blocks obtained from paragraphs. On the other hand, RuleBM is more flexible since it simply exploits blocks obtained from a sliding text window.

It is possible to obtain a better performance with RuleBM if we use contextually meaningful sliding text blocks: For the construction of such cohesive units we may use an automatic text segmentation method [9].

Acknowledgments This work is partially supported by the Scientific and Technical Research Council of Turkey (TÜBITAK) under the grant number 109E006. Any opinions, findings and conclusions or recommendations expressed in this article belong to the authors and do not necessarily reflect those of the sponsor.

References

1. Dankoff, R.: Evliyâ Çelebi in Bitlis: The Relevant Section of the Seyahatn âme. E.J. Brill, Netherlands (1990)
2. Raeder, T., Chawla, N.V.: Modeling a store's product space as a social network. In: Proceedings of the 2009 International Conference on Advances in Social Network Analysis and Mining, pp. 164–169, IEEE Computer Society, Washington DC, USA (2009)

3. Kahraman, S.A., Dağlı, Y.: Evliyâ Çelebi Seyahatnâmesi IV. Kitap. Yapı Kredi Yayınları, İstanbul (2003)
4. Hawking, D., Thistlewaite, P.: Relevance weighting using distance between term occurrences. Technical Report , The Australian National University, Canberra (1996)
5. Agrawal, R., Srikant, R.: Fast algorithms for mining association rules in large databases. In: Proceedings of the 20th International Conference on Very Large Data Bases. VLDB '94, pp. 487–499, Morgan Kaufmann Publishers Inc., San Francisco, CA, USA (1994)
6. Manning, C.D., Raghavan, P., Schütze H.: Introduction to Information Retrieval. Cambridge University Press, New York (2008)
7. Jain, A.K., Dubes, R.C.: Algorithms for Clustering Data. Prentice-Hall, Englewood Cliffs, NJ (1988)
8. Erdős P., Rényi, A.: On random graphs. I. Publ. Math. Debrecen 6:290–297 (1959)
9. Hearst, M.A.: Texttiling: segmenting text into multi-paragraph subtopic passages. Comput. Linguist. **23**(1), 33–64 (1997)

On the Parameterised Complexity of Learning Patterns

Frank Stephan, Ryo Yoshinaka and Thomas Zeugmann

Abstract Angluin (1980) showed that there is a consistent and conservative learner for the class of non-erasing pattern languages; however, most of these learners are *NP*-hard. In the current work, the complexity of consistent polynomial time learners for the class of non-erasing pattern languages is revisited, with the goal to close one gap left by Angluin, namely the question on what happens if the learner is not required to output each time a consistent pattern of maximum possible length. It is shown that consistent learners are non-uniformly *W*[1]-hard inside the fixed-parameter hierarchy of Downey and Fellows (1999), and that there is also a *W*[1]-complete such learner. Only when one requires that the learner is in addition both, conservative and class-preserving, then one can show that the learning task is *NP*-hard for certain alphabet-sizes.

Supported in part by NUS grant numbers R146-000-114-112 and R252-000-420-112.

F. Stephan
Department of Mathematics and Department of Computer Science,
National University of Singapore,
Singapore 117417, Republic of Singapore

R. Yoshinaka
ERATO Minato Project, Japan Science and Technology Agency Hokkaido University,
N-14, W-9, Sapporo 060-0814, Japan

T. Zeugmann (✉)
Division of Computer Science, Hokkaido University, N-14, W-9,
Sapporo 060-0814, Japan
e-mail: thomas@ist.hokudai.ac.jp

E. Gelenbe et al. (eds.), *Computer and Information Sciences II*,
DOI: 10.1007/978-1-4471-2155-8_35, © Springer-Verlag London Limited 2012

1 Introduction

Angluin [1] introduced pattern languages as an example for an interesting class that is learnable in the limit from text. A *pattern* π is a finite string over a finite alphabet Σ and a countably infinite set X of variables, where $\Sigma \cap X \neq \emptyset$. The *language* $L(\pi)$ generated by π is the set of strings which can be obtained by substituting each variable in the pattern by a non-empty string over Σ.

Allowing the empty string as a possible substitution was also studied, yielding the *erasing pattern languages* (cf. e.g., [10, 12]) which are *not learnable from text* [10]. So we follow Angluin [1] and allow only non-empty substitutions. Such pattern languages are called *non-erasing pattern languages.*

We consider here the model of learning in the limit from text (see Sect. 3). Our learners are required to be *consistent* and/or *conservative* (cf. e.g., [9, 15]).

In [1, 2] the class of all non-erasing pattern languages was shown to be learnable in the limit from text and an easy modification of her learning algorithm is consistent and conservative (cf. e.g., [15]) but not polynomial-time computable if $P \neq NP$ (cf. [1], Theorem 3.6). Lange and Wiehagen [8] sacrificed the consistency requirement and provided a polynomial time learner. Zeugmann [14] studied the learning properties of their algorithm from a statistical perspective. Various approaches to learn pattern languages followed [6, 11, 13].

Note that there is no pattern learner which outputs only consistent patterns of maximum possible length unless $P = NP$ (cf. [1]). So what happens if we drop the maximum possible length condition? It is shown that the following holds:

The existence of a polynomial-time consistent learner for the class of non-erasing pattern languages is equivalent to the condition that *CLIQUE* is non-uniformly fixed-parameter tractable, that is, $W[1] \subseteq FPT^n$;
If there is a polynomial-time consistent and conservative learner using a hypothesis space where $H_i = L(\pi)$ for an index i and a pattern π is decidable, then *CLIQUE* is uniformly fixed-parameter tractable, i.e., $W[1] \subseteq FPT$.

There is an oracle A relative to which $FPT = W[Poly]$ and $P \neq NP$. This result gives some evidence that one cannot show that the existence of a polynomial time consistent and class-preserving pattern language learner would imply $P = NP$.

In [1] the word problem of the pattern languages was shown to be in *W[poly]* which is the highest class of the *W*-hierarchy [5] but the exact level of that hierarchy was not located. Polynomial-time consistent and conservative class-comprising learning algorithms for classes where the membership-problem is uniformly polynomial-time computable and some mild other conditions hold were provided in [3, 4]. Below we use the algorithm of [8] instead of the work of [3, 4], but the algorithm of Theorem 4 can also be obtained by their methods.

2 The Complexity of the Pattern-Membership Problem

The *uniform* membership-problem $\{(\pi, w) \mid w \in L(\pi)\}$ is *NP*-complete (cf. [1]) but the membership-problem for a fixed pattern was not checked. So we ask what happens if the underlying numbering of all the pattern languages is *not* the default-numbering but given in a different way. Then certain information could be coded into the hypothesis space and the learner could gain power.

So it is adequate to look at the parameterised complexity (cf. [5]). A parameterised set A is in *FPT* iff there is a recursive function f and a polynomial p such that the question whether $(k, x) \in A$ can be decided in $p(f(k) + |x|)$ time.

A parameterised set S is *fixed-parameter many-one reducible* to a parameterised set T (abbr. $S \leq^n_m T$) iff there is a polynomial p such that for each parameter k there is a parameter k', a factor d_k and a reduction ψ_k such that ψ_k translates every $\langle x, k\rangle$ in time $d_k \cdot p(k + |x|)$ into $\langle x', k'\rangle$ with $\langle x, k\rangle \in S \Leftrightarrow \langle x', k'\rangle \in T$. Moreover, S is *strongly uniformly fixed-parameter many-one reducible* to T (abbr. $S \leq^s_m T$) iff there is a recursive function computing k', the factor d_k and a program for ψ_k from k (cf. [5]). We write $S \equiv^s_m T$ iff $S \leq^s_m T$ and $T \leq^s_m S$.

Let *PMO* denote the membership problem of pattern languages with the parameter k being the number of occurrences of variables in the pattern π, e.g., let π be such that one variable occurs twice and two other variables occur just one time each in π, then $k = 4$. For *CLIQUE* the parameter is the size of the clique requested. The complexity class $W[1]$ can be characterized as those sets which are strongly uniformly fixed-parameter many-one reducible to *CLIQUE*.

Theorem 1 *The problem PMO is W[1]-complete for strongly uniform fixed-parameter many-one reducibility. In particular,* $PMO \equiv^s_m CLIQUE$.

3 Learning Theory

First, we recall notions from learning theory and then apply Theorem 1 to it. The source of information are texts. A *text* t is an infinite sequence eventually containing all words of the target language L and possibly some pause symbols but no non-members. The learner is fed incrementally growing initial segments t_n of t and computes a hypothesis from its input, say i_n, if it has seen precisely the first n words of t. The hypotheses are interpreted with respect to a chosen *hypothesis space* $\{H_i \mid i \in I\}$. The sequence $(i_n)_{n\in\mathbb{N}}$ ($\mathbb{N}$ is the set of all natural numbers) of all created hypotheses has to *converge* to an $i \in I$ such that $H_i = L$.

A learner *learns* L from text if it learns L from all texts for it. A learner learns a class C with respect to a hypothesis space $\{H_i \mid i \in I\}$ from text iff it learns every language from C from text. This model is called *learning in the limit* from text (cf. [7]). Since we consider this model only, we refer to it just as *learning*.

Our learners are also required to be consistent and conservative (cf. [1, 2]). A learner M is *consistent* iff for every input t_n it outputs an i_n such that $range(t_n) \subseteq H_{i_n}$; if no such i_n exists, M outputs a special no-conjecture symbol. We call M *conservative* iff for every two subsequent hypotheses i_n based on t_n and i_{n+k} based on t_{n+k}, $i_n \neq i_{n+k}$, there is an $x \in range(t_{n+k})$ with $x \notin H_{i_n}$.

A hypothesis space $\{H_i \mid i \in I\}$ is *class-preserving* (with respect to the target class C) iff $\{H_i \mid i \in I\} = C$. We call $\{H_i \mid i \in I\}$ *class-comprising* iff $\{H_i \mid i \in I\} \supseteq C$. Every hypothesis space must be class-comprising; ideally it should be class-preserving, but this restricts learnability sometimes (cf. [15]).

We relate our results to Angluin's [1, 2] which are based on descriptive patterns. By *sample* we mean $range(t_n)$. A pattern π is *descriptive* of a sample S iff $S \subseteq L(\pi)$ and for every pattern τ with $S \subseteq L(\tau)$, we must have $L(\tau) \not\subset L(\pi)$. Every learner producing on every input a descriptive pattern learns the class of all non-erasing pattern languages (abbr. *PAT*) and there is an algorithm computing on input any sample S a pattern that is descriptive of S (cf. [1, 2]). If we require learnability with respect to a class-preserving hypothesis space this is the only way to obtain a consistent and conservative learner for *PAT*.

Theorem 2 *Let M be a consistent and conservative learner for PAT with respect to a class-preserving hypothesis space. Then M must output in every step a hypothesis that is descriptive for the content of the text seen so far.*

Theorem 3 *PAT has a consistent polynomial-time learner iff* $W[1] \subseteq FPT^n$.

The next result shows that for consistent and conservative learning with a minimum decidability requirement on the hypothesis space, polynomial-time learnability becomes linked to the more restrictive condition $W[1] \subseteq FPT$.

Theorem 4 $W[1] \subseteq FPT$ *iff there is a consistent and conservative polynomial-time learner for PAT which uses a hypothesis space* $\{H_i \mid i \in I\}$ *such that* $\{(\pi, i) \mid L(\pi) = H_i\}$ *is decidable.*

Next, we use an infinite alphabet Σ, e.g., Σ could be $\mathbb{N}$. Then it is *NP*-hard to make a consistent and conservative learner using a class-preserving hypothesis space. So the weakening from class-preserving to class-comprising hypotheses spaces is an important ingredient for the general polynomial-time learners obtained by [3, 4] for many uniformly polynomial-time decidable classes.

Theorem 5 *Let* $\Sigma = \{0, 1, 2, \ldots\}$. *Now* $P = NP$ *iff there is a consistent and conservative class-preserving polynomial-time learner for PAT over* Σ.

The same result holds if one works with finite alphabets and the learner has to learn the pattern plus the alphabet from the data. Then the learnability problem uniform over finite alphabets is *NP*-complete.

References

1. Angluin, D.: Finding patterns common to a set of strings. J. of Comput. Syst. Sci. **21**, 46–62 (1980)
2. Angluin, D.: Inductive inference of formal languages from positive data. Inform. Control **45**, 117–135 (1980)
3. Case, J., Kötzing, T.: Dynamically delayed postdictive completeness and consistency in learning. In: Freund, Y., Györfi, L., Turán, G., Zeugmann, T. (eds) ALT 2008. LNAI, vol. 5254, pp. 389–403. Springer, Berlin (2008)
4. Case, J., Kötzing, T.: Difficulties in forcing fairness of polynomial time inductive inference. In: Gavaldà, R., Lugosi, G., Zeugmann, T., Zilles, S. (eds) ALT 2009. LNAI, vol. 5809, pp. 263–277. Springer, Berlin (2009)
5. Downey, R.G., Fellows, M.R.: Parameterized complexity. Monographs in computer science. Springer, Berlin (1999)
6. Erlebach, T., Rossmanith, P., Stadtherr, H., Steger, A., Zeugmann, T.: Learning one-variable pattern languages very efficiently on average, in parallel, and by asking queries. Theoret. Comput. Sci. **261**, 119–156 (2001)
7. Gold, M.: Language identification in the limit. Inform. Control **10**, 447–474 (1967)
8. Lange, S., Wiehagen, R.: Polynomial-time inference of arbitrary pattern languages. New Genera. Comput. **8**, 361–370 (1991)
9. Lange, S., Zeugmann, T., Zilles, S.: Learning indexed families of recursive languages from positive data: a survey. Theoret. Comput. Sci. **397**, 194–232 (2008)
10. Reidenbach, D.: A non-learnable class of E-pattern languages. Theoret. Comput. Sci. **350**, 91–102 (2006)
11. Rossmanith, P., Zeugmann, T.: Stochastic finite learning of the pattern languages. Mach. Learn. **44**, 67–91 (2001)
12. Shinohara, T.: Polynomial time inference of extended regular pattern languages. In: RIMS Symposium on Software Science and Engineering, Proc. LNCS, **147**. pp. 115–127, Springer, Berlin (1983)
13. Shinohara, T., Arikawa, S.: Pattern inference. In: Algorithmic Learning for Knowledge-Based Systems, LNAI, **961**. pp. 259–291, Springer, Berlin (1995)
14. Zeugmann, T.: Lange and Wiehagen's pattern language learning algorithm: an average-case analysis with respect to its total learning time. Ann. Math. Artif. Intell. **23**, 117–145 (1998)
15. Zeugmann, T., Lange, S.: A guided tour across the boundaries of learning recursive languages. In: Algorithmic Learning for Knowledge-Based Systems, LNAI, **961**. pp. 190–258, Springer, Berlin (1995)

Using SVM to Avoid Humans: A Case of a Small Autonomous Mobile Robot in an Office

Emi Matsumoto, Michèle Sebag and Einoshin Suzuki

Abstract In this paper we construct a small, low-cost, autonomous mobile robot that avoids humans in an office. It applies an SVM classifier to each of 16 regions of interest in each picture taken by its camera mounted toward the ceiling. Experiments with 1 and 2 subjects with 3 kinds of velocities show encouraging results.

1 Introduction

Avoiding humans is a mandatory capability for an autonomous mobile robot in reality [1, 4–7]. In this paper we exploit the recent advancement of hardware and software and construct a small, low-cost, autonomous mobile robot using SVM [8] as its example.We stick to research with physical robots due to the gap between the reality and simulators [2]. Note also that such a robot is promising as a physical platform for discovery science (http://www.i.kyushu-u.ac.jp/~suzuki/expDS.html), a scientific discipline on any discovery process that is mainly approached by computer science.

E. Matsumoto · E. Suzuki (✉)
Department of Informatics, ISEE, Kyushu University, Fukuoka, Japan
e-mail: suzuki@inf.kyushu-u.ac.jp

E. Matsumoto
e-mail: duc.ubb@gmail.com

M. Sebag
TAO—CNRS & Univ., Paris-Sud, 91405 Orsay, France
e-mail: Michele.Sebag@lri.fr

E. Gelenbe et al. (eds.), *Computer and Information Sciences II*,
DOI: 10.1007/978-1-4471-2155-8_36, © Springer-Verlag London Limited 2012

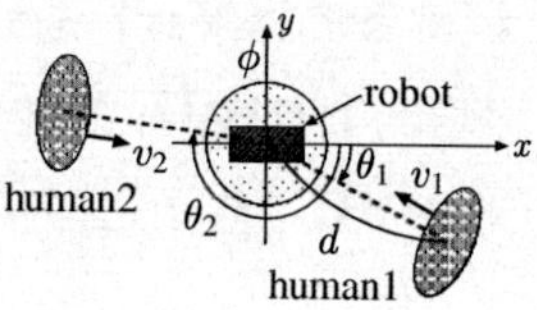

Fig. 1 Problem description

Figure 1 shows a description of the problem of avoiding two humans by a small autonomous robot, which is headed toward $x = \infty$. Humans 1, 2 represented by shaded ellipses approach the robot from directions θ_1, θ_2 with velocities v_1, v_2 and we record the shorter distance d between them when the robot starts moving. Note that avoiding a standing human by a moving robot is a kind of obstacle avoidance problem and is thus easier than this problem. We assume a circle of radius ϕ around the center of the robot, and judge a success if $d \geq \phi$ and a failure otherwise. Figure 2 shows an example of the images taken by the camera of our robot. Note the human around the upper corner on the right, the relatively low resolution, and the fluorescent lamp which obscures the detection.

2 Small Autonoumous Mobile Robot

The robot, shown in Fig. 3, is of width 22.5 cm, length 32.0 cm, height 22.5 cm, and costs about 113K JPY. It is equipped with two cameras, two micro processing units (MPUs), a driving system, three IR sensors, and two pairs of batteries. In our paper, we do not use the camera placed in front, which will be used for navigation and patrolling purposes in other papers. The camera we use is Sanwa Supply CMOS Web camera CMS-V25SETW with maximum resolution 640 × 480 pixels and frame rate 30 fps. It is equipped with a fisheye lens Toda Seiko K-180 to widen its view for an early detection. One of the MPUs, BeagleBoard, which has a processor ARM Corte-A8 600 MHz, memory units 128 MB LPDDR RAM and 256 MB NAND Flash, is used for processing images. It runs an operating system Ubuntu Linux 9.04 ARM version. The batteries are fully charged in 30 min, which allows the robot to function for 60 min.

For each processed image, BeagleBoard sends a command to Arduino with serial communication and Arduino sends control signals to the motors via the control tip. The command is either move forward, turn right, turn left, move backward, or stop. Note that the direction along which a human stands in a picture is always from the center (head) to the border (body), as shown in Fig. 2. We thus use 16 rectangles as shown in Fig. 4 as the regions of interest. An SVM classifier trained on a PC is applied to each of the region of interest and the directions of the escape are specified by the designer in advance.

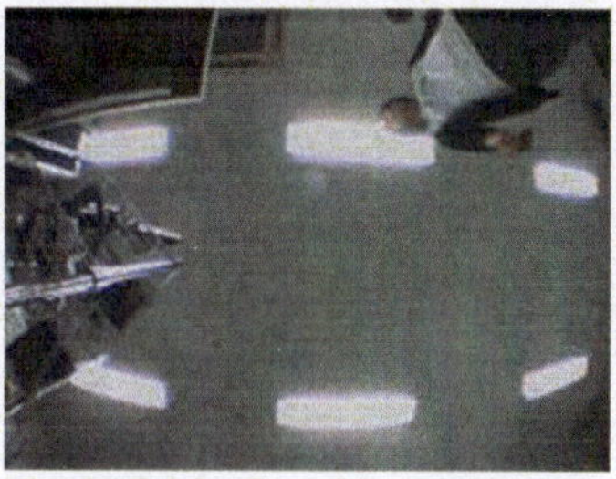

Fig. 2 Example of image

Fig. 3 Sideview of the robot

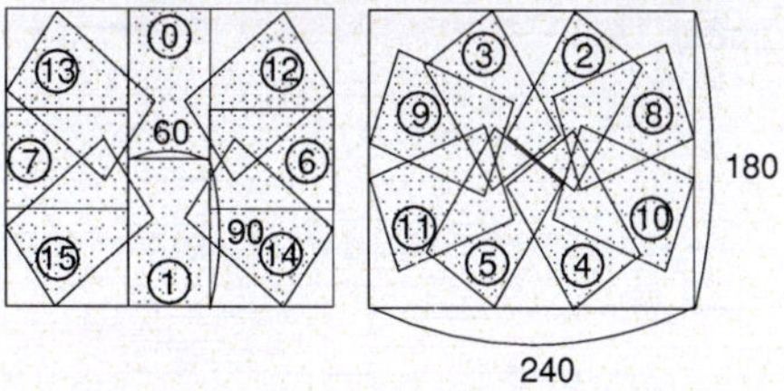

Fig. 4 Regions of interest for the SVM classifier, where the numbers represent the order of the applications

3 Experiments

3.1 Training SVM

We collected 60 images taken by the robot, each containing one person and is reduced to size 240×180. Each region of interest, of which size is 60×90 pixels, was transformed into a 16.2 K-dimensional vector of HSV values for the 5.4 K pixels. We labeled each region of interest to either +1 (with a person) or -1 (without a person) and obtained a training data set which consists of 165 and 795 images of class +1 and -1, respectively.

We used *SVM*^{light} [3] without kernel function because it performed best at the default settings. The accuracy, recall, precision, and F-value are 0.909, 0.529, 0.938, and 0.677, respectively. Note that the higher the recall is, the smaller is the risk of being crushed by a human. On the other hand, a high precision implies a small possibility of useless avoidance. It should be empathized that a perfect performance, i.e., recall = precision = 1, is desirable but not mandatory as the

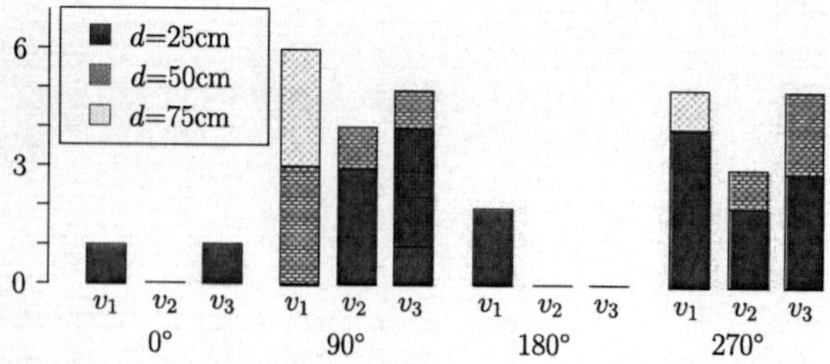

Fig. 5 Numbers of successes in avoiding 1 human

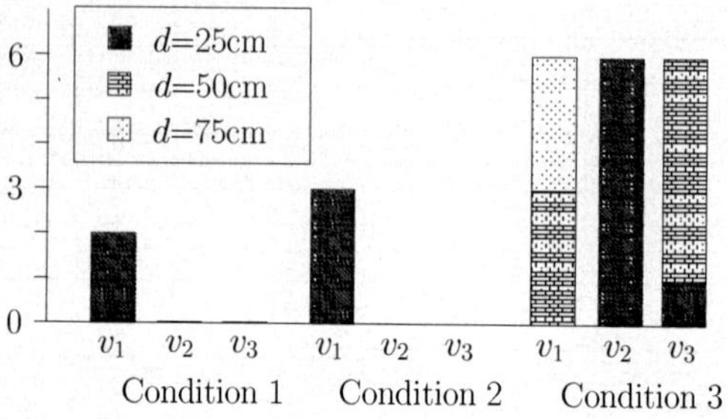

Fig. 6 Numbers of successes in avoiding 2 humans

robot has multiple chances to detect a human. The robot processes one image in about 1/s.

3.2 Experiments on Human Avoidance

We performed the experiments of human avoidance in an office with chairs, desks, white boards, fluorescent lamps, and windows during the daytime. We tested a case of one human (two kinds of subjects) with $\theta = 0, 45, 90, 135, 180, 225, 270, 315(^\circ)$ and another case of two humans (one kind of combination) with $(\theta_1, \theta_2) = (0, 180), (0, 90), (90, 270)(^\circ)$. For both cases, we adopted $\phi = 25$ cm and tested $v_i = 12.50, 18.75, 25.00$ (cm/s) and repeated the experiments 6 times for each condition. For the one-human case we report only the results of a 163 cm-height subject and exclude those with of a 183 cm-subject. The heights of the two subjects in the two-human case are about 160 cm.

A part of the results are shown in Figs. 5 and 6. For the one-human case, we see that $\theta = 90, 270$ show larger numbers of successes than $\theta = 0, 180$. We attribute the reason to the fact that it is easier for the robot to escape by moving forward, i.e., to the direction of $x = \infty$. It makes sense that the cases of v_1 give the best results but it seems contradictory that the cases of v_2 give the worst results. We are trying to explain the latter results by the timings of the image processing.

For the two-human case, conditions 1 and 2, which respectively correspond to $(\theta_1, \theta_2) = (0, 180), (0, 90)$, give bad results while condition 3, i.e., $(\theta_1, \theta_2) = (90, 270)$, give excellent results. They may be explained by the fact that the robot does not have to avoid a person coming from the front ($\theta_1 = 0$) in condition 3.

4 Conclusions

In this paper we have constructed a low-cost, small, autonomous mobile robot that avoids humans in an office. The obtained expertise in terms of the hardware, the software, and the experiments form a precious basis for further research for physical robots that learn and discover in an office. Now in Kyushu University we possess 15 robots of this type and 12 robots of a similar type with a higher computation power. We plan to use them for monitoring elder persons indoor without using a sensor network in the near future.

Acknowledgments A part of this research was supported by Strategic International Cooperative Program funded by Japan Science and Technology Agency (JST) on the Japanese side and Agence Nationale de Recherches (ANR) on the French side.

References

1. Burgard, W., Cremers, A., Fox, D., Hähnel, D., Lakemeyer, G., Schulz, D., Steiner, W., Thrun, S.: Experiences with an interactive museum tour-guide robot. Artif. Intell. **114**(1-2), 3–55 (1999)
2. Jakobi, N., Husbands, P., Harvey, I.: Noise and the reality gap: the use of simulation in evolutionary robotics. In: Proceedings of the ECAL 1995, pp. 704–720, (1995)
3. Joachims, T.: Making Large-Scale Support Vector Machine Learning Practical, pp. 169–184. MIT Press, Cambridge (1999)
4. Nomdedeu, L., Sales, J., Cervera, E., Alemany, J., Sebastia, R., Penders, J., Gazi, V.: An experiment on squad navigation of human and robots. In: Proceedings of the ICARCV 2008, pp. 1212–1218. IEEE (2008)
5. Okada, K., Kagami, S., Inaba, M., Inoue, H.: Walking Human Avoidance and Detection from a Mobile Robot Using 3D Depth Flow. In Proc. ICRA **2001**(3), 2307–2312 (2001)
6. Tamura, Y., Fukuzawa, T., Asama, H.: Smooth collision avoidance in human-robot coexisting environment. In Proc. IROS 2010, pp. 3887–3892 (2010)
7. Tsalatsanis, A., Valavanis, K., Yalcin, A.: Vision based target tracking and collision avoidance for mobile robots. J. Intell. Robotic Syst. **48**(2), 285–304 (2007)
8. Vapnik, V.N.: The Nature of Statistical Learning Theory. Springer, New York (1995)

Part VII
Computer Vision and Image Processing

Palmprint Verification Using SIFT Majority Voting

H. Pasindu Abeysundera and M. Taner Eskil

Abstract In this paper we illustrate the implementation of a robust, real-time biometric system for identity verification based on palmprint images. The palmprint images are preprocessed to align the major axes of hand shapes and to extract the palm region. We extract features using Scale Invariant Feature Transform (SIFT). Classification of individual SIFT features is done through KNN. The class of the hand image is decided by a majority based voting among its classified SIFT features. We demonstrate on the CASIA and PolyU datasets that the proposed system achieves authentication accuracy comparable to other state of the art algorithms.

Keywords Palmprint verification · Biometric identification · SIFT · KNN · Majority voting

1 Introduction

Reliable personal identification promise of biometrics and the growing demand for public security attracted researchers' significant interest to this field. Authentication by biometric verification has found a wide range of application areas from

H. P. Abeysundera · M. T. Eskil (✉)
Department of Computer Science and Engineering,
Pattern Recognition and Machine Intelligence Laboratory,
ISIK University, Istanbul, Turkey
e-mail: eskil@isikun.edu.tr
URL: http://pi.isikun.edu.tr

H. P. Abeysundera
e-mail: pasindu@isikun.edu.tr
URL: http://pi.isikun.edu.tr

E. Gelenbe et al. (eds.), *Computer and Information Sciences II*,
DOI: 10.1007/978-1-4471-2155-8_37,

commercial security applications to criminal investigation systems and sophisticated homeland security systems.

One important criterion of success for biometric verification systems is their ease of use. Among the biometric traits iris is consistently found to be one of the most reliable [1]. Nevertheless fingerprint recognition has gained more acceptance in a variety of application areas due to the convenience it offers to the end user. Likewise, human palm is both convenient to extract, and it embodies many distinctive features. Although this has been known for some time, the palmprint trait has not been studied as thoroughly as the other biometric features.

Our motivation in this study is to exploit the numerous advantages offered by palmprint as a biometric trait. First and most important advantage of the palmprint is that its features can be extracted without too much computational effort. Consequently palmprint acquisition does not require sophisticated equipment. It also has significant practical value due to its resistance to counterfeit.It is seldom that one leaves his/her complete palmprint somewhere unintentionally.

Palmprint contains a sophisticated and a distinctive pattern that inherits adequate traits to substantiate a person's identity. Among the four key hand based biometric traits; fingerprint, hand geometry, palm vein and palmprint, palmprint is alleged to have the potential to attain the authentication accuracy equivalent to fingerprints [2].

Principal lines and texture are the most prominent traits found in a palmprint. Consequently these traits are easily observable even in low resolution images. The curve delineated by the principal line is reported to be utilized in several studies in literature. Huang et al. [3] proposed a technique based on Radon transform to extract principal lines effectively and efficiently. Principal lines were matched by means of a pixel to area ratio. Huang achieved false positive and false negative rates as low as 0.493 and 0.499% in this study.

Zhang [4] employed wavelet transformation in palmprint verification. Their study resulted in a modified complex wavelet structural similarity index in computing the score of an input palmprint for each known subject. The proposed method is fast, invariant to small rotation and distortions, and achieves a genuine acceptance rate of 99.2%.

Chen and Moon [2] suggested combining SIFT with 2D SAX in the feature extraction phase. A weighting method is applied in authentication process.

Our study differs from Chen's in a couple of ways. First, we preprocess the hand images to align them, which enables us to use the (x, y) coordinates of the SIFT features in the verification stage. We also do morphological processing to remove the fingers from the image. We observed that the creases that are on the joints of fingers are the main source of the features that cause imposter matches. Third, we take advantage of majority voting in the feature fusion stage.

In this paper we propose a method that is invariant to rotation and scale and offers satisfactory genuine acceptance and very low false acceptance rates. We make use of spatial coordinates of SIFT features, their magnitudes and orientations in palmprint verification. Scale invariance is achieved inherently by use of SIFT features. Orientation invariance is accomplished by aligning each hand image through Principal Component Analysis. We implemented majority based decision fusion at the level of SIFT features. Also we applied 3-fold cross validation to improve the reliability of our results.

The rest of the paper is organized as follows. Section 2 will describe our proposed method and how we utilize SIFT in extracting palmprint traits. Section 3 presents the experimental results and we conclude our research in Sect. 4.

2 Proposed Method for Palm Verification

2.1 Image Acquisition

We acquired our palmprint images from two databases. Experiments were carried out on each database independently. CASIA Palmprint image database [5] contains 5502 (640 × 480, 96 dpi) palmprint images captured from 312 subjects. For each subject approximately 16 images exist with eight images from each palm. Images were captured for each subject in two sessions. The average interval between the initial and final session was one month.

PolyU palmprint image database [6–9] contains 8000 grayscale images (384 × 284 pixels, 96 dpi) from 400 different subjects. For each subject approximately 20 samples were taken. Similarly images were captured in two sessions, where the mean interval between the initial and final sessions was around two months.

2.2 Preprocessing

PolyU palmprint database contains images of the center region of the palm. Conversely CASIA palmprint database contains full hand images with diverse finger orientations. Our interest lies only on the center region of the palm, since it contains the majority of the discriminative features. The region of interest (ROI) is retrieved by applying morphological operators on palm images.

Preprocessing commences with converting the grayscale images to binary. Before extracting the ROIs all palmprints are aligned using PCA and the eigenvectors of the hand shapes. Next we extract the ROI by applying morphological opening. Figure 1 presents the results of the extraction process.

2.3 Extracting SIFT Features

SIFT was proposed to detect local image features invariant to scaling and translation. It also demonstrates partial invariance to rotation, illumination variations and background noise [10]. One of the most distinctive characteristics of SIFT is

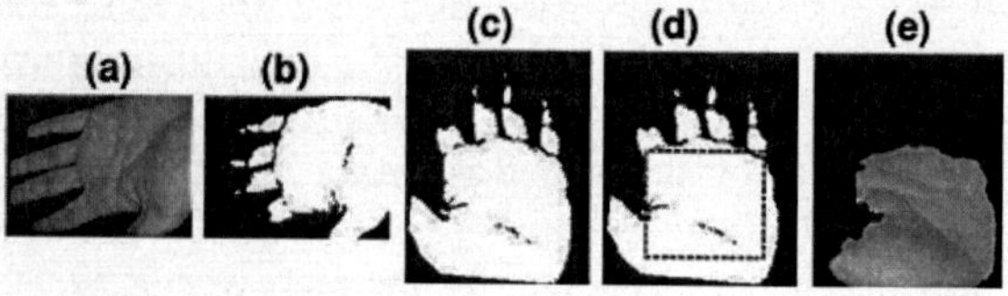

Fig. 1 Image alignment through PCA and ROI extraction. **a** Original image, **b** binary image, **c** aligned image after PCA, **d** region of interest, **e** extracted ROI

its low computational cost. A cascade filtering approach utilizes an initial test to eliminate more expensive computations. The process of extracting SIFT features consists of four key stages.

In the preliminary stage difference of Gaussian filtered images are utilized to discover the scale space extrema. In our study, we applied Gaussian filter five times and obtained four difference of Gaussian (DoG) images for each octave. For each octave, DoG images are cascaded together to obtain n by m by 4 matrices, where n and m are the dimensions of the original image in that octave. The local maxima and minima are detected in this matrix by comparing a cell with its neighbors in a $3 \times 3 \times 3$ cube. After the key points are detected the points that lie on an edge and low contrast points were eliminated.

In the next stage the locations of key points are approximated to subpixel accuracy. Once the key points are accurately localized their orientations are determined with the assistance of a gradient orientation histogram. Gradient magnitude m and orientation Θ are calculated using the following formulae.

$$m(x,y) = \sqrt{(L(x+1,y) - L(x-1,y))^2 + (L(x,y+1) - L(x,y-1))^2} \quad (1)$$

$$\theta(x,y) = atan(\frac{L(x,y+1) - L(x,y-1)}{L(x+1,y) - L(x-1,y)}) \quad (2)$$

Highest peak in the histogram is assigned as the orientation. Finally the image (key point) descriptor is generated. Typically it is a vector composed of 128 values. In the original SIFT architecture features were matched using Euclidean distances. We employed city block distance to enhance the computation speed.

Figure 2a displays some of the SIFT traits retrieved from the palmprints. The directions and magnitudes of vectors illustrate the orientations and magnitudes of SIFT features. Figure 2b shows the matched key points in the retrieved ROI using SIFT. It is observed that most key points are matched accurately.

2.4 KNN Classification and Palm Verification

We extracted 20918 and 73669 SIFT features from CASIA and PolyU datasets respectively. We classified each SIFT feature in the test set using KNN where the number of neighbors is selected as three. In case where all labels of the neighbors

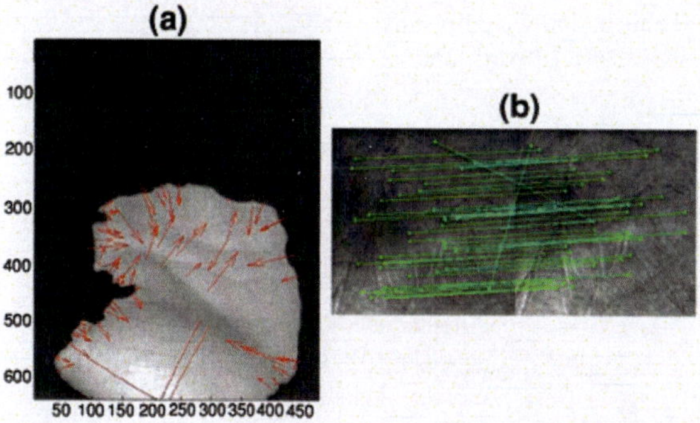

Fig. 2 Palmprint matching. **a** Sample SIFT features on the palm, **b** matching SIFT features in ROI

are dissimilar we employed a nearest neighbor tie-break. A test set observation is labeled according to the mode of the classes of its SIFT features. Finally we used a 3-fold cross validation to improve the reliability of our results.

3 Results

Results we obtained using the mode of classification of features on two data sets is summarized in Table 1 .

Figure 3 shows the Receiver Operating Characteristic (ROC) curve using both datasets. ROC curve is plotted with respect to varying number of features.

We achieved 87.23% true positive rate with around 12.77% false rejections using the mode of the labels of the features using PolyU dataset. Similarly for CASIA dataset 81.79% true positive rate is achieved with about 18.21% false rejections.

The results can be significantly improved by employing two palmprints in the verification process. In this scenario 93% true positive rate with 7% false rejection was achieved for the CASIA dataset. In the PolyU dataset we obtained 96% true accept rate with 4% false negative rate. These improvements in results are due to the fact that we use more features in the verification stage.

4 Conclusion

In this paper we presented a biometric verification method that utilizes palmprints. The proposed system is robust to translation, rotation and scale. Rotation invariance is achieved by aligning the hand shapes using PCA. Invariance to translation, scale and partial invariance to illumination is achieved by the use of SIFT features.

We classified the SIFT features using KNN classification with three neighbors and nearest neighbor tie-break. The final classification of the palmprint image is

Table 1 Individual results obtained for two databases

	CASIA	PolyU
Number of images	5502	8000
Genuine acceptance (%)	81.79	87.23
False acceptance (%)	1.43	0.75

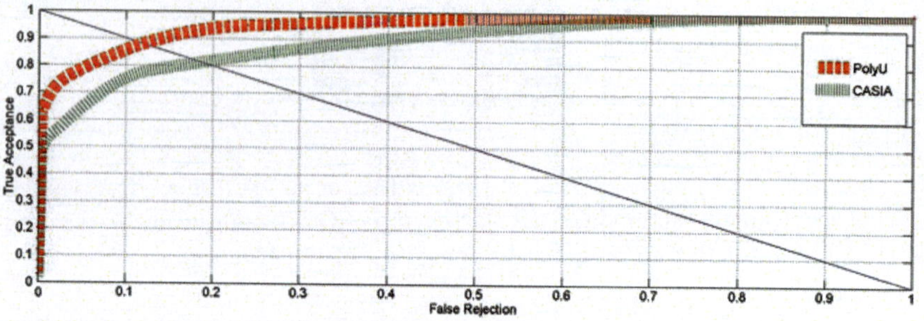

Fig. 3 ROC curve for PolyU dataset and CASIA dataset

done by treating each classified SIFT feature as a classifier and performing a mode-based voting scheme among these classifiers. Observed results reveal that our system can have up to 87.23% true acceptance rate with 0.75% false acceptance rate. In order to improve the reliability of our results we exercised three fold cross validation.

Testing performance, high precision, and invariance to scale, translation and orientation of the palmprint suggests the possibility of the proposed system being used in real time civilian applications and high end security environments.

We are currently concentrating on increasing our genuine acceptance rate. Our plans include improving the preprocessing step and exploring with different matching strategies.

Acknowledgments This research is part of project "Expression Recognition based on Facial Anatomy", grant number 109E061, supported by The Support Programme for Scientific and Technological Research Projects (1001) of The Scientific and Technological Research Council of Turkey (TÜBİTAK). Portions of the research in this paper use CASIA-Iris V3 database collected by the Chinese Academy of Sciences Institute of Automation (CASIA) and the PolyU database collected by the Department of Computing in Hong Kong Polytechnic University.

References

1. Ezhilarasan, M., Kumar, D.S., Santhanakrishnan, S., Dhanabalan, S., Vino, A.: Person identification using fingerprint by hybridizing core point and minutiae feature. (IJCSE) Int. J. Comput. Sci. Eng. **2**, 3075–3078 (2010)

2. Chen, J., Moon, Y.S.: Using sift features in palmprint authentication. In: ICPR'08, pp. 1–4 (2008)
3. Huang, D.S., Jia, W., Zhang, D.: Palmprint verification based on principal lines. Patt. Recognit. **41**(4), 1316–1328 (2008)
4. Zhang, L., Guo, Z., Wang, Z., Zhang, D.: Palmprint verification using complex wavelet transform. In: ICIP(2)'07, pp. 417–420 (2007)
5. CASIA Palmprint Database. http://biometrics.idealtest.org
6. PolyU 3D Palmprint Database. http://www.comp.polyu.edu.h
7. Zhang, D., Lu, G., Li, W., Zhang, L., Luo, N.: Palmprint recognition using 3-D information. Trans. Sys. Man Cyber Part C 39 (September), pp. 505–519 (2009)
8. Wei, L., Zhang, L., Zhang, D.: Three dimensional palmprint recognition. In: SMC'09, pp. 4847–4852 (2009)
9. Li, W., Zhang, L., Zhang, D., Lu, G., Yan, J.: Efficient joint 2D and 3D palmprint matching with alignment refinement. In CVPR'10, pp. 795–801 (2009)
10. Lowe, D.G.: Distinctive image features from scale-invariant keypoints. Int. J. Comput Vision **6**(3), 91–110 (2010)

3D Object Exploration Using Viewpoint and Mesh Saliency Entropies

Ekrem Serin, Candemir Doger and Selim Balcisoy

Abstract This paper introduces a technique to inspect a 3D object in a scene with minimal loss of information. We exploit the concept of the viewpoint entropy and introduce a novel view descriptor called mesh saliency entropy to explore the object by finding a minimal set of camera positions that covers the maximum information. Here we present a greedy choice algorithm which tries to detect a sub-optimal N-best views using the combination of mesh saliency entropy and viewpoint entropy to perceive the information communicated by the object. The main contribution is that the object can be examined with minimal loss of information for which can be used in scientific and medical analysis.

1 Introduction

3D object exploration has been actively studied in recent years [1–3], and has applications in many areas including medical analysis and training, virtual reality and scientific visualizations. The goal is to perceive as much as information available for recognizing objects, detecting regular or non-regular patterns, and executing the required tasks.

The scientific object exploration depends on information and a measure to quantify it. Here we borrow the concept of viewpoint entropy which is introduced

---

E. Serin (✉) · C. Doger · S. Balcisoy
Sabanci University, Istanbul, Turkey
e-mail: ekremserin@gmail.com

C. Doger (✉)
e-mail: cdoger@sabanciuniv.edu

S. Balcisoy (✉)
e-mail: balcisoy@sabanciuniv.edu

E. Gelenbe et al. (eds.), *Computer and Information Sciences II*,
DOI: 10.1007/978-1-4471-2155-8_38,

by Vazquez et.al. [4]. The viewpoint entropy is an information theoretical measure which is used to determine the amount of information from a viewpoint. Viewpoint entropy depends on the model presented by Shannon for a general communication system [5]. We model the scientific object exploration as a communication between user and object. The entropy lets us to quantify the amount of information from the points on viewing-sphere and from which we can select a set that receives the maximum in amount. Mesh saliency [6] is a measure of regional importance for graphics meshes.

In this work we provide an algorithm that uses viewpoint and mesh saliency entropies to select a set of viewpoints to have maximal information coverage presented by the object. Our contributions are, an efficient greedy choice algorithm that selects high coverage of faces, and introduction of a novel approach called mesh saliency entropy. We also discuss about the combination of viewpoint and mesh saliency entropy methods, and evaluation of the entropy as a metric for information coverage. A usability study is conducted to measure the strength of the techniques we provided.

The rest of the paper is organized as follows, in Sect. 2 we discuss about the related work, in Sect. 3 we present *Greedy N-Best View Selection* and *Mesh Saliency Entropy* algorithms and discuss the differences with the methods presented in [4, 7, 8], in Sect. 4 we deliberate on the statistical results, in Sect. 5 we present a usability study and its outcome, and Sect. 6 concludes our work.

2 Related Work

In recent years many methods have been developed for measuring the quality of the views and have tried to describe the optimum point to place a camera on a scene which can be viewed the best way. Unfortunately the translation of term best or good into measures or numbers is not an easy task. Kamada–Kawai [9] were one of the pioneers in defining a good position to place a camera in a 3D scene. They define a parallel projection of a scene to be good, if the number of surface normals orthogonal to the view direction is minimal. The method has several drawbacks, first it does not guarantee that user will see as much details as possible and will fail when comparing equal number of degenerated faces.

Vazquez et al. [4] propose a metric based on the entropy of the scene. They define the best viewpoint as the one with the highest entropy, i.e. the one that sees the maximum of information. They apply the ratio of the projected area of each face to the area covered by the projection of all faces in the scene. Vazquez et al. [4] suggest the technique in 2001 and make improvements in following years. The differences between our approach and techniques from Vazquez et al. [4] will be analyzed in detail in Sect. 3.

Mesh Saliency is also actively studied for viewpoint selection and mesh simplification. Salient features including luminance, pixel colors or geometry are deliberated. Lee et al. [6] propose a geometrical approach for calculation of mesh saliency in 3D models. Their method uses the curvature attribute of the object and Itti et al.'s center-surround mechanism to highlight the regions that are different from their surroundings. Takashi et al. [10] propose a method to locate optimal viewpoints for volumetric objects by decomposing the entire volume into a set of feature components. Bordoloi and Shen [11] use view goodness, view likelihood and view stability concepts to locate viewpoints for volume rendering where viewpoint goodness measure is based on entropy that uses the visibility of the voxels. Bulbul et al. [2] use the concept of saliency and apply it to the animated meshes with material properties.

3 Information Coverage

We selected viewpoint entropy to cover the polygons of the 3D object and introduced *Mesh Saliency Entropy* to have salient points along with the face coverage. Viewpoint entropy and mesh saliency entropy expose the surface area as information to the viewer, which is suitable for scientific visualization tools.

3.1 Viewpoint Entropy

Viewpoint entropy using Shannon Entropy [5] is defined in Vazquez et al. [4], and we use the orthogonal version presented in Vazquez et al. [8]. The techniques to compute the viewpoint entropy using Graphics Processing Unit can be found in Castello et al. [12]. Here we provide a comparative analysis of our approach to the work of Vazquez for solving the best view selection problem. For more details about Vazquez's work we refer the reader to [4, 7, 8]. Vazquez et.al. [4] predict the middle point entropy, add only the highest predicted entropy to the view set, and use spherical triangles for middle point calculations. In our approach we compute each entropy instead of estimating it, we use binary combination of points in view set for sampling, and we employ entropy-weighted midpoints. Differences provide us more viewpoint samples on viewing-sphere, that outputs a viewpoint with higher polygon coverage.

3.2 Greedy N-Best View Selection

Best View Selection algorithm is modified for N-Best View Selection to take the previously covered faces as input and to return the currently covered faces as output. The viewpoint entropy computation is also changed not to include the pixels from already visited faces. It works as continuously calling the *Best View Selection* with supplying the face coverage set in each call. The output of the algorithm is the faces that are covered along with a selected viewpoint and entropy value for that iteration. In each call of the *Best View Selection* the returned faces are added to face coverage set. The algorithm terminates when it can not return any newly covered faces or all faces of the model are covered. Algorithm starts from initial sample points and navigates around the object on each best view selection call to find the best viewpoint, whereas Vazquez et al. [4] perform entropy re-computation only for already computed viewpoints. Our method resembles to finding the best viewpoint of non-visited faces for each iteration and therefore can be called Greedy N-Best View Selection.

3.3 Viewpoint Mesh Saliency Entropy

In our approach we borrow the techniques [6, 13] to calculate curvature based mesh saliency. We use the Gaussian filtered mean curvatures of vertices proposed by Lee et al. [6] using Taubin's procedure to calculate mean curvatures [14]. Let $N(v,\sigma)$ be the set of points within a distance σ for vertex v therefore $N(v,\sigma) = \{x \,|\, ||x-v|| < \sigma, x \text{ is a meshpoint}\}$. Let $G(S(v),\sigma)$ denote the Gaussian-weighted average of the mean curvature.

$$G(S(v),\sigma) = \frac{\sum_{x \in N(v,2\sigma)} S(x) exp(-\frac{||x-v||^2}{2\sigma^2})}{\sum_{x \in N(v,2\sigma)} exp(-\frac{||x-v||^2}{2\sigma^2})} \tag{1}$$

The saliency for vertex v is the absolute difference between coarse and fine scales, where the standard deviation coarse scale is the twice of the fine scale. Then the saliency for vertex v for multiple scales is,

$$M_i(v) = |G(S(v),\sigma_i) - G(S(v),2\sigma_i)| \tag{2}$$

where σ_i is the standard deviation of the Gaussian filter at scale i. After calculation of curvature saliency for five different scales [6] we linearly added those feature maps after the normalization method proposed by Itti et al. [13] hence denote $M(v)$. *Viewpoint Mesh Saliency* entropy is defined as the total saliency entropy of the scene from a selected viewpoint. In each Greedy N-Best View Selection, mesh saliency entropy is calculated for each face and summed up. The saliency entropy for viewpoint p given surface S is defined as,

$$\mathbf{I}(\mathbf{S},\mathbf{p}) = -\sum_{i=1}^{N_f} \frac{\sum_{j=1}^{N_v} M_j}{\sum_{v \in S} M_j} \log_b \frac{\sum_{j=1}^{N_v} M_j}{\sum_{v \in S} M_j} \tag{3}$$

In Eq. 3, M_j stands for the saliency value for vertex v (i.e. normalized Gaussian Mean Curvature given in Eq. 2) which is a positive real number. N_f stands for the number of faces, and N_v for the number of vertices belonging to a face. $\sum_{v \in S} M_j$ denotes the total saliency of the given surface S.

Fig. 1 Hand and Bunny shown from five viewpoints using combined approach. Figures are ordered from the most contribution to the least.

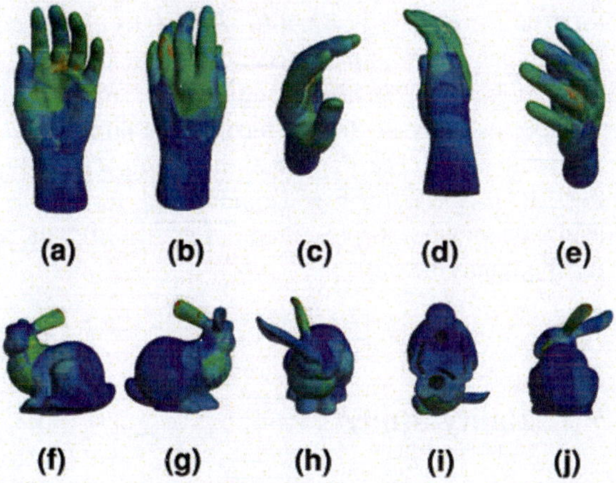

3.4 Combined Approach

Viewpoint Mesh Saliency Entropy is combined with Viewpoint Entropy in Greedy N-Best View Selection to cover both surface area and surface curvature information. Hence the entropy is defined as the information amount on the communication channel, the combined entropy for a given view point p on the surface S can be specified as the product of the two quantities therefore the aggregate information value is,

$$\mathbf{I}(\mathbf{S},\mathbf{p}) = I_f(S,p).I_s(S,p) \tag{4}$$

In Eq. 4, $I_f(S,p)$ denotes the face coverage information hence *Viewpoint Entropy*, and $I_s(S,p)$ denotes the saliency coverage information hence *Mesh Saliency Entropy*. Greedy N-Best View Selection can use either one of $I_f(S,p)$, or $I_s(S,p)$, or $I(S,p)$ quantities during the traversal on bounding sphere for optimal viewpoint search.

4 Results and Statistical Output

We have tried three different setups using Greedy N-Best View Selection. The first one takes only surface area information i.e. viewpoint entropy into account, the second takes the mesh saliency entropy and the final one uses the combination of both for view selection. We observed that the outcome of the combined approach was tend to maximize the face coverage as well as salient points. The snapshots from the output of combined approach are shown in Fig. 1. In Fig. 1a–e a hand model with 18905 faces is displayed using five selected viewpoints. These viewpoints cover 98.75% of faces of the model. The contribution of Fig. 1a to the total face coverage is 42.96%. Figure 1b covers 42.67% of the faces. Figure 1c contributes 6.92%, this ratio is 4.22% for Fig. 1d and 0.98% for Fig. 1e.

In Fig. 1f–j a Stanford Bunny with 16301 faces is presented. The displayed viewpoints cover 98.88% of faces of the bunny model. Figure 1f covers 49.56% of the total faces. The contribution of Fig. 1g is 42.68%, this ratio is 3.36% for Fig. 1h, 1.22% is for Fig. 1i and 0.06% for Fig. 1j . The saliency coverage ratio using the viewpoints presented in Fig. 1 is 99.72% for the hand model and 99.26% for the bunny model.

5 Usability Study

A simple usability study is conducted to measure the effectiveness of our technique and evaluate user tendencies for salient points interests. A group of 15 university students has participated to the study. The task to be completed by users was to place 20 points on the model where they were interested in most. The users were able to freely rotate/orient and zoom in/out the model shown to them. Hand model is displayed with gray color, and users were visually queued by the small red spheres on the surface where they double-clicked.

We conducted two analyses on the user selected points. The first analysis was to find the coverage ratio of the user selected points by the viewpoints provided by our algorithm. When the results were analyzed, we observed that the face coverage of the user selected points was 100% for all participants using viewpoints from our technique. The second analysis was to find the average of the saliency for the selected points. The saliency mean of points selected by users were higher than the surface mean.

Those analysis provided us preliminary feedback about the user interested points coverage by the viewpoints calculated by our algorithm, and a primitive answer for the question that users are mostly interested in salient points on the model presented to them.

6 Conclusion

In this paper we presented a technique to inspect a 3D object in a scene with minimal loss of information where the information is modeled as faces and mesh saliency. We combined the Viewpoint Entropy and *Mesh Saliency Entropy* presented above in *Greedy N-Best View Selection* algorithm to explore the object in 3D scene via minimal set of camera positions. Our experiments and user studies have shown that Shannon's entropy model is a promising way to solve viewpoint related problems by providing a measure to *quantify* the information on the communication channel between the user and visual world in computer.

Acknowledgement This research is supported by Turkish Scientific and Technological Research Council (TUBITAK) research grant 109E022.

References

1. Ji, G., Shen, H.W.: Dynamic view selection for time-varying volumes. IEEE Trans. Vis. Comput. Graph. 1109–1116 (2006)
2. Bulbul, A., Koca, C., Çapin, T.K., Güdükbay, U.: Saliency for animated meshes with material properties. ACM Proceedings of APGV, 81–88 (2010)
3. Mühler, K., Neugebauer, M., Tietjen, C., Preim, B.: Viewpoint selection for intervention planning. EuroVis by Eurographics, 267–274 (2007)
4. Vázquez, P.P., Feixas, M., Sbert, M., Heidrich, W.: Viewpoint selection using viewpoint entropy. VMV Proceedings, 273–280 (2001)
5. Shannon, C.E.: A mathematical thoery of communication. Bell Sys. Tech. J. **287**, 379–423 (1948)
6. Lee, C.H., Varshney, A., Jacobs, D.W.: Mesh saliency ACM. Trans. Graph. **24**, 659–666 (2005)
7. Vázquez, P.P., Sbert, M.: Fast adaptive selection of best views. ICCSA 2003-3, 295–305. Lecture Notes in Computer Science-2669 by (Springer 2003)
8. Vázquez, P.P., Feixas, M., Sbert, M., Llobet, A.: Realtime automatic selection of good molecular views. Comput. Graph. **30**, 98–110 (2006)
9. Kamada, T., Kawai, S.: A simple method for computing general position in displaying three-dimensional objects. Comput. Vis. Graph. Image Process. **41**, 43–56 (1988)
10. Takahashi, S., Fujishiro, I., Takeshima, Y., Nishita, T.: A Feature-driven approach to locating optimal viewpoints for volume visualization. IEEE Visualization, IEEE Computer Society (2005).
11. Bordoloi, U., Shen, H.W.: View selection for volume rendering. IEEE Visualization by IEEE Computer Society (2005)
12. Castelló, P., Sbert, M., Chover, M., Feixas, M.: Techniques for computing viewpoint entropy of a 3D scene. International conference on computational science. Lecture Notes in Computer Science-3992, 263–270 (2006)
13. Itti, I., Koch, C., Niebur, E.: A model of saliency-based visual attention for rapid scene analysis. IEEE Trans. Pattern Anal. Mach. Intell. **20**, 1254–1259 (1998)
14. Taubin, G.: Estimating the tensor of curvature of a surface from a polyhedral approximation. ICCV, 902–907 (1995)

Boundary Descriptors for Visual Speech Recognition

Deepika Gupta, Preety Singh, V. Laxmi and Manoj S. Gaur

Abstract Lip reading has attracted considerable research interest for improved performance of automatic speech recognition (Rabiner, L., Juang, B.: Fundamentals of speech recognition. Prentice Hall, New Jersey (1993)). The key issue in visual speech recognition is the representation of the information from speech articulators as a feature vector. In this paper, we define the lips using lip contour spatial coordinates as boundary descriptors. Traditionally, Principal Component Analysis (PCA), Discrete Cosine Transform (DCT) and Discrete Fourier Transform (DFT) techniques are applied on pixels from images of the mouth. In our paper, we apply PCA on spatial points for data reduction. DCT and DFT are applied directly on the boundary descriptors to transform these spatial coordinates into the frequency domain. The new spatial and frequency domain feature vectors are used to classify the spoken word. Accuracy of 53.4% is obtained in the spatial domain and 54.3% in the frequency domain which is comparable to results reported in literature.

1 Introduction

Automatic speech recognition (ASR) is a promising area of research in the field of Human Computer Interaction (HCI). By incorporating the visual modality with the

---

D. Gupta · P. Singh · V. Laxmi · M. S. Gaur (✉)
Department of Computer Engineering, Malaviya National Institute of Technology, Jaipur, India
e-mail: gaurms@gmail.com

D. Gupta
e-mail: deepika.guptaa19@gmail.com

P. Singh
e-mail: prtysingh@gmail.com

V. Laxmi
e-mail: vlgaur@gmail.com

E. Gelenbe et al. (eds.), *Computer and Information Sciences II*,
DOI: 10.1007/978-1-4471-2155-8_39,

audio signal, we can enhance the accuracy of speech recognition in an audio-challenged environment [10].

Visual features can be broadly classified as pixel based, shape based and appearance based features. To extract pixel based features from the ROI, images containing the mouth region are taken and transforms like Principal Component Analysis (PCA) [7], Discrete Cosine Transform (DCT) [9], DWT (Discrete wavelet transform) [2] and DFT (Discrete Fourier transform) [14] are applied directly on the gray level values of the pixel intensities. These transformed values are then used as a feature vector.

In this paper, we locate 120 points on the outer lip contour to determine shape based features. These points are represented by their *xy* coordinates and are called the *boundary descriptors*. The boundary descriptors are varying in each frame of the video sequence as the word is spoken. Thus, they can be considered as a time-varying signal which changes in every frame. While traditionally, DCT and DFT transformations are applied directly on pixel intensities in the ROI, we have applied DCT and DFT on these coordinates to transform them into the frequency domain. In the spatial domain, PCA is applied on the boundary descriptors. The high energy components available from each transformation are used to form veature vectors for speech identification.

The following of this paper is organized as follows: In Sect. 2, we give some of the related work reported by researchers in this field. Section 3 describes our visual feature extraction method. Our experimental setup is discussed in Sect. 4 followed by the results in Sect. 5 Finally, Sect. 6 concludes the paper.

2 Related Work

Much of the work related to feature extraction using DCT, DFT and DWT is done by applying these techniques directly on the images of the mouth of the speaker. Hong et al. [6] propose a PCA based method to reduce the dimensionality of DCT coefficients. In this work, block-based DCT features are extracted from 256*256 image sequences. Principal Component Analysis is then applied for dimensionality reduction.

Arsic et al. [1] proposes an application of information theoretic approach. Meaningful visual features are selected using mutual information from audio and visual sources. Principal components projections of the mouth region images are obtained and only those cues having the highest mutual information are used for classification. The experiments are carried out in speaker independent scenario on Tulip database consisting first four English digits uttered tow times each by 12 speakers.

Matthews et al. [7] have compared four different visual speech parameterization methods on IBM ViaVoiceTM audio-visual speech database. Direct mouth image region based transforms viz. discrete cosine, wavelet transforms and principal component analysis are applied. A statistical model of shape and appearance (an active appearance model [3]) is also defined to track and obtain model parameters. All parameterisations are compared using hidden Markov models.

Feng and Wang [4] proposes new visual dynamic texture features, based on Dual Tree Complex Wavelet Transform (DTCWT). Canberra distances between lip texture features in adjacent frames are used as visual dynamic features. Experiments are carried out on a Chinese AV database consisting of 78 isolated Chinese words uttered thrice by 20 subjects. 79.83% accuracy has been reported which is 8% more than that achieved by the common Canberra method.

Wang et al. [12] discuss a novel speech segmentation approach combining Otsus method with traditional short-time energy and zero-crossing rate (ZCR). DCT coefficients, extracted from the mouth and Mel frequency cepstral coefficients (MFCC) are used as the visual/audio features respectively. The experiment is carried out on a database consisting of 10 connected Chinese digits spoken by 10 subjects. Each subject utters each word three times.

3 Visual Feature Extraction

The performance of visual speech recognition largely depends on an efficient representation of features. In our experiments, the lips of the speaker are extracted from the images using the Point Distribution Model (PDM) [5]. A PDM is constructed from a training set of images from 20 different speakers. This PDM is allowed to deform to take the shape of the lip in an input image based on intensity analysis. The process of feature extraction is described below:

3.1 Boundary Descriptors

The boundary of the lips is described by the *xy* coordinates of 120 points lying on it. To obtain these points, twelve key points are detected on the boundary depicted by $P1, P2 \dots P12$ (refer Fig. 1). Twenty points are linearly interpolated between each pair of key points resulting in a total of 120 points. The feature vector giving the coordinates of these points is given by $V = [x_1 y_1 x_2 y_2 \dots x_{120} y_{120}]$, having a length of 240. These points describe the boundary of the lips in the spatial domain and are referred to as *boundary descriptors*.

3.2 Boundary Descriptor Transformation

To reduce the dimension of the boundary descriptors, Principal Component Analysis (PCA) is applied on it to yield the *boundary-PCA* vector. The variation of the boundary descriptors in each frame can be visualized as a time-varying signal.

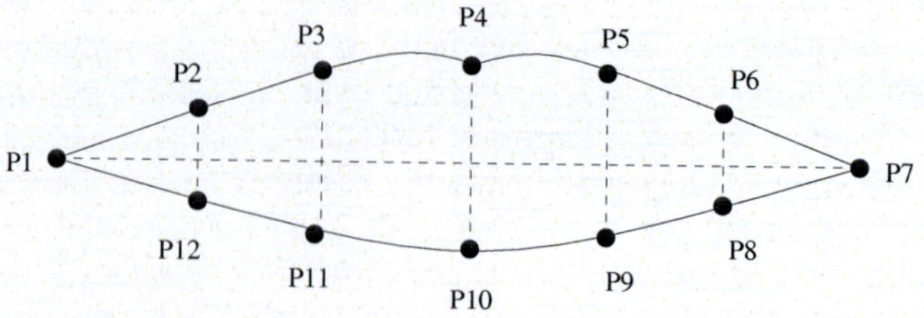

Fig. 1 Boundary descriptors

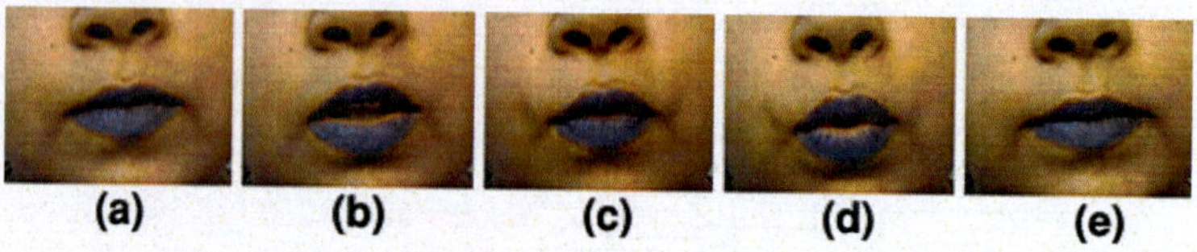

Fig. 2 Example frames from image sequence of utterance of *four*

These are transformed to the frequency domain using 2D-DCT and 2D-DFT. The high energy components are taken to form the *boundary-DCT* and *boundary-DFT* feature vectors. Thus, we obtain three feature vectors: one in spatial domain and two in frequency domain.

4 Experimental Setup

For experimental purposes, we recorded our own audio-visual database in office environment. Since lip segmentation is not the primary focus of our experiments, the lips of the subjects were painted blue for easy segmentation of lips from face. 12 subjects uttered the digits *zero* to *nine*, each digit being repeated five times by each subject. Figure 2 shows a few frames from a sequence of utterance of digit *four* by a speaker.

The images were converted from RGB to HSV image space. The lip contour was extracted using the PDM method described in Sect. 3 120 boundary descriptors were located on the lip contour. PCA was applied on these boundary descriptors to extract significant components. It was observed that seven principal components accounted for 98% variance. These PCA components were taken as the *boundary-PCA* feature vector. DCT and DFT were also applied to the boundary descriptors. 50 high energy components from each transformation were taken to obtain the *boundary-DCT* and *boundary-DFT* feature vectors. Considering speech recognition as a typical pattern recognition problem, we have classified our three feature vectors using multiple data mining algorithms in WEKA [13].

Table 1 TPR, FPR and Precision values for J–48, RF, MLP and SMO

Method	J–48			RF			MLP			SMO		
	TPR	FPR	P(%)	TPR	FPR	P(%)	TPR	FPR	P(%)	TPR	FPR	P(%)
Boundary-PCA	0.41	0.07	40.5	0.53	0.05	53.4	0.19	0.09	19.8	0.16	0.10	15.4
Boundary-DCT	0.47	0.06	43.6	0.54	0.05	54.3	0.22	0.09	22.2	0.18	0.09	17.7
Boundary-DFT	0.42	0.06	42.3	0.53	0.05	54.0	0.21	0.09	21.7	0.18	0.09	17.7

5 Result Analysis

Four classifiers in the WEKA toolkit have been used to classify the three visual feature vectors obtained by applying PCA, DCT and DFT. These are J–48, Random Forest (RF), Multilayer Perceptron (MLP) and Sequential Minimal Optimization (SMO). The *k*–fold cross-validation method has been used for testing with $k = 10$. The advantage of this method is that all observations are used for both training and testing. To evaluate the recognition accuracy of the spoken word, the metrics used are True Positive Rate (TPR), False Positive Rate (FPR) and Precision (P).

From Table 1, we observe that the best recognition accuracy of 54.3% is obtained using the Random Forest classifier. This corresponds to the feature vector using 50 high energy coefficients obtained by applying DCT on the 120 points located on the lip contour. If all the classifier accuracies are compared, DCT features show improvement over DFT and PCA features.

In [6], an accuracy of 68.8% has been reported on classification using features obtained by applying PCA on DCT coefficients. Visual only accuracy is reported to be 50.67% in a speaker independent scenario by Wang et al. [12]. However, these results are obtained on Chinese databases and need to be validated on an English vocabulary. The best accuracy reported in [1] is 89.6%. These results are on a vocabulary of four digits only. Best visual only word error rate (WER) of 58.1% is obtained for DCT transform based features in [7]. Matthews et al. [8] reports that an accuracy of about 40% has been achieved for visual only speech recognition for an English database. Our results are comparable to this.

It can be observed from Table 1 that there is only a marginal difference between the accuracies of DCT and DFT features. DCT features contain only the real values while DFT features are containing the real and imaginary values. Therefore, it can be argued that it is computationally more efficient to use DCT features for classification without compromising on the recognition accuracy.

Application of DCT has been normally done on pixels in the ROI. This involves a large computation and storage effort as all pixel values are transformed. In our experiments, only a few spatial coordinates are being transformed. This reduces computation overheads considerably without decrease in recognition accuracy.

6 Conclusion

This paper uses the concept of transformation of boundary descriptors from spatial domain to frequency domain. To the best of our knowledge, all previous work reported apply DCT and DFT on the ROI. In this paper, we have defined the lip contour using 120 *xy* coordinates. These boundary descriptors are subjected to transforms like Principal Component Analysis, Discrete Cosine Transform and Discrete Fourier Transform which are traditionally applied directly on mouth image pixels. With our approach, we see that the obtained visual speech recognition accuracy is comparable with reported results while the computation effort is reduced as only the lip contour coordinates are being used as the input data. The results motivate us to increase the size of the database and try to determine other visual features which can be an improvement on the achieved accuracy.

Acknowledgment The authors are grateful to the Department of Science & Technology, Government of India, for supporting and funding this project.

References

1. Arsic, I., Thiran, J.P.: Mutual information eigenlips for audio-visual speech recognition. 14th European Signal Processing Conference (EUSIPCO) (2006)
2. Cai, D., He, X., Zhou, K.: Locality sensitive discriminant analysis. International Joint Conference on Artificial Itelligence. pp. 708–713 (2007)
3. Cootes, T.F., Edwards, G.J., Taylor, C.J.: Active appearance models. IEEE Trans. Pattern Anal. Mach. Intell. **23**(6), 681–685 (2001)
4. Feng, X., Wang, W.: DTCWT-based dynamic texture features for visual speech recognition IEEE Asia Pacific Conference on Circuits and Systems (APCCAS 2008). pp. 497–500 (2008)
5. Gupta, D., Singh, P., Laxmi, V., Gaur, M.S.: Comparison of parametric visual features for speech recognition. Proceedings of the IEEE International Conference on Network Communication and Computer (ICNCC, 2011). pp. 432–435 (2011)
6. Hong, X., Yao, H., Wan, Y., Chen, R.: A PCA based visual DCT feature extraction method for lip-reading. International Conference on Intelligent Information Hiding and Multimedia Signal Processing, (IIH-MSP '06). pp. 321–326 (2006)
7. Matthews, I., Potamianos, G., Neti, C., Luettin, J.: A comparison of model and transform-based visual features for audio-visual LVCSR. Proceedings of IEEE International Conference on Multimedia and Expo (ICME 2001). pp. 825–828 (2001)
8. Matthews, I., Cootes, T.F., Bangham, J.A., Cox, S., Harvey, R.: Extraction of Visual Features for Lipreading. IEEE Trans. Pattern Anal. Mach. Intell. **24**(2), 198–213 (2002)
9. Nefian, A.V., Liang, L., Pi, X., Liu, X., Murphy, K.: Dynamic Bayesian networks for audio-visual speech recognition. EURASIP J. Appl. Signal Process. 1274–1288 (2002)
10. Potamianos, G., Neti, C., Huang, J., Connell, J.H., Chu, S., Libal, V., Marcheret, E., Haas, N., Jiang, J.: Towards practical deployment of audio-visual speech recognition. Proceedings of IEEE International Conference on Acoustics, Speech, and Signal Processing (ICASSP 2004) **3**(iii), 777–80 (2004)
11. Rabiner, L., Juang, B.: Fundamentals of speech recognition. Prentice Hall, New Jersey (1993)

12. Wang, X., Hao, Y., Fu, D., Yuan, C.: Audio-visual automatic speech recognition for connected digits. Proceedings of 2nd International Symposium on Intelligent Information Technology Application. pp. 328–332 (2008)
13. University of Waikato.: Open Source Machine Learning Software WEKA. http://www.cs.waikato.ac.nz/ml/weka/
14. Yau, W.C., Kumar, D.K., Arjunan, S.P., Kumar, S.: Visual speech recognition using image moments and multiresolution wavelet images. International Conference on Computer Graphics, Imaging and Visualisation. pp. 194–199 (2006)

Semi-Automatic Adaptation of High-Polygon Wireframe Face Models Through Inverse Perspective Projection

Kristin S. Benli, Didem Ağdoğan, Mete Özgüz and M. Taner Eskil

Abstract Precise registration of a generic 3D face model with a subject's face is a critical stage for model based analysis of facial expressions. In this study we propose a semi-automatic model fitting algorithm to fit a high-polygon wireframe model to a single image of a face. We manually mark important landmark points both on the wireframe model and the face image. We carry out an initial alignment by translating and scaling the wireframe model. We then translate the landmark vertices in the 3D wireframe model so that they coincide with inverse perspective projections of image landmark points. The vertices that are not manually labeled as landmark are translated with a weighted sum of vectorial displacement of k neighboring landmark vertices, inversely weighted by their 3D distances to the vertex under consideration. Our experiments indicate that we can fit a high-polygon model to the subject's face with modest computational complexity.

Keywords Facial expression analysis · Generic wireframe · High-polygon · Face model customization · Adaptation · Initialization · Landmark · Perspective projection

K. S. Benli · D. Ağdoğan · M. Özgüz · M. T. Eskil (✉)
Department of Computer Science and Engineering,
Pattern Recognition and Machine Intelligence Laboratory,
ISIK University, Istanbul, Turkey
e-mail: eskil@isikun.edu.tr
URL: http://pi.isikun.edu.tr

K. S. Benli
e-mail: kristin@isikun.edu.tr

D. Ağdoğan
e-mail: didem.agdogan@isik.edu.tr

M. Özgüz
e-mail: mete.ozguz@isikun.edu.tr

E. Gelenbe et al. (eds.), *Computer and Information Sciences II*,
DOI: 10.1007/978-1-4471-2155-8_40,

1 Introduction

Human face serves as an interface for one of the principal modes of interpersonal communication. Due to this fact facial expression research is very popular in many domains, ranging from psychology to computer science. In computer science, facial expressions are mostly studied in the context of human-computer interaction (HCI). In the most general sense HCI aims to analyze the emotional status of the human operator and use the obtained results in the decision processes of the machine for better adaptation to the user's intentions.

A typical model-based facial expression recognition system incorporates the following five stages; detecting the human face in the input, fitting a 3D model onto face region, tracking the rigid body motion of the head and the deformations of the face, extracting features and classifying the facial expression. As one of the earlier stages, registering the subjects face with a generic 3D face model is critical for the overall success of the system. Any imprecision introduced in this stage will accumulate as error in the later stages. Adaptation of the generic face model must also be flexible enough to accommodate interpersonal variations. This stage is referred to as adaptation, initialization or customization by researchers.

Essa [1] used Moghaddam's [2, 3] view-based and modular eigenspace methods for fitting 3D face model to a face in an image. The positions of the eyes, nose and lips are extracted automatically and the canonical face mesh is deformed and matched with the face image.

Active Shape Models (ASM) approach is proposed by [4]. These models are parametric deformable models. They are used to create models of human hearts, hands and faces. A statistical shape model of the face object is built using a set of training examples. Pose and shape parameters are iteratively modified for a better fit.

Active Appearance Model (AAM) was also proposed by [5]. AAM combines the statistical model of the shape and the gray-level appearance of the object of interest. The synthesized model is projected onto the face image and matching is done iteratively.

Krinidis and Pitas [6] used a semi-automatic approach for fitting the wireframe model to a face image. The face model is a 2D mesh whose elements are springs with stiffness. In the first step they coarsely initialized the wireframe on the face image, and then they manually matched model nodes with the corresponding positions of the face image. Using these correspondences driving force values that will cause the deformation of the wireframe are calculated.

There are still a limited number of studies in the literature that focus solely on the topic of wireframe adaptation, and those studies that address to this topic attempt to solve the fitting problem on low-polygon wireframe models. In this study we propose a semi-automatic algorithm for fitting a generic high-polygon 3D wireframe model to an input face image. Our algorithm produces visually accurate results that are free of collapsing polygons.

2 Semi-Automatic Wireframe Fitting

We propose a semi-automatic model fitting method for wireframe deformation. Our approach starts with manually marking important feature points of the face image and the wireframe model, which will be called landmark points and landmark vertices respectively. We define a landmark point as a point on 2D face image that represents an important feature on the face of a subject such as an eye corner, tip of the chin or the top of the forehead. In our study we marked 36 landmark points on the face image and their corresponding landmark vertices on the wireframe model. These traits are carefully selected in order to sketch the general shape of the subject's face.

Given the image landmark points, we estimate their locations in the 3D coordinate system. We call this process *inverse perspective projection* and detail it in Sect. 3 The original wireframe model is translated and scaled for an initial alignment with the inverse projected landmark points. In the last stage of fitting we calculate the new positions of the unmarked vertices, which ends the customization process.

3 Perspective and Inverse Perspective Projections

We aim to reshape the generic wireframe model based on the facial landmark points selected on the 2D image. In order to calculate translation vectors for the wireframe vertices we need to estimate the 3D coordinates of these facial landmark points. Perspective projection is a method for mapping a 3D object onto 2D camera plane (Fig. 1)

In Eqs. 1 and 2, z_{vp} stands for the z-coordinate of the view (camera) plane and $(x_{prp}, y_{prp}, z_{prp})$ stand for the projection reference point. The projection reference point is chosen on the z-axis to simplify the calculations of perspective projection $(x_{prp} = y_{prp} = 0)$. Then, Eqs. 1 and 2 can be rewritten as in Eq. 3.

$$x_p = x(\frac{z_{prp} - z_{vp}}{z_{prp} - z}) + x_{prp}(\frac{z_{vp} - z}{z_{prp} - z}) \tag{1}$$

$$y_p = y(\frac{z_{prp} - z_{vp}}{z_{prp} - z}) + y_{prp}(\frac{z_{vp} - z}{z_{prp} - z}) \tag{2}$$

$$f_p(x, y, z) = (x_p, y_p) = \begin{bmatrix} x(\frac{z_{prp}-z_{vp}}{z_{prp}-z}) \\ y(\frac{z_{prp}-z_{vp}}{z_{prp}-z}) \end{bmatrix} \tag{3}$$

When the user selects a landmark point on the face, we know the depth of the corresponding vertex in the wireframe. Assuming that the depth of the vertex is fixed, we can take the inverse of perspective projection using Eq. 4.

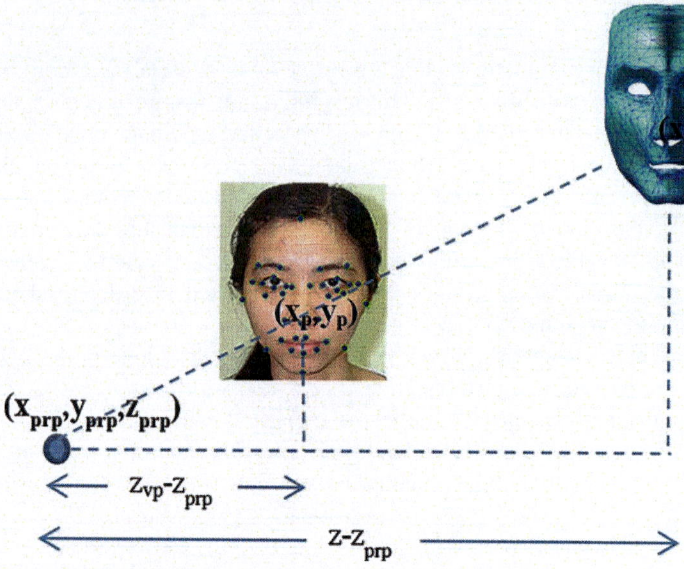

Fig. 1 Perspective and inverse perspective projections

$$f_p^{-1}(x_p, y_p, z) = (x, y, z) = \begin{bmatrix} x_p\left(\frac{z_{prp}-z}{z_{prp}-z_{vp}}\right) \\ y_p\left(\frac{z_{prp}-z}{z_{prp}-z_{vp}}\right) \\ z \end{bmatrix} \tag{4}$$

4 Distance Weighted Nearest Neighbor Algorithm

We found the customized locations of the landmark vertices through inverse perspective projection. For each landmark vertex we can compute the translation vector by subtracting its initial position in original wireframe from the customized position in the deformed wireframe model.

$$\Delta v_i^l = v_{i,orig}^l - v_{i,custom}^l \tag{5}$$

For each non-landmark vertex in the original wireframe we calculated the Euclidean distances between the vertex and the landmark vertices. These distances are to be used to determine the nearest k landmark vertices and their weights in calculating the displacement vector for the non-landmark vertex.

$$d_{i,j} = \left\| v_{i,orig}^l - v_{j,orig}^{nl} \right\| \quad i = 1 \ldots 36, j = 1 \ldots 576 \tag{6}$$

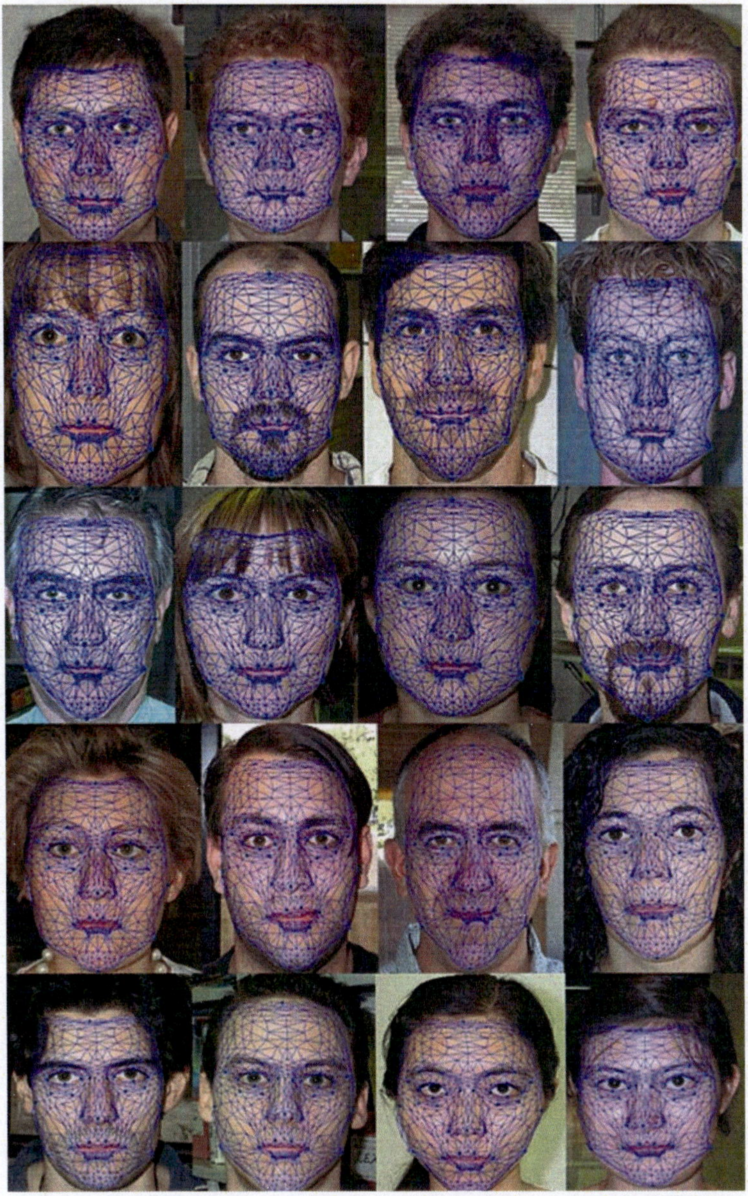

Fig. 2 Test results

Each non-landmark vertex is translated with a sum of translation of k nearest-neighbor landmark vertices, weighted by the inverse of their distances to the vertex.

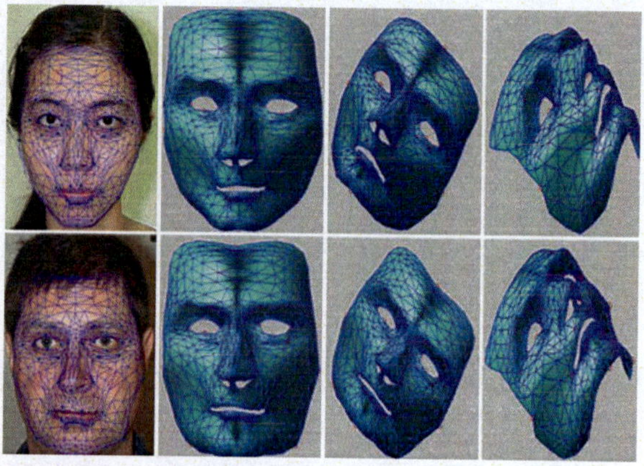

Fig. 3 Test results of wireframe model with different views

$$T_j = \frac{\sum_{i=1}^{k} \frac{\Delta v_i^l}{d_{ij}^2}}{\sum_{i=1}^{k} \frac{1}{d_{ij}^2}} \tag{7}$$

$$v_{j,custom}^{nl} = v_{j,orig}^{nl} + T_j \tag{8}$$

5 Experiments and Results

We used the face database prepared by the California Institute of Technology Computational Vision group [7]. Our test images contain neutral faces that are oriented towards camera. Experimental results are shown in Figs. 2 and 3 .

Our generic wireframe model consists of 612 vertices. In the first step of the study we manually marked 36 landmark vertices on the wireframe model. For each test image, we marked the corresponding landmark points. We then applied 3D translation and scaling operations to the wireframe model for initial alignment. The customized and final coordinates of the landmark vertices are found by taking the inverse perspective projection of the facial landmark points. The coordinates of non-landmark vertices in the customized model are computed using the distance weighted nearest neighbor algorithm. The new positions of the non-landmark vertices are calculated by weighting the displacement vectors of closest 10 landmark vertices on the wireframe model.

6 Conclusion

In this study we focus on registering a 3D generic wireframe model with an input face image. We proposed a semi-automatic wireframe fitting algorithm for customizing a high-polygon wireframe model. It is currently not possible to quantify the precision of our algorithm since our input images are 2D. We are currently working on repeating these experiments using 3D databases. By registering vertices in the customized wireframe model with 3D face points, we will be able to quantify our error and take steps to improve our algorithm and reduce the error rate.

Acknowledgments This research is part of project "Expression Recognition based on Facial Anatomy", grant number 109E061, supported by The Support Programme for Scientific and Technological Research Projects (1001) of The Scientific and Technological Research Council of Turkey (TÜBİTAK).

References

1. Essa, I.A.: Coding, analysis, interpretation, and recognition of facial expressions. IEEE Trans. Pattern Anal. Mach. Intell. **19**, 757–763 (1998)
2. Moghaddam, B., Pentland, A.: Face recognition using view-based and modular eigenspaces. In: Automatic Systems for the Identification of Humans, vol. 2277. SPIE (1994)
3. Pentland, A., Moghaddam, B., Starner, T.: View-based and modular eigenspaces for face recognition. In: Computer vision and pattern recognition conference. IEEE Comput. Soc. pp. 84–91 (1994)
4. Cootes, T.F., Cooper, D., Taylor, C.J., Graham, J.: Active shape models—their training and application. In: Comput. Vis. Image Understand. **61**(1), 38–59 (1995)
5. Cootes, T.F., Edwards, G.J., Taylor, C.J.: Active appearance models. In: Burkhardt, H, Neumann, B. (eds.) Proceedings of the European Conference on Computer Vision 1998, vol. 2, pp. 484–498, Springer (1998)
6. Krinidis, S., Pitas, I.: Facial expression synthesis through facial expressions statistical analysis. In: European Signal Processing Conference (EUSIPCO06), Florence, Italy (2006)
7. Caltech Face Database. http://www.vision.caltech.edu/html-files/archive.html

A Fuzzy Metric in GPUs: Fast and Efficient Method for the Impulsive Image Noise Removal

María G. Sánchez, Vicente Vidal, Jordi Bataller and Josep Arnal

Abstract The implementation of image correction algorithms on the CUDA platform is a relatively new field. Although the platform is easy to program, it is not easy to optimize the applications due to the number of decisions that have to be made. This paper reports an optimization study on the use of the CUDA platform to remove impulsive noise in images using fuzzy metric and the concept of peer group. The texture memory is used to speed up the access to data. In order to get the maximum bandwidth on the GPU memory, a strategy based on storing each pixel in 4 bytes is proposed.

Keywords Noise removal in images · Parallel systems · GPU · CUDA

M. G. Sánchez
Departamento de Sistemas y Computación, Instituto Tecnológico de Cd. Guzmán, 49100 Cd. Guzman, Jal, Mexico
e-mail: msanchez@dsic.upv.es

V. Vidal · J. Bataller
Departamento de Sistemas Informáticos y Computación E.P.S. Gandia, Universidad Politécnica de Valencia, 46730 Grao de Gandia, Valencia, Spain
e-mail: vvidal@dsic.upv.es

J. Bataller
e-mail: bataller@dsic.upv.es

J. Arnal (✉)
Departamento de Ciencia de la Computación e Inteligencia Artificial, Universidad de Alicante, 03071 Alicante, Spain
e-mail: arnal@dccia.ua.es

E. Gelenbe et al. (eds.), *Computer and Information Sciences II*,
DOI: 10.1007/978-1-4471-2155-8_41,

1 Introduction

In recent years, the incorporation of GPUs (Graphics Processing Units) in graphic cards has achieved significant improvements in computational speed, offering a high parallel processing level. For this reason, developments based on this hardware have become increasingly widespread, not only for graphic implementations but also for general purpose applications. The most commonly used programming platform for these graphic cards is CUDA (Compute Unified Device Architecture) [1]. Its multiple fields of application include medicine, astrophysics, computational chemistry and signal processing, among many others [2, 3]. Whilst it is relatively easy to use the CUDA platform to program the GPU, the problem lies in the difficult task of optimizing application performance, due to several hardware restrictions and the multiple types of memories included. Thus, it is necessary to carry out a specific study in order to identify the best approach to using the resources offered by CUDA in each case. Such a study is the subject of the present paper, applied in this case to noise removal in images. We propose a parallel algorithm to remove impulse noise in images based on the fuzzy metrics [4] and the concept of peer group $\mathcal{P}$ [5]. Many algorithms have been proposed for correcting impulsive noise, for instance those mentioned in [6, 7]. At the hardware level, there are studies regarding image correction implementations [8], in which the hardware is configured for that purpose. In the present study, we implemented a parallel algorithm based on the algorithms proposed in [6]. The process of noise removal was divided into two steps: erroneous pixel detection, and the elimination of these pixels. We analyze the access time to global memory and use the texture memory in order to reduce it. A strategy for getting the maximum bandwidth of the GPU memory is proposed.

This paper is organized as follows: In Sect. 2, the GPU is described together with CUDA programming models. Section 3 illustrates the noise removal method employed and its implementation on CUDA. Experimental results are shown in Sect. 4, and lastly, the conclusions are presented in Sect. 5.

2 GPU and CUDA Programming Models

Physically, the GPU contains a set of multiprocessors and a physical memory that can be used in different ways. The main use is as global shared memory among the GPU multiprocessors. However, this memory also permits its several areas to be used in other modes: as local memory, as texture and as a constant read-only area. Internally, each multiprocessor has four kinds of memories [1]: a set of registers per processor, a shared memory, a constant cache memory and a texture cache. The large variety of memories and their different features complicate the task of achieving optimal performance in programs using CUDA. One of the main issues that must be considered in order to obtain efficient programs is the coalescence of accesses to global memory. Global memory bandwidth is used more efficiently when the simultaneous memory accesses by threads in a half-warp can be coalesced into a single memory transaction of 32, 64, or 128 bytes [1]. Initially, data are available in the GPU global memory but if coalescence problems arise, it becomes necessary to consider whether it would be advisable to copy data from the global memory to the shared memory of each multiprocessor in order to process them. Another option is to use textures, if part of the information is read-only and the information contained is organized dimensionally in arrays. Since a GPU has various multiprocessors, it is necessary to group the threads (a piece of code being run in parallel) in blocks to make clear the allocation of the execution of a block to a specific multiprocessor of the GPU. This feature permits to synchronize the threads only if they belong to the same block. The set of blocks is called a grid. When a block of threads ends and there are still data to be analyzed, a new block is assigned to the multiprocessor that remains available. As can be appreciated, in the design of a CUDA program is essential to decide at each moment the best location, among the different available memories, for the input and output data, and how to access them, performing the necessary copies at the appropriate time.

3 Image Noise Removal and Implementation Strategies with CUDA

In the present study, the process of image noise removal was divided into two steps. The first step was to detect erroneous pixels and the second, to eliminate them. For the detection stage, the fuzzy metric between the vectors of the color image x_i and x_j was used, which is given by the following function:

$$M(x_i, x_j) = \prod_{l=1}^{3} \frac{\min\{x_i(l), x_j(l)\} + k}{\max\{x_i(l), x_j(l)\} + k}, \tag{1}$$

where $(x_i(1), x_i(2), x_i(3))$ is the color vector for the pixel x_i in space color RGB. In [6], k = 1,024 was shown to be an appropriate setting, and this was therefore the value that was used in the present study. Fuzzy metric is employed in peer group $\mathcal{P}(x_i, d)$,

$$\mathcal{P}(x_i, d) = \{x_j \in W : M(x_i, x_j) \geq d\}, \tag{2}$$

where x_i is the central pixel in a window W with size $n \times n$, $n = 3, 5, 7, \ldots$, and $0 \leq d \leq 1$ is the distance threshold. The peer group associated with the central pixel of W is a set consisting of the central pixel x_i and its neighbors belonging to W, whose distance from x_i exceeds d. After several tests, $d = 0.95$ proved to be a good value for d. In the detection process described in Algorithm 1, the kernel was configured so that each thread processed one item of pixel data. The thread corresponding to the pixel x_i analyzes the $n \times n$ pixels of W (in the present study, $n = 3$ was considered), calculates the peer group and if this satisfies the cardinality

$m+1$, i.e. $\#\mathcal{P}(xi,d) \geq m+1$, the central pixel is diagnosed as uncorrupted; if not, it is diagnosed as corrupted. In the correction step described

Algorithm 1 Detection of Erroneous Pixels

Require: *m, d, W.*
Ensure: The image with the pixels classified as corrupt or not.

Calculate $\mathcal{P}(xi,\ d)$:
    Each thread calculates:
        distance = M $(x_i,\ x_j)$
        if distance $\geq$ d **then**
            pixel $x_j \in \ \mathcal{P}(x_i\ i,\ d)$
        endif
if (# $\mathcal{P}(xi,\ d) \geq m$ +1) **then**
    x_i *is declared as non-corrupted pixel*
else
    x_i *is declared as corrupted pixel*
end if

in Algorithm 2, we used an equal number of threads for kernel 1. The threads corresponding to pixel x_i apply filtering by substitution where this has been labeled as corrupted in detection step. The replacement value is determined by creating a window *W* for the pixel x_i and calculating the well-known AMF (Arithmetic Mean Filter) [6] on corrupted pixels *W*. Threads with uncorrupted pixel values continue as before. The arithmetic mean was used instead of the median for two main reasons: first, because in the calculation of the median, comparison operations are required and these operations are not recommended on the GPU. Second, because the computational cost of the arithmetic mean is lower than the median. To determine the number of threads per block that

Algorithm 2 Elimination of Erroneous Pixels

Require: *n*, threads corresponding to the pixels labeled as corrupted.
Ensure: Filtered Image.

Each thread defines the row and column corresponding to the central pixel x_i.
Each thread builds its windows W of pixels.
Calculate the AMF of uncorrupted pixels W.
Thread corresponding to the pixel x_i replaces value obtained by the AMF.

best fits the application, a heuristic study concluded that 64 $\times$ 64 threads per block gave lowest computational costs. As can be observed in Sect. 2, a series of choices must be considered in order to implement an algorithm on CUDA. In the present case, two strategies were employed:

1. Storing the image with 3 channels per pixel or 4 channels per pixel.
2. Accessing data through the texture memory or not.

Deciding whether to store the image in RGB (three channels) or RGBA (three channels + padding), is necessary since the RGB format uses 3 bytes and thus does not achieve access coalescence. Moreover it is not a multiple of 32, 64 or 128 bits that matches a correct access pattern to write or read to the memory. However, if a padding byte is added, even using a time for it, the accesses will fit in blocks of 4 bytes and performance may improve. Furthermore, the addition of a fourth byte can be used to indicate pixel status: corrupted, uncorrupted or undiagnosed. In addition, we evaluated the improvements deriving from the use or not of texture memory. In our case, we evaluate whether access to this memory improves access time to data. For this research, two textures were used: one in the first phase of detection, another in the elimination phase. In all cases, these were used with the purpose of reading x_i neighboring pixels in the fuzzy and peer group calculations.

4 Experimental Results

This section presents the results obtained for the implementations discussed in Sect. 3. The CPU used was a Mac OS X Intel Xeon Quad-Core processor at 2 $\times$ 2.26 GHz with 8GB memory. The GPU was an NVIDIA GeForce GT 120 with 512 MB of memory and 4 multiprocessors. Our implementation used C language and single-precision calculations. The lenna image [6] was employed, with RGB format square dimensions 128, 256, 512, 1024, 2048 and 4096 pixels and 5 and 10% noise impulse. For each algorithm, we designed both the CPU serial code and the GPU parallel code and then we compared execution time. We have taken into account the preparation time and data transfer time between CPU and GPU and the calculation time on the GPU. In the first experiment (strategy 1 of Sect. 3) we use the char data type to store each pixel component, using 3 bytes per pixel, and we compare with the option of storing the pixel in 4 bytes. The data access was done through a normal pointer directly to global memory. In the second experiment (strategy 2 of Sect. 3) we reserve space in memory using 4 bytes per pixel and we access data through the texture memory. Table 1 shows that the 4 bytes storage option obtained better execution times than the 3 bytes storage option. Moreover, it can be seen that the access to the data through the texture memory improves significantly the global memory option. These performance was obtained for the two choices of percentage of image noise. When the image size increases the reduction rate in execution time increases.Working with 4 bytes per pixel we get more bandwidth when accessing data of the GPU memory and we can use it to classify the pixel as corrupt or not, and then it is not needed aditional memory space for this purpose. Moreover, more time reduction is obtained if the texture memory is used because we access through its cache. The results regarding relative GPU time spent by each kernel are shown in Table 2. As can be seen, the

Table 1 Processing time in milliseconds for the implementations

Size	5% noise						10% noise					
	3 bytes per pixel. Data access directly to global memory		4 bytes per pixel. Data access directly to global memory		4 bytes per pixel. Data access through texture		3 bytes per pixel. Data access directly to global memory		4 bytes per pixel. Data access directly to global memory		4 bytes per pixel. Data access through texture	
	GPU	Total(GPU and CPU)	GPU	Total(GPU and CPU)	GPU	Total(GPU and CPU)	GPU	Total(GPU and CPU)	GPU	Total(GPU and CPU)	GPU	Total(GPU and CPU)
128	1.44	2.03	1.11	1.34	0.47	0.87	1.48	2.07	1.15	1.38	0.51	0.90
256	5.62	6.85	4.18	4.65	1.63	2.39	5.82	7.07	4.34	4.82	1.78	2.56
512	31.78	35.79	16.67	18.11	6.29	8.54	32.69	36.70	17.25	18.69	6.68	9.10
1024	130.45	145.95	68.09	73.48	24.91	33.08	134.85	150.30	70.61	75.99	27.54	35.72
2048	569.35	630.91	306.20	327.09	104.48	136.14	593.74	655.02	316.54	337.47	115.35	147.00
4096	2536.16	2779.93	1250.89	1333.77	419.92	545.59	2653.68	2897.51	1300.98	1383.85	464.69	590.29

Table 2 Computational times on GPU for kernels with 5% impulsive noise

Strategy						
	3 bytes/Global memor		4 bytes/Global memor		4 bytes/Texture memor	
Image Size	Detection	Elimination	Detection	Elimination	Detection	Elimination
128	0.92	0.52	0.73	0.37	0.28	0.19
256	3.70	1.92	2.78	1.40	0.92	0.70
512	23.40	8.38	11.20	5.47	3.60	2.69
1024	96.80	33.65	46.21	21.88	14.07	10.84
2048	427.66	141.69	215.91	90.29	57.82	46.66
4096	1937.16	599.00	886.58	364.31	232.93	186.99

Table 3 Speedup for different image sizes with 5% impulsive noise

SIZE	TEXTURES GPU	CPU	SPEEDUP
128	0.87	8.17	9.39
256	2.39	31.84	13.32
512	8.54	126.96	14.87
1024	33.08	491.20	14.85
2048	136.14	1961.51	14.41
4096	545.59	7895.66	14.47

computational cost in the elimination phase is lower than in the detection. This is due to the fact that in the detection phase all the threads have to check if the pixel is corrupted or not using the process of peer group and fuzzy metric, whereas in the elimination phase only corrupted pixels have to do the filtering of the arithmetic mean. The strategy of 4 bytes with texture reflects the least computational time for all sizes. Table 3 shows the speedup when comparing the sequential version running on CPU and the parallel version on GPU using 4 bytes with textures.

5 Conclusions

In this paper we have described a study conducted in order to determine the best method for implementing image correction processing in RGB format with impulsive noise on a GPU using a CUDA platform. This processing was divided into two steps: noise detection and noise elimination. For detection, the fuzzy measure and the concept of peer group were used. In the correction stage, corrupted pixel values were replaced by calculating the mean of those neighbors not labeled as corrupted in the detection step. The experiments show that if the accesses coalesced and an access pattern of read/write in words of 32, 64, of 128 bytes is satisfied, the results improve significantly. Furthermore, we have also shown that the use of textures for accessing data in global memory reduces significantly the data access time. Likewise, through the use of textures, we have obtained outstanding results in speedup, compared to sequential version of the implementation. For future works, we plan to spread the processing load between multiple GPUs and evaluate the performance of large images.

Acknowledgments This work was funded by the Spanish Ministry of Science Innovation (Project TIN2008-06570-C04-04). María would also like to acknowledge DGEST-ITCG for the scholarship awarded through the PROMEP program (Mexico).

References

1. NVIDIA Programming guide version 2.3.1, http://www.nvidia.es/page/home.html
2. Ruiz, A., Ujaldón, M., et al.: The GPU on biomedical image processing for color and phenotype analysis. In: 7th IEEE International Conference on Bioinformatics and Bioengineering, pp. 1124–1128 (2007)
3. Yang, Z., Zhu, Y., Pu, Y.: Parallel image processing based on CUDA. In: International Conference on Computer Science and Software Engineering, pp. 198–201 (2008)
4. Morillas, S., Gregori, V. et al.: Local Self-Adaptive Fuzzy Filter For Impulsive Noise Removal in Color Images. Sci. Direct Signal Process. **88**, 390–398 (2008)
5. Smolka, B., Chydzinski, A.: Fast detection and impulsive noise remolval in color images. Real-Time Imag. **11**, 389–402 (2005)
6. Camarena, J.G., Gregori, V., Morillas, S., Sapena, A.: Fast detection and removal of impulsive noise using peer group and fuzzy metrics. J. Vis. Commun. Image Represent. **19**, 20–29 (2008)
7. Gonzalez, R.C., Woods, R.E.: Digital image processing. Person Education, Upper Saddle River (2008)
8. Gutiérrez-Ríos, J., Brox, P., Fernández-Hernández, F., Baturone, I., Sánchez-Solano, S.: Fuzzy motion adaptive algorithm and its hardware implementation for video de-interlacing. Appl. Soft Comput. **11**, 3311–3320 (2011)

Improved Secret Image Sharing Method By Encoding Shared Values With Authentication Bits

Guzin Ulutas, Mustafa Ulutas and Vasif Nabiyev

Abstract Authentication capability and stego image distortion issues of secret sharing schemes are addressed and improved by the proposed scheme in this paper. Reduced fake stego block authentication probability and improved stego image Peak Signal to Noise Ratio (PSNR) are achieved by encoding both the share and authentication bits by XOR. Employing only eight bit for both share and authentication reduces the probability of authenticating a fake stego block to 1/256 with 44 dB average PSNR stego images. Experimental results indicate improvements compared to similar studies reported in the literature.

Keywords Secret image sharing · Steganography · PSNR

1 Introduction

Shamir proposed a polynomial based technique to share a secret among a group of participants in 1979 [1]. Thien and Lin used Shamir's approach to share a secret image among *n* participants in 2002 [2]. Lin et al. proposed to use Steganography to hide share images into meaningful cover images [3]. Their method also introduced an authentication mechanism for stego images. Two issues, distortion caused by truncation and the use of a simple parity bit for authentication, are

G. Ulutas (✉) · M. Ulutas · V. Nabiyev
Department of Computer Engineering,
Karadeniz Technical University, Trabzon, Turkey
e-mail: guzin@ieee.org; gulutas@ktu.edu.tr

M. Ulutas
e-mail: ulutas@ktu.edu.tr

V. Nabiyev
e-mail: vasif@ktu.edu.tr

E. Gelenbe et al. (eds.), *Computer and Information Sciences II*,
DOI: 10.1007/978-1-4471-2155-8_42,

reported in Yang et al.'s work in 2007 [4]. In 2008, Chang et al. proposed a new authentication technique based on the CRT to improve the authentication capability [5]. A new secret image sharing scheme encoding both share and authentication bits by XOR to both enhance authentication and visual quality of the stego images is proposed in this paper. Each stego block accommodates eight bit to both represent eight bit shared value and eight bit authentication value. Any modification of a stego block would result in a modified share since share and authentication bits are related by XOR. The proposed method is able to identify alterations of stego images after the secret image is revealed by the reconstruction procedure. However, using only eight bit of the block to accommodate both share and authentication value improve stego images' PSNR by 4 dB on the average as indicated in the results. Authentication capability of the proposed method is also improved compared to similar works since eight authentication bits reduces the probability of a cheat to 1/256.

2 Sharing Scheme with Packed Share and Authentication Bits

The proposed method can be defined as a sharing algorithm and a retrieving algorithm. The sharing algorithm is used to share a secret image among n participants for a (k,n) scheme. It also employs steganography to hide shared values into cover images. The retrieving algorithm needs at least any k of the n stego images from the participants. It retrieves shared values from stego images and Lagrange's interpolation technique is used to reconstruct secret pixel values from shared values. The details of both algorithms are given below. The dealer D should select a unique serial numberx_i, $i \in \{1, \cdots, n\}$ for each participant. Pixels of the secret image S with intensity values in $[251-255]$range are truncated to 250 before the sharing algorithm. Let the intensity of the gray level secret image S of size $N \times M$ be $S = \{s_{ij} | s_{ij} \in [0-250], 1 \le i \le N, 1 \le j \le M\}$.

1. Repeat the following steps for $i = 1, 2, \cdots, N$
2. Repeat the following steps for $j = 1, 1+k, 1+2k \cdots, \lfloor M/k \rfloor \cdot k - (k-1)$

   2.1. Obtain the current section from the secret image$(s_{ij}, s_{ij+1}, \cdots, s_{ij+k-1})$

   2.2. Construct a polynomial as$F(x) = (s_{ij} + s_{ij+1}x + \cdots + s_{ij+k-1}x^{k-1})$mod 251.

   2.3. Repeat the following steps for $t = 1, 2, \cdots, n$

      2.3.1. Evaluate the polynomial to compute the shared value $F(x_t)$.

      2.3.2. Obtain the corresponding cover block $\begin{bmatrix} X^t = C^t_{uv} & Y^t = C^t_{uv+1} \\ Z^t = C^t_{u+1v} & W^t = C^t_{u+1v+1} \end{bmatrix}$ at tth cover image where (u, v) is $(2 \cdot i - 1, \lfloor j/k \rfloor \cdot 2 + 1)$.

      2.3.3. Compute authentication value using keyed hash function $H(\cdot)$.

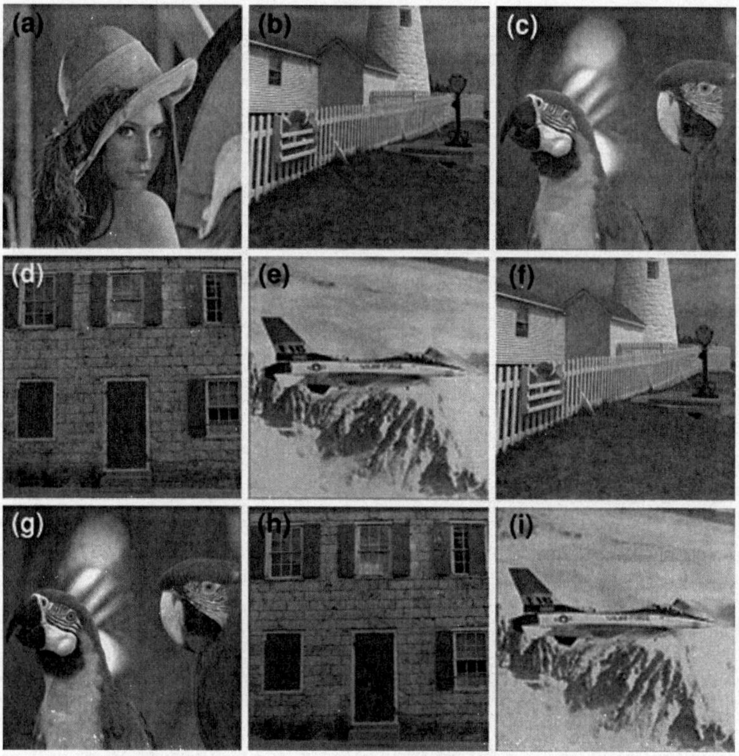

Fig. 1 **a** Secret image of size 256 × 256 **b–e** Gray level cover images **f–i** Stego images

$$A^t_{uv} = \langle H((X^t - (x^t_7, x^t_8)) \| (Y^t - (y^t_7, y^t_8)) \| (Z^t - (z^t_7, z^t_8)) \| (W^t - (w^t_7, w^t_8))) \rangle_8$$

2.3.4. Compute the encoded embedding value E^t_{uv} to be stored into corresponding cover block from the share $F(x_t)$ and authentication hash A^t_{uv} values by $E^t_{uv} = A^t_{uv} \oplus F(x_t)$.

2.3.5. Partition embedding value E^t_{uv} into two bit chunks to be embedded into corresponding cover pixels' last two bits.

The proposed method modifies only eight bits of cover blocks to accommodate both share and authentication values using XOR encoding whereas similar studies use $8 + b$, eight bit share and b, $b \in [1 - 4]$ bit authentication. Separate share and authentication bits impair PSNR and limit the authentication capability of those methods due to increased number of modified bits per block. In other words, the proposed method do not need extra bits to represent authentication value which both enhance the visual quality of stego images while ensuring eight bit authentication capability. Previous methods used maximum four bits for authentication purposes. The retrieving algorithm reconstructs shares and reveals the secret.

Let k different stego images be denoted by $ST^1, ST^2, \cdots, ST^k$. The steps of the retrieving algorithm processing k stego image block by block are given below.

1. Repeat the following steps for $i = 1, 3, \cdots, 2N$
2. Repeat the following steps for $j = 1, 3, \cdots 2M$

   2.1. Repeat the following steps for $t = 1, 2, \cdots, k$

   2.1.1. Obtain a stego block $\begin{bmatrix} \tilde{X}^t = ST^t_{ij} & \tilde{Y}^t = ST^t_{ij+1} \\ \tilde{Z}^t = ST^t_{i+1j} & \tilde{W}^t = ST^t_{i+1j+1} \end{bmatrix}$ from tth stego image.

   2.1.2. Compute eight bit authentication value $\tilde{A}^t_{ij}$ from block as $\tilde{A}^t_{ij} = \langle H((\tilde{X}^t - (\tilde{x}^t_7, \tilde{x}^t_8)) \| (\tilde{Y}^t - (\tilde{y}^t_7, \tilde{y}^t_8)) \| (\tilde{Z}^t - (\tilde{z}^t_7, \tilde{z}^t_8)) \| (\tilde{W}^t - (\tilde{w}^t_7, \tilde{w}^t_8))) \rangle_8$

   2.1.3. Extract the embedded value E^t_{ij} by concatenating last two bits of pixels of the current stego block.

   2.1.4. Decode the shared value $F(x_t)$ using $F(x_t) = \tilde{A}^t_{ij} \oplus E^t_{ij}$.

   2.2. Use Lagrange's interpolation technique on $F(x_1) \cdots F(x_k)$ to compute k coefficients of the polynomial corresponding to secret pixel values values.

Computed authentication value in step 2.1.2 would indicate modifications of the stego image pixels since decoded shared value by the authentication bits would result in incorrect coefficients. By utilizing XOR function, the method not only improves stego image quality but also prevents construction of the correct secret pixel values, if stego image pixels are modified during transmission or storage.

3 Experiments and Discussion

The two metrics of secret image sharing schemes, visual quality and authentication capability, of the proposed method are tested during the experiments. The secret image "Lena" of size 256 × 256 and cover images of size 512 × 512 given in Fig. 1a–e respectively are used in the experiments. Proposed method shares the secret among four participants for $k = 3$ for as the first experiment. Generated stego images have approximately 44 dB PSNR as shown in Fig. 1f–i. Authentication capability of the proposed method is tested by the second experiment. First stego image obtained in the first experiment is replaced by a fake stego image given in Fig. 2a. The least significant two bits of the fake stego image pixels are replaced by the least significant two bits of the original stego image pixels in order to preserve the share information. The reconstructed secret image given in Fig. 2b does not reveal any information about the secret image. Improved authentication capability of the method correctly identifies alterations on any one of the stego images. Using only eight bit for both hiding shared and authentication values improves both the PSNR and authentication capabilities compared to similar works by others reported in the literature. eight bit authentication capability

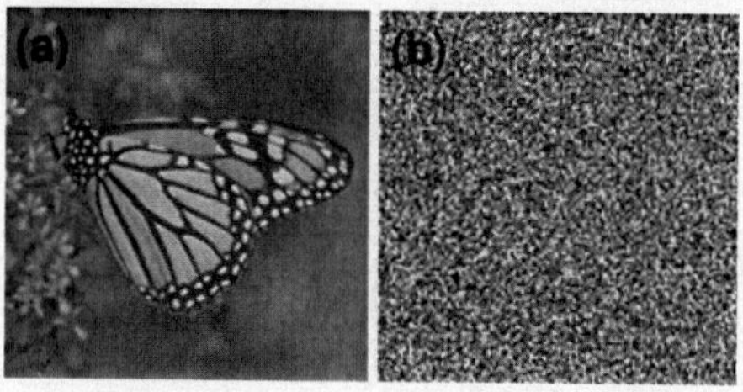

Fig. 2 **a** Fake stego image **b** Reconstructed secret image with a fake stego image

Table 1 Properties of the proposed method and other methods

	[2]	[3]	[4]	[5]	Proposed method
Num. of auth. Bits	–	1	1	4	8
Avg. PSNR	39 dB	41 dB	38 dB	37 dB	44 dB

provided to verify the stego blocks during the retrieving algorithm reduces the probability of authenticating a modified block to 1/256. Table 1 lists a feature comparison of the proposed method and related works. Proposed method has higher PSNR and improved authentication capability compared to similar works.

References

1. Shamir, A.: How to share a secret. Commun. ACM. **22**, 612–613 (1979)
2. Thien, C.-C., Lin, J.-C.: Secret image sharing. Comp. Graph. **26**, 765–770 (2002)
3. Lin, C.-C., Tsai, W.-H.: Secret image sharing with steganography and authentication. J. Syst. Softw. **73**, 405–414 (2004)
4. Yang, C.-N., Chen, T.-S., Yu, K.H., Wang, C.-C.: Improvements of image sharing with steganography and authentication. J. Syst. Softw. **80**, 1070–1076 (2007)
5. Chang, C.-C., Hsieh, Y.-P., Lin, C.-H.: Sharing secrets in stego images with authentication. Pattern Recogn. **41**, 3130–3137 (2008)

Part VIII
Agent Based Systems

Intelligent Navigation Systems for Building Evacuation

Gokce Gorbil, Avgoustinos Filippoupolitis and Erol Gelenbe

Abstract Intelligent navigation systems can significantly improve the evacuation of civilians in urban emergencies by providing dynamic directions to the evacuees as the hazard spreads. In this paper, we propose two distributed and adaptive systems for building evacuation. The first system, called *intelligent evacuation system (IES)*, is composed of a wireless network of static decision nodes (DNs) installed in the building which provide movement decision support to evacuees by directing them to exits via less hazardous routes. The second system, called *opportunistic communications based evacuation system (OCES)*, is based on mobile communication nodes (CNs) carried by civilians, which form an opportunistic network to exchange information on the situation and provide directions to civilians. We evaluate our systems' performance with simulation experiments of a three-floor office building. Our results indicate that both systems can greatly improve the outcome of an emergency.

Keywords Building evacuation · Intelligent systems · Opportunistic communications · Emergency navigation

G. Gorbil (✉) · A. Filippoupolitis · E. Gelenbe
Department of Electrical and Electronic Engineering,
Imperial College London, SW7 2BT London, UK
e-mail: g.gorbil@imperial.ac.uk

A. Filippoupolitis (✉)
e-mail: afil@imperial.ac.uk

E. Gelenbe
e-mail: e.gelenbe@imperial.ac.uk

E. Gelenbe et al. (eds.), *Computer and Information Sciences II*,
DOI: 10.1007/978-1-4471-2155-8_43,

1 Introduction

To address the issues of safe and quick evacuation of buildings under dynamic conditions, we propose the use of emergency evacuation systems which use sensing and communications to safely evacuate people when some major hazard, such as fire or explosion, is detected. We specifically propose two different evacuation systems and provide simulation results to show that significant improvements can be obtained through their use. In this paper, we evaluate the IES and OCES as separate systems: only one of them would be installed in the building for evacuation support at any time.

2 Description of Proposed Systems

We assume that the building is represented as a graph $G(V,E)$. Vertices V are locations where civilians can congregate, such as rooms, corridors and doorways, and edges E are physical paths that can be taken by the civilians for movement. Edges have multiple costs: edge length $l(i,j)$, which is the physical distance between vertices $i,j \in V$, $h(i,j)$, which is the hazard intensity along this edge, and $L(i,j) = l(i,j) \cdot h(i,j)$, which is the "effective" edge length, a joint metric combining physical distance and hazard for shortest path calculations. Note that $h(i,j) = 1$ when there is no hazard on the edge and its value increases with the value of the observed hazard.

We also assume that the building graph is known for a building. Both the IES and OCES are supported by fixed *sensor nodes (SNs)* installed in the building, where each SN monitors a graph edge. Each SN has a unique device ID, a location tag that corresponds to the area (i.e. edge) it monitors, and very short-range (1–2 m) wireless communication capability so it can relay its measurements to other entities in the system, such as DNs in the IES and CNs in the OCES. We assume that SNs are very simple, battery-operated devices, with little to no generic computing power, low memory capacity and restricted energy constraints. Because of these limitations, SNs *do not* perform any data storage, processing or decision making. Each measurement is stored until it is over-written by a newer measurement. When a DN or CN requests the current measurement from an SN, the SN sends its $h(i,j)$ value to the requester.

2.1 Intelligent Evacuation System

Our proposed intelligent evacuation system (IES) [1] consists of static *decision nodes (DNs)*, which are installed at specific locations inside the building (at each graph vertex). Each DN has short-range wireless communication capability, some local processor and memory, and a dynamic visual panel to present directions to civilians. The main role of a DN is to compute the best direction towards a building exit and communicate this to the evacuees in its vicinity. Hazard information is provided to DNs by their adjacent SNs, and this information is further propagated among DNs based on the distributed decision algorithm as presented below. Each DN, positioned at vertex u, stores the following information:

- the effective edge lengths to neighbors: $L(u,n)$, $\forall n \in V \mid (u,n) \in E$
- the effective lengths of the paths to an exit for all neighbors: $L(n,e)$, $\forall n \in V \mid (u,n) \in E$ and e is a building exit,
- the effective length of the shortest path (SP) from u to an exit e: $L(u,e)$,
- the next suggested DN d (i.e. the next hop along the SP from u to an exit).

The distributed decision algorithm, given in Algorithm 1, is executed periodically by each DN and updates both the executing DN's neighbors (i.e. their $L(n,e)$ values) and the DN itself (its $L(u,n)$, $L(u,e)$ and d values). Note that $L(u,n)$ values are updated based on measurements received from the SNs. The algorithm is based on principles developed in [2, 3], and inspired by the distributed shortest path algorithm [4] and adaptive routing techniques such as the Cognitive Packet Network [5]. Its output is the next hop (i.e. DN) along the SP to the nearest building exit. As edge costs are a combination of physical distance and hazard intensity, the IES directs the civilians towards the exits while avoiding dangerous areas in the building. Prior to executing the algorithm for the first time, exit DNs set their $L(u,e)$ values to 0, while all other DNs set it to ∞. Note that since DNs only use local information, the IES is a distributed system that does not require global information.

Algorithm 1 Distributed decision algorithm for the IES. A DN updates its suggested direction via communication with its adjacent DNs and SNs.

procedure UPDATEDN (u)

  for all $n \in V \mid (u,n) \in E$

    Send $L(u,e)$ to n
    Get $h(u,n)$ from SN for (u,n)
    $L(u,n) \leftarrow l(u,n) \cdot h(u,n)$

  end for
  $L(u,e) = min\{L(u,n) + L(n,e), \forall n \in V \mid (u,n) \in E\}$
  $d = argmin\{L(u,n) + L(n,e), \forall n \in V \mid (u,n) \in E\}$

end procedure

2.2 *Opportunistic Communications Based Evacuation System*

The opportunistic communications based evacuation system (OCES) is composed of *mobile communication nodes (CNs)* carried by civilians. In the OCES, we assume that each civilian is equipped with a pocket- or hand-held device, with storage and processing capacity that would be equivalent to a mobile phone or similar unit, capable of short range (up to 10 m) wireless communication. CNs form a network in an opportunistic manner as devices come into contact as a result of the vicinity of other humans and their mobility. Opportunistic communications (oppcomms) are characterized by the "store-carry-forward" paradigm [6] where CNs carry messages in local storage and then forward it to others when they get in communication range. Thus, a message is delivered to its destination via successive opportunistic contacts. Because the opportunistic network (oppnet) can be disconnected for long periods of time, CNs may need to carry messages for long durations and delivery of messages is not guaranteed.

Oppcomms are used to disseminate *emergency messages (EMs)* containing information on the hazard (i.e. location and intensity) among CNs. A CN obtains hazard observations from SNs in its vicinity, which are then translated by the CN into EMs that include the CN ID, locations (e.g. edges), intensities and timestamps of the hazard observation(s). An EM is disseminated among all CNs in the OCES, meaning each EM is sent network-wide. The first hazard observation or EM received by a CN acts as an alarm, indicating that there is a hazard and the civilian should evacuate the building. Each received EM is used to update the edge costs stored locally by a receiving CN, and triggers re-calculation of its local evacuation SP. The evacuation SP from the current CN location to the nearest building exit is calculated using Dijkstra's SP algorithm. Since effective edge lengths ($L(i,j)$ values) are used in SP calculation, the "shortest" path minimizes exposure to the hazard while also minimizing travel distance.

A CN uses its evacuation SP to provide a navigation service to its civilian by guiding her towards the next hop (i.e. graph vertex) on the SP. CNs use the SNs to find their location in the building. CNs request the location tag from their nearby SNs as they move within the building, and each near-by SN sends back a *localization message (LM)* which contains its location (or the monitored area, i.e. edge). CNs can then find out where they are in the building based on these LMs. The actual position of a CN is therefore approximated by its inferred location (vertex) on the building graph. *Epidemic routing (ER)* [7] is used for the dissemination of EMs in the oppnet. We have found that ER is very suitable for the OCES due to its flooding-based approach which closely matches how EMs should be disseminated, and its high message delivery ratio and low message latencies [8], which are critical in emergency communications. In order to store EMs, CNs employ *timestamp-priority queues*, where EMs with the earliest creation timestamps are dropped from the queue when it is full.

3 Experimental Evaluation

We have evaluated the IES and OCES with simulation experiments in the case of a fire in a three-floor office building using the distributed building evacuation simulator (DBES) [9], which is a multi-agent discrete-event simulation platform

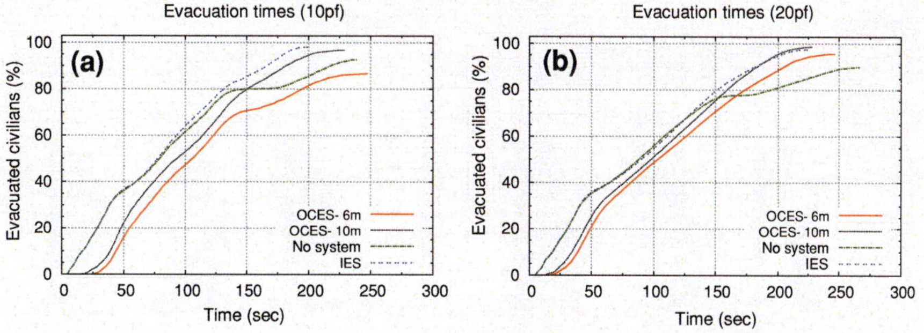

Fig. 1 Ratio of evacuated civilians vs. simulation time. **a** 10 civilians per floor **b** 20 civilians per floor

that can be coupled with real-life networks, such as WSNs [10]. The two building exits are located on the ground floor. Before a civilian is notified of the fire, she follows a probabilistic mobility model which simulates inter-office movement. After being alerted of the fire, civilians follow directions provided by the IES or the OCES. We assume that in the IES experiments the execution period of the distributed decision algorithm is 100 ms. All communication entities are assumed to have wireless capabilities similar to IEEE 802.15.4-2006. We also assume that entities on different floors cannot communicate. Each data point in the presented results is an average of 50 simulation runs; each run represents a different distribution of civilians over all vertices of the building. The spread pattern for the fire is based on a probabilistic model inspired by [11]. Where appropriate, experimental results are presented with their 95% confidence intervals. In all experiments the fire starts at time 0 at the intersection of two corridors (near the staircases) on the second floor.

We compare our proposed systems under different scenarios: (i) In the *IES with alarm (IES)* scenario, all civilians are alerted of the fire via a central alarm as soon as the fire starts and they then follow dynamic directions provided by the DNs to evacuate the building. (ii) In the *OCES without alarm* ($\mathbf{OCES^-}$) scenario, we assume that there is *no central alarm* in the building, so civilians are alerted and guided only by the OCES. (iii) The *OCES with alarm* ($\mathbf{OCES^+}$) scenario assumes a central alarm as in the IES scenario, but civilians are guided by the OCES. (iv) The *no system (NoSys)* scenario simulates the case where there is no evacuation support system in the building other than a fire alarm.

Figure 1a shows that OCES^- performs worse than NoSys when population density and CN communication range are both low. In this case, CNs are disconnected from the rest of the oppnet for a considerable amount of time, which means that civilians are alerted late of the fire and do not receive up-to-date evacuation guidance as the fire spreads. However, OCES^- performance improves with a longer communication range (i.e. 10 m) and surpasses NoSys in terms of evacuated civilians even with a low population density. In the 10pf case, the IES performs the best in terms of both evacuated civilians and evacuation time. With a

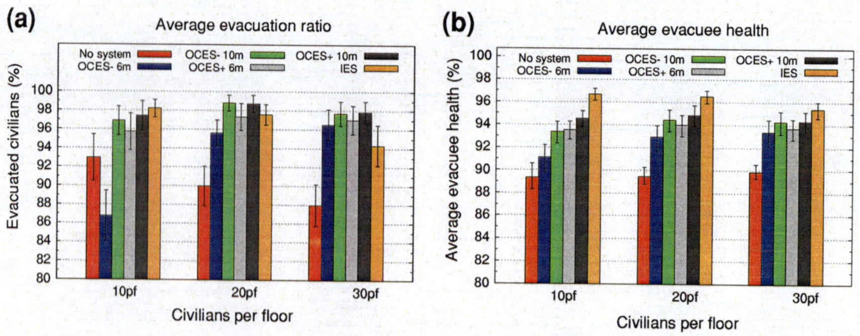

Fig. 2 Ratio of evacuated civilians and average evacuee health vs. population density. **a** Evacuation ratio **b** Evacuee health

higher population density (20pf), $OCES^-$ performance improves and as population density and communication range increase, $OCES^-$ performs better than IES. This is due to increased connectivity in the oppnet with increasing density and range, while IES performance gets slightly worse with increasing density due to physical congestion during evacuation (e.g. at the staircases). $OCES^-$ is not affected by congestion as much since civilians start evacuating at different times due to the lack of a central alarm. In Fig. 2, trends similar to the above discussion are observed. The effect of the central alarm in $OCES^+$ (compared to $OCES^-$) is most apparent at low population densities (10pf), and the alarm loses its effectiveness as OCES performance improves with increasing density and/or communication range.

4 Conclusions and Future Work

We have proposed and described two intelligent navigation systems for building evacuation in cases of dynamic hazards such as a fire. We have evaluated our systems using simulation experiments of a three-floor building in case of a fire. Our results indicate that IES provides a good overall performance, while OCES performs poorly in low population densities when a short communication range is used but surpasses IES in denser populations or with higher communication ranges. In future work, we aim to investigate how IES and OCES can be deployed simultaneously to improve evacuation, for example when there are failures in the system.

References

1. Filippoupolitis, A., Gelenbe, E.: A distributed decision support system for building evacuation. In: Proceedings of the 2nd IEEE International Conference on Human System Interaction. pp. 323–330, May 2009
2. Gelenbe, E.: Sensible decisions based on QoS. Comput. Manag. Sci. **1**(1), 1–14 (2004)

3. Gelenbe, E., Seref, E., Xu, Z.: Simulation with learning agents. Proc. IEEE. **89**(2), 148–157 (2001)
4. Humblet, P.A.: Another adaptive distributed shortest path algorithm. IEEE Trans. Commun. **39**(6), 995–1003 (1991)
5. Gelenbe, E., Lent, R., Xu, Z.: Measurement and performance of a cognitive packet network. Comput. Netw. **37**(6), 691–701 (2001)
6. Pelusi, L., Passarella, A., Conti, M.: Opportunistic networking: Data forwarding in disconnected mobile ad hoc networks. IEEE Commun. Mag. **44**(11), 134–141 (2006)
7. Vahdat, A., Becker, D.: Epidemic routing for partially-connected ad hoc networks. Technical Report CS-2000-06, Duke Univerdity. Department of Computer Science (Apr 2000)
8. Song, L., Kotz, D.F.: Evaluating opportunistic routing protocols with large realistic contact traces. In: Proceedings of the 2nd ACM Workshop on Challenged Networks (CHANTS'07), pp. 35–42, ACM (2007)
9. Dimakis, N., Filippoupolitis, A., Gelenbe, E.: Distributed building evacuation simulator for smart emergency management. Comput. J. **53**(9), 1384–1400 (2010)
10. Filippoupolitis, A., Hey, L., Loukas, G., Gelenbe, E., Timotheou, S.: Emergency response simulation using wireless sensor networks. In: Proceedings of the 1st International Conference on Ambient Media and Systems (Ambi-Sys'08). Institute for Computer Sciences, Social-Informatics and Telecommunications Engineering (ICST), Brussels, Belgium, pp. 21:1–21:7 (2008)
11. Elms, D., Buchanan, A., Dusing, J.: Modeling fire spread in buildings. Fire. Technol. **20**(1), 11–19 (1984)

Automatic Population of Scenarios with Augmented Virtuality

Oscar Ripolles, Jose Simo and Roberto Vivo

Abstract In this paper we propose a new augmented virtuality framework where the synthetic scenario is populated with the information coming from the real-world. Our proposal is based on a smart camera which processes the images to detect objects. With this information, our framework relies on the collisions of the optical rays with the scenario to locate the detected objects. Our solution is also capable of handling situations where objects are piled on others.

Keywords Augmented virtuality · Surveillance · Computer vision

1 Introduction

Video surveillance is an increasingly active research area, and different researchers have proposed systems for visual surveillance using mixed reality. Most surveillance solutions are based on augmented reality and very few systems offer augmented virtuality applications, where the virtual environments are *augmented* with data and media coming from the real world.

In this paper we present a new augmented virtuality framework using automatic video interpretation. The framework that we propose combines a computer graphics application with a smart camera which is currently being developed under

O. Ripolles (✉) · J. Simo · R. Vivo
Universitat Politècnica de València, Inst. Universitari d'Automàtica i Informàtica Industrial, Camino de Vera s/n, València, Spain
e-mail: oripolles@ai2.upv.es

J. Simo
e-mail: jsimo@ai2.upv.es

R. Vivo
e-mail: rvivo@ai2.upv.es

E. Gelenbe et al. (eds.), *Computer and Information Sciences II*,
DOI: 10.1007/978-1-4471-2155-8_44,

the SENSE project.[1] This smart camera performs image analysis onboard and outputs a stream of XML-coded information containing the details on the detected objects [1]. In this way, our aim is to develop a fully synthetic environment where the augmentation is only used to indicate the characteristics of the detected objects and their location. The graphics application uses the information coming from the smart camera to infer the 3D characteristics of each detected object. Previous solutions have typically used bounding boxes to locate the objects [2, 3], which result in very simple approximations. In our proposal we use the silhouette information to calculate the size of the objects in their three dimensions. Once the 3D scene is populated, the user can vary the view-point as desired to observe the scene from a better perspective.

This paper is structured as follows. The following section introduces previous solutions for the simulation of real scenarios in 3D. Section 3 details the proposed framework and describes the algorithms. Section 4 presents the results that we have obtained and, finally, the conclusions and the main ideas for future work are given in Sect. 5.

2 Previous Work

There are two main lines of research regarding augmented virtuality for surveillance. On the one hand, some authors propose combining 3D environments with real videos. In this sense, 3D scenarios are augmented by locating the real cameras on them and adding a textured quad on this position to render the video sources. These solutions aim at offering situational awareness so that the user has a clearer understanding of where the cameras are located [3, 4].

On the other hand, some authors propose using computer graphics techniques to estimate the position of the objects [2, 3]. This process is depicted in Fig. 1, where the synthetic camera (point A) is located and oriented as the real-world camera, while the image plane is located and oriented so that it matches exactly the perspective view from the synthetic camera. Authors commonly use very rough 3D scenarios and create a simple textured polygon at point C to represent the object with an image from the video source [2, 3, 5]. From a different perspective, other authors propose training the systems to detect a low range of objects. Then, when simulating a scene, the algorithms perform a classification of the objects by finding the combination of type of object, height and orientation that best fits the detected bounding box in the 2D image [6]. With a similar idea, Osawa et al. use the information of previous frames to produce a set of synthetic images by varying the position of the objects in the 3D scenario until the objects detected in the real and the synthetic images are the same [7].

[1] SENSE stands for Smart Embedded Network of Sensing Entities (Project 033279), developed under a grant from European Sixth Framework Programme

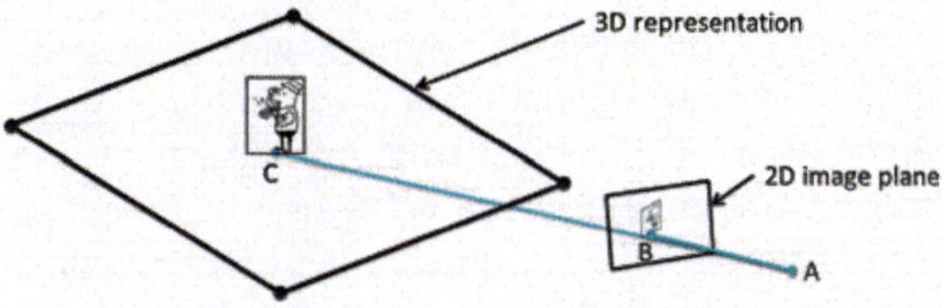

Fig. 1 Example of object location by casting a ray from A (*optical center*) through B (*object location on the image plane*) intersecting the floor in C (*calculated 3D position*)

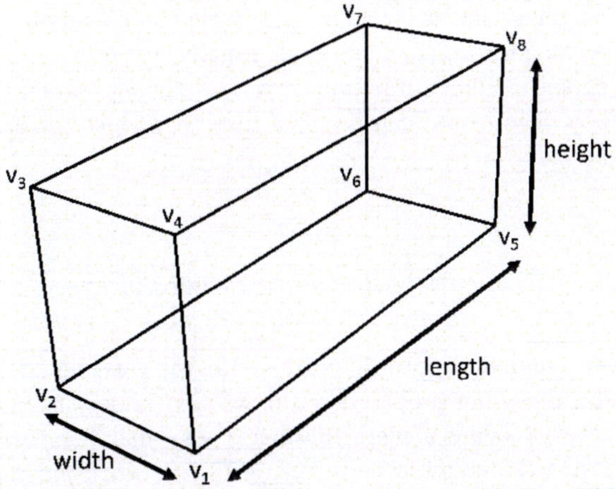

Fig. 2 Cuboid, the geometrical shape used to represented the detected objects

3 Augmenting the Virtual Environment

To convert the 2D information obtained from the camera into accurate 3D data, we follow an approach similar to the one presented in Fig. 1. The main difference with previous approaches is that we use multiple points from the silhouettes to obtain a much better approximation. For the visualization system we have chosen OpenSceneGraph as rendering engine [8]. For the correct performance of the proposed solution, the 3D scenario must be modeled with the actual sizes of the real-world and the camera must be correctly registered. Within the synthetic environment, each detected object is rendered as a cuboid (see Fig. 2).

Depending on the information retrieved from the camera, the system is capable of modeling the objects with more or less detail. For every detected object we are always able to use the information of their bounding box. Nevertheless, if the detected silhouette meets some requirements the framework is also capable of using the silhouette information to model the object with a higher accuracy.

3.1 Using the Bounding Box to Model the Object

The information on the bounding box that the smart camera outputs can be used to approximate the location and shape of the detected object. The algorithm to follow in this case is simple. Firstly, we start by casting two rays towards the scenario through the two bottom corners of the bounding box. The two collisions give us the location of the object within the scenario and the projected width. Secondly, we create a quad centered between the two collisions and oriented accordingly. The width of the box is slightly increased and the height is given a high value to avoid errors in the following step. The third step consists in casting two more rays through the two top corners of the bounding box. These collisions indicate the height of the detected object, which is calculated as the average of the values obtained from each collision. Finally, we must infer a value for the length of the cuboid, as this third dimension of the object cannot be assessed from the bounding box. The length can be fixed for every object or proportional to its width and/or height.

3.2 Using the Silhouette to Model the Object

Although the approach presented in the previous section was suitable for all cases, the algorithm proposed in this section is only adequate for objects oriented with respect to the camera in such a way that perspective information is available. Using the notation presented in Fig. 2, when the casted ray towards vertex v_1 is parallel to the edge (v_2, v_1) and perpendicular to the edge (v_5, v_1) (or vice-versa), it is not possible to infer further details on the cuboid as we lack information on the width or the length of the object.

The idea of our second augmentation algorithm is to select the silhouette points that can correspond to the vertices of the cuboid:

- from the points located near the left side of the silhouette, select the one at the bottom to represent v_2
- from the points located near the right side of the silhouette, select the one at the bottom to represent v_5
- from the points located near the bottom side of the silhouette, select the one at the center to represent v_1
- from the points located near the left side of the silhouette, select the one at the top to represent v_3
- from the points located near the right side of the silhouette, select the one at the top to represent v_8

The first three points are projected on the 3D scenario and then used to calculate the position (we consider v_1 to be the location), the width and the length of the cuboid. The only dimension to find out is the height. We create two cuboids, one located at edge (v_2, v_6) and the other one on edge (v_5, v_6). We project the two latter points (v_3 and v_8) and calculate the height as the mean of the values obtained from these two collisions.

Table 1 Results obtained when varying the orientation or the location

Angle(°)	Real Position (x,y)	Calculated position (x,y)	Width	Height	Length
15	50, 25	50.45, 27.23	2.95	4.22	14.07
30	50, 25	51.22, 26.59	3.39	4.09	13.92
45	50, 25	51.02, 25.97	3.86	4.33	13.52
60	50, 25	51.89, 24.45	3.49	4.10	12.93
75	50, 25	49.03, 26.33	3.75	3.78	11.75
45	50, 25	50.46, 24.77	3.81	4.15	13.84
45	50, 40	50.72, 36.98	3.75	3.75	13.72
45	80, 25	82.86, 23.24	4.54	4.14	12.52
45	80, 40	83.72, 39.15	3.27	4.15	12.72

3.3 Solving Ambiguities

Our framework is capable of working with complex environments with objects piled on others. A problem can appear when parts of an object are hanging off, as we could obtain erroneous values when projecting the corners of the bounding box or the silhouette points. One of the advantages of working with OpenSceneGraph is the fact that the engine is capable of indicating the object that the ray collides to. With this information, it is easy to retrieve the height of the different objects the rays collide and select the highest one (provided the collision happens on top of the objects; if not, errors may be introduced). This height is then used to locate an imaginary plane where to perform the calculations of the location and the shape of the object using the algorithms presented before.

4 Results

In this section we present some results that have been obtained using our augmented virtuality application, which has been developed using OpenSceneGraph, which is written entirely in C++ and OpenGL.

We want to start by testing how the accuracy of the detection depends on the orientation of the object with respect to the camera. Table 1 presents the results obtained when detecting a stapler which measures $3.67 \times 4.55 \times 14.4$ centimeters (width $\times$ height $\times$ length) using the silhouette algorithm. In this table we vary the location and the orientation, being the latter referred to the angle between the ray that goes through v_1 and the edge (v_1, v_5).

Firstly, in the results depicted in the first half of Table 1 we vary the angle, and we can see how the lower the angle is, the higher the accuracy for calculating the length, as we have less perspective information. Similarly, the higher the angle is, the higher the precision for calculating the width. It is worth mentioning that, angles 0° and 90° have not been considered as the silhouette algorithm cannot be applied. Secondly, the results of the second half test the accuracy when the staple is always oriented at approximately 45°. These results prove how the calculations

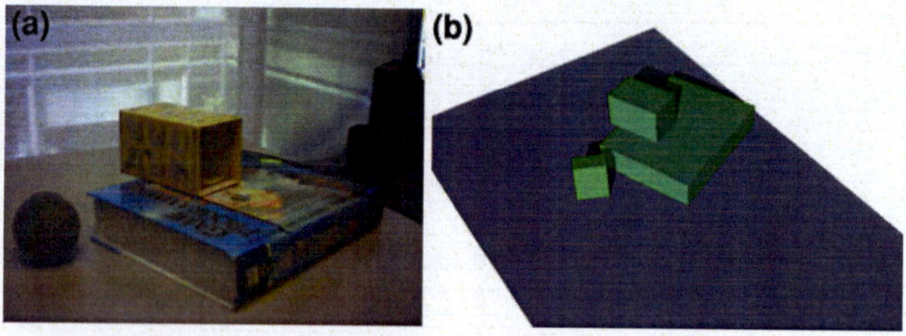

Fig. 3 Synthetic environment obtained from a real scenario. **a** Scenario with three objects **b** Image from a different perspective

are more precise when the object is closer to the camera, as the algorithm has more information on the silhouette.

From a different perspective, Fig. 3 offers a more complex scenario, where the box and the book are modeled with the silhouette algorithm, while the ball is modeled using the bounding box as no perspective information can be obtained from its silhouette. Finally, we must underline that the box has been located in such a way that some points of the silhouette hang off the book. In this case, as some rays collide on the book and some on the floor plane, the algorithm selects the height of the book to locate a plane with which continues the modeling process.

5 Conclusions

In this paper we have presented an augmented virtuality framework based on a smart camera. We have shown how our approach is capable of locating the objects on the scenario with high accuracy. This paper offers important improvements over previous solutions, as it uses the silhouette to better adjust the shape of the objects. Once the synthetic scene is populated, the user can modify the camera position and orientation as desired, obtaining new perspectives that can be useful for surveillance tasks.

As future work, our main line for future work implies extending the vision algorithms of the SENSE camera to offer features like classification, speed or color to improve the quality of the simulation performed in the graphics application. For example, having information on trajectories can enable prediction if there were delays in the visual feedback loop. From a different perspective, we must say that the use of a multi-camera framework would imply altering our algorithms but would definitely improve our simulation.

Acknowledgments This work has been funded by the Spanish Government (TIN2009-14103-C03-03, TSI-020400-2009-0133 and DPI2008-06737-C02-01/02) and by the European Union (ITEA2 IP08009).

References

1. Benet, G., Simo, J.E., Andreu, G., Rosell, J., Sanchez, J.: Embedded low-level video processing for surveillance purposes. In: 3rd International Conference on Human System Interaction (HSI) (2010)
2. Sebe, I., Hu, J., You, S., Neumann, U.: 3D video surveillance with augmented virtual environments. In: ACM Workshop on Video Surveillance, pp. 107–112 (2003)
3. Rieffel, E., Girgensohn, A., Kimber, D., Chen, T., Liu, Q.: Geometric tools for multicamera surveillance systems. In: Conference on Distributed Smart Cameras (2007)
4. de Haan, G., Scheuer, J., de Vries, R., Post, F.: Egocentric navigation for video surveillance in 3D virtual environments. In: IEEE workshop on 3D User Interfaces, pp. 103–110 (2009)
5. Fleck, S., Busch, F., Biber, P., Strasser, W.: 3D surveillance a distributed network of smart cameras for real-time tracking and its visualization in 3D. In: Conference on Computer Vision and Pattern Recognition Workshop (CVPRW06), p. 118 (2006)
6. Ziga, M., Brmond, F., Thonnat, M.: Fast and reliable object classification in video based on a 3D generic model. In: Proceedings of the International Conference on Visual Information Engineering (VIE2006), pp. 26–28 (2006)
7. Osawa, T., Wu, X., Wakabayashi, K., Koike, H.: 3D human tracking for visual monitoring. NTT. Tech. Rev. **5**, 1–8 (2007)
8. Osfield, R., Burns, D.: OpenSceneGraph. http://www.openscenegraph.org (2011)

Support for Resilient Communications in Future Disaster Management

V. M. Jones, G. Karagiannis and S. M. Heemstra de Groot

Abstract Disasters are often accompanied by damage to critical infrastructure, including (wireless) communications infrastructure. Our solution for emergency communications is based on advanced networks: Generalized Access Networks (GANs), Body Area Networks (BANs) and Vehicular Networks, to support dynamic, resilient communication services for disaster management.

Keywords Disaster management · Health monitoring · Wireless communications · GAN · BAN · Vehicular networks · MANETS · Resilience · QoS

1 Introduction

Wireless communications have been widely deployed in disaster management. Experience around the world however shows that communication and coordination continue to present major challenges. Problems are compounded if communications are disrupted through destruction of infrastructure. Digital solutions for emergency communications have not resolved all the issues; problems such

V. M. Jones (✉) · G. Karagiannis
Faculty of Electrical Engineering Mathematics and Computer Science,
University of Twente, Drienerlolaan 5, 7522 NB Enschede, The Netherlands
e-mail: v.m.jones@utwente.nl

G. Karagiannis
e-mail: g.karagiannis@utwente.nl

S. M. H. de Groot
Twente Institute for Wireless and Mobile Communications,
Institutenweg 30, 7521 PK Enschede, The Netherlands
e-mail: sonia.heemstra.de.groot@ti-wmc.nl

S. M. H. de Groot
Delft University of Technology, Postbus 5, 2600 AA Delft, The Netherlands

E. Gelenbe et al. (eds.), *Computer and Information Sciences II*,
DOI: 10.1007/978-1-4471-2155-8_45,

as network congestion, loss of connectivity and indoor and underground coverage continue, resulting in difficulties for first responders at the scene and for incident commanders. We present a solution to support dynamic, resilient communication services for disaster management. The work is based on the MOSAIC project's Major Incident Scenario [1, 2].

2 Proposed Solution

In our solution, emergency services personnel and vehicles are equipped with Body Area Networks (BANs) and Vehicular Networks which comprise nodes that connect to form mobile ad hoc networks (MANETS) to support intra- and inter-service communications. MANETS may provide the only communication possibility for emergency services by providing islands of communication whereby personnel at the scene can communicate. Further, these ad hoc networks should discover and communicate with any surviving telecommunication infrastructure, connecting over damaged infrastructure networks with disaster and emergency services' coordination centres. Such a solution must support resilience mechanisms providing fast connectivity restoration, resulting in a self-healing communication environment. The advanced networking technologies needed to realise this solution are described below.

Body Area Networks (BANs). A BAN is a body worn network of communicating devices (e.g. sensors, actuators). A BAN may communicate wirelessly with external networks. In 2001 BANs were proposed to support Virtual Trauma Teams [3]; a casualty attended by an ambulance team would be fitted with a trauma patient BAN which would measure vital signs and transmit them to the hospital. Each paramedic would be equipped with a paramedic BAN effecting A/V communication with the hospital. These BANs were implemented and trialled during the IST MobiHealth project. The MobiHealth trauma trials involved a single casualty and one ambulance team. In IST MOSAIC this scenario was extended to cover a major incident, where multiple teams from different emergency services cooperate during first response. The futuristic MOSAIC scenario shows wearable microelectronics incorporated into the uniforms of the emergency services personnel to support biosignal monitoring, positioning and A/V communication with control centres. Firefighter BANs would also include environmental sensors and paramedic BANS would also be able to discover/query casualties' BANs, link BAN data with the EMR and enable telepresence and an augmented reality experience for hospital staff. The BANs communicate with the emergency vehicles to provide an ad hoc communications network at the scene.

*Wireless Vehicular Net*works range from vehicular area networks (VANs), for communication between persons, equipment and devices located in the vicinity of the vehicle, to ad-hoc networks for inter-vehicle communication and vehicle to infrastructure communications. Vehicular networks make use of different wireless technologies (e.g WiFi IEE802.11, WiMAX, Bluetooth and Zigbee, GPRS, or UMTS/W-CDMA), as well as domain-specific technologies such as TETRA and satellite links. IEEE802.11p [4] is the main wireless technology for vehicle to vehicle communication and the two main standards are IEEE WAVE [5] and ISO CALM [6]. In a disaster we envision a large hybrid ad hoc and infrastructure network providing ad hoc communication between vehicles of the different emergency services, vehicle to infrastructure, and vehicle to emergency services personnel. In addition, a wide range of sensing systems (such as rapidly deployed sensors to detect toxic gases and high temperatures in the incident area, including cameras and robots with sensor and information processing units) will be part of the combined network supported by the vehicular networks.

Generalized Access Networks (*GANs*) allow ubiquitous connectivity to ad hoc networks. The Generalized Access Network (GAN) proposed provides an enhancement to the 3GPP General Access Network [7] using a radio access method which utilizes unlicensed spectrum on an all IP-based based broadband transport network. This is a B4G architecture, including mobile and wireless access networks, based on a flexible, seamless all-IP infrastructure with enhanced interworking features and global roaming for all supported access technologies. The supported radio access methods include wireline and radio access methods and are based on existing radio technologies using licensed and unlicensed spectrum, sharing the same network infrastructure among different operators and supporting appropriate levels of security and QoS. The approach is based on separating physical networks into a number of overlay networks to support, for example, separation of multiple mobile virtual operators and the formation of service networks and Virtual Private Networks (VPNs). The system incorporates: *a GAN operator* with a common infrastructure supporting multiple radio interface technologies through which mobile hosts and several forms of ad hoc networks are attached, e.g. Mesh networks, Wireless Sensor networks, BAN and PAN; *different forms of ad hoc communication*, involving emergency services personnel and vehicles; *multiple radio technologies* for communication with remote centres using the different GAN system and radio interfaces. Any existing radio technology can be supported, e.g. TETRA, Zigbee, Wibree, Bluetooth, WLAN, GSM/GPRS, UMTS/W-CDMA, CDMA2000, UMTS/HSPA, WiMax, LTE or UMB. A GAN should be able to satisfy the following critical requirements: (1) Security support; (2) Resilience support; (3) Mobility and load distribution support; (4) Network management support; (5) Ad hoc networks for emergency management support. Some solutions that can be used within the GAN to satisfy these requirements are: (a) for support of strict QoS requirements in combination with the requirement on supporting different communication modes it is recommended to use MPLS in combination with RSVP-TE for point to multipoint traffic engineering label switched paths and MPLS multicast label; (b) for support of mobility requirements it is recommended to use the generic PMIP with multicast support; (c) for support of security, it is recommended to use IPSec combined with security solutions applied for MPLS.

3 Conclusions and Discussion

The challenges relating to emergency communications are being tackled by governments, industry, standards bodies and NGOs, and by many international fora, conferences and researchers. The IASC Working Group on Emergency Telecommunications addresses regulatory, operational and technical aspects of telecommunications for disaster relief. The Major Incident Medical Management and Support procedures have been adopted as a military standard in the UK, Netherlands, Italy and Sweden and are used in NATO training. Many research projects focus on the technical challenges surrounding communication and related ICT issues. The success of emergency response depends on access to timely and reliable information, however networks and systems are constrained by extreme operating conditions. The Hyperion projects investigate adaptive agent-based information management systems for enhancing quality and resilience of information made available to defense users [8]. Simulation tools, important for training responders in different emergency situations, may also be used as a means to evaluate alternative actions during operation. Examples are tools for real-time evaluation of optimal evacuation routes [9] and an on-line decision support system [10]. In [11] a multi-agent simulator is integrated with a real wireless sensor network. Robots equipped with wireless communications to create ad hoc communication networks to assist trapped victims are proposed in [10]. In this paper we address part of the problem by proposing a combination of advanced networking technologies to support *communication and reliable transfer of information* during first response. We believe the proposed solution supports the identified requirements by enabling ubiquitous connectivity to ad-hoc networks, resilience in the face of damaged infrastructure, mobility and load distribution and adaptive network management. Major and critical challenges remain; they include: ensuring security of the emergency communications services, especially crucial in case of terrorist attack, and quality assuring correct system behaviour in such dynamic adaptive distributed systems.

References

1. Jones, V., Saranummi, N.: MOSAIC vision and scenarios for mobile collaborative work related to health and wellbeing. ICE 2005, 11th International Conference on Concurrent Enterprising, University BW Munich, Germany, 20–22 June 2005, AMI@Work Forum Day: Towards Ambient Intelligence at Work, Vision 2010. In: Proceedings of the 1st AMI@work communities Forum Day 2005, Towards Ambient Intelligence at Work 2010, Munich, Germany, 22 June 2005, Marc Pallot and Kulwant S Pawar (eds.) ISBN 13 978 0 85358 225 0
2. Jones, V., Karagiannis, G., Heemstra de Groot, S.: Ad hoc networking and ambient intelligence to support future disaster response. In: Proceedings. ASWN 2005, 5th Workshop on Applications and Services in Wireless Networks, 29 June–1 July 2005, Paris, pp. 137–146, IEEE. ISBN 2-9156-18-08-9

3. Jones, V.M., Bults, R.G.A., Konstantas, D.M., Vierhout P.A.M.: Healthcare PANs: personal area networks for trauma care and home care. Presented at fourth international symposium on wireless personal multimedia communications (WPMC). Aalborg, Denmark (2001)
4. IEEE P802.11p: IEEE standard for information technology—telecommunications and information exchange between systems—local and metropolitan area networks—specific requirements Part 11: Wireless LAN Medium Access Control (MAC) and Physical Layer (PHY) Specifications Amendment 6: Wireless Access in Vehicular Environments, June 2010
5. IEEE 1609.3: IEEE trial—use standard for wireless access in vehicular environments (WAVE)—networking services (2007)
6. ISO 21217: Intelligent transport systems—communications access for land mobiles (CALM)—architecture, 1st edn., Stage: 40.60, TC 204, ISO 21217 (2010)
7. Karagiannis, G., Jones, V.M., Heemstra de Groot, S.M.: Support of Future Disaster Response Using Generalized Access Networks (GANs). Proceedings of Information Technology in Biomedicine, ITAB 2006, International Special Topic Conference on Information Technology in Biomedicine. IEEE Engineering in Medicine and Biology Society, pp. 26–28, Oct 2006
8. Ghanea-Hercock, R., Gelenbe, E., Jennings, N.R., Smith, O., Allsopp, D.N., Healing, A., Duman, H., Sparks, S., Karunatillake, N.C., Vytelingum, P.: Hyperion: next-generation battlespace information services. Comput. J. **50**(6), 632–645 (2007)
9. Dimakis, N., Filippoupolitis, A., Gelenbe, E.: Distributed building evacuation simulator for smart emergency management. Comput. J. **53**(9), 1384–1400 (2010)
10. Filippoupolitis, A., Loukas, G., Timotheou, S., Dimakis, N., Gelenbe, E.: Emergency response systems for disaster management in buildings. In: Proceedings of the NATO Symposium on C3I for Crisis, Emergency and Consequence Management, NATO Research & Technology Organisation, Bucharest, Romania, May 2009
11. Filippoupolitis, A., Hey, L., Loukas, G., Gelenbe, E, Timotheou, S.: Emergency response simulation using wireless sensor networks. In: Proceedings of the 1st International Conference on Ambient Media and Systems, pp. 1–8, Quebec City, Canada, Feb 2008. ICST, Brussels, Belgium

Agent-Based Modeling of a Price Information Trading Business

Saad Ahmad Khan and Ladislau Bölöni

Abstract We describe an agent-based simulation of a fictional (but feasible) information trading business. The Gas Price Information Trader (GPIT) buys information about real-time gas prices in a metropolitan area from drivers and resells the information to drivers who need to refuel their vehicles. Our simulation uses real world geographic data, lifestyle-dependent driving patterns and vehicle models to create an agent-based model of the drivers. We use real world statistics of gas price fluctuation to create scenarios of temporal and spatial distribution of gas prices. The price of the information is determined on a case-by-case basis through a simple negotiation model. The trader and the customers are adapting their negotiation strategies based on their historical profits. We are interested in the general properties of the emerging information market: the amount of realizable profit and its distribution between the trader and customers, the business strategies necessary to keep the market operational (such as promotional deals), the price elasticity of demand and the impact of pricing strategies on the profit.

1 Introduction

In this paper we describe an agent-based simulation of a fictional (but feasible) information trading business. The Gas Price Information Trader (GPIT) buys information about real-time gas prices in a metropolitan area from drivers, and resells this information to drivers who need to refuel their vehicles.

S. A. Khan · L. Bölöni (✉)
Department of EECS, University of Central Florida, Orlando, USA
e-mail: lboloni@eecs.ucf.edu

S. A. Khan
e-mail: skhan@eecs.ucf.edu

E. Gelenbe et al. (eds.), *Computer and Information Sciences II*,
DOI: 10.1007/978-1-4471-2155-8_46,

Similar systems have been proposed previously [1, 2], online webpages tracking gas prices currently exist (the ad-supported GasBuddy and GasPriceWatch in the United States, the government-run FuelWatch in Western Australia). These are related in spirit and operation to a large number of applications proposed in the field of urban computing/citizen computing, which can be similarly construed as information trading. While feasibility has been repeatedly demonstrated, many of these systems have not been, in general, deployment successes. We argue that the low deployment is due to the fact that voluntary participation can be only maintained for projects with emotional and political impact. Instead of relying on voluntary participation, our system relies on the self interest of the participants. We assume that the business is stable only if all the participants profit financially over the long run. We are interested in the overall market dynamics of the system. To illustrate the operation of the system, let us consider an example. A driver travels from work to home in Orlando. Near Lake Burkett, the car signals low fuel. The driver has several nearby choices for refueling: some of them are on his planned path, while others require short detours. He contacts the GPIT and requires information about the cheapest option in the vicinity. The GPIT obtains this information from its local database which contains information it has acquired ahead of time from drivers. It will provide this information for a price, which will be negotiated in real time.

From the point of view of the cost structure of its business, the GPIT is similar to a software company: it has an initial cost to acquire the information, but subsequently, it can sell it an arbitrary number of times, without incurring further costs (as long as the database is accurate). What makes the GPIT business challenging is the difficulty in determining the utility of the information to the user, the very large variation of the utility from user to user, and, as we shall see, the variation of the utility over time periods.

These considerations require us to consider a system where the cost of information is determined on a case by case basis, by a short negotiation process between the user and the GPIT. Not all negotiations will be successful. In fact, there will be clients who will effectively drop out from the system. This will happen if the clients estimate the utility of the information provided by the GPIT to be zero (considering the driving habits and location of the client). The subset of clients who will regularly purchase information depends on the pricing and negotiation strategy of the GPIT. Specific long term pricing strategies such as promotional offers are also an important part of the system.

To understand the behavior of this system, we need a *detailed simulation*. In paper we use real world geographic data, lifestyle-dependent driving patterns and vehicle models to create an agent-based model of the drivers. We use real world statistics of gas price fluctuation to create scenarios of temporal and spatial distribution of gas prices. The price of the information is determined on a case-by-case basis through negotiation. The trader and the customers are adapting their negotiation strategies based on their historical profits.

We are interested in the general properties of the emerging information market: the amount of realizable profit and its distribution between the trader and customers, the business strategies necessary to keep the market operational (such as promotional deals), the price elasticity of demand and the impact of pricing strategies on the profit.

2 Negotiating the Price of Information

To save on the cost of fuel by choosing the optimal gas station, the customer contacts the GPIT, sends its location, planned path, and requests information about the optimal gas station at the same time making a price offer $x^1_{c\rightarrow s}$. The GPIT calculates this information and replies with a counter offer $x^2_{c\leftarrow s}$. The negotiation $\mathcal{X}^m_{c\leftrightarrow s}$ will consist of a series of exchanged offers $\{x^1_{c\rightarrow s}, x^2_{s\leftarrow c} \ldots x^n_{c\leftrightarrow s}\}$. If the kth negotiation was successful, it will conclude with a deal v^k. In this case, the client will pay v^k dollars to the GPIT.

The customer's negotiation strategy. The first offer made by the client is at u_{min}, it will uniformly concede with a concession rate $\delta \in (0, 0.5)$. The conflict offer set by the customer is twice the initial offer and is normally distributed around the initial offer, i.e., $u^c_{max} = 2 \cdot u^c_{min} + \mathcal{N}(\mu, \sigma^2)$ where $\mu = 0$ and $\sigma^2 = 1$. Therefore, the cost of a deal is given as

$$cost(\mathcal{X}^k_{c\leftrightarrow s}) = v^k = \begin{cases} x^n_{c\leftrightarrow s} & \text{if } x^n_{c\rightarrow s} \geq u^s_{min} \vee x^n_{c\leftarrow s} \leq u^c_{max} \\ 0 & \text{otherwise} \end{cases}$$

At the start of each negotiation the initial offer is dependent upon the previous r negotiations with the GPIT:

$$u^c_{min,k} = \left(m \cdot \frac{\sum_{n=1}^{r} v^{k-n}}{r} \right) + u^c_{min,k-1} \cdot (1-m) \tag{1}$$

where $m = 0.125$ bounds the initial offer to increment within one fourth of previous negotiated deal. The limitation of the amount of history is justified by the experimental fact that while recent negotiation experience can be a good predictor of the current negotiation, older negotiations are not.

The trader's negotiation strategy. The GPIT uses exponential smoothing to give more weight to the recent deals. The initial offer from GPIT in start of kth negotiation is

$$u^{s,k}_{max} = m \cdot \phi^k(r) + u^{s,k-1}_{max}(1-m) \tag{2}$$

where $m = 0.125$ and $\phi^k(r)$ is the exponential moving average of previous r deals:

$$\phi^k(r) = \alpha \cdot v^{k-1} + (1-\alpha)\phi^{k-1}(r) \tag{3}$$

where, there control factor is chosen as $\alpha = 2/(r+1)$ providing more weight to the recent negotiation.

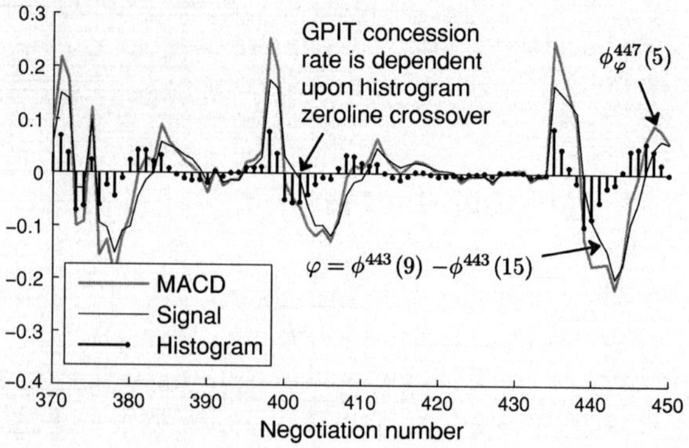

Fig. 1 Estimation of customer concession policy using MACD

To model the gradual change in the customers' concession, the GPIT uses *Moving Average Convergence Divergence* $\phi_{\varphi}^{k}(r)$ (MACD). This indicator provides a histogram with the help of which GPIT estimates the concession behavior of the customer. The histogram is obtained by computing the difference of two EMAs (of different weights) against an EMA of the difference. The signal φ, i.e., difference between two EMAs is given as:

$$\varphi = \phi(9) - \phi(15) \tag{4}$$

Evaluating φ with brackets of 9 and 15 takes-in consideration the current negotiations as well negotiations preceding it and these values have been determined experimentally to provide a good match of the customer behavior (in stock market analysis the values of 9, 12 and 26 day moving averages are used). The concession step of GPIT δ_k^c is dependent upon φ and is given as Fig. 1

$$\delta_k^c = \begin{cases} 0.15 & \text{if} \quad \varphi < \phi_{\varphi}^{k}(5) \\ 0.25 & \text{else} \end{cases} \tag{5}$$

3 Environment Modeling

The environmental conditions under which the system operates have been modeled based on real world data (and extrapolations of it) and at a relatively high level of detail.

Modeling geography. We have chosen to model a relatively large area of East Orlando. This area has a number of employers in education and high tech industry, as well as residential, shopping and recreation areas. It is a "driving city": public transportation options are virtually non-existent, the distances between areas are large and the zoning restrictions frequently prevent the building of mixed use areas. The simulator was directly contacting the Google Maps web service, parsing the returned KML files, and caching them locally for reuse.

Modeling gas prices. We have acquired our gas prices from the website GasBuddy.com, which works as an urban portal for people who can participate in posting and reviewing gas prices. Based on these prices we were able to generate a number of fictional scenarios. To extrapolate the values we used a technique to scale the spatial variance of the gas prices with an arbitrary factor, and added a Gaussian noise component.

Modeling lifestyle. The lifestyle of East Orlando residents consists of well distinguished time periods spent at distinct locations, and driving between them. The clients were assumed to have an 8-to-5 work schedule on weekdays, not going to work on weekends, and a randomly distributed set of other activities in the remaining time.

Scenario generation. The simulation of the economic system starts with a baseline scenario containing the movement patterns of the clients, over the course of a year, assuming that no refueling is needed. We assume that the variations of the price of the fuel do not impact the in-city driving patterns of the users. Running the economic model on this scenario will add the refueling times, occasionally inserting small detours necessary to find the cheapest gas station.

The generated scenario is implemented as a series of events which represent driving activities happening at specific locations and time. Each event is described by (a) the participating agent, (b) the event type, (c) the location, (d) the timestamp and (e) the distance travelled since the previous event.

4 Experimental Study and Results

We performed a series of simulations to analyze the properties related to the information market and the GPIT, observing (a) the total income for the server, (b) the agreement price and (c) the number of successful negotiations. Our first series of experiments verify if the dynamic price formation mechanism operates as expected. Our expectation is that the strategies deployed by the customers and the trader interact in such a way that they will agree on a price level where both the customers and the trader will profit over the long run. Figure 2(*left*) shows that the histogram of the agreed price for successful negotiations had a relatively wide distribution around the value of \$3/information. This is consistent with the variation of the gas prices and the potential savings of the customers filling their 15 gallon gas tanks.

Figure 2 (*right*) shows the ratio of successful negotations for a fixed initial offer by the trader ranging in the interval $u^s_{max} = [\$0..\$7]$. The conflict offer was set to $u^c_{min} = u^c_{max}/3$, and the concession rate was fixed at $\delta^c = 0.25$.

Figure 3 shows the profits of the trader obtained using a fixed initial offer. We observe the *rachet effect* [3] where even at higher prices some customers are willing to pay for the price of information yet the total profit is very low. If we compare the profits that are obtained using dynamic pricing as seen in Fig. 2, we see that it is not feasible for the GPIT to operate using non-variable prices.

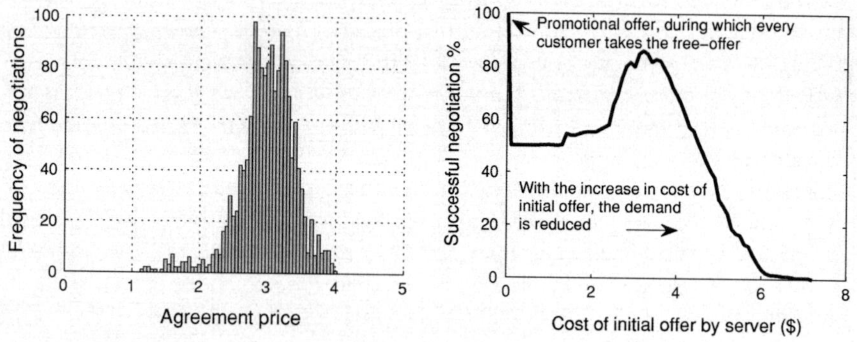

Fig. 2 Dynamic pricing analysis (*left*) and price elasticity analysis (*right*)

Fig. 3 Total profit versus fixed initial offer cost

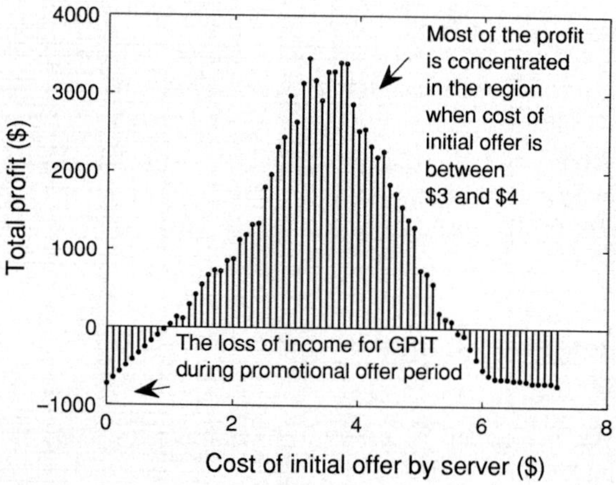

5 Related Work

Sim [4] discusses that variable adapting behavior should incorporated for dynamic-market. An agent based urban economic market with neighbourhood-variance and other spatial properties is presented in [5]. In [6, 7] a model for automated bidders and sellers with performance analysis is presented. Further it has been discussed that how network performance [8] and other parameters, e.g., arrival rates of bids or decision rates of sellers [9], affects the different parties involved.

References

1. Bulusu, N., Chou, C., Kanhere, S., Dong, Y., Sehgal, S., Sullivan, D., Blazeski, L.: Participatory sensing in commerce: using mobile camera phones to track market price dispersion. In: Proceedings of the UrbanSense (2008)

2. Dong, Y., Kanhere, S., Chou, C., Bulusu, N.: Automatic collection of fuel prices from a network of mobile cameras. Distributed Computing in Sensor Systems, pp. 140–156 (2008)
3. Acquisti, A., Varian, H.: Conditioning prices on purchase history. Mark. Sci. **24**(3), 367–381 (2005)
4. Sim, K.: A market–driven model for designing negotiation agents. Comput. Intell. **18**(4), 618–637 (2002)
5. Parker, D., Filatova, T.: A conceptual design for a bilateral agent-based land market with heterogeneous economic agents. Comput. Environ. Urban. Syst. **32**(6), 454–463 (2008)
6. Gelenbe, E.: Analysis of automated auctions. Computer and Information Sciences–ISCIS, pp. 1–12 (2006)
7. Gelenbe, E., Györfi L.: Performance of auctions and sealed bids. Comput. Perform. Eng. pp. 30–43 (2009)
8. Gelenbe, E.: Analysis of single and networked auctions. ACM Trans. Internet. Technol. (TOIT) **9**(2), 8 (2009)
9. Gelenbe, E., Velan, K.: An approximate model for bidders in sequential automated auctions. Agent Multi-Agent Syst. Technol. Appl. 70–79 (2009)

Modeling Lane Preferences in Agent-Based Multi-Lane Highway Simulation

Yi Luo and Ladislau Bölöni

Abstract Simulation of highway driving was traditionally the domain of virtual physics based models. Yet, traffic simulation is incomplete without considering the drivers' conscious strategic and tactical behavior. These aspects can be naturally simulated through an agent-based driver model. In this paper, we describe a model of the strategic lane preferences of the drivers, with a special attention to the optimal lane positioning for a safe exit. Our experiments show that the simulated traffic of Orlando's Highway 408 matches well with the real world traffic data. The increased simulation detail can be applied to crash prediction and the control of intelligent transportation system devices, such as variable speed limits.

1 Introduction

Existing microscopic traffic simulation models describe the behavior of the car and driver in terms of mathematical formulas which are often expressed in formulas analogous to those describing various physical phenomena: forces, viscosity, potential fields and so on. We will call these *virtual physics* models.

Our work is centered on improving the accuracy of microscopic highway simulation through agent based modeling of the *conscious* aspect of the driver behavior. For the starting point of the contributions described in this paper see [1]. We separate the behavior of the driver into two simulation modules:

Y. Luo · L. Bölöni (✉)
Dept of EECS, University of Central Florida, Orlando, USA
e-mail: lboloni@eecs.ucf.edu

Y. Luo
e-mail: yiluo@mail.ucf.edu

E. Gelenbe et al. (eds.), *Computer and Information Sciences II*,
DOI: 10.1007/978-1-4471-2155-8_47,

The virtual physics model models the physics of the vehicle as well as those aspects of the driver which are either reflexive (such as emergency braking) or learned to the point of becoming sub-conscious (such as lane following and keeping a constant distance from the car in front).

The agent model models the conscious cognition of the human driver. This includes both strategic planning (which exit to take, which lane to prefer for long distance driving) and tactical (the decision to join a convoy or overtake a slow moving car).

In this paper we concentrate on a single aspect of strategic behavior, the planning, decision and execution of lane changes.

2 Virtual Physics-Based Lane Change Models

Our baseline model extends the traditional car following model with the lane change model described by Kesting et al. [2]. This model assumes that lane changes happen instantaneously: for a shift to the left lane, a vehicle which has been previously in the middle lane, at time t disappears from the middle lane and appears in the left lane. This opens the possibility that a car, coming from behind in the new lane with a higher speed can not break sufficiently quickly and collides with the lane changing car. The model assumes that it is the responsibility of the lane changing car to ensure that the rear left vehicle $j-1$ has sufficient buffer distance such that it can decelerate before hitting the lane changing vehicle

$$\hat{a}_{j-1}(t) \geq -b_{max} \tag{1}$$

If this condition is not satisfied, the vehicle concludes that it is not safe to change lanes.

The notion of *politeness* models the fact that the driver might consider the accelerations of the other vehicles as well when taking a decision to change the lane. The politeness parameter p specifies how much does the vehicle discount the other vehicles' desired acceleration compared to its own. A value $p=0$ indicates an impolite, fully selfish driver which does not care about other drivers (however, it still considers the safety criteria). The vehicle i will decide to change the lane if the following inequality is verified:

$$\left(\hat{a}_i + p \cdot (\hat{a}_{j-1} + \hat{a}_{i-1}\right) - \left(a_i + p \cdot (a_{i-1} + a_{j-1})\right) \geq \Delta p_{th} \tag{2}$$

where Δp_{th} is the politeness threshold. The left hand side is the difference between the new accelerations $\hat{a}_i$, $\hat{a}_{j-1}$ and $\hat{a}_{i-1}$ if the vehicle i successfully changes into the target lane and the old accelerations a_{i-1} and a_{j-1} if it doesn't change lane. The intuition is that the vehicle favors to change lane only when the advantage of the action is greater than the disadvantage it exerts to its neighboring vehicles. However, because the vehicle i can not obtain the parameters (T, v_0, a, b)

for its successor $i-1$ and $j-1$, the utility of lane change can only be calculated by vehicle i's own parameters.

3 Strategic Lane Change Behavior

Strategic behavior considers, beyond the short term measurable advantages of a certail lane, the utility of the line for the long term goals of the driver. The first idea would be to use the left hand side of Formula 2 as the utility metric. This value, however, can be negative: it's range is $[-C, C]$ where

$$C = (a + b_{max})(1 + p) \tag{3}$$

We need, however, a strictly positive utility metric for the further definitions. To achieve this, we add C to the formula. Thus the utility of the current, left and right lanes will be defined as:

$$\begin{aligned} U_c &= \Delta p_{th} + C \\ U_l &= \left(\hat{a}_i + p \cdot (\hat{a}_{j-1} + \hat{a}_{i-1}\right) - \left(a_i + p \cdot (a_{i-1} + a_{j-1})\right) + C \\ U_r &= \left(\hat{a}_i + p \cdot (\hat{a}_{h-1} + \hat{a}_{i-1}\right) - \left(a_i + p \cdot (a_{i-1} + a_{h-1})\right) + C \end{aligned} \tag{4}$$

The preference model modifies the virtual physics model by allowing the long term planning process to assign the preference value $W_c \in [0.0, 1.0]$ to the lanes of the road. The virtual physics model will consider the *weighted utilities* of the lanes $U_c^w = W_c \cdot U_c$. When all the lanes have the same preference, the behavior reverts to the basic virtual physics model.

The preference weights are directly associated to the lanes of the highway, yet the vehicle needs to make decisions *one lane change at a time*. Thus the vehicle occasionally needs to accept a decrease in utility in order to reach a preferred lane after more lane changes.

To resolve this problem, we define the lane change preferences as follows. W_c is the preference of the vehicle's current lane. W_l and W_r are the *maximum* of all the preferences to the left and right of the vehicle, respectively.

Let's now consider some examples of the use of the preferences by the agent:

(i) When *entering* the highway, the agent will set the preference of its terminating entrance lane to zero. This will cause it to move to the highway's continuing lanes as soon as it is safe (see Fig. 1a).

(ii) When *driving* on the highway, the vehicle will assign higher preference to the lanes it prefers driving on. The preference gradients will be, however, milder. This allows the preference to be overridden if significant advantage is to be gained - or if the tactical maneuver requires it (see Fig. 1b).

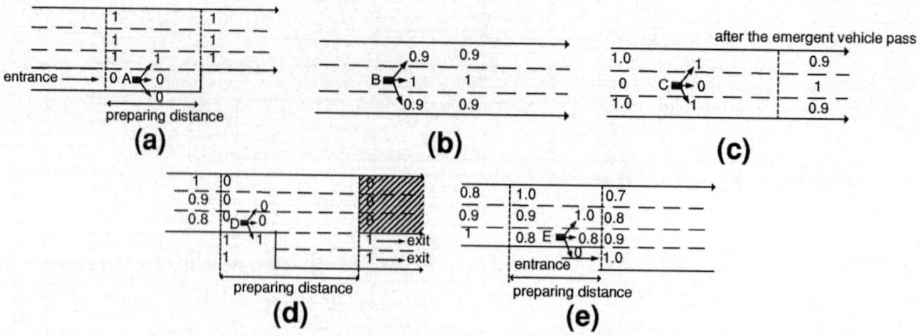

Fig. 1 The agent tries to evaluate the preferences of lane changes. **a** Entering **b** Driving **c** Give a way **d** Exit **e** Avoid busy lane

(iii) If the vehicle needs to *"give a way"* to a police car or emergency vehicle, it will set the specific lane(s) to zero preference, forcing it to move to one of the non-zero preference lane as soon as it is safe. Once the emergency vehicle has passed, the vehicle reverts to its previous preferences (see Fig. 1c).

(iv) If the vehicle prepares to *exit*, it will modify the lane preferences to prefer the exit lane. Note that this does not mean that the vehicle will immediately change to the exit lane, as a number of other safety conditions need to be satisfied for each lane change(see Fig. 1d).

(v) *Avoiding entering lanes.* Let us consider a vehicle which is on inter-city routes prefers to drive on the rightmost lanes. These lanes, however become extremely busy before and after exits with cars which are entering and exiting the highway. Thus many drivers prefer to the left side of the road around the exit to their default preferences after the merge finished (Fig. 1e).

To model the behavior of human drivers we implemented a speed adaptation technique. The safety condition for lane change is more likely to be satisfied if the vehicle modifies the speed of the desired destination lane. Thus, under situations of *forced lane change*, the vehicle will change its desired speed to the current speed of the neighboring vehicle in the destination lane. If the vehicle needs to cross several lanes (as the case of the exit) it will change its desired speed in steps, always adapting it to the speed of the next destination lane. Once the forced lane change situation is terminated, the desired speed of the vehicle reverts to the one dictated by the virtual physics model.

3.1 *Case Study: Preparing for Exit or Merge*

Exiting the highway is a maneuver which requires several consecutive forced lane changes. It was found that about 10% of the crashes occuring on highways are *sideswipe crashes* while about 11% of them are *angle crashes* [3]. Both types are associated with lane changes (the reminder of the crashes are mostly rear-end crashes). Modeling the mechanics of this process is of a major importance as it can predict traffic situations with high crash risk.

Probability of successful lane change. The need to prepare in advance for exit is due to the fact that a driver who intends to perform a lane change might not be able to execute it for a certain amount of time.

The difficulty of the lane change depends on the local density of the vehicles in the target lane D_i and the average speed difference between the vehicle and the neighboring vehicles in the target lane ΔV_i. An experienced driver can estimate $Pr(t, D_i, \triangle V_i)$—the probability that it can successfully change lanes in time t for a specific value of the density and speed difference.

Probability of successful exit. If the vehicle is currently n lanes away from the exit lane, it will need to successfully execute n lane changes before exit. The driver needs to start its exit preparations at such a time/distance ahead so that it can safely exit. Let us denote with $Pr(t_s, D_i, \triangle V_i)$ the probability of a single lane change which is finished at time t when the next lane i has density D_i and speed difference $\triangle V_i$. In general, if the agent tries to change from lane i to j in time n, the probability that it can succeed is

$$Pr(i,j,n) = \begin{cases} \sum_{t=1}^{n-(j-i)+1} Pr(t, D_{i+1}, \triangle V_{i+1}) Pr(i+1, j, n-t) & i<j \\ \sum_{t=1}^{n-(i-j)+1} Pr(t, D_{i-1}, \triangle V_{i-1}) Pr(i-1, j, n-t) & i>j \\ 1 & i=j \end{cases} \quad (5)$$

4 Experimental Study

For the experimental study we implemented the virtual physics and agent models in our simulator. The agent model also includes a number of tactical behavior components not discussed in this paper (such as communication through signaling), which ensures a higher accuracy and realism of the overall simulation. The experiments have been performed on a detailed, lane-by-lane model of a 22.13 mile stretch of Highway 408. Inflow and outflow information was acquired from the statistics of the expressway authority.[1] The vehicle inflow was modeled as a Poisson traffic, matching the specified average inflow rate. For the following experiments we will study two different types of vehicle behavior with the same virtual physics model but different agents. The SIG agent does not change the speed of the vehicle when trying to change lane. In contrast, the VAR agent is changing its desired speed to match the destination lane.

Rate of exit misses function of the exit preparation distance. In this experiment, we study the rate of the exit misses (or, in a different interpretation, of the

[1] http://www.expresswayauthority.com/Corporate/aboutStatistics/HistoricalTraffic.aspx

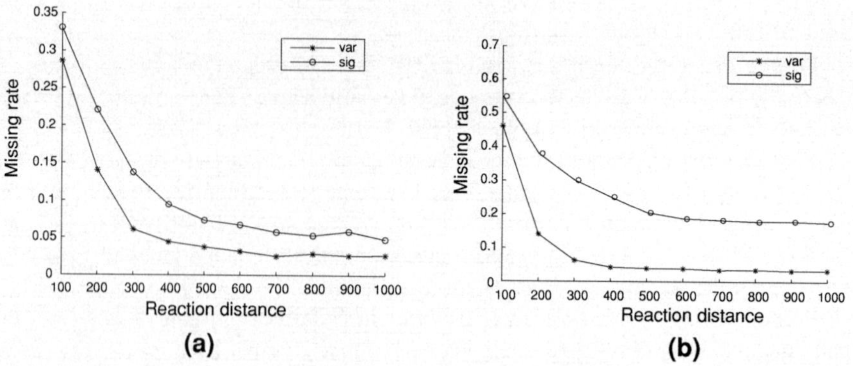

Fig. 2 The rate of exit misses function of the preparation distance with normal inflow and outflow rate 2, and during rush hour 2. **a** Normal **b** Rush hour

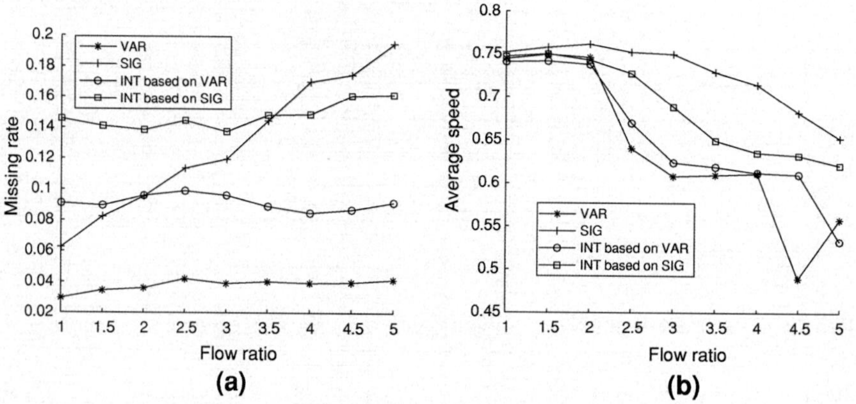

Fig. 3 The rate of exit misses and average speed in the function of flow ratio on Highway 408. **a** Rate of exit misses **b** Average speed

dangerous exits) in function of the distance where the vehicles start their preparation for exit by changing their lane preferences to prefer the exit lane (as in Fig. 1d).

Figure 2a shows rate of exit misses for the SIG and VAR agents for regular traffic on Highway 408. We find that for agents the miss rate decreases with the distance, but in general the VAR agent has a lower miss rate.

Figure 2b shows the same measurements for rush hour traffic (with the inflow and outflow increased five times). The conclusions from the normal traffic situation extend to this scenario as well. The rate of exit misses of the VAR agent did not change significantly, on the other hand the miss rate of the SIG agent is much higher, and it cannot be reduced below about 20% even with early preparation.

We conclude that adapting the speed to the target lane is a major requirement for safe driving under high traffic conditions. While this might appear as a commonsense advice for an experienced driver, it is an observation which does not appear in purely virtual physics based models, yet it emerges naturally when that model is augmented with an agent-based conscious behavior simulator.

Rate of exit misses function of inflow ratio. Figure 3 plots the exit misses as well as the average speed function of the inflow ratio. We compare four strategies: SIG and VAR with a fixed preparation distance of 600 m, and their "intelligent" variants with adaptive preparation distance INT-SIG and INT-VAR. We find that, as expected, the adaptive strategies have a more "flat" diagram, allowing us to choose our preferred compromise between performance and risk.

References

1. Luo, Y., Bölöni, L.:Towards a more accurate agent-based multi-lane highway simulation. In: Proceedings of International Workshop on Agents in Traffic and Transportation (ATT10), in conjunction with the 9th Joint Conference on Autonomous and Multi-Agent Systems (AAMAS 2010), pp. 13–20. May 2010
2. Kesting, A., Treiber, M., Helbing, D.:General lane-changing model MOBIL for car-following models. Transp. Res. Rec. J. Transp. Res. Board. **1999**(−1), 86–94 (2007)
3. Pande, A., Abdel-Aty, M.:Assessment of freeway traffic parameters leading to lane-change related collisions. Elsevier J. Accid. Anal. Prev. **38**, 936–948 (2006)

Chromosome Coding Methods in Genetic Algorithm for Path Planning of Mobile Robots

Adem Tuncer and Mehmet Yildirim

Abstract In this study, various chromosome coding methods are analyzed for genetic algorithm to solve path planning problem of mobile robots. Path planning tries to find a feasible path for mobile robots to move from a starting node to a target node in an environment with obstacles. Genetic algorithms have been widely used to generate an optimal path by taking the advantage of its strong optimization ability. Binary, decimal and orderly numbered grids coding methods are used to create chromosomes in this study. Path distance, generation number and solution time parameters are observed and compared for the three coding methods under the same conditions. Results showed that the solution time is directly affected by chromosome coding method.

Keywords Genetic algorithm · Path planning · Mobile robot

1 Introduction

Recently, interest of researchers on autonomous vehicles increases with technological developments. There are many studies in the literature about autonomous vehicles. One of the main subject on autonomous vehicle is path planning. Path planning tries to find a feasible path for mobile robots to move from a starting node to a target node in an environment with obstacles [1].

Path planning environment can be either static or dynamic. For the static environment, the whole solution must be found before starting execution.

A. Tuncer (✉) · M. Yildirim
Networked Control Systems Laboratory, Technical Education Faculty,
Kocaeli University, 41380 Umuttepe/Kocaeli, Turkey
e-mail: adem.tuncer@kocaeli.edu.tr

E. Gelenbe et al. (eds.), *Computer and Information Sciences II*,
DOI: 10.1007/978-1-4471-2155-8_48, © Springer-Verlag London Limited 2012

However, for the dynamic or partially observable environments re-plannings are required frequently in each step and it needs more planning update time.

There are so many methods that have been developed to overcome path planning problem for mobile robots. Each method differs in their effectiveness depending on the type of application environment and each one of them has its own strengthness and weaknesses. Compared to traditional search and optimization methods, such as calculus-based and enumerative strategies, the evolutionary algorithms are robust, global and generally more straightforward to apply in situations where there is little or no a priori knowledge about the problem to solve [2].

In the last decade, genetic algorithms have been widely used to generate the optimal path by taking the advantage of its strong optimization ability [3]. Genetic algorithms have been recognized as one of the most robust search techniques for complex and ill-behaved objective functions. The basic characteristic that makes the GA attractive in developing near-optimal solutions is that they are inherently parallel search techniques [4]. They can search all working environment simultaneously in a parallel manner and so they can reach a better solution more quickly.

Solution time is critical issue for path planning of mobile robots on the dynamic environment. It is directly proportional to the generation number in genetic algorithms. Decrease in the length of chromosome leads to decrease in generation number or solution time. In this study, different chromosome coding methods are analyzed for genetic algorithm to solve path planning problem of mobile robots. Grid-based environment model is used, since the calculation of distance and the representation of obstacle are easier with grid-based representation. Binary, decimal and orderly numbered grids coding methods are used to create chromosomes. Path distance, generation number and solution time parameters are observed and compared for the three coding methods under the same conditions.

2 Path Planning with Genetic Algorithms

2.1 *Representation of Environment and Chromosome*

Many path planning methods use a grid-based model to represent the environment space. It has been determined that calculation of distance and representation of obstacle are easier with grid-based representation. The grid-based environment space is represented in two ways, by the way of coordinate plane [4–6] or by the way of orderly numbered grids [1, 7, 8]. Coordinates can be represented with both the binary or decimal numbers. Figure 1 shows the binary, decimal and orderly numbered grids representation of the path planning environment. The shadowed grid on the environment shows the infeasible obstacle area and blank grid shows the feasible area where mobile robots can move freely.

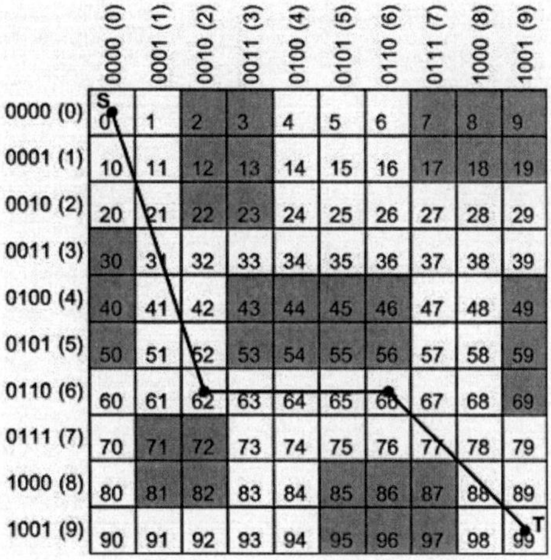

Fig. 1 Binary, decimal and orderly numbered representations of environment

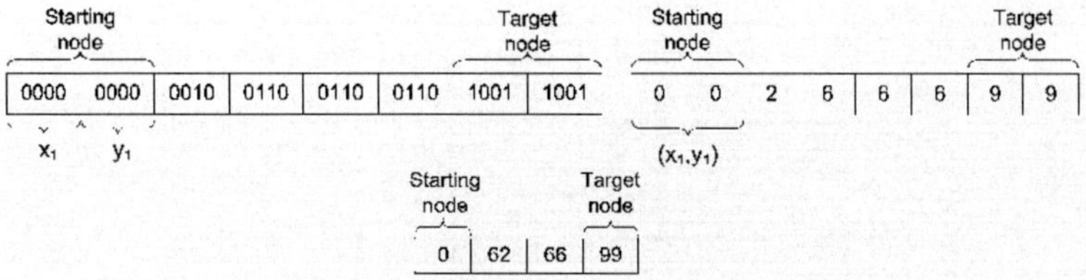

Fig. 2 Binary, decimal and orderly numbered coding of chromosomes

A chromosome represents a candidate solution [9] for the path planning problem. A chromosome or a path consists of a starting node, a target node and the hopping nodes which mobile robot across on them. These nodes or steps in the path are called as genes of the chromosome. Different coding methods are used to create chromosomes (Fig. 2), depending on the representation method of the environment. Binary coded string method [4, 10, 11] is used in general, however decimal coded string method is also used [1, 7, 8] and it is thought as to be more flexible. Decimal coding needs less computational overhead in time and space.

2.2 Initialization of Population

The initial population is generally generated randomly. Some of the generated chromosomes may include infeasible paths which intersect an obstacle. An optimal or near optimal solution can be found by genetic operators, even though the initial population includes infeasible paths.

In the study, genetic algorithm is being run with both randomly generated and feasible initial population. For the same environment and 10 runs, the average

Table 1 Comparison of the initial population methods

Initial population method	Random	Feasible
Fitness value	28.34	28.29
Number of generation	81	36
Solution time(s)	0.87	0.49

fitness value, number of generation and solution time of each method are given in Table 1. It is determined that starting the genetic algorithm with the feasible initial population is fairly beneficial, as also stated in the [7, 8].

2.3 Fitness Function and Selection

The purpose of the path planning problem is to find an optimal path between a starting and a target node. Optimal path may be the shortest, the least time and energy requiring path to trip on it. Generally, in the path planning problems, objective function is considered as a shortest path. In this study, objective function value of a chromosome used in genetic algorithm is given in Eqs. 1 and 2.

$$f=\begin{cases}\sum_{i=1}^{n-1} d(p_i,p_{i+1}), & \text{for feasible paths}\\ \sum_{i=1}^{n-1} d(p_i,p_{i+1})+penalty, & \text{for infeasible paths}\end{cases} \tag{1}$$

$$d(p_i,p_{i+1}) = \sqrt{(x_{(i+1)}-x_i)^2 + (y_{(i+1)}-y_i)^2} \tag{2}$$

Where; f is fitness function, p_i is the ith gene of chromosome, n is the length of chromosome and d is the distance between two nodes.

Objective function value is defined as the sum of distances between nodes in a path. If there is an obstacle between two nodes, a penalty is added to objective function value. The penalty value should be greater than the maximum path length on the environment.

In this study, the rank based fitness assignment is used instead of the proportional assignment method. This prevents a few better chromosomes to be dominant in the population. At the last step, chromosomes are selected according to fitness values and then put into mating pool to produce new chromosomes.

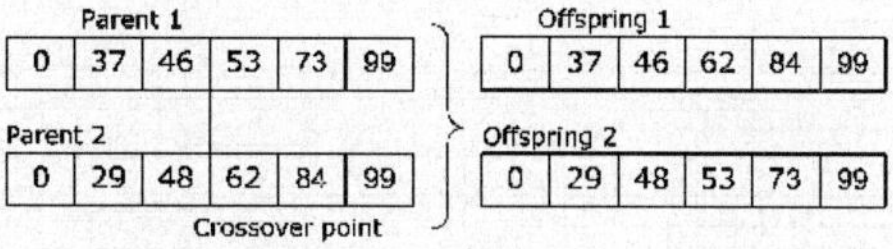

Fig. 3 Single-point crossover

2.4 Crossover and Mutation Operators

Generally, crossover combines the features of two parent chromosomes to form two offspring. Single-point crossover operator is used in this study (Fig. 3). The genes of two chromosomes after the crossover point are swapped. All candidate chromosomes in the population are subjected to the random mutation after the crossover operation. This is a random bit-wise binary complement operation or a random small change in gene, depends on the coding of chromosomes, applied uniformly to all genes of all individuals in the population with a probability of mutation rate. The mutation operation expands the search space to regions that may not be close to the current population, thus ensuring a global search [8]. Mutation operation increases the diversity of the population and avoids the premature convergence.

3 Experiments and Performance Evolution

Solution time is critical issue for path planning of mobile robots on the dynamic environment. An optimal new path must be determined until the robot move from present node to next node. Minimization of the solution time may be more important than reducing the path distance. Solution time is directly proportional to the generation number in genetic algorithms. Decrease in the length of chromosome, that means decrease in number of genes in the chromosome, leads to decrease in generation number or solution time.

In order to point out the success of chromosome coding methods, binary, decimal and orderly numbered grids coding methods are analyzed with two experiments. Path distance, generation number and solution time parameters are observed and compared for the three coding methods under the same conditions.

Experiment 1: Path planning problem is solved by genetic algorithm for the environment shown in Fig. 4a. The environment consists of 16 $\times$ 16 grids and has eight obstacle regions that are shadowed on the figure. Binary, decimal and orderly numbered grids coding methods are used one by one to create chromosomes in the genetic algorithm. The parameters of the genetic algorithm are like these; the population size is taken as 60, crossover probability is taken as 1.0 and mutation probability is taken as 0.1. After 10 runs of genetic algorithm for each method, the average values of path distance, generation number and solution time are given in Table 2. According to the table, decimal coding method finds the optimal

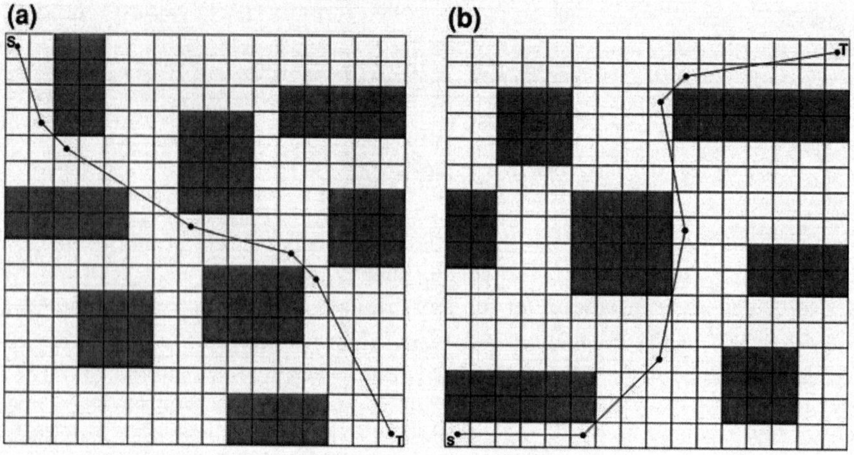

Fig. 4 **a** Environment for *experiment* 1 consists of eight obstacle regions. **b** Environment for *experiment* 2 consists of seven obstacle regions

Table 2 Results for *experiment* 1

	Binary	Decimal	Numbered
Path distance	28.40	25.07	26.30
Generation number	153	75	54
Solution time(s)	2.7	1	0.7

Table 3 Results for *experiment* 2

	Binary	Decimal	Numbered
Path Distance	31.9	29.6	30.6
Generation Number	101	44	32
Solution Time(s)	1.5	0.55	0.47

(shortest) path and orderly numbered grids coding method finds a near optimal path more quickly than others.

Experiment 2: Path planning problem is solved again for the environment shown in Fig. 4b. The environment has seven obstacle regions that are shadowed on the figure. The parameters of the genetic algorithm are taken as same as the Experiment 1. After 10 runs of genetic algorithm for each method, the average values of path distance, generation number and solution time are given in Table 3. According to the table, decimal coding method finds again the optimal path and orderly numbered grids coding method finds a near optimal path more quickly than others.

4 Conclusion

Path planning tries to find a feasible path for mobile robots to move from a starting node to a target node in an environment with obstacles. In this study, binary, decimal and orderly numbered grids coding methods are used one by one to create chromosomes in the genetic algorithm. After 10 runs of genetic algorithm for each method, the average values of path distance, generation number and solution time are observed and compared for the three coding methods under the same conditions. According to the results of two experiment, orderly numbered grids coding method finds a near optimal path most quickly. Decimal coding method finds the optimal path in a reasonable time when compared to the solution time of orderly numbered grids. However, binary coding is the worst of three methods concerning with the path distance, generation number and solution time. Results showed that the solution time is directly affected by chromosome coding method. Decrease in the length of chromosome leads to decrease in generation number or solution time.

References

1. Hu, Y., Yang, S.X.: A knowledge based genetic algorithm for path planning of a mobile robot. In: Proceedings of the 2004 IEEE, International Conference on Robotics and Automation, vol. 5, pp. 4350–4355 (2004)
2. Tu, J., Yang, S.X.: Genetic Algorithm based path planning for a mobile robot. In: Proceedings. ICRA '03. IEEE International Conference on Robotics and Automation, vol. 1, pp. 1221–1226 (2003)
3. Al-Taharwa, I., Sheta, A., Al-Weshah, M.: A mobile robot path plannig using genetic algorithm in static environment. J. Comput. Sci. **4**, 341–344 (2008)
4. Elshamli, A., Abdullah, H.A., Areibi, S.: Genetic algorithm for dynamic path planning. Canadian Conference on Electrical and Computer Engineering **2**, 677–680 (2004)
5. Manikas, T.W., Ashenayi, K., Wainwright, R.L.: Genetic algorithms for autonomous robot navigation. Instrumentation and Measurement Magazine, IEEE **10**, 26–31 (2007)
6. Naderan-Tahan, M., Manzuri-Shalmani, M.T.: Efficient and safe path planning for a mobile robot using genetic algorithm. CEC '09. IEEE Congress on Evolutionary Computation, pp. 2091–2097 (2009)
7. Yao, Z., Ma, L.: A static environment-based path planning method by using genetic algorithm. International Conference on Computing, Control and Industrial Engineering (CCIE), vol. 2, pp. 405–407 (2010)
8. Li, Q., Zhang, W., Yin, Y., Wang, Z., Liu, G.: An improved genetic algorithm of optimum path planning for mobile robots. ISDA '06. Sixth International Conference on Intelligent Systems Design and Applications, vol. 2, pp. 637–642 (2006)
9. Gelenbe, E., Liu, P., Lainé, J.: Genetic algorithms for route discovery. IEEE Transactions on Systems, Man, and Cybernetics. Part B: Cybernetics **36**, 1247–1254 (2006)
10. Sugihara, K., Smith, J.: Genetic algorithms for adaptive motion planning of an autonomous mobil robot. In: Proceedings, IEEE International Symposium on Computational Intelligence in Robotics and Automation, CIRA'97, pp. 138–143 (1997)
11. Nagib, G., Gharieb, W.: Path planning for a mobile robot using genetic algorithms. ICEEC '04, International Conference on Electrical, Electronic and Computer Engineering, pp. 185–189 (2004)

Part IX
Security and Safety

Client-Based CardSpace-OpenID Interoperation

Haitham S. Al-Sinani and Chris J. Mitchell

Abstract We propose a novel scheme to provide interoperability between two of the most widely discussed identity management systems, namely CardSpace and OpenID. In this scheme, CardSpace users are able to obtain an assertion token from an OpenID-enabled identity provider, the contents of which can be processed by a CardSpace-enabled relying party. The scheme, based on a browser extension, is transparent to OpenID providers and to the CardSpace identity selector, and only requires minor changes to the operation of the CardSpace relying party.

1 Introduction

To mitigate identity-oriented attacks, a number of identity management systems (e.g. CardSpace, OpenID, etc.) have been proposed. An identity provider (IdP) in such systems supplies a user agent (UA) with an authentication token that can be consumed by a relying party (RP). Whilst one RP might support CardSpace, another might only support OpenID. To make these systems available to the largest possible group of users, interoperability between such systems is needed.

We consider CardSpace-OpenID interoperation because of OpenID's wide adoption; complementing this, the wide use of Windows, recent versions of which incorporate CardSpace, means that enabling interoperation between the two

This author is sponsored by the Diwan of Royal Court, Sultanate of Oman.

H. S. Al-Sinani (✉) · C. J. Mitchell
Information Security Group, Royal Holloway,
University of London, Egham, UK
e-mail: Haitham.Al-Sinani.2009@rhul.ac.uk;
Haitham.Al-Sinani.2009@live.rhul.ac.uk

C. J. Mitchell
e-mail: C.Mitchell@rhul.ac.uk

E. Gelenbe et al. (eds.), *Computer and Information Sciences II*,
DOI: 10.1007/978-1-4471-2155-8_49,

systems is likely to be of significance for large numbers of identity management users and service providers. CardSpace-OpenID interoperation is particularly attractive since both schemes support user attribute exchange. In addition, supporting interoperation on the client could be practically useful, because IdPs/RPs might not accept the burden of supporting two identity systems unless there is a significant financial incentive. Client-based interoperation means that the performance of the server is not affected, since the overhead is handled by the client.

2 CardSpace and OpenID

2.1 CardSpace

CardSpace is Microsoft's implementation of a digital identity metasystem, in which digital identities are represented to users as Information Cards (or InfoCards). There are two types of InfoCards: personal (self-issued) cards and managed cards, issued by remote IdPs. Personal cards are created by users themselves, and the claims listed in such cards are asserted by the self-issued identity provider (SIIP) that co-exists with the CardSpace identity selector (or just the selector) on the user machine [1]. The integration scheme uses CardSpace personal cards to make information provided by OPs available to CardSpace RPs.

The personal card protocol operates as follows, in the case where the RP does not employ a security token service (STS).

1. UA → RP. HTTP/S request: GET (login page).
2. RP → UA. HTTP/S response. A login page is returned containing the CardSpace-enabling tags in which the RP security policy is embedded.
3. User → UA. The RP page offers the option to use CardSpace; selecting this option activates the selector, which is passed the RP policy. If this is the first time that this RP has been contacted, the selector will display the identity of the RP and give the user the option to either proceed or abort the protocol.
4. Selector → User. The selector highlights InfoCards matching the policy.
5. User → Selector. The user chooses (or creates) a personal card. The user can preview the card to ensure that they are willing to release the claim values.
6. Selector ⇌ SIIP. The selector sends a Request Security Token (RST) to the SIIP, which responds with a Request Security Token Response (RSTR).
7. UA → RP. The RSTR is passed to the UA, which forwards it to the RP.
8. RP → User. The RP validates the token, and, if satisfied, grants access.

The PPID is an identifier linking an InfoCard to an RP. When a user first uses a personal card at a particular RP, CardSpace generates a card-site-specific PPID by combining the card ID[1] with data taken from the RP certificate, and a

---

[1] During card creation, a card ID and master key are created and stored.

card-site-specific signature key pair by combining the card master key with data taken from the RP certificate. The PPID could be used on its own as a shared secret to authenticate a user to an RP. However, it is recommended that the associated (public) signature verification key should be used to verify the signed security token to provide a more robust authentication method [1].

2.2 OpenID

OpenID is a decentralised user authentication scheme supporting remote single sign-on to multiple websites using a single digital identity. Two 'major' OpenID versions have been released: OpenID 1.1 [2], and OpenID 2.0 [3]; fortunately v2.0 is backward compatible with v1.1. OpenID 2.0 uses two types of user identifier: URLs and XRIs. A user could adopt a self-owned URL, or register a (new) URL at an OP. OpenID operates as follows (the use of TLS/SSL is recommended).

1. UA → RP. HTTP/S request: GET (login page).
2. RP → UA. HTTP/S response. A login page is returned containing a form.
3. User → UA. The user enters their OpenID identifier into the OpenID form.
4. RP: OP Discovery. The RP uses the identifier to discover the user's OP.
   - *HTML-based discovery (OpenID 1.1/2.0).* The RP requests an HTML document identified by the user's OpenID URL; such a document contains the information necessary to discover the required OP.
   - *XRDS-based discovery (OpenID 2.0).* The RP requests an XRDS document containing the data necessary to discover the OP. If the user's identifier is an XRI, the RP will retrieve an XRDS document identified by the XRI; if it is a URL, the RP will use Yadis to retrieve an XRDS document, and, if this fails, the RP will revert to HTML-based discovery.
5. RP $\rightleftharpoons$ OP (optional). The RP and OP agree a shared secret key to be used for a specified period of time to MAC-protect exchanged messages. This (back-channel) request-response process, known as the 'association' mode, requires the two parties to be able to store the secret.
6. RP-OP Interaction. The RP and OP can communicate in either 'checkid_immediate' mode, involving direct RP-OP communications without user interaction, or 'checkid_setup' mode, where the user is interactively involved. The 'checkid_setup' mode is more commonly used; if 'checkid_immediate' mode fails, the scheme reverts to 'checkid_setup' mode. If using 'checkid_immediate' mode, the RP directly sends the OP an OpenID authentication request, and the OP directly replies with an OpenID authentication response; step 9 then takes place.However, if using 'checkid_setup' mode, the RP redirects the user to the OP with an OpenID authentication request, and step 7 follows.
7. OP $\rightleftharpoons$ User. If necessary, the OP authenticates the user. If successful, the OP constructs an OpenID assertion token, including user credentials/attributes,

a freshly-generated nonce, a current time-stamp, and a MAC computed on the token. If a shared key was agreed in step 5, the OP uses it to generate the MAC; otherwise the OP employs an internally-generated MAC key. The OP requests permission to send the assertion token to the requesting RP.

8. OP → UA → RP. The OP redirects the user back to the RP with a positive or negative OpenID authentication response.
9. RP → User. The RP verifies the MAC-protected OpenID authentication response, and, if satisfied, grants access. If a shared secret was previously agreed (see step 5), the RP uses its copy to verify the MAC. If not, the RP must make an extra request to the OP to verify the MAC, typically via a TLS/SSL channel. This request-response process is known as the 'check_authentication' mode, and is adopted in the integration scheme.

3 The Integration Scheme

The parties involved are a CardSpace-enabled RP, a CardSpace-enabled UA (e.g. a suitable web browser), an OP, and a browser extension implementing the protocol described below. The scheme has the following operational requirements.

- The user must have accounts with the CardSpace RP and the OP (thus the OP can authenticate the user). The RP and the user must trust the OP.
- Prior to use, the user must create a personal card, referred to here as an IDcard, containing the following data items in specific fields (the choice of which is implementation-specific): the user's OpenID identifier; a predefined sequence of characters (e.g. 'OpenID') used to trigger the browser extension and indicate which OpenID version to use; and the OP URL.
- The CardSpace RP must not employ an STS; instead, it must express its policy using HTML/XHTML, and selector-RP interactions must be based on HTTP/S via a browser (a simpler and probably more common scenario). This is because the scheme uses a (JavaScript-based) browser extension, and is incapable of managing the communications with an STS.
- The RP must accept an unsigned SAML token which includes OP-asserted attributes and the signed RSTR containing the card-RP-specific PPID.

The novel protocol operates as follows. Steps 1, 2, and 4–7 are the same as steps 1, 2, and 3–6, respectively, of the personal card protocol given in Sect. 1.

3. Browser extension → UA. The extension performs the following steps.
   (a) It scans the login page to detect whether the RP website supports CardSpace; if so, it proceeds, otherwise it terminates.
   (b) It examines the RP policy to check whether the use of personal cards is acceptable. If so, it proceeds; otherwise it terminates, giving CardSpace the opportunity to operate normally.

(c) It keeps a local copy of any RP-requested claims.
(d) It modifies the policy to include the claim types employed in the IDcard.

8. Selector → Browser Extension. Unlike in the 'standard' case, the RSTR does not reach the RP; instead the extension intercepts it and temporarily stores it. It then determines the communication protocol (HTTP or HTTPS) in use.[2] If the RP uses HTTP, the extension uses the RSTR's contents to construct an OpenID authentication request,[3] which it forwards to the appropriate OP, having discovered its address from the RSTR. On the other hand, if the RP uses HTTPS, the browser extension:

   (a) asks the user to enter his/her OpenID identifier and uses the supplied OpenID identifier to perform OP discovery (see Sect. 2;) and
   (b) constructs an OpenID authentication request (precisely as in the HTTP case), which it forwards to the discovered OP.
   In both cases the format of the OP authentication request will depend on the version of OpenID being used. In both cases the more commonly used OpenID 'check_setup' mode is adopted (the 'check_immediate' mode is not supported as it requires back-channel RP-IdP communication).

9. OP ⇌ User. If necessary, the OP authenticates the user. If successful, the OP requests permission to send the OpenID token to the RP return-page.
10. OP → UA → RP. The OP redirects the UA back to the RP return-page with a positive or negative OpenID authentication response[4] depending on whether or not the user granted permission in step 9.
11. Browser Extension → UA. The extension verifies the MAC-protected OpenID token by interacting with the OP using the 'check_authentication' mode via a TLS/SSL channel. If successful, it constructs a CardSpace-compatible SAML token,[5] and forwards it to the RP. If unsuccessful, the extension terminates.
12. RP → User. The RP verifies the SAML token (including verifying the RSTR signature, PPID, nonce, time-stamps, etc.), and, if satisfied, grants access.

---

[2] The protocol operates slightly differently depending on whether the RP uses HTTP or HTTPS. This is because, if HTTPS is used, the selector will encrypt the RSTR using the site's public key, and the browser extension does not have access to the corresponding private key. Hence, it will not know whether to trigger the integration protocol, and will be unable to obtain the user's OpenID identifier; such issues do not occur if HTTP is used since the selector will not encrypt the RSTR.

[3] This will indicate the RP-requested attributes which are to be asserted by the OP.

[4] The RP will receive the OP-issued token unchanged (embedded in the URL); however it is assumed that the RP will ignore it because of its inability to process the token.

[5] Observe that this (unsigned) SAML token will contain the user attributes as asserted by the OP and the digitally-signed SIIP-issued RSTR (which contains the PPID).

4 Security Analysis

The unsigned browser extension-generated SAML token (referred to as the 'user token') includes the PPID, the OP-asserted user attributes, the signed SIIP-issued RSTR, and (optionally) the MAC-protected OP-issued token. The RP compares the SIIP-asserted PPID (and the public key) in the user token with its stored values and verifies the digital signature. The RP can optionally verify the MAC in the OP-issued token, which is embedded unchanged in the user token; for the RP to verify the MAC, the extension must skip the verification process and the RP must interact with the OP via the 'check_authentication' mode.

It is infeasible for a malicious entity to fabricate a user token to masquerade as a legitimate party since it will not have access to three key token components: the PPID; the SIIP-signed RSTR, which is only issued if the appropriate InfoCard is selected on the correct platform; and the MAC-protected OP-issued token, which is only issued if the genuine user has been authenticated by the OP. In addition, nonces and time-stamps are used to prevent replay attacks, and RPs can also employ IP address validation. The use of SSL/TLS on the OP-client and RP-client channels is strongly recommended.

The selector identifies the RP to the user and indicates whether or not they have visited that particular RP before; if this RP is being visited for the first time, CardSpace requests the user's permission to proceed. This helps to support mutual authentication since the user and the RP are both identified to each other (a security gain over 'native' OpenID, which does not identify the RP).

The scheme mitigates the risk of phishing, because the redirect to the OP[6] is initiated by the browser extension and not by the RP, i.e. the RP cannot redirect the user to an OP of its choosing. By contrast, in OpenID a malicious RP could redirect a user to a fake OP, which might capture the user credentials.

Finally, the scheme allows the user attributes to be remotely stored at the OP; this has potential security advantages over storing the attributes locally on the user machine, as is currently the case with CardSpace SIIP-issued attributes.

5 Related Work

A somewhat similar scheme [4] has been proposed to support CardSpace-Liberty interoperation. However, unlike the scheme proposed here, the CardSpace-Liberty integration scheme is not transparent to the IdPs, does not support the exchange of identity attributes, and does not support HTTPS-enabled websites.

Kim et al. [5] have proposed an OpenID authentication method using an identity selector. The scheme uses a specially modified identity selector to enable OpenID authentication, unlike our scheme which uses an unmodified selector.

---

[6] In HTTP mode the OP address is retrieved from the IDcard as entered by the user.

Microsoft and OpenID have announced plans (http://www.guardian.co.uk/technology/blog/2007/feb/07/openidgetsab) to enable a level of interoperation. A stated aim of this effort is to reduce the risk of phishing in OpenID by enabling an OpenID user to employ CardSpace when authenticating to an OP. The scheme proposed here inherently protects against phishing (see Sect. 4), and also supports the use of CardSpace to authenticate to OPs.

6 Conclusions

We have proposed a means of interoperation between two leading identity systems, namely CardSpace and OpenID. CardSpace users are able to obtain an assertion token from an OpenID provider, the contents of which can be processed by a CardSpace-enabled relying party. The scheme is transparent to OpenID providers and identity selectors, uses a browser extension, and requires only minor changes to a CardSpace relying party. It uses the selector interface to integrate OpenID providers with CardSpace relying parties, using personal cards.

The integration scheme takes advantage of the similarity between the OpenID and the CardSpace frameworks, and this should help to reduce the effort required for full system integration. A full version of this paper, including a description of a prototype implementation, is available [6].

References

1. Mercuri, M.: Beginning Information Cards and CardSpace: From Novice to Professional. Apress, New York (2007)
2. Recordon, D., Fitzpatrick, B.: OpenID Authentication 1.1. http://openid.net/specs/openid-authentication-1_1.html (2006)
3. OpenID Community: OpenID Authentication 2.0—Final. http://openid.net/specs/openid-authentication-2_0.html (2007)
4. Al-Sinani, H.S., Alrodhan, W.A., Mitchell, C.J.: CardSpace-Liberty integration for CardSpace users. In: Klingenstein, K., Ellison, C.M. (eds.) Proceedings of the 9th Symposium on Identity and Trust on the Internet, (IDtrust'10), Gaithersburg, Maryland, USA, 13–15 April 2010, pp. 12–25. ACM, New York, NY, (2010)
5. Kim, S.H. et al.: OpenID Authentication Method Using Identity Selector. Patent Application Publication, United States, Publication Number US 2009/0249078 A1 (2009)
6. Al-Sinani, H.S., Mitchell, C.J.: CardSpace-OpenID integration for cardspace users. Technical Report: RHUL–MA–2011–12, Department of Mathematics, Royal Holloway, University of London (2011)

An Artificial Immune Classifier Using Pseudo-Ellipsoid Rules

Aris Lanaridis and Andreas Stafylopatis

Abstract We suggest in this papers a fuzzy rule-based classification algorithm based on an Artificial Immune Network. The network evolves a dynamically-determined number of pseudo-ellipsoid rules to cover the patterns of the problem, focusing both on the quality and the diversity of the evolved rules. The proposed classifier is compared against C4.5 and GEP on a number of benchmark problems from the UCI Repository, yielding satisfactory results.

Keywords Artificial immune systems · Artificial immune networks · Fuzzy rule classifiers · Fuzzy ellipsoid rules · Pattern classification

1 Introduction

Fuzzy rule classifiers using ellipsoid regions have been extensively researched in the literature of pattern classification, producing very satisfying results [1–3]. However, the calculations of high-dimensional ellipsoids are computationally expensive, imposing several restrictions on algorithms based on such rules, such as requiring clustering algorithms to be employed, determining the number of rules in advance, or demanding a priori knowledge of the data to be classified.

We suggest in this paper a fuzzy classification system based on an Artificial Immune Network. The algorithm uses fuzzy rules based on a simplified generalization of the ellipse in high-dimensional spaces, using these pseudo-ellipsoid

A. Lanaridis (✉) · A. Stafylopatis (✉)
Department of Electrical and Computer Engineering Intelligent Systems Laboratory, National Technical University of Athens, Athens, Greece
e-mail: aristeides@gmail.com

A. Stafylopatis
e-mail: andreas@cs.ntua.gr

E. Gelenbe et al. (eds.), *Computer and Information Sciences II*,
DOI: 10.1007/978-1-4471-2155-8_50, © Springer-Verlag London Limited 2012

surfaces to cover the problem space. The immune network evolves in a way that focuses on the diversity of the rules, dynamically determining the number of rules required to cover all the patterns to be classified.

2 An Artificial Immune Network for Pattern Classification

The Artificial Immune Network proposed here attempts to create a rule-based classification system. The proposed method is based on the usual principles of Artificial Immune Networks (we refer the reader to [4] for a thorough overview), with one important modification. As is the case in the natural immune system, the objective of the algorithm is not the recognition, but the elimination of the antigens. Only after this objective has been achieved we measure the classification capability achieved by the network throughout the process.

To incorporate this modification into the algorithm, we assign to each antigen a factor h representing its health, assumed to lie in the interval $[0, h_{max}]$. We note that, the term 'health' is used here in the same sense that it is used in computer games, that is it represents the amount of damage the antigen can withstand. The health of antigen is also an indicator of its strength. Stronger antigens impose an greater threat on the immune system, and consequently the network assigns greater importance to them.

The set of antigens represent the dataset of the problem, with each antigen representing a pattern belonging to a specific class. The antibodies represent fuzzy rules, each dedicated to recognizing antigens of a specific class, randomly selected among the classes of the problem. The antibodies evolve in a way that maximizes the total membership of positive (belonging to the same class as the antibody) antigens, while minimizing the total membership of negative (belonging to any other class) antigens. The strength of the antigens is also taken into account, with stronger antigens weighing more to the total membership function.

The evolution of each antibody is based on the clonal selection principle. That is, for each antibody a number of clones are created and mutated, and the best clone replaces the original antibody. The process is repeated for a given number of generations, after which the new antibody is added to the network. To preserve the quality and diversity of the population, each time a new antibody is added to the network the quality and similarity of all antibodies is compared. Antibodies that present significantly inferior quality or high similarity to other antibodies are removed.

The network continues its evolution until the health of all antigens falls below a specific threshold h_{min}. The calculation of the antigen health is based on the antibodies currently present in the network. Each antibody weakens the health of each positive antigen to a degree equal to the membership function, while it has no effect on negative antigens.

3 Fuzzy Pseudo-Ellipsoid Rules

The antibodies in the network represent fuzzy rules, based on a generalized form of the ellipse, used to cover the problem space. An ellipse is the locus of all points of a plane whose distances to two fixed points add to the same constant. Each of these points is called a *focus* of the ellipse, while the constant is called *radius*. If we denote the two foci f_1 and f_2 and the radius r, the ellipse consists of all the points x of the plane for which the quantity

$$d(x) = \frac{\|x - f_1\| + \|x - f_2\|}{r} \tag{1}$$

is $d(x) = 1$. The set of points satisfying the above equation in an n-dimensional space with $n > 2$ define a quadric surface in this space. This surface does not constitute an ellipsoid in the strict sense of the term. However, it is based on a generalized form of the ellipse, and will be thus called pseudo-ellipsoid for the rest of this paper. To create a fuzzy pseudo-ellipsoid based on the above definition, we define a membership function

$$\mu(x) = \exp\left(-d(x)^f\right) \tag{2}$$

where $f \in [0, +\infty]$ is a factor defining the steepness of the membership function. For $f \to +\infty$ the function reduces to a crisp one, while for $f \to 0$ it is uniform with value equal to the membership of the points on the surface of the ellipsoid. This factor is allowed to vary among different rules.

The value of the membership function also represents the extent to which the antibody interacts with a specific antigen. That is the antibody reduces the health of all positive antigens (until the health drops to zero) by a value equal to the value of the membership function, while it has no effect on antigens of other classes, that is

$$h'_i = \begin{cases} \max(h_i - \mu_i, 0), & a_i \in P \\ h_i, & a_i \in N \end{cases} \tag{3}$$

4 Mutation and Evaluation of Antibodies

The mutation process is based on the common non-uniform mutation operator defined by Michalewicz [5]. To apply this operator, each attribute x has to be confined in an interval $[x_{min}, x_{max}]$. To simplify the calculations, we first normalize all the patterns in the dataset in the $[-1, 1]$ range. The normalization does not affect the function of the classifier in any way, since all parameters will be scaled accordingly.

After the normalization, each attribute of the foci will also be in $[-1, 1]$. For the radius of the ellipse, we use a factor $\alpha \in [0, 1]$ which produces a linear map from $[0, 1]$ to $[r_{min}, r_{max}]$, that is

$$r = r_{min} + \alpha(r_{max} - r_{min}) \tag{4}$$

and apply mutation on this factor. In the above equation, the minimum possible value of the radius is the distance between the two foci $r_{min} = \|f_1 - f_2\|$ (recalculated each time the foci are moved), and the maximum is set to the maximum possible distance in the normalized problem space, that is $r_{max} = 2\sqrt{n}$. Finally, the steepness of the membership function f is limited to the interval [2, 6], since its shape does not change significantly for values outside this interval.

To evaluate the rules of the classifier we define a fuzzy variant of the crisp rule evaluation metric known as M-estimate [7]. For the fuzzy rule, if P, T are the sets of positive and total patterns in the dataset, the coverage p, t of positive and total patterns covered will be

$$p = \sum_{i \in P} \mu_i, \qquad t = \sum_{i \in T} \mu_i \tag{5}$$

We also need to incorporate in the above quantities the health factor h assigned to each antigen, so that the network becomes less concerned with antigens as they weaken. By incorporating this factor into the rule coverage and replacing in the original M-estimate, the evluation metric becomes

$$e = \frac{\sum_{i \in P} \mu_i h_i + M\frac{|P|}{|T|}}{\sum_{i \in T} \mu_i h_i + M} \tag{6}$$

5 Antibody Death

As explained above, an essential aspect of Artificial Immune Networks is that they regulate their population by generating new antibodies and removing existing ones. The birth of new antibodies is aimed at confronting the invading antigens, while the death of existing antibodies is aimed at maintaining the quality and diversity of the population. Specifically, antibodies die when they are either significantly lower in quality than the rest of the population, of too similar to other antibodies. To achieve this we apply a common outlier detection criterion, namely the upper and lower inner fences of a distribution [6].

To remove antibodies of significantly lower quality, each time an antibody is added to the network we evaluate all the antibodies using equation (6). To compare the rules on equal basis we regard the health of all antigens to be $h_i = h_{max}, \forall i$. Having evaluated all the rules, we create the distribution $\mathbf{e} = [e_1, e_2, \ldots, e_{\|B\|}]$, where B is the set of antibodies currently present in the network, and remove the antibodies whose quality is below the lower inner fence of the distribution of $\mathbf{e}$.

In a similar manner, rules that are too similar to existing rules must be removed. If $\mu_i = [\mu_{i1}, \mu_{i2}, \ldots, \mu_{im}]$ is the vector containing all membership values for the ith rule, we define the similarity between the ith and the jth antibody as

$$s_{ij} = \frac{\mu_i \cdot \mu_j}{\|\mu_i\| \cdot \|\mu_j\|} \tag{7}$$

For the ith antibody of the network we find the most similar one, that is the antibody for which the similarity value is

$$M_i = \max_{1 \leq j \leq \|B\|,\, j \neq i} s_{ij} \tag{8}$$

We repeat the process for all antibodies, creating the distribution of maximum similarity pairings $\mathbf{M} = \left[M_1, M_2, \ldots, M_{\|B\|}\right]$.

If the similarity between two rules exceeds the upper inner fence of the above distribution, one of them must be removed. Let M_i be a such similarity value, where M_i represents the similarity between the ith rule and the kth rule, where $s_{ik} = M_i$. In this case either the ith or the kth rule must be removed. We remove the one with the smaller evaluation value. The process is repeated for all M_i exceeding the upper fence of the distribution.

6 Evolution of the Network

Let $A = \{a_1, \ldots, a_m\}$ be the set of antigens, with each antigen representing a pattern which is to be classified. The attributes of the antigens are normalized to the interval $[-1, 1]^n$, where n is the number of attributes of each antigen. Let $CA = \{ca_1, \ldots, ca_m\}$ be the classes of the antigens. Regarding the initial health h_{max} of the antigens, it is set to 1 for all antigens. The network is initialized to the empty set, and a new antibody is created, with each of its parameters randomly initialized in their respective domains, specifically, $f_1, f_2 \in [-1, 1]^n, f \in [2, 8], \alpha \in [0, 1]$, and cb_i randomly selected among the classes of the problem.

After the initialization, N_C clones (exact copies) of the antibody are created and mutated. The rate of the mutation depends on the current generation t as defined by the operator, while the bounds of the mutation are the same as those used to initialize each parameter. Among the mutated clones produced the one that maximizes the evaluation metric given by (6) is selected to replace the original antibody. The procedure is repeated for T generations, after which the antibody is added to the network.

Having added the newly evolved antibody to the network, all antibodies of the network are re-evaluated, with the health of all antigens considered to be 1, and the similarity of between all pairs of antibodies is calculated. The antibodies that are

Table 1 Values of the parameters used to train the classifier

Parameter	Value		
(a) Distance Metric			
N_C	100		
T	20		
M	$\sqrt{noinstances}$		
h_{min}	$1/e$		
b	2		
(b) Distance Metric			
Dataset	C4.5	GEP	AIS
Breast Cancer	94.7 ± 1.5	$\mathbf{96.7 \pm 1.8}$	96.08 ± 1.22
Iris	93.9 ± 8.1	95.3 ± 4.6	$\mathbf{96.67 \pm 4.08}$
Pima Indians	74.8 ± 4.7	69.7 ± 3.8	$\mathbf{75.72 \pm 2.09}$
Wine	91.6 ± 8.1	92.0 ± 6.0	$\mathbf{96.63 \pm 2.32}$

considered outliers, in terms of either low quality or high similarity to other antibodies are removed, as described in Sect. 5, producing the current set of antibodies B. At this point, the health of each antigen is calculated by

$$h_j = \max\left(0, 1 - \sum_{i=1}^{\|B\|} \mu_{ij} \cdot I(i,j)\right) \quad \text{where} \quad I(i,j) = \begin{cases} 1, & cb_i = ca_j \\ 0, & cb_i \neq ca_j \end{cases} \tag{9}$$

According to equation (9) the health of each antigen is reduced by a quantity equal to its membership to each antibody of the same class, until its health drops to zero. The calculation is repeated for all antigens, producing the distribution $h = [h_1, \ldots, h_m]$. If, after removing the outliers above the upper inner fence, the health of all antigens is below h_{min}, the evolution of the network stops. In any other case a new antibody is added and the procedure is repeated.

When the evolution of the training has been completed, each pattern is assigned to the class of the rule to which it exhibits the largest membership value. Specifically, an antigen a_j will be assigned to the class cb_k of the kth antibody, where the value of k is given by $k = \arg\max_{1 \le i \le \|B\|} \mu_{ij}$.

7 Experimental Results

In this section, the classifier described above is tested on a number of benchmark problems from the UCI repository [8], and its performance is compared against two typical rule-based classifiers, C4.5 and Gene-Expression Programming (GEP). The system is trained using the parameters listed in Table 1a. To assess the classification accuracy of the proposed method, we applied 5-fold cross-validation to each dataset, and the process was repeated 5 times. The results given in Table 1b correspond to the mean and the standard deviation of the resulting 25 experiments.

As illustrated by the experiments, the classification accuracy of the algorithm is satisfactory, surpassing the competition in all problems with the exception of *Breast Cancer* where it was slightly surpassed by GEP. In the future we will study the effect of the training parameters to a greater extent, and we will try alternative rule evaluation criteria, in an attempt to further improve the classification accuracy achieved. Finally we will test the efficiency of the method of different pattern recognition problems.

References

1. Yao, L., Weng, K.S.: An evolutionary fuzzy classifier with adaptive ellipsoids. In: Proceeding of IEEE international conference on Systems, Man, and Cybernetics (2006)
2. Feng, H.M., Wong, C.C.: Fewer Hyper-Ellipsoids Fuzzy Rules Generation Using Evolutional Learning Scheme. Cybern. Syst. **39**(1) (2008)
3. Yao, L., Weng, K.S., Huang, C.D.: Fuzzy classifier with evolutionary design of ellipsoidal decision regions, World Academy of Science, Engineering and Technology (2005)
4. Garrett, S.M.: How do we evaluate artificial immune systems? Evolutionary Computation, vol. 13(2), pp. 145–177. MIT Press, Cambridge (2005)
5. Michalewicz, Z.: Genetic Algorithms + Data Structures = Evolution Programs
6. Ott, R.L., Longnecker, M.T.: An introduction to statistical methods and data analysis, 5th edn. Duxbury Press, California (2010)
7. Furnkranz, J., Flach, P.A.: An analysis of rule evaluation metrics. In: Proceedings of the twentieth international conference on machine learning. Washington, D.C. (2003)
8. Frank, A., Asuncion, A.: UCI Machine Learning Repository, http://archive.ics.uci.edu/ml, University of California, Irvine, School of Information and Computer Sciences (2010)

Contextually Learnt Detection of Unusual Motion-Based Behaviour in Crowded Public Spaces

Ognjen Arandjelović

Abstract In this paper we are interested in analyzing behaviour in crowded public places at the level of *holistic motion*. Our aim is to learn, without user input, strong scene priors or labelled data, the scope of "normal behaviour" for a particular scene and thus alert to novelty in unseen footage. The first contribution is a low-level motion model based on what we term *tracklet primitives*, which are scene-specific elementary motions. We propose a clustering-based algorithm for tracklet estimation from local approximations to tracks of appearance features. This is followed by two methods for motion novelty inference from tracklet primitives: (a) an approach based on a non-hierarchial ensemble of Markov chains as a means of capturing behavioural characteristics at different scales, and (b) a more flexible alternative which exhibits a higher generalizing power by accounting for constraints introduced by intentionality and goal-oriented planning of human motion in a particular scene. Evaluated on a 2 h long video of a busy city marketplace, both algorithms are shown to be successful at inferring unusual behaviour, the latter model achieving better performance for novelties at a larger spatial scale.

1 Introduction

Public spaces such as squares and shopping centres are unpredictable and challenging environments for computer vision based inference. Not only are they rich in features, texture and motion, but they also continuously exhibit variation of both high and low frequencies in time: shopping windows change as stores open and close, shadows cast by buildings and other landmarks move, delivery lorries get parked intermittently etc. The focus of primary interest, humans, also undergo

O. Arandjelović (✉)
Trinity College, Cambridge, CB2 1TQ, UK
e-mail: ognjen.arandjelovic@gmail.com

E. Gelenbe et al. (eds.), *Computer and Information Sciences II*,
DOI: 10.1007/978-1-4471-2155-8_51,

extreme appearance changes. Their scale in the image is variable and usually small, full or partial mutual occlusions and occlusions by other objects in the scene are frequent, with further appearance variability due to articulation, viewpoint and illumination.

2 Low-Level Building Blocks

Our approach can be broadly described as bottom-up. We first extract trajectories of apparent motion in the image plane. From these a vocabulary of elementary motions is built, which are then used to canonize all observed tracks. Inference is performed on tracks expressed in this fixed vocabulary of motion primitives. In this section we address low-level problems related to motion extraction and its filtering, and the learning of motion primitives vocabulary.

2.1 Motion Extraction

As the foundation for inference at higher levels of abstraction, the extraction of motion in a scene is a challenging task and a potential bottleneck. The key difficulty stems from the need to capture motion at different scales, thus creating the compromise between reliability and permanence of tracking. Generally speaking, the problem can be approached by employing either holistic appearance, or local appearance in the form of local features. Holistic, template based methods capture a greater amount of appearance and geometry, which can be advantageous in preventing tracking failure. On the other hand, in the presence of frequent full and partial occlusions, these methods are difficult to auto-initialize and struggle with the problem of gradual bias drift as the tracked object's appearance changes due to articulation, variable background and viewpoint, etc. All of these difficulties are very much pronounced in the scenario we consider, motivating the use of local features.

Focusing on computationally efficient approaches, we explored several methods for detecting interest points. Recently proposed Rosten-Drummond fast corner features and Lowe's popular scale-space maxima were found to be unsuitable due to lack of permanence of their features: for an acceptable total number of features per frame, features that were detected at some point in time remained undetected in more than 80% of frames. Success was achieved by adopting a simple method of tracking small appearance windows, in a manner similar to Lucas and Kanade. Image region corresponding to a window in frame F_i, is localized in subsequent frame F_{i+1} by finding the translation which minimizes the observed error between the two regions:

$$\mathrm{d}_i^* = \arg\min_{\mathrm{d}} \int_{\mathrm{x}\in\mathcal{W}} \left[F_{i+1}(\mathrm{x} + \frac{1}{2}\mathrm{d}) - F_i(\mathrm{x} - \frac{1}{2}\mathrm{d}) \right] d\mathcal{W} \qquad (1)$$

This criterion is similar to that originally formulated by Lucas and Kanade, with the difference that the expression is symmetrized in time (i.e. with respect to F_i and F_{i+1}). Further robustness in comparison to the original method is also gained by performing iterative optimization in a multiscale fashion, whereby d_i^* is first estimated using smoothed windows and then refined by progressively less smoothing. The best features to track were selected as those corresponding to the 2×2 gradient matrices with the largest magnitude eigenvalues.

Trajectory filtering. Following the basic extraction of motion trajectories, we filter out uninformative tracks. These are tracks which are too short (either due to low feature permanence or due to occlusion of the tracked region) or which correspond to stationary features (possibly exhibiting small apparent motion in the image plane due to camera disturbances, such as due to wind). We accept a track $\{(x_1, y_1), \ldots, (x_N, y_N)\}$ if $N \geq 30$ (i.e. it lasts for at least 30 frames, or 1.2 s at 25 fps) and:

$$\frac{1}{N-1}\sum_{i=1}^{N}\left[(x_i - \bar{x})^2 + (y_i - \bar{y})^2\right] \geq \sigma_{min}^2 \quad \bar{x} = \sum_{i=1}^{N} x_i/N \quad \bar{y} = \sum_{i=1}^{N} y_i/N$$

2.2 Tracklet Motion Primitives

People's motion trajectories in a scene can exhibit a wide range variability. However, not all of it is relevant to the problem we address. For example, motion of interest is corrupted by noise and at a short scale modulated by articulation. To reduce the effects of confounding variables, we represent all tracks using the same vocabulary of elementary motion primitives, inferred from data.

Inferring primitives. We construct the vocabulary of motion primitives by clustering *tracklets*—local, linear approximations of tracks. We extract a set of tracklets t_i from a feature track $T = \{(x_1, y_1), \ldots, (x_N, y_N)\}$ by first dividing the track into overlapping segments $\{(x_{s(i)}, y_{s(i)}), \ldots, (x_{e(i)}, y_{e(i)})\}$ such that:

$$\Delta D = \sum_{j=s(i)}^{e(i)-1} \left\| \begin{pmatrix} x_j \\ y_j \end{pmatrix} - \begin{pmatrix} x_{j+1} \\ y_{j+1} \end{pmatrix} \right\|,$$

where ΔD is the characteristic scale parameter of the corresponding tracklet model. The ith tracklet is then defined by a numerical triplet consisting of its location and orientation $t_i = (\hat{x}_i, \hat{y}_i, \theta_i)$, where:

$$\hat{x}_i = \frac{\sum_{j=s(i)}^{e(i)} x_j}{e(i) - s(i) + 1} \quad \hat{y}_i = \frac{\sum_{j=s(i)}^{e(i)} y_j}{e(i) - s(i) + 1} \quad \hat{\theta}_i = \tan^{-1}\frac{y_{e(i)} - y_{s(i)}}{x_{e(i)} - x_{s(i)}} (\mathrm{mod}\ \pi)$$

All tracklets extracted from training data tracks are clustered using an iterative algorithm. In each iteration, a new cluster is initialized with a yet unclustered tracklet as the seed. The cluster is refined further in a nested iteration whereby tracklets are added to the cluster under the constraint of maximal spatial and directional distances, respectively ΔQ and $\Delta\Theta$, from both the seed and the cluster centre. Cluster centre is then set equal to the mean of the selected tracklets and the procedure repeated until convergence.

Note that the equivalence of directions θ and $\theta \pm \pi$ introduces some difficulty in the estimation of the cluster centre orientation. Specifically, it is *not* appropriate to average member directions using modulo π arithmetic. First, note that the problem is not always well posed, i.e. that it does not always have a unique solution. Thus, we require that $\forall i.\Delta\hat{\theta}_i < \pi/4$, where:

$$\Delta\hat{\theta}_i = \min\Big\{(\hat{\theta}_i - \theta_c)(\text{mod } \pi), |\hat{\theta}_i - \theta_c|\Big\}.$$

This condition ensures that the range of directions in a cluster is sufficiently constrained that the mean direction is unambiguously definable. It is sufficient that $\Delta\Theta < \pi/4$ for this to be the case, which is certainly true in this paper, as a directional spread of over $\pi/2$ within a cluster would produce meaningless tracklet groupings. An algorithm for computing the cluster angle (provided that a unique solution exists) is described in the extended version of this paper.[1]

Expressing tracks using primitives. A track is expressed in a particular tracklet model by diving it into overlapping segments, computing the location and direction of each segment as before and, finally, associating each segment $t_i = (\hat{x}_i, \hat{y}_i, \theta_i)$ with the most similar tracklet $\mathfrak{T}_{j(i)} = (X_{j(i)}, Y_{j(i)}, \Theta_{j(i)})$:

$$t_i \longrightarrow \mathfrak{T}_{j(i)} : j(i) = \arg\min_k \epsilon(\mathfrak{T}_j, t_i),$$

$$\text{where } \epsilon(\mathfrak{T}_j, t_i) = (X_{j(i)} - \hat{x}_i)^2 + (Y_{j(i)} - \hat{y}_i)^2 + \left[\frac{\Delta Q}{\tan(\Delta\Theta)}\tan(\Delta\hat{\theta}_i)\right]^2.$$

This functional form ensures that the angular contribution to distance is infinite for orthogonal tracklets and approximately linear for small $\Delta\hat{\theta}$, whereas the proportionality factor ensures that relative scaling of spatial and angular distances corresponds to the spread of cluster tracklets.

3 Two Tracklet Based Motion Models

In the previous section we described our approach to extraction and representation of motion in a scene. We now turn to the problem of learning a motion model from these representations and applying it to infer novelty in unseen data.

[1] Extended paper with additional detail is available on the author's website.

3.1 First Order Markov Chain Ensemble

The first model we introduce in this paper utilizes an ensemble of K complementary first order Markov chains models to learn "normal" behaviour in a scene. The idea is that each model learns behaviour on a different characteristic spatial scale. While this idea is now new, it should be noted that our approach is different in that the ensemble we construct is (in general) not a hierarchial one—states describing behaviour on longer scales do not consist of sequences of lower scale states. Rather, each model is built independently by extracting tracklets and the corresponding motion primitives using different characteristic scales. The kth model thus comprises learnt priors $P(\mathfrak{T}_i^{(k)})$ and transition probabilities $P(\mathfrak{T}_j^{(k)}|\mathfrak{T}_i^{(k)})$ for tracklet primitives at the corresponding scale.

Consider a particular novel feature track T. In our model, the track is expressed independently in each of the K vocabularies of tracklet primitives:

$$T \longrightarrow T_k = \left\{\mathfrak{T}_1^{(k)}, \ldots, \mathfrak{T}_{M_k}^{(k)}\right\} \text{where } M_1 < M_2 < \ldots M_K,$$

where T_1 corresponds to the smallest scale of interest and T_K the largest. Each of the chains can then be used to compute the log-likelihood estimate corresponding to its scale, which we average to normalize for differing track lengths:

$$R_k(T) = \frac{1}{M_k}\left[\log P(\mathfrak{T}_1^{(k)}) + \sum_{i=2}^{M_k} \log P(\mathfrak{T}_i^{(k)}|\mathfrak{T}_{i-1}^{(k)})\right].$$

Thus, the task of deciding if motion captured by T sufficiently conforms to behaviour seen in training is reduced to inference based on log-likelihoods $R_k(T)$. This would not be a difficult problem if both positive and negative training data (i.e. both unusual and normal motion patterns) was available, or if log-likelihoods corresponding to different models were directly comparable. However, a representative amount of positive training data is difficult to obtain in this case. Furthermore, although each $R_i(T)$ is normalized for track length, the range of variation of its value is dependent on the model's characteristic scale. This is a consequence of lower entropy (generally) of larger scale models, which have fewer states (tracklet primitives). To solve this problem, we transform the average log-likelihoods of all models to conform to the same cumulative distribution function of the lowest scale (highest entropy) model:

$$R_k \longrightarrow \hat{R}_k = C_1^{-1}[C_k(R_k)]$$

where $C_k(R)$ is the cumulative distribution function of the average log-likelihood of the kth model, estimated from the training data set: $C_k(R) = \int_{-\infty}^{R} p_k(r)dr$. The conformance of the track to the overall multiscale model is then computed as the minimal conformance over different scales: $\rho_1(T) = \min_k \hat{R}_k$.

3.2 Pursuit-Constrained Motion Model

In the previous section we described an approach to learning the range of normal motion in a scene which treats a feature trajectory as a sequence of states corresponding to extracted tracklet primitives. To make the parameter estimation practically tractable, the sequence of states was modelled as a first order Markov chain which inherently restricts the scope of the model to aggregating single-step behaviour. Progressively less spatially constrained behavioural characteristics were captured by multiple tracklet models, each with a different characteristic scale.

What this approach does not exploit is the structure of observed motion governed by the of *intentionality* of persons in the scene (whether they are on foot or using a vehicle). While it is certainly the case that if unlimited data was available the described purely statistical model would eventually learn this regularity, this insight can help us achieve a higher degree of generalization from limited data which is available in practice. Our idea is based on the simple observation that people perform motion with the aim of reaching a particular goal and they generally plan their it so as to minimize the invested effort, under the constraints of the scene (such as the locations of boulders and paved areas, or places of interest such as shopping windows). Consequently, we concentrate on learning the distribution of traversed trajectory lengths between two locations in a scene, rather than the exact paths taken between them (a far greater range of possibilities). Unlike in Sect. 1, we now express a feature track T as a sequence of tracklet primitives *only* in the vocabulary of the smallest scale:

$$T \longrightarrow T_1 = \left\{ \mathfrak{T}_1^{(1)}, \ldots, \mathfrak{T}_{M_1}^{(1)} \right\}.$$

For each pair of tracklets $\mathfrak{T}_i$ and $\mathfrak{T}_j$ $(i,j = 1,\ldots,M_1)$ from the sequence we compute the corresponding distance L_{ij} between them along the path. Since the tracklet primitives were estimated using a single scale model with the characteristic scale parameter ΔD_1, by construction this distance is $L_{ij} = (i-j)\Delta D_1/2$. The track is thus decomposed into $M_1(M_1-1)/2$ triplets $(\mathfrak{T}_i, \mathfrak{T}_j, L_{ij})$. We wish to estimate $p(\mathfrak{T}_i, \mathfrak{T}_j, L_{ij})$. By expanding the joint probability as:

$$p(\mathfrak{T}_i, \mathfrak{T}_j, L_{ij}) = P(\mathfrak{T}_i) P(\mathfrak{T}_j|\mathfrak{T}_i) p(L_{ij}|\mathfrak{T}_i, \mathfrak{T}_j),$$

we can see that the first two terms—the prior probability of the ith primitive and the probability of $i \rightarrow j$ transition—can be learnt in a similar manner as for the Markov chain based model described previously. On the other hand, the last term corresponding to the distribution of possible path lengths between the ith and jth primitive, is computed by modelling it using a normal distribution:

$$p(\mathfrak{T}_j|\mathfrak{T}_i, L_{ij}) = \mathcal{N}(L|\bar{L}_{ij}; \sigma_{ij}).$$

We estimate its parameters—the mean $\bar{L}_{ij}$ and standard deviation σ_{ij}—using transitions between primitives observed in the training data set. Finally, the conformance of a novel track to the learnt motion model is computed as the log of the lowest probability transition within the observed motion: $\rho_2(T) = \min_{i,j} \log p(\mathfrak{T}_j, \mathfrak{T}_i, L_{ij})$.

4 Evaluation

Using a stationary camera placed on top of a small building overlooking a busy city marketplace we recorded a video sequence which we used to evaluate the proposed methods. This footage of the total duration of 1 h:59 m:40 s and spatial resolution 720×576 pixels contains all of the challenging aspects used to motivate our work: continuous presence of a large number of moving entities, frequent occlusions, articulation and scale changes, non-static background and large variability in motion patterns.

4.1 Results

After training each of the proposed methods using the 113,700 extracted tracks, we computed the corresponding histograms of conformity measures ρ_1 and ρ_2. From these we automatically selected thresholds for novelty detection, R_1 and R_2, such that 0.05% of training tracks produce lower conformities. An examination of these tracks revealed that the two algorithms generally identified the same tracks as being the least like the rest of the training set. A common aspect which can be observed between them is that they correspond to motions which include sharp direction changes *at locations in the scene where there is little reason for them.* This motion not only novel by the construction of our model but it also conforms to our intuition about what constitutes unusual behaviour. Note the scene-specific, contextual aspect of the learnt motions: many extracted tracks contain sharp turns (e.g. at the end of the row of marketplace stalls or at the corner of the buildings) which are not deemed unusual because the constraints of the scene made such turns (comparatively) frequent in the training data.

Next, to evaluate how our algorithms cope with unseen data, by clicking on the image of the scene we generated a series of synthetic tracks which a human might consider unusual in the context of the marketplace in question. Here, the two methods produced different results. The discrepancy can be explained by observing that in the ensemble approach, a trade-off is made between the precision of motion localization by tracklet primitives and the ability to capture behaviour at a larger scale. Such compromise does not exist in the proposed pursuit-constrained model.

Lastly, since the conformance measure ρ_2 is effectively dependent only on a single pair of primitives (those which are explained the worst by the path length model), we could visualize these for a series of tracks in which novelty was detected. This is useful firstly as a way of ensuring that the model is capturing meaningful information and secondly to draw attention not only to a particular behaviour on the whole but its particular feature.

Grid Security Loopholes with Proposed Countermeasures

Nureni Ayofe Azeez, Tiko Iyamu and Isabella M. Venter

Abstract Grid computing is an emergent computing innovation which offers endless access to computing infrastructure across various organizations (academia and industry). This technology allows the aggregation of various computer systems for usage by different users to run applications. The information stored on these computers may be vulnerable. According to research on attribute based access control for grid computing, security mechanism to authorize and authenticate users, before accessing information on a grid system, is not totally adequate. The issue of security in grid technology has not been fully addressed even though it is a precondition for optimizing grid usability. In this paper some of the security attacks on a grid system were explored and reasonable countermeasures are proposed. An architectural model to prevent any form of attacks explained, is presented.

Keywords Authorization · Grid · Attack · Sharing · Circumvent · Sensitive · Model · Access · Control

N. A. Azeez (✉) · I. M. Venter
Department of Computer Science, University of the Western Cape,
Private Bag X17, Bellville 7535, South Africa
e-mail: 3008814@uwc.ac.za; nurayhn@yahoo.ca

I. M. Venter
e-mail: iventer@uwc.ac.za

T. Iyamu
Department of Informatics, Faculty of Information and Communication,
Tshwane University of Technology, Pretoria South, South Africa
e-mail: iyamuT@tut.ac.za

E. Gelenbe et al. (eds.), *Computer and Information Sciences II*,
DOI: 10.1007/978-1-4471-2155-8_52,

1 Introduction

For any grid participant to enjoy the full benefits of resource sharing, resource distribution and resource aggregation across various domains, assurance has to be given that the data that will be shared is safe from any form of attack. In a distributed environment, such as the grid, which involves users from various domains, data manipulation, alteration and falsification are possible. Security breaches cause problems that can be devastating therefore participants in distributed environments sometimes fear sharing resources. This paper attempts to explain what causes security lapses [3] in a grid-based environment and propose various measures to circumvent some of these anomalies.

2 Grid Classification Based on Topology

Grid computing can be classified into four basic groups based on its topology. The classifications that will be discussed are : clusters; intra-grids; extra-grids and inter-grids [11].

2.1 Cluster: is the smallest form of grid both in scope and size [1]. It is a combination of various servers which can generate computing power similar to what is obtainable in an offline (standalone) system. This form of grid computing was developed to solve problems in a unit or same department and it is most often implemented on campus intranets.

2.2 Intra-grid: Is a combination of various clusters. This form of grid computing is also referred to as a campus grid. It allows resources to be shared across various departments and units. An intra-grid allows resources to be shared across various organizations (with the same policies) without any need to address the security and policy management relating to global grids. Fig. 1.

Cluster grid Intra grid Extra grid Inter grid

2.3 Extra-grid: is also referred to as "Partner grids" or "Extraprise grids". It is a combination of two or more intra-grids with various security domains. This form of grid is geographically distributed between various establishment, companies and organization. Virtual Private Networks (VPN) are used for implementing this type of grid in order to make resources easily available [1] to users.

2.4 Inter-grid: is the final phase of grid evolution. It is also known as a global grid. It is a combination of intra-grids and cluster grids joined together by the internet. This form of grid is often used in academia.

3 Security Attacks on a Grid System

It has been recognized by researchers that all forms of distributed systems are vulnerable and cannot be secured completely. Some of the security measures currently being used are inflexible and not scalable [8].

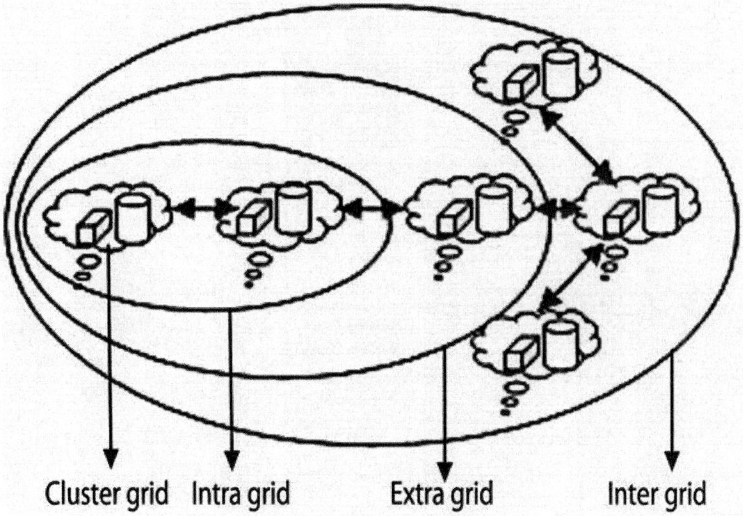

Fig. 1 Grid classification based on topology

The most common security attacks on grid systems can be divided into the following categories:

- Passive attacks
- Active attacks
- Dictionary attacks

3.1 Passive attacks: Passive security attacks ranges from [2] secret monitoring of transmissions, such as electronic mails, file exchange on any distributed system, client–server transmission or eavesdropping. Passive attacks involve a thorough analysis of traffic as well as exposing and releasing of message content.

- Release of message contents: This form of attack can be best explained as shown in Fig. 2. Electronic mail messages, telephone conversations and files being transferred may contain sensitive and confidential information [22]. To limit the vulnerability of the information, it is important to prevent intruders to learn about and understand the contents of the transmission [21].
- *Traffic analysis:* This is a form of security attack [3] whereby messages are intercepted and examined in order to get information from patterns in communication process. This form of attack can be carried out even when the message is encrypted and is difficult to be decrypted. The higher the number of messages intercepted and examined the more deduction and inference can be made from the traffic. Traffic analysis (often carried out in military intelligence) is a serious concern for computer security [15].

***3.2 Active attacks*:** This type of attack attempts to change, remove, or destroy data being transferred from one system to another system on a grid network. Some

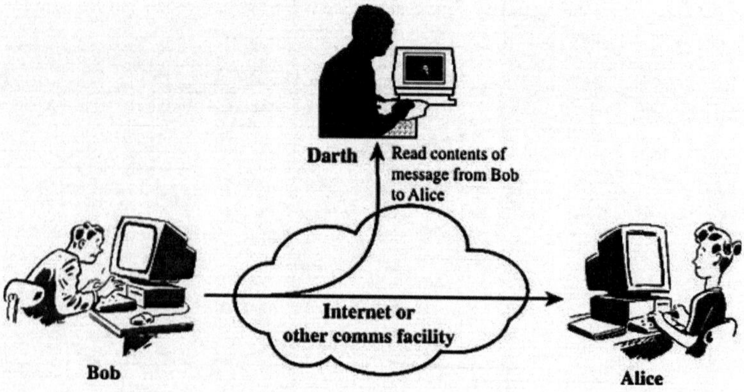

Fig. 2 Release of message contents (Passive attack) *Source*: [22]

of the types of active attacks are denial of service, masquerade, replay and modification [21]. It can be prevented with the aid of common and popular security mechanisms such as firewalls or encryption techniques. The following section gives a brief explanation of some of the examples of active attacks.

- **Denial of Service (DoS):** This is a type of active security attack, which halts the system on a grid so that it fails to respond to users' demands. To accomplish this objective, attackers will send a very large quantity of data to the access point [16] so that it becomes extremely difficult to respond appropriately.
- **Impersonation:** In this case the behaviour of an authorized grid user is mimicked and copied by the attacker. With this approach, attackers can easily modify and change information to the detriment of a legal user. Man-in-the middle attack is the commonest form of impersonation.
- **Disclosure:** Sensitive information across the grid might be disclosed by a compromised machine on the grid to another machine that is not authorized to have access to such information. Sometimes classified information is unwittingly compromised or divulged by a user and thus the content is exposed[2].
- **Unauthourized access:** This type of security attack occurs when a person, who is permitted to access particular information, gains access to it and plays around with it as if he/she is the rightful owner of the information. When this happens the sensitivity of the information will be lost and the data can be manipulated without beingy challenged [12].
- **Replay attack:** Replay attack can also be likened to a "Man-in-the-middle" form of attack. This form of security attack allows the data packets to be intercepted and replayed (resent) to the server. For example, it is possible for a client to send a password and a username that is encrypted to a server to gain access to grid information. A hacker may use monitoring software to intercept such a message and replay/resend it. Hence, such a hacker will have the same rights as the rightful owner of such a message.

3.3 Dictionary attack: This method is used by attackers to break a security measure put in place for a system or a server, by trying all possible passwords with the intention of gaining access. Dictionary attacks allow spammers [4] to randomly send mail across to various addresses using a combination of some popular domain names with the intention of getting information across to a large percentage of e-mail users [5]. For example a dictionary attack may begin with the following e-mail addresses john@uwc.ac.za, john1@uwc.ac.za, john2@uwc.ac.za, john3@uwc.ac.za. This may continue till all possible combinations of both the numbers and variable has been tried and exhausted [5].

4 Security Requirements in a Grid Environment

Security requirements for a grid were defined by The International Organization for Standardization (ISO) and the International Telecommunication Union (ITU) (ITU-T Provision X.805 and X.800) [7].

4.1 Authourization: For any organization to allow its resources to be jointly shared between all parties involved there is need for authorization—thus who should have access to any particular resources and who should not. Permission is only given to authorized nodes on the network [2]. Globus toolkit gridmap files, Community authorization service (CAS) and Virtual organisation membership service (VOMS) are some of the authorization measures used in grid computing [6].

4.2 Authentication and access control: Impersonation has been identified as a big threat in a grid environment. Authentication is important to purposely prevent resources from being illegal accessed. The main purpose of authentication [9] is solely to confirm that the user is he who he claims to represent and not any other person. Authentication is usually carried out with the use of a password a username. A digital certificate, verified by a Certificate authority [3], is generally accepted to be the best way to ensure authentication on the internet.

4.3 Availability: Irrespective of security attacks, data must be readily available across the network to satisfy the demands of grid computing users at any point in time. Data availability means that data is available at all times. In a grid environment, data availability is usually achieved through redundancy; which has to do with how data is stored and how such data can be reached [7].

4.4 Data Confidentiality**:** The purpose of data confidentiality is to protect data from being divulged to the wrong or an unintended party [24]. Two steps can be used to achieve data confidentiality; data encryption and data decryption. Two types of cryptography that can be used to provide data confidentiality [17] is: symmetric and asymmetric cryptography.

4.5 Data integrity**:** In the banking industry, military operation as well as aviation industry [10] data modification by unauthorised [23] persons is very serious. Data integrity assumes that data in a grid environment is removed, updated, modified, deleted, edited and transmitted only by an authourised [19] person.

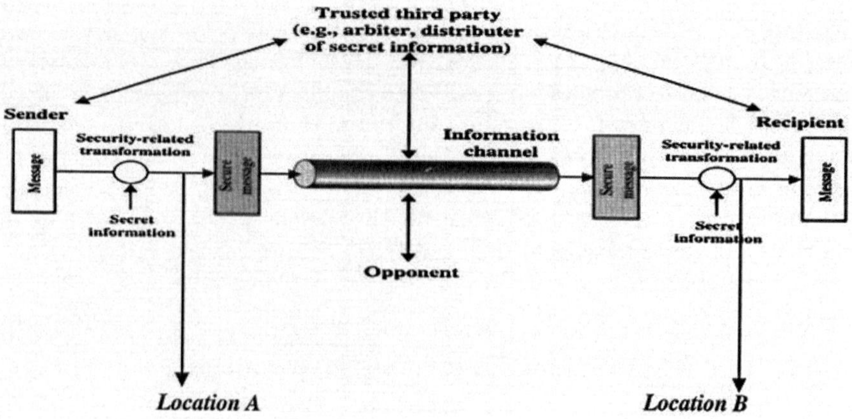

Fig. 3 Secured grid model

4.6 Non repudiation: Since many transactions take place on the internet, this security service protects parties involved in a transaction from denying that a particular transaction took place [13]. Non-repudiation therefore ensures that both the receiver and the sender cannot renege or deny that a message has been sent or received [18]. This security measure is usually achieved through the use of digital signatures and certificates. Timestamps which contain the date and time can also be used.

4.7 Privacy: The main purpose of privacy [20] is to ensure that information being shared on the grid system is protected. Every grid user wants his sensitive information to be completely secured from misuse and abuse. According to Alan Westin, the definition of privacy is given as follows [14]:

> Privacy is the right of individuals to determine for themselves when, how, and to what extent information about them is communicated to others

5 Proposed Secured Grid Model

As shown in the Fig. 3, when a message is to be sent from *location A* to *location B* across a grid to a rusted third party, both the sender and the receiver are expected to cooperate in order to ensure the smooth exchange of information across the grid. A reliable information channel ensures the smooth passage of the message from source to destination. To secure the information and protect the message flow from being tampered with, confidentiality, authenticity and integrity needs to be ensured. How can such a secured transmission be put in place? In this model, an encryption key is used to protect the message before transmission and unprotect it immediately after it is received at *location B*. It will thus be difficult for an intruder to unencrypt an intercepted message. Additional code, based on the message contents, is used to trace and identify the identity of the sender.. This is an added protection measure.

6 Conclusion

Since the confidentiality as well as integrity of information in a distributed system should be sacrosanct, it is important to ensure that appropriate mechanisms are put in place to ensure that the information is protected. Various grid security loopholes have been identified which could compromise a grid-based environment. The countermeasures suggested should go a long way in protecting sensitive information. Although the suggested model has not been tested as yet, it is believed that when implemented it should address many of the security challenges within the grid.

References

1. Alfawair, M., Aldabbas, O., Bartels, P., Zedan, H.: Grid Evolution. International, Conference on Computer Engineering and Systems (ICCES'07) (2007)
2. Ali, A.-B., Hussein, Z., François, S.: Access Control Mechanism for Mobile Ad Hoc Network of Networks (MANoN). Software Technology Research Laboratory De Montfort University, Leicester (2009)
3. Ayofe A.N., Osunade, O: Towards ameliorating cybercrime and cybersecurity. Int. J. Comput. Sci. Inf. Secur. USA (IJCSIS) **3**(1), 153
4. Azeez, N., Azeez, R., Sulaimon, B., Atanda, F.: A framework for computer aided investigation of ATM fraud in Nigeria. Pac. J. Sci. Technol. USA **11**(1), 356–361 (2010)
5. Benny, P., Tomas, S: Securing passwords against dictionary attacks. Accessed 18 Feb 2011, from www.pinkas.net/PAPERS/pwdweb.pdf
6. Chadwick, D.: Authorisation in Grid computing Elsevier. University of Salford, UK (2005)
7. Fran, B., Geoffrey, F., Anthony, J.G.: Grid Computing: Making the Global Infrastructure a Reality. Wiley and Sons Ltd, West Sussex (2005)
8. Laccetti, G.G.S.: A framework model for grid security. ScienceDirect- Future Gener. Comput. Syst. **23**, 702–713 (2007)
9. Hogue, M., Avery, V.: Novel strategies to speed-up query response. Res. J. Inform. Technol. **2**, 11–20 (2010)
10. Husam, K., Taha, O., Nasser, S.: Attribute-based authorization for grid computing. international conference on intelligent systems, modelling and simulation, IEEE Computer Society. USA pp. 70–74. (2010)
11. Ian Foster, C.K., Kesselman, C.: Globus: A metacomputing infrastructure toolkit. 11. Int. J. Supercomput. Appl. High Perform. Comput. **11**(2), 115–128 (1997)
12. IASTED: International Association of Science and Technology for Development. www.iasted.com. URL: http://www.iasted.org/conferences/2004/Innsbruck/pdcn.htm. Accessed 2 Mar 2011 (2004)
13. Jianying, Z., Robert H., D., Feng, B.: Evolution of fair non-repudiation with TTP.In: ACISP '99 Proceedings of the 4th Australasian Conference on Information Security and Privacy. Springer, UK (1999)
14. Lalanal, K., Hal, A: Access Control is an Inadequate Framework for Privacy Protection MIT Computer Science and Artificial Intelligence lab, Cambridge. URL: www.w3.org/2010/api-privacy-ws/papers/privacy-ws-23.pdf. Accessed 19 Feb 19 2011 (1990)
15. McClure, S., Scambray, J., Kurtz, G.: Hacking Exposed: Network Security secrets and solutions. McGraw Hill, New York (2003)
16. Mohteshim, H.: Passive and Active attacks against Wireless LANS. URL: www.designmpire.com/mohteshim.com/projects/anp.pdf. Accessed 2 Mar 2011 (2005)

17. MSDN.: Data Confidentiality. Accessed 12 Feb 2011, from http://msdn.microsoft.com/en-us/library/ff650720.aspx(2005)
18. Onieva, J.A., Zhou, J.: Secure multi-party non-repudiation protocols and applications. Institute for Infocomm Research, Singerpore. Springer (Security and Cryptology), 43:20. URL: http://portal.acm.org/citation.cfm?id=1481008. Accessed 28 Feb 2011 (2008)
19. Prasannakumari, V.: A robust tamperproof watermarking for data integrity in relational databases. Res. J. Inform. Technol. **1**, 115–121 (2009)
20. Rasheed, M., Ghazali, O., Norwawi, N.: Server scanning worm detection by using intelligent failure connection algorithm. Res. J. Inform. Technol. **2**, 228–234 (2010)
21. Welch D., Lathrop S.: Wireless security threat taxonomy. In: Proceedings of the 2003 IEEE workshop on information assuarance United States Military Academy West Point. NY URL: http://www.eecs.usma.edu/webs/people/lathrop/publications/wirelessthreat_IEEE.pdf. Accessed 8 Jan 2011 (2003)
22. William, S.: Network Security Essentials: Applicationsand Standards. Pearson Education Inc, (2002)
23. Zanjani, M., Sakhaee, N., Shahbaznezhad, H.: Mechanisms of customer knowledge management in e-commerce websites. Res. J. Inform. Technol. **1**, 86–93 (2010)
24. Zhi-Dong, S., Fei, Y.: Grid system integrated with trusted computing platform. IEEE **2**, 619–625 (2006)

Detection of Broken Link Fraud in DSDV Routing

Rajbir Kaur, Vijay Laxmi and Manoj Singh Gaur

Abstract In this paper we present a technique to defend against Broken Link Fraud, a denial of service attack against Destination Sequenced Distance Vector. Each node in the network keeps track of the frequency of next hop for each destination. Next hop is blacklisted if frequency exceeds a threshold. Our method does not rely on expensive cryptographic techniques making it suitable for energy constrained Mobile Ad hoc NETworks. Simulation results demonstrate effectiveness of our method in blacklisting misbehaving node, resulting in improved Packet Delivery Ratio.

keywords Broken Link Fraud · Denial of service · DSDV · MANET

1 Introduction

Destination Sequenced Distance Vector (DSDV) [1] is a proactive protocol to enable communication between mobile nodes in Mobile Ad hoc NETwork (MANET). It is essential to study attacks in routing protocols [2] and incorporate provisions for detection to minimize their impact. In this paper we present an inexpensive countermeasure for Broken Link Fraud (BLF) [3].

R. Kaur · V. Laxmi · M. S. Gaur (✉)
Department of Computer Engineering, Malaviya National Institute of Technology, Jaipur, India
e-mail: gaurms@mnit.ac.in; gaurms@gmail.com

R. Kaur
e-mail: rajbir@mnit.ac.in

V. Laxmi
e-mail: vlaxmi@mnit.ac.in

E. Gelenbe et al. (eds.), *Computer and Information Sciences II*,
DOI: 10.1007/978-1-4471-2155-8_53,

2 Related Work

Wang et al. [4] study security properties of DSDV. They have not studied impact of large distance fraud on DSDV. Smith et al. [5] propose digital signatures to provide authentication and integrity of routing messages which does not protect against short/long distance fraud (LDF). SEAD proposed by Hu et al. [6] does not prevent a misbehaving node from advertising a distance longer than the one it has received. Wan et al. [7] state that LDF can only be used to launch passive attacks (e.g., selfishness). BLF uses LDF to launch active attacks that severely hampers network performance.

3 Methodology for Detection and Countermeasure

Dynamic topology in MANETs ensure that neighbors of a node change with time; a persistent neighbour is indication of presence of malicious node. Improper behavior may be modelled as same next hop neighbor for period longer than some threshold. Detection of malicious node is done in two phases:

Phase 1: Each node keeps track of the frequency of next hop for a given destination for *observation time*. Neighbour is labelled malicious if its frequency exceeds threshold.

Phase 2: Nodes broadcast identity of suspected node and assimilate information received from other nodes. Node labelled as suspected by maximum nodes is designated as malicious and blacklisted.

Algorithm 1 Algorithm for detection of malicious node

```
begin
    N : Network Size
    t_current: Current time,
    t_observe = 10s: Observation time,
    t_assim = 5s: Processing time
    τ = 75: Threshold value
    Γ_freq: Number of times an event was observed
    while t_current < t_observe do
        numCount[nxtHop]++;
        Γ_freq++;
    i = arg(max[numCount]);
    if (i*100)/Γ_freq > τ then
        attackerId = i;
        numCount[N] = 0;
    pkt.packetHeader = attackerPkt;
    pkt.srcID = nodeID;
    pkt.attackerId = attackerId;
    /* collect information from other nodes */
    while t_current < t_observe + t_assim do
        attackerIdResult[srcID] = pkt.attackerId;
```

for $i = 0$ *to* N **do**
  numCount[attackerIdResult[i]]++;
$malNode = \arg(\max[numCount])$;
if $t_{current} > t_{assim}$ **then**
  Do not accept packets coming from *malNode*;

---

4 Simulation

We use *NCTUns-5.0* [8] for simulation. Packet Delivery Ratio (PDR) (ratio of packets received to packets sent) is used as metric. Table 1 shows the simulation parameters used. To incorporate our methodology, we have made changes as illustrated in Algorithm 3 to DSDV module.

Simulation Result and Discussion: Figure 1 shows the PDR for different simulation periods. In the graph we plot curves for four scenarios: with/without attack and with/without countermeasure. In attack, PDR is very low. When countermeasure is applied, there is an improvement of 20 to 25% in PDR.

It takes $\approx$ 17 s for countermeasure to converge. During this, the attacker claims itself as next hop for the destination. Routing tables at most nodes have attacker as next hop. Incorrect routing tables cause a packet to be discarded after expiry of TTL (time-to-live). As a result, PDR drops. For each simulation time, We have averaged 10 readings by assuming different nodes as attackers. We also observe from Figure 1 that the performance of modDSDV in no attack situation is comparable to that of normal DSDV. For a small and relatively less dynamic topology, proposed method may blacklist a 'honest' node. This requires searching for an alternate path which may be less efficient than the previous one in terms of hop count. Slight drop in PDR is observed in such cases. This explains slightly low performance of modDSDV. Figure 2 shows false positive rate with different values of observation time. Probability of false positives is low even if the observation time is 5 s. Figure 3 shows PDR for different values of observation time. Lowering observation time does not have any appreciable effect on PDR. We infer that observation time between 7–10 s can also be chosen. Figure 4 shows overhead associated with our approach.

$$\text{Overhead} = \frac{\text{No. of broadcasts}}{(\text{No. of broadcasts + No. of packets transmitted})} \quad (1)$$

Proposed countermeasure does not significantly increase broadcast packets indicating that there is no flooding of channel bandwidth.

Table 1 Simulation Parameters

Number of nodes	16
Signal range of each node	200 m
Mobility model	Random direction waypoint model
Traffic type	UDP traffic
Number of packets sent from source to destination	1 packet/s
Simulation time	100 s

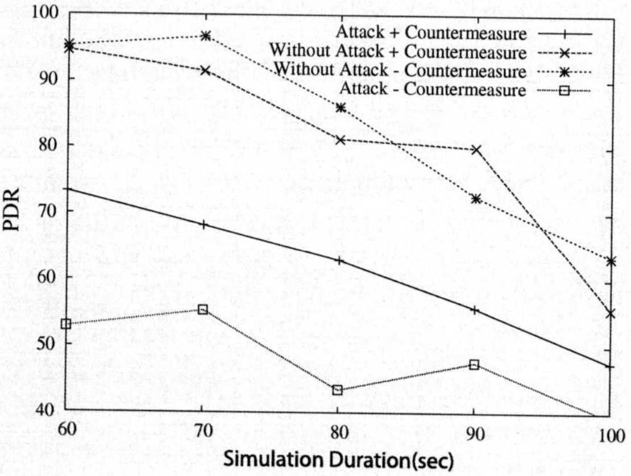

Fig. 1 PDR improves with modDSDV

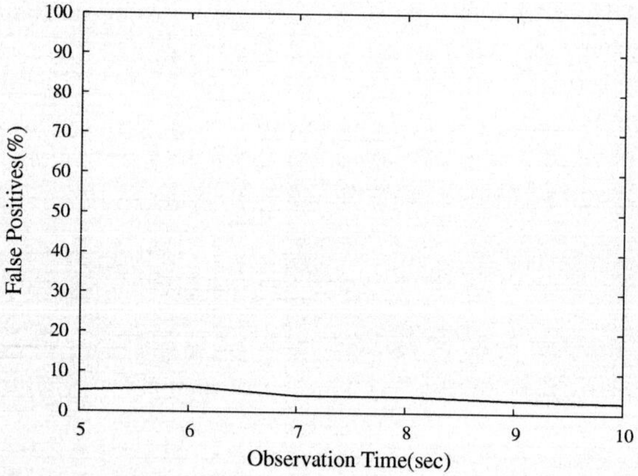

Fig. 2 False positives for different $t_{observe}$

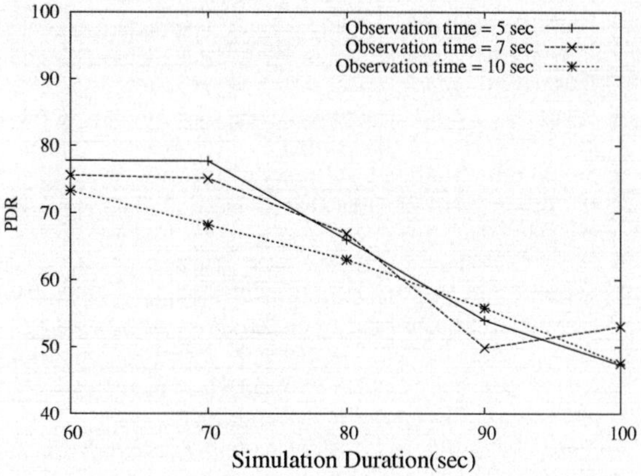

Fig. 3 PDR for different observation times

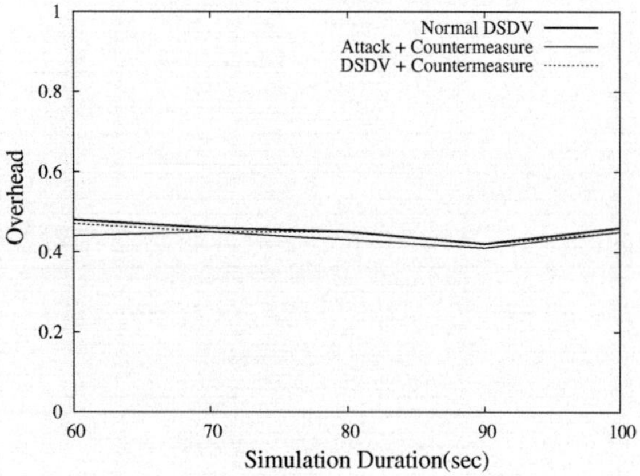

Fig. 4 Overhead of proposed countermeasure

5 Conclusion and Future Work

In this paper we have proposed a detection and countermeasure for BLF. Simulation results prove effectiveness of the approach to detect attacker. Probability of false positives is negligible. Our approach does not use expensive computations as by cryptographic methods. In future, we will refine the method to override underlying assumptions.

References

1. Perkins, C.E., Bhagwat, P.: Highly dynamic destination-sequenced distance-vector routing for mobile computers. SIGCOMM Comput. Commun. Rev. **24**(4), 234–244 (1994)
2. Kaur, R., Gaur, M.S., Suresh, L., Laxmi, V.: DOS attacks in MANETs: Detection and Countermeasures. In: Santanam, R., Sethumadhavan, M., Virendra, M. (eds.) Cyber Security, Cyber Crime and Cyber Forensics: Applications and Perspectives, pp. 124–145. IGI Global, US (2011). doi:10.4018/978-1-60960-123-2.ch010

3. Kaur, R., Gaur, M.S., Laxmi, V.: A novel attack model simulation in DSDV routing. In: 4th International Conference on New Technologies, Mobility and Security (NTMS'11), Paris, France (2011)
4. Wang, W., Lu, Y., Bhargava, B.: On security study of two distance vector routing protocols for mobile Ad Hoc networks. IICPCC, pp. 179–186, March 2003
5. Smith, B.R., Murthy, S., Garcia-Luna Aceves, J.J.: Securing distance-vector routing protocols. In: Proceedings of the 1997 Symposium on Network and Distributed System Security (SNDSS'97), IEEE Comput. Soc, p. 85, Washington, DC, USA (1997)
6. Hu, Y.C., Johnson, D.B., Perrig A.: SEAD: secure efficient distance vector routing for mobile wireless Ad Hoc networks. In: Proceedings of the Fourth IEEE Workshop on Mobile Computing Systems and Applications, pp. 3–13 (2002)
7. Wan, T., Kranakis, E., Van Oorschot, P.C.: Securing the destination-sequenced distance vector routing protocol (S-DSDV). In: Information and Communication Security, LNCS, vol. 3269, pp. 311–318. Springer, Heidelberg (2004)
8. Wang, S.Y., Lin, C.C.: NCTUns 5.0: A network simulator for IEEE 802. 11(p) and 1609 Wireless Vehicular Network Researches. IEEE International Symposium on Wireless Vehicular Communication, September 2008

Strategies for Risk Facing in Work Environments

MariaGrazia Fugini, Claudia Raibulet and Filippo Ramoni

Abstract Risk management in work environments requires the introduction of mechanisms able to identify the causes and the indications, which precede accidents in order to avoid them whenever this is possible. Most of the accidents are announced by risk events, which may be identified and managed before their evolution into accidents through preventive strategies. We propose to address risks by proposing risk management strategies where the risk is explicitly defined and revealing and protection devices generate risk events managed through conditions and set of actions.

Keywords Risk management · Work environments · Risk events · Risk strategies databases

1 Introduction

The advances available in IT, service-based software, telecommunications and sensors networks, work safety garments may be exploited successfully in the identification and management of risky situations which usually precede and announce accidents. In this paper, we describe a *Risk Management System (RMS)*

M. Fugini (✉) · F. Ramoni
Dipartimento di Elettronica e Informatica, Politecnico di Milano, Via Ponzio, 34/5, 20133 Milan, Italy
e-mail: fugini@elet.polimi.it

F. Ramoni
e-mail: filippo.ramoni@yahoo.it

C. Raibulet
Dipartimento di Informatica, Sistemistica e Comunicazione, Università degli Studi di Milano-Bicocca, Viale Sarca 336, Ed. U14, 20126 Milan, Italy
e-mail: raibulet@disco.unimib.it

E. Gelenbe et al. (eds.), *Computer and Information Sciences II*,
DOI: 10.1007/978-1-4471-2155-8_54, © Springer-Verlag London Limited 2012

whose software and hardware components are able to monitor and analyze risk events, and to plan and execute risk facing strategies.

The *RMS* is specified according to a MAPE (Monitoring, Analysis, Planning, Execution) loop [3]. In case of risky situations, the sensors and the software services attached to objects used to tag the environment and the workers are able not only to signal the danger but also to compute and put in place preventive or corrective mechanisms able to diagnose and possibly repair the risky situation. In previous work [4], we have explored how technologies can provide information about the individual who wears the computer sensors and services and the world which this individual interacts with. We have also studied design issues [7] of an *RMS* in work areas, investigating what a risky situation is and how it can be modeled and simulated.

Risks in dangerous environments are usually solved technically through the use of sensors, ad-hoc devices and suitable software for empowering the individuals' lives. The related solutions available in the literature are based on sensors (and sensor networks), and various types of monitoring devices, among which wearable devices [4, 9]. In [5], emphasis is on the use of key decision structuring steps and analytical tools to help ensure the systematic treatment of fact-based and value-based risk knowledge claims. The overall problem of risks and safety, with particular attention to the way of communicating risks to people and companies is discussed for instance in [8]. Both articles address our problems of education to risk through monitoring and simulation techniques. Research areas related to our work are the area of sensors [1, 2], and of wearable objects in ambient intelligence [10].

The paper is organized as follows. Section 2 introduces the basic issues of our *RMS*. Section 3 describes risk facing strategies. Conclusions and further work are given in Sect. 4.

2 Risks and the Risk Management System

Among the most common risks which may arise in work environments (such as building sites, industrial plants, or underground areas), we consider the loss of gas.

The loss of gas may lead to an explosion, if not timely revealed. The *causes* of the loss of gas may be of various types such as a consequence of a deterioration of the gas conducts in time or the interaction of persons with tools/machineries in the nearby of gas pipe lines. The loss of gas may be *gradual*, in time. In this case, the loss is hardly identifiable directly by the persons/workers because they get used to the gas smell. Initially, in this case, the *risk level* is low and it *increases in time* if it is not revealed and addressed. On the other hand, if the loss of gas occurs suddenly and powerfully, then the risk level is high and a risk or even an emergency may occur immediately.

Moreover, if in the nearby of the gas loss, persons are at work, than the risk level is higher. If workers are performing activities that may produce sparks (e.g., digging), than the level increases, too. This example of risk can be identified through gas sensors. Depending on the risk level, there may be various strategies to face this risk: close the gas provisioning, turn electricity off, evacuate all the persons/workers in the interested area, and/or repair the gas conduct.

After identifying the reason for a high risk value, the *RMS*, which continuously receives data from sensors for monitoring, performs the analysis, trying to find a correlation between the cause (e.g., high concentration of gas in the air) and a possible risk facing strategy. The *actions* which are part of the strategy and which can take place to decrease the risk in this case are:

1) *instruction* for general valve: close the general gas pipe;
2) *message* to person in the room (such as, "leave the room quickly");
3) *message* to the team leader: problem with gas in the environment.

Upon application of these actions, the *risk index* of a person is likely to decrease.

In case the gas level increases, the risk for the person increases too, and the *risk index* exceeds the emergency threshold. The analysis phase of MAPE realizes that the source of risk is the gas concentration, and a different strategy has to be selected, e.g., a strategy with stricter time constraints, such as the following:

1) *message* to the persons in the environment: leave the environment quickly;
2) *instruction* for general valve: close the central gas pipe;
3) *message* to the team leader: serious problem with gas in the environment.

If no persons are in the environment, in our approach the *risk index* is not computed.

We have implemented a prototype *RMS* to make available to simulate the data flow from sensors and devices (sensor networks, RFiDs, antennas, tags, and so on) and the risk reaction process in terms of alarms, actions, and strategies.

The *RMS* is aware of which persons is the environment and detects the change of persons' locations, determines the variation of environmental parameters, determines if a risk exists, and generates the necessary data for risk management. The simulation is centered around the worker actions, to focus on the Human-Environment interactions by simulating the actions typical of persons in a work environment. Workers' actions are the driver of both the *RMS* interaction with the simulation and of the generation of events for risk computation and prevention. The simulator is described in [6].

3 Risk Strategies

A *strategy* is defined as a set of *Actions* which can take place to decrease the risk. Strategy 1 beneath is a possible strategy, composed of actions to be executed in sequence. Words in italic denote basic action types, followed by the parameter of such action in ( ) and boldface, denoting to whom the action is directed and which environment elements are involved, and by an *instantion* of the action terms, appearing in < >.

Strategy 1—Risk Level *1*

{a_1 {*instruction* (for **central valve**): < *close*, central gas pipe > ;
/* this is a prescription action */
a_2 *message* (to **persons in the room**): < leave room within 5 min > ;
/* this is a message for risk facing */
a_3 *message* (to **persons in adjacent rooms**): < leave rooms within 10 min > ;
/* this is a message for risk facing */
a_4 *message* (to **team leader**): < leakage in pipe n. 756 room R1 to be solved in the next 2 h > .
/* this is an explanation message sent by the *RMS* analysis module giving textual explanation of fault e.g. on PDA*/
}

In order to have an *RMS* which is useful also for learning risk management procedures, both from end-users (workers) and administrators, the *RMS* is conceived as endowed with a Business Intelligence module. This module is structured according to a Case-based reasoning approach and works as in Fig. 1.

The Cases database contains ECA rules describing the *RISK CASE* and the strategies to face it:

RISK CASE = <EVENT, CONDITION, STRATEGY>

Events are expressed as risk types (e.g., HEALTH, FIRE, GAS, TOOLS HANDLING, and so on). *Conditions* are environment or context-related predicates describing the risk situation. They contain the following elements (each is optional):

- Risk value (LOW, MEDIUM. HIGH, VERY HIGH)
- Presence of persons (yes/no)
- Time constraints (a reaction has to be applied within *x* time units)
- Alarm constraints (if alarms have been activated)
- Required skills for interventions (a fire specialist is required, or a team leader must be present in the risk physical area, etc.)
- Devices (hardware and software devices, sensors, tags to be activated/ deactivated).

The strategy is specified in the ECA rule as a set of *Actions* to be executed to react to the *Event*. The *Strategy* has a name, a description, a priority level, denoting if the strategy applies during a risk, a combination possibility, describing if the strategy can be used *together* or *as an alternative* to other strategies (and in this case, which ones). It contains a set of Actions to be executed in sequence or in parallel (depending on the gravity of the risk and on the inherent possibility to handle parallel interventions such as "activate an alarm" and "switch power off").

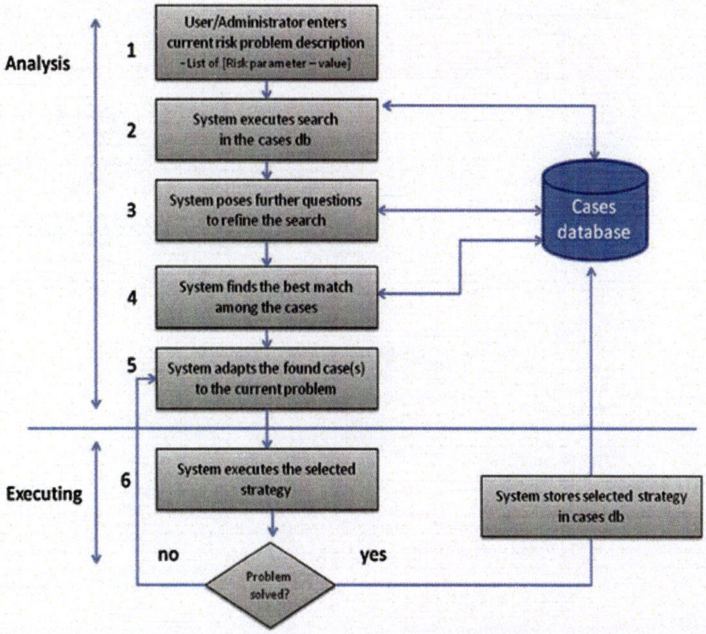

Fig. 1 Reasoning about risk

The person(s) expected to execute the action can also be specified (e.g., the fire brigade chief only, or "at least 3 people together"). These are conditions attached to the Actions. Each Action can be shared by various strategies. The Cases database is structured as a relational database with the schema reported in Fig. 2:

A Case can be queried by the DBMS on the basis of RISK TYPES, PERSONS, GRAVITY OF THE RISK, etc., according to predefined indexes. These indexes can evolve during the system lifetime in that the system is tuned to more refined versions according to observation of the effects of both search processes (if they are successful immediately or need many queries to be posed to the user to find a match) and of the effectiveness of the applied strategy. The observation of the success of application of a given (set of) strategy(ies) is the final part of the MAPE loop, namely the observation of the effects of the strategy on the risk. If the risk is mitigated in less than n time units, then the strategy can be defined as VERY SUCCESSFUL and stored in the db with a high success score. Other criteria to evaluate the goodness of a strategy (or a combination of strategies) can be the number of persons who were safely guided out of the risky status, or how many tools and machinery were safely guarded during the risk management procedure. For a strategy to be stored in the Cases database, a score must be defined as a result of both an automatic evaluation ("yes" branch in Fig. 1) and upon manual analysis by the *RMS* managers.

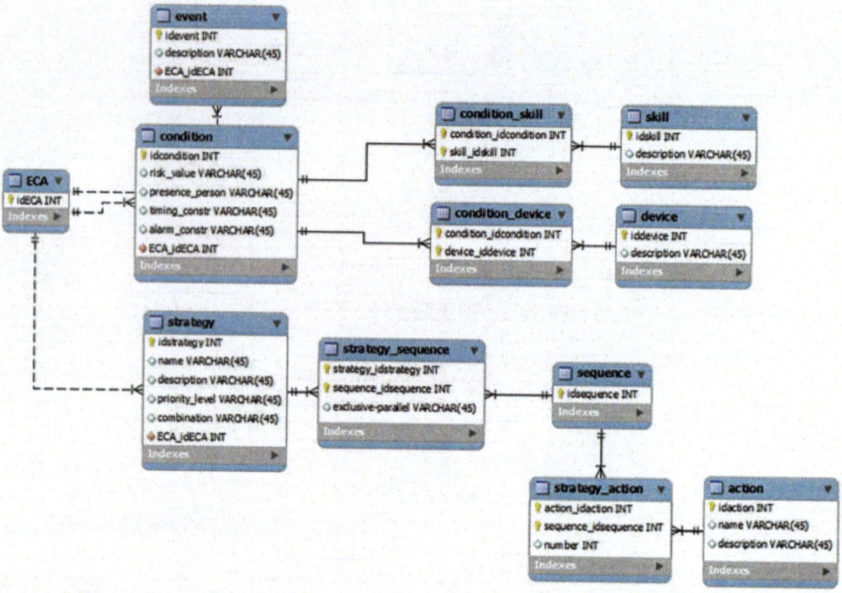

Fig. 2 Risk Cases database schema

4 Conclusions and Further Work

In this paper we have presented an overview of our approach to risk management in work environments. We have presented risk facing strategies. The novelty consists in the explicit design of risks and their facing strategies according to a risk prevention approach useful to implement a training/learning tool for risk managers. We are currently implementing a set o strategies within the java-based prototype simulator and are refining the risk computation functions to achieve a more precise reporting on risks and on the application of strategies.

Acknowledgments This work has been partially supported by the TeKNE Project and by the S-Cube European Network of Excellence in Software Services and Systems.

References

1. Botts, M., Percivall, G., Reed, C., Davidson, J.: Sensor web enablement: overview and high level architecture, OGC® sensor web enablement: overview and high level architecture, In: GeoSensor Networks, LNCS, vol. 4540/2008, pp. 175–190. Springer, Berlin (2008)
2. Chen, N., Di, L., Yu, G., Min, M.: A flexible geospatial sensor observation service for diverse sensor data based on web service. ISPRS J. Photogramm. Remote Sens. **64**, 234–242 (2009)
3. Cheng, B.H.C., de Lemos, R., Giese, H., Inverardi, P., Magee, J.: Software Engineering for Self-Adaptive Systems. LNCS 5525. Springer, Heidelberg (2009)

4. Conti, G.M., Rizzo, F., Fugini, M.G., Raibulet, C., Ubezio, L.: Wearable services in risk management. In: 2009 IEEE/WIC/ACM International Joint Conferences on Web Intelligence and Intelligent Agent Technologies, Web2Touch Workshop, Milan, pp. 563–566 (2009)
5. Failing, L., Gregory, R., Harstone, M.: Integrating science and local knowledge in environmental risk management: a decision-focused approach. Ecol Econ **64**(1), 47–60 (2007)
6. Fugini, M.G., Raibulet, C., Ubezio, L.: Risk characterization and prototyping. In: Proceedings of the 10th International Conference on New Technologies of Distributed Systems (NOTERE 2010), Tozeur, Tunisia, May 31–June 2, pp. 57–64 (2010)
7. Fugini, M.G, Raibulet, C., Ramoni, F.: A prototype for risk prevention and management in working environments. In: Information Technology and Innovation Trends in Organizations (ITAIS'2010), Naples, Italy, October 8–9 (2010)
8. Lundgren, R.E., McMakin, A.H.: Risk Communication a Handbook for Communicating Environmental Safety and Health Risks, 4th edn. Wiley, Hoboken (2009)
9. Pederson, T., Surie, D.: Towards an Activity-Aware Wearable Computing Platform Based on an Egocentric Interaction Model in Ubiquitous Computing Systems, pp. 211–227. Springer, Tokyo (2008)
10. Southworth, F.: Multi-criteria sensor placement for emergency response. Appl. Spatial Anal. Policy **1**(1), 37–58 (2008)

Part X
Modelling and Simulation

Numerical Integration Methods for Simulation of Mass-Spring-Damper Systems

Mete Özgüz and M. Taner Eskil

Abstract The dynamics of a face are often implemented as a system of connected particles with various forces acting upon them. Animation of such a system requires the approximation of velocity and position of each particle through numerical integration. There are many numerical integrators that are commonly used in the literature. We conducted experiments to determine the suitability of numerical integration methods in approximating the particular dynamics of mass-spring-damper systems. Among Euler, semi-implicit Euler, Runge-Kutta and Leapfrog, we found that simulation with Leapfrog numerical integration characterizes a mass-spring-damper system best in terms of the energy loss of the overall system.

Keywords Numerical integration methods · Physically-based facial modeling · Animation · Semi-implicit Euler · Runge-Kutta · Leapfrog

1 Introduction

In mass-spring-damper systems, a face is composed of mass points and springs that connect these points. The charm of physically-based modeling lies in the simplicity of its formulation. The first muscle based face model is done by Platt and Badler [2]. Their skin model had one layer. Terzopoulos and Water modeled facial tissue as trilayer [1].

M. Özgüz (✉) · M. T. Eskil
Department of Computer Science and Engineering,
Pattern Recognition and Machine Intelligence Laboratory,
ISIK University, Istanbul, Turkey
e-mail: mete.ozguz@isikun.edu.tr; meteo158@gmail.com
URL: http://pi.isikun.edu.tr

M. T. Eskil
e-mail: eskil@isikun.edu.tr
URL: http://pi.isikun.edu.tr

E. Gelenbe et al. (eds.), *Computer and Information Sciences II*,
DOI: 10.1007/978-1-4471-2155-8_55,

Our face model is composed of one layer triangular mesh for skin, one layer triangular mesh for skull and three types of anatomically-motivated facial muscles; linear, sheet and sphincter. The details of these muscle types can be found in [3]. Every vertex on the skin has a mass m and every edge between vertices is a spring with spring constant k and damping constant b.

2 Simulation Results

We designed an experiment to show the energy loss for the numerical integration methods mentioned above. Experiments are conducted on a 10×10 grid.

Steady state of grid is disrupted by displacing dynamic points at (3, 7), (4, 7), (5, 7), (6, 7) and (7, 7) to (3, 8.5), (4, 8.5), (5, 8.5), (6, 8.5) and (7, 8.5) respectively. For each time step in the simulation the kinetic, potential and total energy of the system are calculated. There is no muscle force acting on the vertices once the simulation is started.

In Fig. 1, when the vertices are released at time zero, system's potential energy is at its maximum. Alternatively, if vertices pass through their steady state positions simultaneously, the kinetic energy of the system reaches its maximum. If there is energy loss, next peak in potential energy will have a smaller magnitude and this loop continues until the total energy becomes zero. Figure 1a shows these oscillations of the potential and kinetic energy for the Euler integration with position refinement [3] while the total energy diminishes.

The energy loss with the semi-implicit Euler method can be seen in Fig. 1b. In this figure, we see that the kinetic and potential energies converge to a stable energy level while the total energy becomes approximately constant. This is because of the phase shift in the oscillations of the vertices. In a snapshot in the later stages of the simulation, there will be vertices that are momentarily stationary (zero kinetic energy), there will be vertices in motion under an equilibrium of spring forces (zero potential energy), and there will be vertices in transition. The energy loss is getting noticeable over 100 s.

Leapfrog surpasses both Euler integration with position refinement and semi-implicit Euler in terms of the conservation of energy . As observed in Fig. 1c, the energy loss is not noticeable even at 100th s when Leapfrog numerical integration is used.

Leapfrog offers higher robustness against energy loss while keeping the execution time the same with semi-implicit Euler method. Further tests showed significant improvement in conservation of energy when Leapfrog is used.

Our experiments so far imply that Runge-Kutta 4 (RK4) and Leapfrog have the same performance in terms of conservation of energy. To distinguish them further, we experimentally determined the largest time step that the simulation can endure before getting unstable (Table 1). For each integration method, simulations were repeated 100 times starting time step of 0.01 s and incrementing it by 0.01 until it reaches 1 s.

With the largest achieved time step value of 0.91, RK4 nearly doubles Leapfrog. RK4's extra steps in integration have a direct impact on execution time

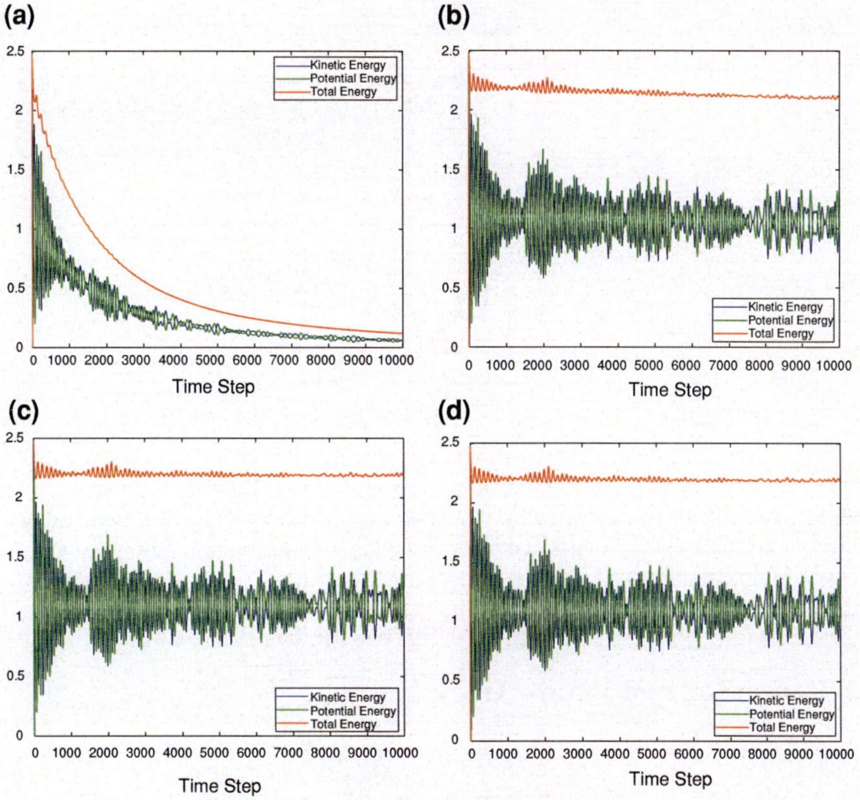

Fig. 1 Change in potential, kinetic and total energies (Δt = 0.01, 100 s)

Table 1 Largest achieved time steps

Integration method	Largest time step	Collision
Zhang	0.41	Not observed
Semi-implicit Euler	0.45	Not observed
Leapfrog	0.55	Observed
Runge-Kutta	0.91	Not observed

but as the time steps become larger, it may be possible to decrease the overall execution time as long as the error introduced is tolerable. This means that RK4 can be preferable over Leapfrog in certain applications.

To determine which integration method will be used, we did a last test to show energy loss of all numerical integration methods when time step Δt is 0.1 s instead of 0.01 s.

Fig. 2 clearly shows that RK4 can not conserve energy as well as Leapfrog when time step is increased. It must be noted that this time step is one ninth of its achieved largest time step. On the other hand, Leapfrog is still quite robust against

Fig. 2 Total energy loss (Δt = 0.1, 1000 s)

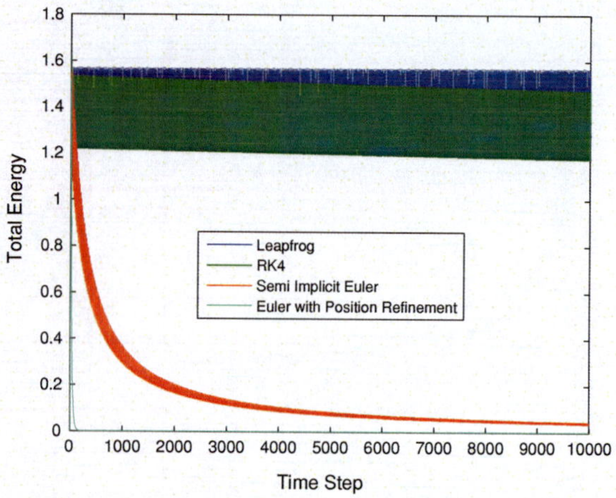

energy loss. With its largest achieved time step of 0.55 s, which is quite advantageous for real-time animation, its performance in conservation of energy and its ease of implementation; Leapfrog integration steps ahead as the preferable choice.

3 Conclusion and Future Work

Conservation of energy is an important characteristic that suggests the accuracy of the simulation. It is particularly important in catching the subtle facial movements for realistic simulation of facial expressions.

We found in our experiments that the Leapfrog method conserves the overall energy of the system best and it is quite robust against instability for increased time steps. This implies that Leapfrog is preferable for both higher realism and real-time simulations.

Acknowledgments This research is part of project "Expression Recognition based on Facial Anatomy", grant number 109E061, supported by The Support Programme for Scientific and Technological Research Projects (1001) of The Scientific and Technological Research Council of Turkey (TÜBİTAK).

References

1. Terzopoulos, D., Waters, K.: Physically-based facial modeling, analysis, and animation. J. Vis. Comput. Animat. 1(2), 73–80 (1990)
2. Platt, S.M., Badler, N.I.: Animating facial expressions. IEEE IEEE Comput. Graph. **13**, 245–242 (1981)
3. Zhang, Y., Prakash, E.C., Sung, E.: Efficient modeling of an anatomy-based face and fast 3D facial expression synthesis. Comput. Graph. Forum **22**(2), 159–169 (2003)

Multi-Layered Simulation Architecture: A Practical Approach

Okan Topçu and Halit Oğuztüzün

Abstract This paper presents a practical approach to design federate architectures by applying a well-known architectural style—layered architecture. Applying the layered architecture to specify the architecture of an HLA-based simulation provides a good separation of concerns, which are the interface, the simulation logic, and the HLA-specific communication. Thus, the layered simulation architecture allows the simulation developers to focus on each concern separately and gives them the freedom to implement each layer in a different programming language, and to encapsulate the repetitive and low-level implementation details of HLA.

1 Introduction and Background

Software engineering has a long history, built by academic and industrial efforts, towards developing better software. While building software, the engineers and the programmers generally face to solve the same type of design and implementation problems, which are affected not only by the user requirements, but also the technological advances. Observing the problems and solutions yielded architectural styles and design patterns, which provide good abstracted templates. Additionally, the need for building quality software promoted the use of architectural patterns. This need is no exception to computer simulations.

Today, the dominating framework and standard for distributed simulation applications is High Level Architecture (HLA) [1–3]. It is a distributed

O. Topçu (✉) · H. Oğuztüzün
Dumlupınar Bulvarı, ODTÜ, 06800 Ankara, Turkey
e-mail: okantopcu@gmail.com

H. Oğuztüzün
e-mail: oguztuzn@ceng.metu.edu.tr

E. Gelenbe et al. (eds.), *Computer and Information Sciences II*,
DOI: 10.1007/978-1-4471-2155-8_56,

simulation framework emphasizing the interoperability and reuse in simulation components (i.e. federate applications in HLA terminology) in a distributed simulation (i.e. federation). The architectural design of an HLA-based distributed simulation can be grouped into two broad categories: (1) the architectural design of a federation (called federation architecture) and (2) the architectural design of each federate (called federate architecture) within a federation. Note that neither the HLA standard nor the accompanying Federation Development and Execution Process (FEDEP) recommendation [4] address the issue of how to architect a federate. FEDEP recommends developing designs for new federates or tailoring the existing ones in order to comply with "federation agreements" if necessary. But, it does not provide guidance to federate design and architecture. In many large scale simulation projects we are familiar with, new federate applications have to be built. Thus, there is a need for guidance for designing federate architectures. Although federation architectures for some projects have been reported in the literature [5–7], there is a lack of reports on federate architectures.

The federation architecture deals with modeling the structure and the observed interactions among federates and the environment [8]. The federated architecture focuses the internal structure of a federate. Either generated or coded manually, the federate internal structure can be enhanced by the good practices in architecture and design patterns in order to develop quality simulations.

In this study, the federate architecture is formalized in a multithreaded and layered way to take the advantage of modern software architectures and patterns. This article contributes to the literature on the design of simulation software by elucidating a layered federate architecture for simulation design.

2 Layered Federate Architecture

A typical design of a federate conforms to the layered architecture. Layering [9] is an encapsulation of the assemblies by providing separation of concerns of three types of tasks: The user interaction (graphical output, the user input, and synchronization of data between view and simulation), the simulation logic and computation (model), and the communication which provides the federation-wide data (objects). The architecture mainly includes a presentation layer, a simulation layer, and a communication layer (see Fig. 1). The federation specific data structures are found in the federation foundation library (FFL).

The separation of algorithms and the visual presentation of their results is a point worthy of discussion. Consider, for example, a weather condition, such as fog. The algorithm that models the dynamics of fog dispersion should be implemented in the simulation layer, while the visualization of the fog dispersion should be handled in the presentation layer, and updates on the fog status should be sent out via the communication layer.

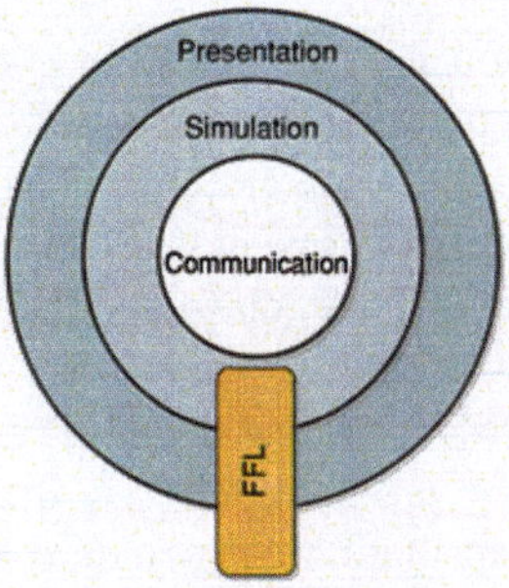

Fig. 1 Layers in a federate application

The *presentation layer* (also called the *user interface layer*) includes the pure presentation (view), input, and (if involved) interaction with user. The main components are the graphical user interface (GUI) and the user interface controller. The UI controller binds the user interface widgets with simulation data, manages and handles the user events, and iterates the simulation flow (i.e. the main simulation loop). A federate can have more than one GUI based on different architectures (e.g. web-based vs. windows-based) and/or libraries in order to support various working environments. The architecture is technology-independent to the extent possible, but it can be implemented with currently available technologies. The presentation layer is separately developed for each federate.

The *simulation layer* (also called *processing layer*) includes the computation (the simulation) and the local data (the federate-specific objects). It defines the federate behavior. The simulation layer is application (federate) specific. The simulation layer classes directly include or extend the generic classes found in the communication layer.

The *communication layer* deals with the HLA runtime infrastructure level communication in order to access the federation-wide data (namely, the objects and interactions exchanged in federation) by using an RTI abstraction component for .NET (RACoN). RTI is middleware broker software that manages the federation execution and object exchange through a federation execution. The presented federate architecture is not dependent on any particular RTI abstraction layer. Developing an abstraction layer (wrapper) over RTI is a popular approach in HLA-based development projects [10–12] as it offers more maintainable, easier to use, and portable methods. RACoN is used to wrap the native C++ interface to obtain a .NET interface to provide language independency (only for .NET languages). It also makes transparent the use of both the various HLA specifications (e.g. DoD HLA 1.3 specification [13] and IEEE 1516–2000 standard [3]) and the various vendor specific RTI implementations for the same specification (e.g. Portico, DMSO 1.3 NG) to the developer. The RACoN component is federation independent and can be used in all kinds of federates. But, its current limitation is that it only supports a subset of federate interface specification of HLA 1.3 and HLA 1516-2000 standards.

The Federation Foundation Library provides common data types, structures, GUI components, and the federation object model (FOM)[1] to the federate applications. In addition, it provides cross cutting concerns (e.g. exception handling, internationalization, and security) for each layer. This library encapsulates the federation-specific data structures (i.e. classes and data types where each federate application uses this library in order to construct its local data structures (i.e. simulation object model (SOM) in HLA jargon). The FFL supports the code reuse. Thus, using the FFL classes promote reduced development effort in federate development, and better encapsulation and standardization in common interests of the federates. The FFL also supports the code maintenance by keeping the common code in a central library. Consequently, a change in FFL reflects in all the federates using it. Note that our approach is not the layered architecture in the pure sense due to the FFL.

3 Discussion and Conclusions

This paper presents a multi-layered simulation architecture adopting the layered architecture style for HLA federates. The architecture and its supporting framework have been used in a variety of small-scale projects, but that body of work is outside the scope of this paper. It has been observed that adopting the layered architecture style provides a good separation of the concerns, which are the interface, the simulation logic, and the HLA-specific communication.

As a future work, the multi-layered simulation architecture can be extended to take advantage of the recent advances such as multicore processors and graphical processing units.

References

1. IEEE 1516 Standard for Modeling and Simulation (M&S) High Level Architecture (HLA)–Framework and Rules, 21 Sept 2000
2. IEEE 1516.2 Standard for Modeling and Simulation (M&S) High Level Architecture (HLA)–Object Model Template Specification, 21 Sept 2000
3. IEEE 1516.1-2000 Standard for Modeling and Simulation (M&S) High Level Architecture (HLA) – Federate Interface Specification, 21 Sept 2000
4. IEEE 1516.3 Standard for IEEE Recommended Practice for High Level Architecture (HLA) Federation Development and Execution Process (FEDEP), 23 Apr 2003
5. Molla A., Sarıolu K., Topçu O., Adak M., Oğuztüzün H.: Federation Architecture Modeling: A Case Study with NSTMSS", 07F-SIW-052. In: Proceedings of 2007 Fall Simulation Interoperability Workshop (SIW), Orlando, Florida, USA, 16–21 Sept 2007

[1] FOM mainly describes the format and the structure of data (i.e. objects) and events (i.e. interactions) that can be exchanged in the federation execution.

6. Canney Shane A.: Virtual Maritime System Architecture Description Document Issue 2.00. Defense Science and Organisation Australia Document Number 00034, 10 July 2002
7. Reading R., Örnfelt M., Duncan J.: Results and lessons learned from a multi-National HLA development supporting simulation based acquisition, 02S-SIW-045. In: Proceedings of 2002 Spring Simulation Interoperability Workshop (SIW), 2002
8. Topçu O., Adak M., Oğuztüzün H.: A metamodel for federation architectures. ACM Trans Model Comput Simul (TOMACS). **18**(3), article no. 10, pp. 10:1–10:29, DOI:10.1145/1371574.1371576, July 2008
9. Microsoft.: Microsoft Application Architecture Guide: Patterns & Practices, 2nd edn. 2009
10. Savaşan H.: The RToolkit: An Open Source Object Oriented Distributed Simulation Framework. In: Proceedings of Fall Simulation Interoperability Workshop, 2008
11. Open and Distributed Simulation Platform, EODiSP, http://www.pnp-software.com/eodisp/index.html, last accessed 17 Mar 2011
12. Cengiz S, Oğuztüzün H.: A COM Component Set for HLA Interface Specification, Fall 2002 SIW Simulation Interoperability Workshop, 2002
13. Defense Modeling and Simulation Office-DMSO.: High Level Architecture Run-Time Infrastructure RTI 1.3-Next Generation Programmer's Guide Version 6, 2002

Variable Threshold Based Cutting Method for Virtual Surgery Simulations

Ömer Çakir, Fatih Çakir and Oğuzhan Çakir

Abstract We propose a variable threshold based cutting method for tetrahedralized deformable objects used in virtual surgery simulations. Our method uses cut-nodes and cut-edges scheme to determine cutting path with variable threshold improvement. Static threshold may cause ambiguity in cut-node selection when the threshold is so large that one of two vertices of the triangle to be cut can be selected as a cut-node. In addition, small threshold value results in many tetrahedrons while a smooth cut and large threshold value leads to a more jagged incision with less small tetrahedrons. Thus, both small and large threshold values have advantages and disadvantages. Simulation results show that our variable threshold based cutting method offers relatively better results than static threshold based one in terms of advantages and disadvantages mentioned above.

Keywords Physically based modeling · Animation · Cutting simulation

Ö. Çakir (✉)
Department of Computer Engineering, Karadeniz Technical University, 61080 Trabzon, Turkey
e-mail: cakiro@ktu.edu.tr

F. Çakir · O. Çakir
Department of Electrical and Electronics Engineering, Karadeniz Technical University, 61080 Trabzon, Turkey
e-mail: fcakir@ktu.edu.tr

O. Çakir
e-mail: cakir@ktu.edu.tr

E. Gelenbe et al. (eds.), *Computer and Information Sciences II*,
DOI: 10.1007/978-1-4471-2155-8_57,

1 Introduction

In recent years virtual surgery has become a very active thread of research. One of the major parts of virtual surgery is cutting and several methods have been presented before. Bielser et al. [1] pioneered interactive tetrahedral mesh cutting for surgical simulation or volume visualization. This work used a tetrahedron based mass-spring model and was later included in a framework for open surgery simulation [2], including collision detection and haptic interaction. Tetrahedron cuts are restricted to a small number of possible cases, but the predefined decomposition schemes result in a large number of new sub-elements. Ganovelli et al. [3, 4] extend the method for multi-resolution mesh representations, and improve the decomposition schemes with respect to number of elements. Mor et al. [5] adapt these schemes to model progressive cutting by dragging nodes along a blade. Bielser et al. [6] present a state machine where element decomposition is recursively changed depending on the movement of the cutting tool inside the tetrahedron, allowing for progressive and high-resolution cutting operations. Nienhuys et al. [7] modeled interactive cutting by restricting cut paths to element boundaries.

Denis Steinemann implements a similar approach to [7] to find the sets of cut-nodes and the set of cut-edges for cutting path selection. He uses static threshold and offers threshold to be variable in his PhD dissertation [8], but does not give sufficient information setting on how to define threshold variable. In this paper we propose a method for a threshold value for each of the edges of all tetrahedrons belonging to the same object based on edge lengths. Selected thresholds are variable because every edge has different length. We will explain our proposed method in detail in Sect. 2.

2 Proposed Cutting Method

Utilized cutting algorithm is based on determining cut-nodes and cut-edges. If the cutting surface is sufficiently close to (smaller than a threshold d) a node, this node is selected as a cut node and duplicated. Otherwise cutting surface splits cut edge into two new edges. This is illustrated in Fig. 1(a).

The cutting algorithm mentioned causes cut-node ambiguity when the threshold d is so large that it is possible to select any near vertex of the triangle as cut-node. This is illustrated in Fig. 1(b) (topmost) with bold edge. Our method uses the threshold d variable by dividing every edge into three equal parts. If cutting surface passes through one of the two parts that are close to a node, this node is selected as a cut node and duplicated. Otherwise it passes through the middle part and splits cut edge into two new edges. This is illustrated in Fig. 1(b) (middle and undermost).

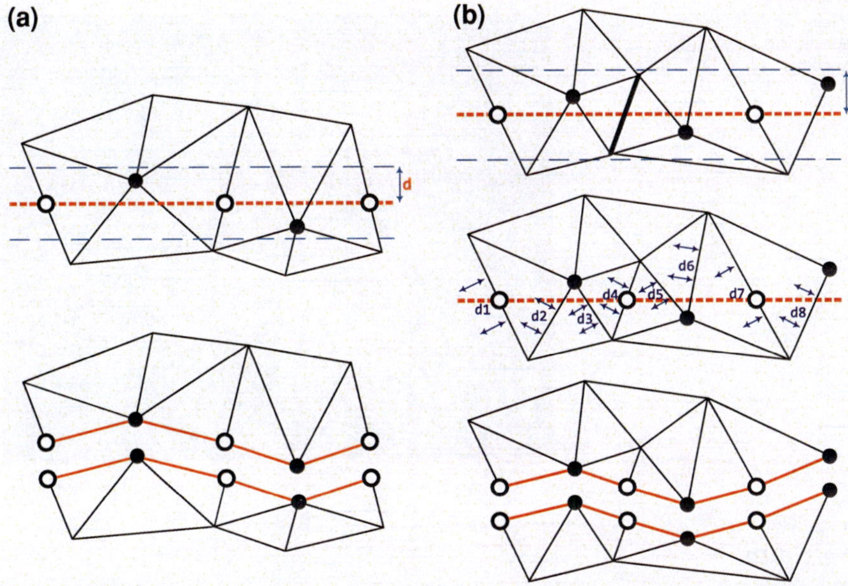

Fig. 1 **a** Cut nodes and cut edges scheme [8] **b** Cut-node ambiguity problem in **bold** edge and solution with the proposed method

3 Simulation Method

Finite Element Method (FEM) is used as a deformable body simulation method. FEM part of the simulation program solves the dynamic equilibrium equation and updates positions utilizing implicit Euler integration. The rotational part R of tetrahedrons are updated using "Stiffness Warping" method [9] to conserve volume during the simulation. The concept of stiffness warping consists of rotating the deformed positions back into the undeformed frame, calculating the elastic force in this frame, and than rotating the result back to the deformed frame using (1) or (2),

$$f_{ext} = RK(R^{-1}x - x_u) \tag{1}$$

$$= RKR^{-1}x - RKx_u \tag{2}$$

where f_{ext} is the external force, K is the element's stiffness matrix, x and x_u are the positions of the four vertices of the tetrahedron in deformed and undeformed states, respectively.

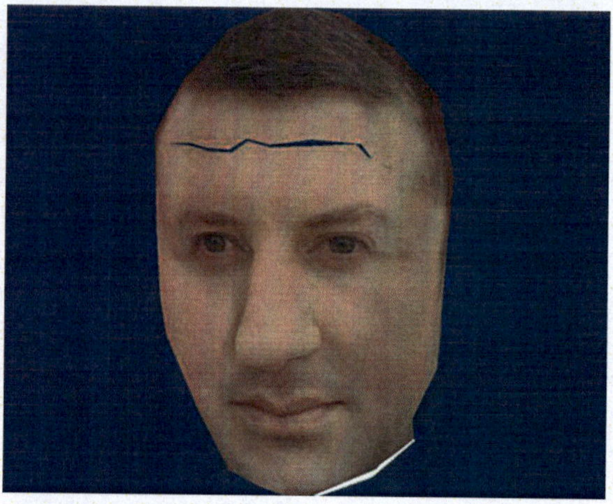

Fig. 2 A screen shot from the cutting simulation program

4 Results

The cutting simulation was performed on a Intel Core i7 2.66 GHz PC with a Geforce GTX460 GPU. The simulation program runs at about 60 frames per second at 1920 × 1080 screen resolution in release mode compilation. Fig. 2 shows a screen shot from the simulation program. A screen-captured video of the simulation program can be seen at this page: http://ceng2.ktu.edu.tr/~cakir/iscis2011.html.

The simulation program uses a head model that was modeled with MAYA. The model consists of 482 polygons and 1321 tetrahedrons. Tetrahedrons were generated using NETGEN.

5 Conclusions and Future Work

The proposed method removes cut-node ambiguity and the simulation program runs in real-time but FEM part of the simulation still runs on CPU. Conjugate gradient part of the FEM could be executed on GPU. CUDA programming language supports conjugate gradient calculations on GPU. Simulation program was written using DirectX 10 API. CUDA supports DirectX via DirectCompute so in the future conjugate gradient calculations will be performed with DirectCompute.

The head model has no skull and muscles. The skull and the muscles will be added to the model to achieve a more realistic soft tissue deformation.

References

1. Bielser, D., Maiwald, V.A., Gross, M.H.: Interactive cuts through 3-dimensional soft tissue. Proc. Eurograph. **18**, 31–38 (1999)
2. Bielser, D., Gross, M.: Interactive Simulation of Surgical Cuts. Proc. Pac. Graph. **8**, 116–125 (2000)
3. Ganovelli, F., Cignoni, P., Montani, C., Scorpigno., R.: A multiresolution model for soft objects supporting interactive cuts and lacerations. Proc. Eurograph. **19**(3), 271–282 (2000)
4. Ganovelli, F., Cignoni, P., Montani, C., Scorpigno., R.: Enabling cuts on multiresolution representation. Vis. Comp. **17**(5), 274–286 (2001)
5. Mor, A.R., Kanade, T.: Modifying soft tissue models: progressive cutting with minimal new element creation. CVRMed Proc. **19**, 598–607 (2000)
6. Bielser, D., Glardon, P., Techner, M., Gross, M.: A state machine for real-time cutting of tetrahedral meshes. Proc Pac. Graph. **22**, 377–386 (2003)
7. Nienhuys, H.W., van der Stappen, A.F.: Combining finite element deformation with cutting for surgery simulations. Proc. Eurograph. **19**, 274–277 (2000)
8. Steinemann, D.: Interactive simulation of topological changes on deformable objects. PhD thesis No. 18071, Department of Computer Science, ETH Zürich (2008)
9. Müller, M., Dorsey, J., McMillan, L., Jagnow, R., Cutler, B.: Stable real-time deformations. In: Proceedings ACM SIGGRAPH, San Antonio, Texas, pp. 49–54 (2002)

Model Based Approach to the Federation Object Model Independence Problem

Mehmet Fatih Uluat and Halit Oğuztüzün

Abstract One of the promises of High Level Architecture (HLA) is the reusability of simulations. Although HLA supports reusability with the Object Model Template, the Federation Object Model (FOM) Independence Problem arises when the developer desires to reuse a legacy federate application within a federation with a different FOM. Usually, federate code modifications become necessary. There have been attempts to alleviate this problem and they offer help to some extent, but they fall short of providing a flexible and complete mapping mechanism even when only Object Management is considered. In this work, a model based approach that addresses Declaration, Object and Federation Management services are demonstrated. The transition process is arranged into three distinct phases, namely, modeling, automatic code generation and component construction.

1 Introduction

High Level Architecture (HLA) is as a common software architecture that provides a general framework within which developers can structure their distributed simulation applications [1]. In 2001, HLA became an IEEE (Institute of Electrical and Electronic Engineers) standard [2–4]. The main motivation behind HLA is to support reusability and interoperability of simulations. These are facilitated

---

M. F. Uluat (✉)
Ankara Teknoloji Geliştirme Bölgesi Bilkent, STM Inc,
Cyberpark E Blok 5.Cad. No: 6/A, Çankaya, Ankara, Turkey
e-mail: f_uluat@yahoo.com

H. Oğuztüzün
Department of Computer Engineering, Middle East Technical University,
Inonu Bulvari, Ankara, Turkey
e-mail: oguztuzn@ceng.metu.edu.tr

E. Gelenbe et al. (eds.), *Computer and Information Sciences II*,
DOI: 10.1007/978-1-4471-2155-8_58,

through Object Model Template (OMT) [3] and Federate Interface Specification (FIS) [2]. The OMT provides a standard mechanism to define and document the form, type and structure of data that will be shared among the members of a distributed simulation. The HLA FIS describes the runtime services offered to the members by the Run Time Infrastructure (RTI) [2], which is the software that implements this specification. RTI provides the services and functions necessary to support a HLA compliant distributed simulation. In HLA parlance, the whole distributed simulation is termed as a federation and its members as federates. Federates can be thought of as the basic functional units of an HLA federation. A federate typically simulates a real-world entity, such as an aircraft, or carries out some utility function, such as monitoring.

HLA supports reusability, particularly by means of the Federation Object Model (FOM) and the Simulation Object Model (SOM). A SOM specifies the data that a federate can provide and the data that it requires. A FOM specifies the data that is exchanged among the member federates. All members are supposed to agree on the FOM at development time. At execution time, RTI will enforce that all data exchanges abide by the FOM. Problems related with reusability of federates became prevalent with the expanding adoption of HLA. When the federate developer wants to use her legacy federate within another federation having a different FOM, she faces the FOM Independence Problem (FIP), also known as the FOM agility problem. A federation developer can construct a new FOM based on existing SOMs. For large federations, this process often involves negotiations among federation developers and federate developers. Once a FOM is decided, the developers of existing federates are likely to be called upon to modify their federates so that they can fulfill their responsibilities under the new FOM. These modifications must accurately reflect their respective SOMs as well. The necessity of such modifications to a legacy federate to reuse it under a new FOM is the essence of FIP.

The paper is organized as follows. FIP and its impact on federation development are described in the next section. Some well-known approaches are outlined in section III. The proposed approach is expounded in section IV. Then, a case study is presented in section V and conclusions are drawn in section VI.

2 Problem

FIP is encountered when an existing federate is reused as a member of a new federation with a different FOM. Ideally, no modifications to the federate source code would be needed. However, the existing HLA OMT mechanism alone is not enough make this possible. The brute force solution is to modify the federate source code according to the new FOM. However, at this point, the federate cannot join the previous federation as a result of the modifications. Code modifications might be affordable for small federates with simple functionality, but they could be infeasible for complex legacy federates. Nevertheless, multiple-FOM-related

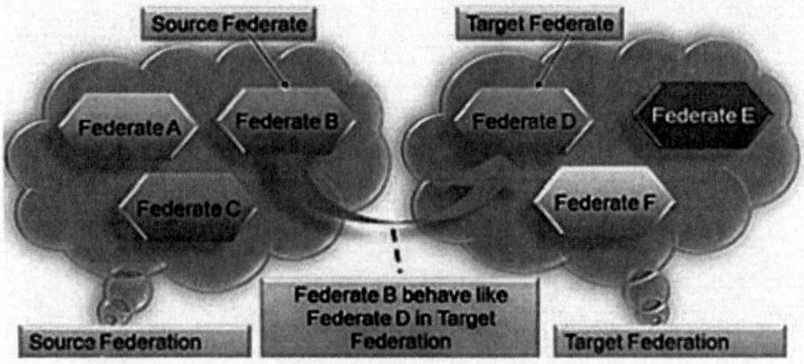

Fig. 1 Target and source federation/federate relationship

versions of the federate application must be maintained which exacerbates the configuration management issue.

The FIP involves two federations, designated as Source and Target Federation. The Source Federation has the Source Federate as a member. The variant of the Source Federate adapted for a new role in the Target Federation is called Target Federate. The FOMs of these two are named as the Source and Target FOMs which is illustrated in Fig. 1.

To utilize the Source Federate in Target Federation, the developer needs to define, at least, the relationship between the Source and Target FOM's object attributes and interaction parameters. To draw the boundaries of problem, the kinds of discrepancies that might occur between two FOMs are examined first. The most obvious ones are those resulting from the syntactic changes in the names of Attributes and Parameters. A more important discrepancy arises when the same concept is represented with different data types or data structures. Suppose, for instance, in Source Federation, "Temperature" is expressed in Celsius and is represented as a floating point. In Target Federation, it is still in Celsius but it is represented as an integer. For the floating point and integer case, federate code modification may not be necessary as long as they have same sizes, but if it is defined as a user defined type or complex data type, code modification is inevitable. Another source of discrepancy is the multiplicity of representation schemes. For instance, suppose in the Source Federation, Temperature is represented as an integer quantity in units of Celsius, and in Target Federation it is represented as a real number type in Fahrenheit. In this case, in addition to data type conversions, conversion between Celsius and Fahrenheit is also needed, which requires further code modifications.

The most problematic discrepancies are those rooted in semantics. Although the concepts used in different federations might be closely related, the precise semantics and representation might differ drastically from Source to Target Federation. For instance, suppose a legacy high fidelity Aircraft federate is reused in a C2 federation as a low fidelity Track federate. Although both federates are employing the location concept, the way they represent location might be very different. All these discrepancies seem to necessitate federate code modifications. The proposed solution, however, can bridge all these discrepancies without changing a single line of federate code.

3 Related Work

Having defined the problem, related approaches tackling the FIP are now summarized. The most two well-known approaches comes from MÄK [5, 6]. The first approach is FOM-specific Code Generation Approach where a middleware API is generated for a specific FOM. The federate code is being developed using this API. Thus, the developer does not need to deal with the details of the HLA API. However, when the FOM changes, the API needs to be regenerated, and the existing federate code needs to be adapted according to this new API manually. The second approach is called FOM-Configurable Fixed-API Middleware Approach. It provides a mechanism to develop the federate code once and use it without any modifications for new federations. As stated in [5], in this approach a fixed FOM-Configurable API is provided and when the FOM changes a new mapper, which is usually packaged as a Dynamic Linked Library (DLL), is generated and used without changing code.

Another approach is represented by Real-time Platform Reference Federation Object Model (RPR FOM), which let the stakeholders (including developers and users) agree on a common FOM and use it for a variety of HLA-based projects [7]. Although this might solve the problem to some extent, in reality, there are diverse kinds of applications and specialized concepts which are very difficult and sometimes impossible to capture with only one FOM, or any fixed number of FOMs. Even for common concepts, such as geographical coordinates, there are a number of representations that result from different standards or specific constraints.

There is also an ongoing study [8] to enhance agreement on FOM prior to federation execution for interoperability, but it is not yet standardized.

Having examined these approaches, it is concluded that all of them alleviate FIP to some extent, but they all fall short of providing a flexible and complete mapping mechanism. On a minor note, each one is based on some propriety API deviating from the HLA standard. These two concerns are the main motivation of this paper.

4 Model Based Approach to FIP

The solution proposed for FIP adopts the Model Driven Engineering (MDE) approach. This approach aims at developing software from domain-specific models via a series of transformations. MDE, treats models as first class entities like source code. The relevance of the model-driven engineering approach in modeling and simulation has already been well recognized [9, 10].

In this work, a particular manifestation of MDE, Model Integrated Computing (MIC), is adopted [11, 12]. For all metamodeling and modeling activities, the Generic Modeling Environment (GME) software is used [13]. To attack FIP,

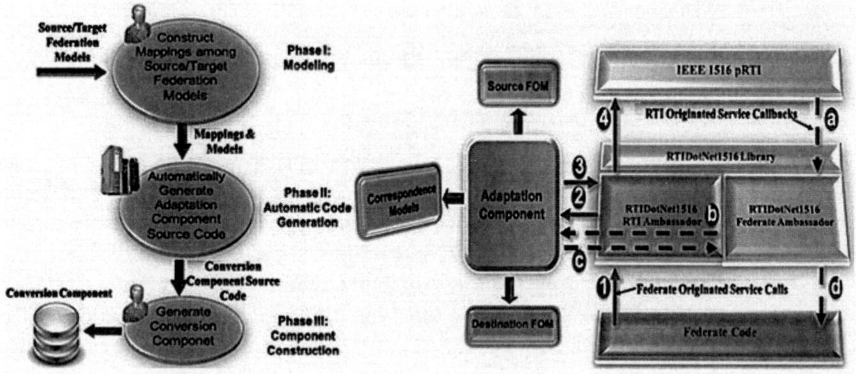

Fig. 2 **a** The three phases of the proposed approach. **b** Overview of the software components involved in the solution to FIP

a modeling capability for the HLA domain is needed and for that a base metamodel is required. The HLA Object Model Metamodel (HOMM) [14, 15] forms the basis for the modeling activities in this work.

The proposed approach is organized into three sequential phases: modeling, automatic code generation and component construction. The three-phased solution is illustrated in Fig. 2a, where the main activities performed in each phase, along with their inputs and outputs are shown. The steps that need user involvement are indicated by a person icon on the upper left part of the circle. Figure 2b shows an overview of the components proposed and their relationship with each other. In a basic federation, only the native RTI and federate application codes are involved, which are shown as boxes at the upper and bottom parts. The other parts represent software components resulting from the present study (i.e. RTIDotNet1516 Library, Adaptation Component). Whenever a federate is needed to join a new federation with different FOM, only the Adaptation Component which is shown at the left part is changed, no source code or library is required to be modified. Figure 2b also shows the corresponding paths for both federate-initiated calls (marked as numbers) and RTI-initiated callbacks (marked by letters). The federate-initiated calls are shown by '1', and then these calls are directed into Adaptation Component to check if there are any mapping for incoming attribute/parameters values, shown by '2'. After necessary actions are taken in the Adaptation Component, the obtained attribute/parameter values are fed into RTI, shown by '3' and '4'. The similar path is followed for federate-initiated callbacks, simply in reverse direction and through Federate Ambassador.

The modeling phase consists of the construction of FOMs in compliance with FOM Independence MetaModel (FIMM), Correspondence Models and the verification of these models. The Correspondence Model is a representation of a mapping from Source FOM Attributes/Parameters to Target ones, which are modeled to conform to FIMM. These mappings are created between two or more of them according to the semantics and representations of the Attributes/

Parameters. The collection of all the Correspondence Models is called an Adaptation Model. Thus, an Adaptation Model needs to be created for each particular instance of FIP. In contrast to existing approaches using tables or special programming notations for mapping, the mapping itself is a model.

The main issue addressed in the modeling phase is how to specify mappings and how to formulate the adaptations that carry out these mappings at federation execution time. There are four possibilities. The simplest case is a one-to-one mapping from the Attributes/Parameters of a single Object/Interaction Class in the Source Federate to the corresponding Attributes/Parameters of a single Object/ Interaction Class in the Target one. For this type of mapping, when an RTI service is executed in the Source Federate, the same is executed in the Target one. The second formation is a one-to-many mapping in which converters process one or more Attributes of a single Object Class in the Source Federate and generate multiple RTI service calls for each Object Class with corresponding Attributes in the Target one. When multiple RTI services are executed in the Target Federation for more than one Attribute, there will be only one Federate callback for the corresponding Attributes of the Object Class in the Source Federate. Many-to-one mapping is similar to one-to-many, but in reverse direction. The most complicated case is many-to-many mapping where all source and target federation originated RTI calls should be handled according to the Correspondence Model. All information necessary for these calls are obtained from the Correspondence Models.

An important issue in the Modeling Phase is model verification where the purpose is to prevent the developer from constructing inappropriate or potentially problematic Correspondence Models. When the developer commits her models, they are checked and diagnostics are produced.

The automatic code generation phase is started after the developer completes the construction of the Correspondence Models and performed via a plug-in mechanism [13]. Following verification, the interpreter walks over the constructed models and gathers necessary information for automatic code generation. Finally, the source code of Adaptation Component with the Adaptation Method templates is generated automatically.

The third and the final phase is the construction of an Adaptation Component from automatically generated codes in the previous phase. It is used for adaptations determined in the mapping phase and handling of RTI services during the federation execution. In this phase, templates generated with input and output arguments are filled by developer and then component is generated through a simple build process.

5 Case Study

In this section, the proposed approach is illustrated through a case study with two federations having different FOMs: Combat and Command and Control (C2) Federation. Combat Federation, is responsible for forming a high fidelity combat environment for performing various air defense scenarios. It consists of two federates: F16 and Environment federates. First one is in charge of high-fidelity simulation of the behaviors of an F16 aircraft. Second one is in charge of providing necessary environmental data, such as temperature, date, time and humidity to federation. C2 Federation is responsible for carrying out C2 scenarios on a low fidelity representation of a battlefield. It consists of three federates. The Track federate provides data related to entities that are to be detected by Radars. The Radar federate simulates the radar behavior by processing the data provided by Track federate according to the radar model employed. Finally, the Meteorology federate provides meteorological data that is necessary for performing simple C2 scenarios.

In this case study, most complicated mapping scenario which is many-to-many mapping is detailed in following lines. The legacy F16 and Environment federates are going to be reused as the Track, Radar and Meteorology federates in C2 Federation. Obviously, there is a semantic disagreement between the F16 and Radar federates, which stems from the way that these federates publish the data. F16 federate publish position data very frequently; however Radar needs to analyze these data, process it and publish it as trajectory information which contains previously received data with a given interval. For implementation of this capability, the Adaptation Method needs to utilizes a mechanism to determine whether the adapted data is passed or it is held. With utilization mentioned, the F16 data is used with the employed radar model and processed track data is provided to the federation through trajectory data. As a benefit of this, the C2 Federation developer can use the F16 federate both as a Track and Radar federate.

Performance implications of the proposed solution are of interest. We present quantitative measurements of the overhead that was incurred due to the added intermediate steps. In fact, there are two points where overhead is introduced: one is additional RTI/middleware calls and the other one is adaptation methods. Here, the latter overhead is directly related to the adaptation logic.

To obtain related measurements, the critical path was determined and the time spent along that path was assessed. For this purpose, the Aircraft and Environment federates of the case study were used. To focus on the overhead stemming from the proposed solution the adaptation method definitions were simplified so that their execution times would be negligible. Time spent by the RTI is not considered. The measurements are obtained for federate initiated "updateAttributeValues", "sendInteraction" service calls and RTI-initiated "reflectAttributeValues", "receiveInteraction" service callbacks. These are chosen because they are among the most heavily used service calls/callbacks. The measured overhead per service call and callback results are illustrated in Table 1. These values were obtained from average measurements of 10000 trials. The callbacks took more time than the service calls as federate ambassador related actions are needed.

The test configuration employed is as follows: Windows XP SP3 PC, equipped with Intel Core 2 Duo 2.26 GHZ CPU and 2 GB RAM. It can be concluded that the layering overhead that the proposed solution brings should be negligible for most applications.

Table 1 Obtained measurements

Service Call/Callback	Average Execution Time Overhead (in microseconds)
Update attribute values Service Call	5,519
Reflect attribute values Service Callback	15,814
Send Interaction Service Call	8,796
Receive Interaction Service Callback	13,755

Regarding the implementation,.NET 2.0 technologies are chosen. A reason for this choice is the lack of.NET adoption in the HLA arena, and our desire to evaluate it as a side benefit. In fact, The RTIDotNet1516 library, which is essentially an RTI abstraction layer, is developed for this purpose as a side product of this study.

Note, however, that by its very nature, our approach does not rely on any particular technology, be it.NET, J2EE or CORBA. When it comes to demonstrate and validate the approach, however, a particular implementation technology has to be utilized.

6 Conclusion

In this work, a model based solution is put forth to solve the FIP within the confines of Federation, Declaration and Object Management services. The problem, which is frequently encountered in reusing HLA federate applications, is defined and previous approaches are reviewed by examining what they provide or lack. Then the proposed solution is described by grouping related activities into three clearly scoped phases.

There are some limitations of the presented solution. First of all, the Time, Data Distribution and Ownership Management services are beyond the current scope. Second is the infeasibility of a fully automatic solution. The current approaches are required to define, at least, the algorithmic relationship among the Source and Target Federation by hand. In other words, the computation code associated to a mapping cannot be generated automatically from a given FOM. For a sensible mapping, human intervention is needed.

Based on experience acquired, it is believed that this approach can successfully be integrated into forthcoming model-driven HLA application development tools.

References

1. IEEE Std. 1516.: IEEE Standard for Modeling and Simulation (M&S) High Level Architecture (HLA)-Framework and Rules. IEEE, 2000
2. IEEE Std. 1516.1.: IEEE Standard for Modeling and Simulation (M&S) High Level Architecture (HLA)-Federate Interface Specification. IEEE, 2000

3. IEEE Std. 1516.2.: IEEE Standard for Modeling and Simulation (M&S) High Level Architecture (HLA) - Object Model Template Specification", IEEE, 2000
4. IEEE Std. 1516.3.: IEEE Recommended Practice for High Level Architecture (HLA) Federation Development and Execution Process (FEDEP). IEEE, 2003
5. Granowetter, L.: Solving the FOM-Independence Problem, 1999 Spring Simulation Interoperability Workshop (SIW) 1999
6. MÄK Technologies,"VRLink",http://www.mak.com/products/vrlink.php,last Accessed 29 March 2011
7. Simulation Interoperability Standards Organization, http://www.sisostds.org/ DigitalLibrary.aspx?EntryId = 30818,GRIM_RPR-FOM_1-0v2.pdf, Last Accessed 29 March 2011
8. Möller, B., Löfstrand B., Karlsson M.: An Overview of the HLA Evolved Modular FOMs, 2007 Spring SIW, 2007
9. Tolk A.: Metamodels and Mappings – Ending the Interoperability War, 2004 Fall SIW, 2004
10. Parr, S., Keith-Magee, R.: Making the Case for MDA, 2003 Fall SIW, 2003
11. Nordstrom, G., Sztipanovits, J., Karsai, G., Ledeczi, A.: Metamodeling–rapid design and evolution of domain-specific modeling environments. In: IEEE ECBS'99 Conference, Tennessee (1999)
12. Karsai, G., Agrawal, A., Ledeczi, A.: A metamodel-driven MDA process and its tools. In: WISME, UML 2003 Conference, San Francisco, 2003
13. Ledeczi, A., Maroti, M., Bakay, A., Karsai, G., Garrett, J., Thomason, C., Nordstrom, G., Sprinkle, J., Volgyesi, P.: "The Generic Modeling Environment", IEEE International Workshop on Intelligent Signal Processing (WISP'2001). Budapest, Hungary , 2001
14. Cetinkaya, D., Oguztuzun, H.: A Metamodel for the HLA object model. In: 20th European Conference on Modeling and Simulation (ECMS), pp. 207–213 (2006)
15. Topçu, O, Adak, M, Oğuztüzün, H.: A metamodel for federation architectures. ACM Trans Model Comput Simul **18**(3), 10:1–10:29 (2008)

Part XI
Architecture and Systems

Secondary Bus Performance in Retiring Cache Write-Backs to Memory

John W. O'Farrell, Rakshith Thambehalli Venkatesh and Sanjeev Baskiyar

Abstract Processor bus access latencies and queuing delays continue to remain a major bottleneck for communication between the CPU and other devices even with the use of multi-level caches. Write buffer induced stalls can be significant in data intensive applications. A number of writeback strategies have been proposed with a goal of reducing the cache miss penalty, but the focus has not been on these write buffer induced stalls. In this paper, we evaluate the use of a secondary bus that handles memory writes, connecting the level 2 on-chip cache to the main memory. Simulation results on the SPEC CPU 2006 benchmarks show that the addition of this bus results in program execution speed-up of up to 19% and reduction in the queuing delay on the main system bus by more than 95% even during high traffic on the system bus. This secondary bus is useful in real time applications and with programs having substantial I/O and graphics processing operations.

1 Introduction

Generations of technology scaling has led to a wide speed gap between the processor and the memory. As processor speeds continue to scale according to Moore's law [1], DRAMs have not been able to keep-up. The processor

J. W. O'Farrell · S. Baskiyar (✉)
Department of Computer Science and Software Engineering,
Auburn University, Auburn, AL 36849, USA
e-mail: baskisa@auburn.edu

J. W. O'Farrell
e-mail: ofarrell@auburn.edu

R. T. Venkatesh
Department of Electrical and Computer Engineering,
Auburn University, Auburn, AL 36849, USA
e-mail: rzt0004@auburn.edu

E. Gelenbe et al. (eds.), *Computer and Information Sciences II*,
DOI: 10.1007/978-1-4471-2155-8_59,

is connected to the main memory and other peripherals via a high bandwidth bus called the 'main bus' or the 'system bus'. The main bus carries a significant amount of traffic to and from the processor and tends to get clogged with data when programs with large workloads are executed on it. Caches within the processor employ various coherency strategies such as 'write-through' and 'write-back' to maintain data consistency with the next higher level in the memory hierarchy. Both of the write techniques employ a 'write buffer' [2] to prevent the next-level write access latencies from affecting the program execution time. The dirty cache lines have to be retired to the memory once the buffer is full, and retirement is usually made in bursts. The buffer initiates the write-back and waits until all the contents are emptied into the memory before serving the L2 cache miss request by the processor, adding to the latencies. In addition to processor-memory accesses there is bus contention by I/O devices that communicate with the processor through the I/O controller or memory controller.

In this chapter, we measure the effectiveness of a secondary channel [3, 4] between the processor and main memory that would carry the write-back traffic and help in improving the congestion on the system bus. The secondary channel helps retire the write-back buffer entries to memory in a timely fashion, thereby preventing memory access stalls due to the buffer being full. A control mechanism, in addition to the memory controller, would enable the secondary bus to snoop the main bus and initiate memory write-back when the latter is not being used for memory operations or busy with I/O transactions between the CPU and the I/O devices.

2 Background

When the write-back buffer is full, any cache miss for the processor results in wait cycles, until a cache block is evicted. This is because when the buffer gets filled, a burst-write of the write-back data happens and the wait time for the processor is significant as the buffer contents have to be written to the memory over the main bus, which also caters to the subsequent read traffic from the memory. Bus contentions by multiple agents on the main bus and queuing delays for the read traffic are a major bottleneck for program execution.

An observation of the write back rates for various SPEC CPU 2006 benchmarks shown in Fig. 1 indicates that the number of *writes* to the memory, in the benchmarks, can comprise up to 30% of the total memory accesses on the main bus. Furthermore, the burst write following a buffer full may create write traffic, which depending on the buffer size, could add large queuing delays to any read traffic. Simply using dual-ported memories is unlikely to reduce queuing delays. Although dual-ported memories may allow simultaneous read/write accesses, the requests will get queued, as the bus is half duplex [5], [6], [13]. Many techniques have been proposed for preventing the program execution stalls that occur as a result of the system bus contention. Miss Status Holding Registers (MSHRs) [7]

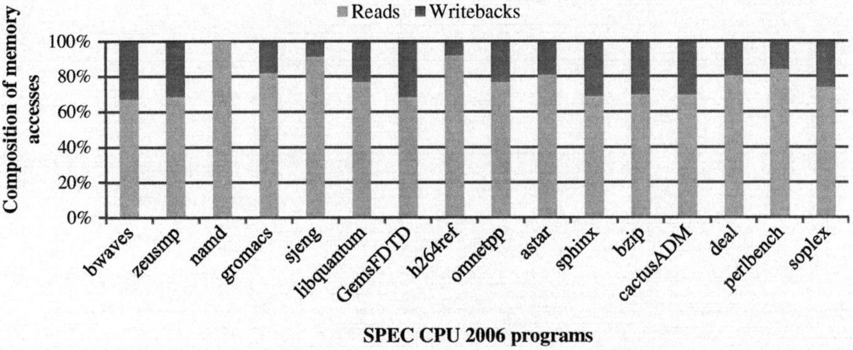

Fig. 1 Memory access requests for reads and writes per 100 million instructions

and Victim Buffers proposed by Jouppi [8] are used for non-blocking direct mapped caches. A technique called Eager Writeback was proposed by Lee, Tyson and Farrens [9] to address the write-back congestion by focusing on efficient main bus bandwidth utilization. This technique tries to retire the write-back buffer lines early to reduce future access conflicts on the main bus. The secondary bus design provides benefits over and above this technique and even in presence of I/O.

3 Architecture of the Secondary Bus

The motivation behind the secondary bus architecture design is to first provide a separate path from write buffers to main memory so that stalls due to the buffer becoming full can be avoided. The design of the secondary bus based architecture is shown in Fig. 2. As seen in Fig. 2, the secondary bus supports the main bus during write-backs and I/O transactions. This is made possible by the secondary bus controller, which snoops the main bus and identifies bus cycles that are not involved in memory accesses. These cycles will be used to retire the dirty cache lines to the memory over the secondary bus. The signals ‘s1’ and ‘s2’ are sent in accordance with the states of the main bus to arbitrate accesses.

A simple main bus state diagram is shown in Fig. 3 [3]. In state T3, the main bus is idle and is waiting for an address strobe to move to the next state. If the address supplied is for an I/O operation, the main bus would move to state T2. When the main bus is busy with I/O, it will be in state T2 and upon completion of I/O, it will return to T3. In state T1 the bus is busy with memory accesses [14]. In states T2 and T3, write-back entries can be retired through the secondary bus. The controller is designed so that s1 in state T1 is a connection enabling signal and s2 is a disabling signal. In states T2 and T3 the reverse is true.

There can be situations where the main bus state would change from T3 to T1 or from T2 to T1 via T3 in the middle of a cache write-back commit via the

Fig. 2 Architecture with the secondary bus

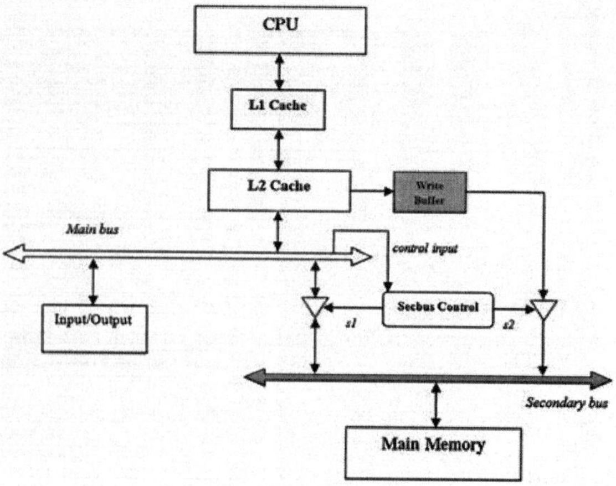

Fig. 3 Simplified bus state diagram for the main bus

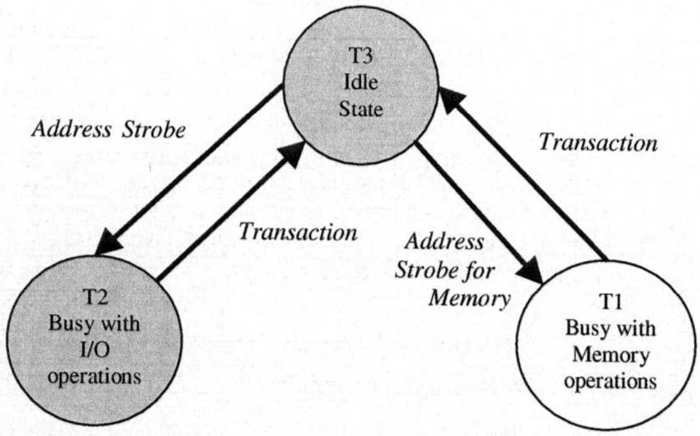

secondary bus. To handle this situation, the first option is to queue the main bus request until the write operation completes and the write-back buffer is empty. The second option aborts the burst write-back and enables the main bus to access memory. The second option of giving the main bus access priority was adopted in our design so as not to add to the queuing delay on the bus and for consistency with existing architectures. We note that the state transitions of the secondary bus depend on the state of the main bus.

The simulation was conducted on the SimAlpha simulator [10] of the SimpleScalar tool set that simulates an Alpha 21264 processor. The SPEC CPU 2006 benchmark suite was used for evaluating the design. The use of 'simpoints' [11] to simulate the programs at specific instruction milestones helped reduce the simulation time to a few days. The specifications used for the simulations have been given in Table 1. The SimAlpha simulator was modified to incorporate the bus controller mechanism and a writeback buffer capable of holding 256 cache lines Fig. 3.

Table 1 Simulation specifications

Processor Parameter	Specifications
Processor speed	3 GHz
Level 1 Data Cache	8 way, 32 KB, virtual-index virtual-tag
Level 1 Instruction Cache	8 way, 32 KB, virtual-index virtual-tag
Level 2 Cache	8 way, 2 MB, physical-index physical-tag
Number of MSHRs per Cache	8
Write mechanism for level 1 Cache	Victim buffer
Write mechanism for level 2 Cache	Writeback buffer
Main Bus (Front side bus)	600 MHz, 8B wide, 10 cycles of arbitration latency
Secondary bus	600 MHz, 1B wide, 10 cycles of arbitration latency

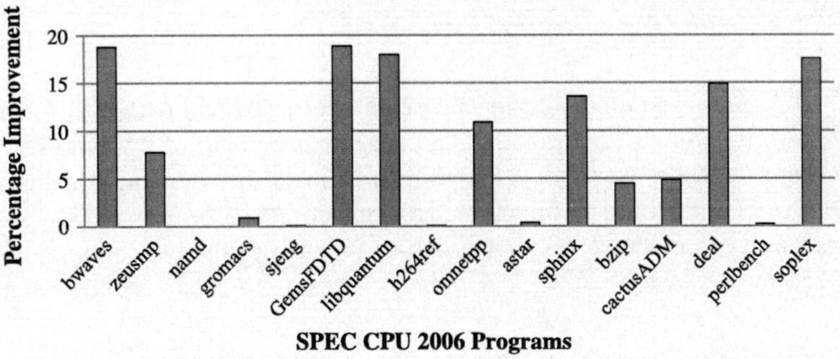

Fig. 4 Improvement in processor throughput across SPEC CPU 2006 benchmark programs

4 Simulation Results and Analysis

4.1 Speed-up

As seen in Fig. 4, speed-ups of up to 19% were achieved with the addition of the secondary bus. With the presence of the secondary bus, the main bus never had to wait for the dirty write-back traffic to be written to the main memory whenever it requested data due to a L2 cache miss. Results also show that the speed-up depends on how much the program strains the memory hierarchy. Programs having a very large working set could benefit compared to the ones using smaller caches and working sets such as *gromacs*, *sjeng*, and *h264ref*. Programs such as *bwaves, zeusmp, gemsFDTD* and *sphinx* were highly write-back intensive, with nearly 30% of the traffic on the main bus being the writeback traffic.

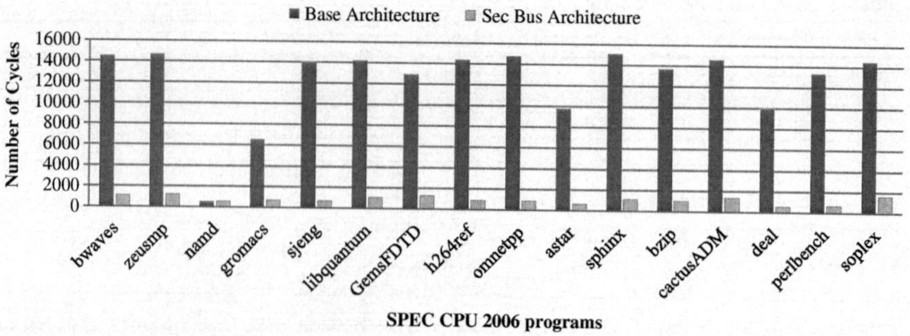

Fig. 5 Max cycles taken by any instruction

4.2 Worst Case Execution Time (WCET)

The WCET of a program or task is dependent on the critical path in the program. Results of our simulations show that the maximum number of cycles taken by an instruction to execute was reduced significantly with the secondary bus architecture compared to the base architecture. This reduction can be seen in Fig. 5, where the maximum cycles taken by an instruction for each of the benchmark programs have been shown. We see up to 95% reduction in the worst case number of cycles.

4.3 Queuing Delay

When the main bus is being used to service requests then any newly arriving requests are queued. In real-time systems these queuing delays can become significant, resulting in unexpected latencies and hard deadlines being missed. For instance, if there were an I/O device requesting the use of the processor during a memory read/write it would have to wait until the main bus is free. Almost all programs showed great reduction in the queuing delays, with an average reduction of nearly 99%. The presence of the secondary bus helped in making the main bus less prone to bus contention. We can predict that the percentage reduction would somewhat diminish as the I/O traffic becomes a larger portion of the main bus bandwidth (since then the write traffic becomes a smaller fraction of the total traffic).

5 Conclusion

A novel idea to reduce latencies due to bus contentions on the main system bus between the CPU and the chipset has been evaluated here. The technique of bus redundancy and a policy for efficient bus bandwidth management through data traffic distribution has shown to give significant performance improvements compared to the existing architecture. Queuing delays and worst-case instruction execution times were considerably reduced due to the sharing of the load on the main bus by the secondary bus. This outcome was verified across a range of the SPEC CPU 2006 benchmarks which were comprised of both CPU and memory intensive workloads to test the architecture.

Acknowledgments We are grateful to the Alabama Supercomputing Authority [12] for giving us access to the supercomputer clusters for running our simulations. This research was supported by grants from DARPA/AFRL Grant FA8750-09-1-0163 (PI-Dr. SanjeevBaskiyar) and NSF CCF-0411540 (PI-Dr. Sanjeev Baskiyar).

References

1. Moore, G.E.: Cramming more components onto integrated circuits. Electronics **38**, 114–117 (1965)
2. Chu, P.P., Gottipati, R.: Write buffer design for on-chip cache. In: Proceedings of the IEEE International Conference on Computer Design: VLSI in Computers and Processors, pp. 311–316 (1994)
3. Baskiyar S, Wang, C.: A secondary channel between cache and memory for decreasing queuing delay. US provisional patent application no 61/003:542 (2007)
4. O'Farrell, J., Baskiyar S.: Improved real-time performance using a secondary bus. In: ISCA Proceedings of Computers and Their Applications, Honolulu, (2010)
5. The Essentials of the Intel Quickpath Architecture, Online. Available: http://www.intel.com/intelpress/files/Intel%28r%29_QuickPath_Interconnect_Electrial_Architecture.pdf
6. Intel Atom Processor N270 for Embedded computing, Online. Available: http://download.intel.com/design/intarch/prodbref/320218.pdf
7. Kroft, D.: Lockup-free instruction Fetch/Prefetch Cache organization. In: Proceedings of the 8th Annual Symposium on Computer Architecture, pp. 81–87. (1981)
8. Jouppi, N.P.: Cache write policies and performance. In: Proceedings of the 20th Annual International Symposium on Computer Architecture, pp. 191–201. (1993)
9. Lee, H.H.S., Tyson, G., Farrens, M.: Eager Writeback–A technique for improving bandwidth utilization. In: Proceedings of the 33rd annual ACM/IEEE International symposium on Microarchitecture, Monterey, California, pp. 11–21 (2000)
10. Desikan, R., Burger, D., Keckler, S.W.: Measuring experimental error in microprocessor simulation. In: Proceedings of the 28th Annual International Symposium on Computer Architecture, pp. 266–277 (2001)
11. Ganesan, K., Panwar, D., John, L.K.: Generation, validation and analysis of SPEC CPU2006 simulation points based on branch, memory and TLB characteristics. In: Proceedings of the SPEC Benchmark Workshop on Computer Performance Evaluation and Benchmarking, Section: Modeling and Sampling Techniques, Austin, pp. 121–137 (2009)
12. Alabama Supercomputing Authority, http://www.asc.edu/
13. Kumar, R., Zyuban, V., Tullsen, D.M.: Interconnections in multi-core architectures: Understanding mechanisms, overheads and scaling. ACM SIGARCH Comput. Archit. News **33**(2), 408–419 (2005)
14. Intel Embedded Pentium Processor Developer Manual, Online. Available: http://www.intel.com/design/intarch/manuals/273204.htm

Runtime Verification of Component-Based Embedded Software

Hasan Sözer, Christian Hofmann, Bedir Tekinerdoğan and Mehmet Akşit

Abstract To deal with increasing size and complexity, component-based software development has been employed in embedded systems. Due to several faults, components can make wrong assumptions about the working mode of the system and the working modes of the other components. To detect mode inconsistencies at runtime, we propose a "lightweight" error detection mechanism, which can be integrated with component-based embedded systems. We define links among three levels of abstractions: the runtime behavior of components, the working mode specifications of components and the specification of the working modes of the system. This allows us to detect the user observable runtime errors. The effectiveness of the approach is demonstrated by implementing a software monitor integrated into a TV system.

1 Introduction

An evident problem in the embedded systems (ES) domain is the increasing software size and complexity. As a solution, component-based development has been recognized as a feasible approach to improve reuse and to ease the creation of

This work has been carried out as part of the TRADER project under the responsibility of the Embedded Systems Institute. This project is partially supported by the Netherlands Ministry of Economic Affairs under the Bsik program.

H. Sözer (✉)
Özyeğin University, Istanbul, Turkey
e-mail: hasan.sozer@ozyegin.edu.tr

C. Hofmann · M. Akşit
University of Twente, Enschede, The Netherlands

B. Tekinerdoğan
Bilkent University, Ankara, Turkey

E. Gelenbe et al. (eds.), *Computer and Information Sciences II*,
DOI: 10.1007/978-1-4471-2155-8_60, © Springer-Verlag London Limited 2012

variants of products [1]. Hereby, usually each component has to deliver a set of well defined services in a set of working modes. Components can correctly work together in the integrated system only if their working modes are consistent with each other; however, several faults can lead to mode inconsistencies at runtime.

We observed that mode inconsistencies between components can cause severe errors that lead to user-perceived failures. To detect and recover such errors, dedicated fault tolerance mechanisms are required. Instead of tolerating faults, one may try to avoid them by adopting theorem proving and model checking techniques at design time. Although these techniques have showed their value for many practical applications, the existing tools do not scale-up easily. Moreover, some faults may simply remain undetected during design and/or new faults may be introduced during the implementation.

In our approach, we define 3 levels of abstractions: the runtime behavior of the components, the working mode specifications of components and the specification of the working modes of the system. We establish explicit links among these levels. This allows us to detect runtime errors caused by inconsistent working modes of components. The effectiveness of the approach is demonstrated with a software monitor integrated into a TV system.

The remainder of this paper is organized as follows. Section 2 introduces the problem using an industrial case. Our solution approach is described in Sect. 3. Section 4 proposes diagnosis and recovery techniques. Section 5 discusses the effectiveness and the limitations of the solution approach. In Sect. 6, related previous studies are summarized. Finally, the paper is concluded in Sect. 7.

2 Industrial Case

In this section, we illustrate the problem using digital TV (DTV) as an industrial case. The DTV software is composed of many components working in coordination [1]. Each component has a set of working modes. These modes should be mutually consistent to provide the functionality that is required by a working mode of the system. If the synchronization between the component's working modes is lost (by loosing a notification message, data corruption etc.) inconsistent behavior occurs and component interactions no longer work in the anticipated way.

Consider for example the *Teletext Page Handler* component, which is responsible for requesting and acquiring a teletext page. Another component, *Display Manager* renders teletext pages on the screen. If *Display Manager* correctly assumes that the TV is in Teletext mode whereas *Teletext Page Handler* assumes that the TV should display the video stream, the combined behavior leads to a failure: no Teletext page is rendered leaving the user with a blank screen.

Due to the large number of components and the cost sensitivity of the ES domain, it is not feasible to check mode consistencies at the system level. Therefore, we propose to detect mode inconsistencies at component level as follows.

Table 1 Mapping between the component modes and the 4 system working modes

Application manager		Txt page handler		Display manager	
Mode	Map	Mode	Map	Mode	Map
On screen display	1000	On	111	On screen display	1000
Txt	0111	Off	1000	Txt full screen	0100
Off	0000	subtitle	0111	Txt left-half screen	0010
TV	1000			Default screen	1000

3 Error Detection

Our approach is based on models of the working modes of the system and its components. We map the models describing the component working modes to the implementation of the corresponding component in order to observe the component modes at runtime. We also map all component modes to the system modes. This makes the inter-dependencies between component modes explicit. The mappings between modes specify the mutual consistency condition, which is checked by monitoring the system at runtime. An error is detected whenever an inconsistency has been observed. In the following, we will discuss a prototype implementation of this approach in more detail. Then, we generalize this implementation and provide a formal definition of the mode consistency condition.

3.1 Implementation: A Prototype

The number of errors that can be detected by our approach depends on the number of working modes of the system that is considered and the number of component modes that are monitored. We developed a prototype and integrated it into a real TV system, where 4 system working modes are considered and 3 components are monitored. System working modes are *TV*, *Txt*, *dual screen*, and *transparent teletext*. The monitored components are *Teletext Page Handler*, *Display Manager*, and *Application Manager*. The *Application Manager* component controls the execution of applications in the system. Each component mode is represented by a bit vector, $\langle b_0 b_1 b_2 b_3 \rangle$, where each bit corresponds to a system working mode, e.g., b_2 corresponds to *dual screen*. The mapping between the system working modes and the component modes is shown in Table 1.

Application Manager, *Teletext Page Handler* and *Display Manager* components can be in one of 4, 3 and 4 different modes, respectively. In total, there are $3 \times 4 \times 4 = 48$ different mode combinations. AND ing the bit vector representations of the component modes leads to the value 0 in 40 of the cases. In 8 cases the result is non-zero, i.e., there is at least one bit where all of the component modes have the value 1. Therefore, our prototype can detect 40 error cases. The error detection mechanism polls the system periodically, where the current state values of the components are ANDed and an error is issued if the result is 0.

We injected 4 different faults in the TV system. These lead to mode inconsistencies between the *Teletext Page Handler* and *Display Manager* components, which eventually end up in lock-up failures in the Teletext functionality. We systematically activated the faults with a key combination from the remote control. In all cases the detection mechanism was able to detect the errors (notified by blinking the TV status LED) even before we observed the associated failure. This allows to (possibly) recover from an error before a failure is perceived by the user.

3.2 Generalization of the Prototype

In this section, we generalize and formalize our approach. We define a finite set of components C, where for each component $C_i \in C$, there exists a set of working modes M_i. Furthermore, there exists a set of working modes of the system M_S. For each mode couple *(m,s)* *s.t.* $m \in M_i$ and $s \in M_S$, we define a mapping function,

$$map(m,s) = \begin{cases} 0, & s \Rightarrow A(\neg mW\neg s) \\ 1, & otherwise \end{cases}$$

where $s \Rightarrow A(\neg mW\neg s)$ is a Computational Tree Logic [2] formula denoting that when s occurs, m should never occur until the mode of the system changes. For every mode $m \in M_i$ of a component C_i, we define a bit-vector $v_{i,m}$ of length $|M_S|$. Each bit in $v_{i,m}$ refers to a mode s in M_S and its value is assigned according to the mapping function. The consistency condition is defined as $\bigwedge_{0 \leq i < |C|} v_{i,m}$, where $v_{i,m}$ corresponds to the current mode of component C_i. There exists an error if this condition is evaluated to 0, meaning that there is no consistent assumption of each component about the current mode of the system.

3.3 Performance Overhead

To measure the performance overhead introduced by the error detection mechanism, we used an existing feature of the system that measures the load in terms of CPU cycles. We have made load measurements during two scenarios; watching TV (*TV*) and reading a Teletext page (*TXT*). For each scenario, we calculated the maximum, minimum and average CPU load of the system with and without the error detection mechanism. The results are presented in Table 2.

In Table 2, we see that the CPU load during the *TV* scenario did not differ at all. For the *TXT* scenario, the average CPU loads was increased by 1,2% on average. This shows that the overhead introduced by the error detection mechanism can vary depending on the usage scenario. In the case of the *TV* and *TXT* scenarios, the overhead is at acceptable levels. In general, our approach provides several advantages in terms of simplicity and efficiency. Error checking is performed with a single AND operation over bit-vectors and a space of size $(|C| \times |S_G|)$ bits must

Table 2 CPU load of the system with and without error detection

Scenario	Without error detection			With error detection		
	Min.(%)	Avg.(%)	Max.(%)	Min.(%)	Avg.(%)	Max.(%)
TV	34	36.9	51	34	36.9	51
TXT	39	42.9	46	41	43.4	47

be allocated only. Also note that the modeling effort is limited to assigning binary values that indicate the mode compatibility.

4 Diagnosis and Recovery

Error detection is the main focus in this paper. Another essential step of fault tolerance is recovering from the detected errors. We have recently developed a local recovery framework [3] for this purpose. Local recovery is an effective approach, in which the recovery procedure takes actions concerning only the erroneous components. To make local recovery possible, an additional diagnosis step should be introduced, which identifies the components that do not have a consensus on the current system mode. We can apply a voting mechanism to pinpoint such components in $O(|S_G| \times |C|)$ time as shown in Algorithm 4.

Algorithm 1 Diagnosis Procedure

```
systemmode ← ø
maximumvotesum ← 0
for j = 0 → |S_G| do
    votesum ← 0
    for i = 0 → |C| do
        votesum ← votesum + v_{i,m_current}[j]
    end for
    if maximumvotesum < votesum then
        systemmode ← j
        maximumvotesum ← votesum
    end if
end for
for i = 0 → |C| do
    if v_{i,m_current}[systemmode] ≠ 1 then
        mark the component C_i
    end if
end for
```

5 Discussion

We assigned the monitor to the lowest priority task, which can proceed after all the other tasks become idle. This provides a safe point in time to perform the error checking: (i) the monitor does not intervene with other tasks and functions, (ii) the system reaches to a stable state before the mode information is collected, and (iii) the introduced performance degradation is negligible. The only drawback of this approach is that error detection might be late. Error checking may never have a chance to execute in case the system is continuously busy or deadlocked. Such errors can be detected by other mechanisms like watchdog [4].

6 Related Work

There have been several proposals regarding formal specification of behavior [5–7]. Behavior protocols [8] and contracts [9] have been mainly used to formalize component interaction and utilized for design-time verification.

The scheme proposed by Thai et al. in [10] detects errors by checking consistency of states. The decision about whether there exists an error or not is made statistically. The outcome is according to the ratio between checks that passed and the total number of checks executed. Our approach is deterministic in the sense that an error is issued whenever a check does not pass.

Classification schemes have been provided for on-line monitoring [11] and real-time system monitoring [12]. We can classify our work as a monitoring approach for *fault tolerance*. It is based on *time-driven* sampling of component modes and *built-in event interpretation* that triggers *recovery* actions.

7 Conclusion

We pointed out a problem associated with component-based software that constitutes a challenge for reliability of ES. Either because of implicit assumptions at the design level or faults introduced during the implementation, mode inconsistencies can occur between components, which end up with the failure of the system. Such errors can be detected and recovered, (possibly) before a failure is observed. In this paper, we proposed an error detection mechanism that can detect mode inconsistencies at runtime. It can be adopted independent of the utilized component technology. We implemented a prototype of our solution and integrated it into a TV system. We obtained promising results.

References

1. van Ommering, R.C., et al.: The Koala component model for consumer electronics software. IEEE Comput. **33**(3), 78–85 (2000)
2. Emerson, E.A., Clarke, E.M.: Using branching time temporal logic to synthesize synchronization skeletons. Sci. Comput. Program. **2**(3), 241–266 (1982)
3. Sozer, H., Tekinerdogan, B., Aksit, M.: FLORA: a framework for decomposing software architecture to introduce local recovery. Softw. Pract. Exper. **39**(10), 869–889 (2009)
4. Huang, Y., Kintala, C.: Software fault tolerance in the application layer. In: Lyu, M.R. (ed.) Software Fault Tolerance, pp.231–248. John Wiley & Sons, Chichester (1995)
5. Peters, D.K., Parnas, D.L.: Requirements-based monitors for real-time systems. IEEE Trans. Softw. Eng. **28**(2), 146–158 (2002)
6. Zulkernine, M., Seviora, R.: Towards automatic monitoring of component-based software systems. JSS ACBSE Special Issue **74**(1), 15–24 (2005)
7. Diaz, M., Juanole, G., Courtiat, J.: Observer—a concept for formal on-line validation of distributed systems. IEEE Trans. Softw. Eng. **20**(12), 900–913 (1994)
8. Plasil, F., Visnovsky, S.: Behavior protocols for software components. IEEE Trans. Softw. Eng. **28**(11), 1056–1076 (2002)
9. Berbers, Y. et al.: CoConES: an approach for components and contracts in embedded systems. LNCS **3778**, 209–231 (2005)
10. Thai, J., et al.: Detection of errors using aspect-oriented state consistency checks. In: ISSRE, pp. 29–30 (2001)
11. Schroeder B.: On-line monitoring: a tutorial. IEEE Comput. **46**(25), 72–78 (1995)
12. Schmid, U.: Monitoring distributed real-time systems. Real-Time Syst. **7**(1), 33–56 (1994)

Model-Based Software Integration for Flexible Design of Cyber-Physical Systems

K. Ravindran

Abstract The paper discusses the design of complex embedded systems around *intelligent physical worlds* (IPW). Here, an IPW is an embodiment of control software functions wrapped around the raw physical processes to perform the core domain-specific adaptation activities. The IPW exhibits an intelligent behavior over a limited operating region of the system—in contrast with the traditional models where the physical world is basically dumb. To perform over a wider range of operating conditions, the IPW interacts with an intelligent computational world (ICW) to patch itself with suitable control parameters and rules/procedures relevant in those changed conditions. The modular decomposition of an application into IPW and ICW lowers the overall software complexity of building embedded systems.

1 Introduction

A cyber-physical system (CPS) allows the computational processes to interact with the physical world processes to impact how the latter is structured and designed. We elevate the definition of an embedded system by eliminating the hardware-centric boundaries of physical processes. An application $\mathcal{B}$ that is traditionally viewed as a non-embedded system because of its heavy software leaning can now be brought into the fold of CPS with a notion of *intelligent physical world* (IPW) A_p. Here, A_p can be an embodiment of diverse software functions, with the embedded hardware instantiating the raw physical processes (RPP). The RPPs are dumb physical component ensembles through which a system interacts with its external environment (e.g., steering to turn a car on the road, network link to transport data packets, and conveyor belt to move assembled parts).

K. Ravindran (✉)
City College of CUNY and Graduate Center, New York, NY 10031, USA
e-mail: ravir222@yahoo.com; ravi@cs.ccny.cuny.edu

E. Gelenbe et al. (eds.), *Computer and Information Sciences II*,
DOI: 10.1007/978-1-4471-2155-8_61,

The sub-system $\mathcal{A}_p$ consists of a software wrapper that controls RPP in such a way to infuse a self-contained and intelligent behavior. From a programming standpoint, the RPP is abstracted as a function $g(I, O)$ that takes an input I and responds with an output O. Here, O depicts an observation of the state of physical processes, with the time-scale of response hidden as part of the abstraction. For example, $g(I, O)$ may be the end-to-end path in a data network, with I and O denoting a packet transmission and its delivery respectively. As another example, $g(I, O)$ may be the motor in an industrial control system, with I and O denoting the electrical signal and rotational speed respectively. Our idea is to extend $g(I, O)$ into a coherent intelligent physical world A_p that is self-aware and can repair itself (in a limited way) from the damages caused by environment conditions. A_p is augmented by an intelligent computational world (ICW) A_c that manages the overall operations of A_p. Their composition is denoted as: $\mathcal{B} \equiv A_p \oplus A_c$, where A_p is wrapped around $g(I, O)$ and the operator '$\oplus$' depicts the inter-module flow of signals between A_p and A_c: which includes a *management-oriented feedback* from A_p to A_c. The signal flow is at a meta-level abstracting a 'monitor-and-control' type of interaction (M&C), while the $A_c - A_p$ concrete interactions are determined by their programming boundaries.

An example of A_p is a smart home that sets the heating and cooling parameters based on the occupancy, ambient conditions, comfort level, and the like. Here, A_c may be a Home Comfort & Security Company outsourced with the task of managing the intelligent home remotely by setting the right parameters and operating procedures (say, different procedures for winter operations and summer operations). In a target tracking system as another example, the radar units reporting the images of objects in a terrain to a data fusion center may also notify the terrain characteristics to enable the choice of a right set of image processing algorithms (say, to meet the processing and communication resource constraints). Here, A_p is the group of radar units and A_c is the fusion center.

The off-loading of domain-specific core adaptation functions into A_p enables the infusion of new functionalities and features in applications with less software complexity. The ease of verification and testing of such modularly structured systems lowers the development cost of distributed control software for complex systems. This is in contrast with a traditional embedded system (TES) which employs an asymmetric control relationship with the physical world: i.e., only the computational processes initiate the M&C interaction with $g(I, O)$ but not vice versa. Figure 1 illustrates the difference between CPS and TES. We focus on the software engineering advantages of CPS arising from this difference.

2 Our CPS-Based View of Embedded Systems

In a TES, only the computational processes interact with the physical world $g(I, O)$. Whereas, a CPS delegates some intelligence to the physical world. So, our notion of CPS underscores a certain degree of self-awareness in A_p, which enables A_p to communicate with A_c to coordinate supervisory control by A_c.

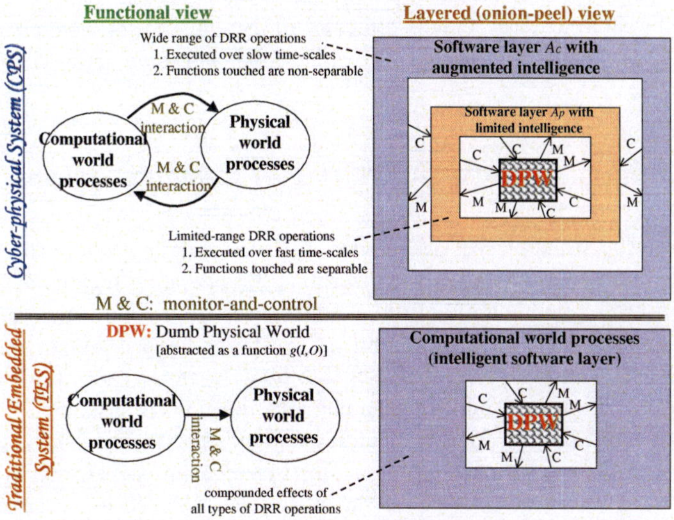

Fig. 1 Traditional embedded systems *versus* Cyber-physical systems

2.1 Existing Design Approaches Cast Through Our CPS View

Computational intelligence in the physical world requires the components to be self-aware, i.e., a component needs to be able react to its external environment—and possibly repair itself. Such an ensemble of self-aware components in A_p need to work together to provide a coherent interface to A_c. In this light, existing works on embedded control systems [1, 2] use an integrated structure that assigns intelligence for adaptation and reconfiguration only to the computational world, which exercises control on the physical world to cause effects in the external environment. In a TES-based design, an application $\mathcal{B}$ is thus a composition:

$$\mathcal{B} \equiv \mathcal{B}'' \oplus g(I,O),$$

where $\mathcal{B}''$ refers to the computational processes (implemented in software) that interface with the RPP function $g(I, O)$. The composition $\oplus$ depicts a M&C type of interaction, with $\mathcal{B}''$ invoking $g(.)$ with an actuator signal I and accepting an output response O. The TES structure assigns intelligence to $\mathcal{B}''$, with the latter interfacing with $g(I, O)$ through signaling hooks to actuate the trigger mechanisms, thereby moving the RPP move from one operating point to another.

We thus have: $A_p(cps) \equiv A'_p \oplus g(I,O)$ as an augmented physical world that exhibits intelligence over a limited operating region. Here, A'_p is the software function wrapped around the RPP $g(I, O)$ such that $\mathcal{B}'' \equiv A'_p(cps) \oplus A_c(cps)$, where $A_c(cps)$ infuses a broader intelligence in the operations of $\mathcal{B}$.

Due to the state-machine complexity of the system as a whole, the TES-based integrated approach does not lend itself well for a seamless addition/removal of automated system features, entails difficulty in incremental software changes, and makes the maintenance of system software a labor-intensive activity.

2.2 CPS-Based Modular Structuring of Complex Systems

We employ the principles of *piece-wise linearity* and *separability* of functions describing the system model [3], to determine the operating regions of A_p where the system-level computations of future trajectories (in a control-theoretic sense) are simpler and fall within the ambit of local intelligence. When the system behavioral changes satisfy linearity/separability, A_p can repair itself. The self-repair can be via a local built-in mapping function that is instantiated with the parameters supplied by A_c for that operating region. An example is the adjusting of TCP flow control window size based on small changes in packet round-trip delay (RTT) over the transport network. On the other hand, if the behavioral changes are larger taking the system into non-linear regions, A_p notifies an exception to A_c for the latter to adjust the parameters for the new operating region: say, by using *domain-specific policy* functions. In the TCP example, the protocol itself may be changed to aggressively adjust the window size when the RTT swings are large. A_c is wired with policies and rules to evaluate the linearity and separability conditions, and then patch A_p with suitable parameters and procedures.

A_p operates over a much faster time-scale than A_c. This is because the control loop in A_p is self-contained to react to the smaller changes that typically occur frequently in the external environment. Whereas, A_c steps in only when larger changes occur in the external environment—which are less frequent (e.g., a network suffering a DOS attack, a car tire losing air due to a puncture). A_p embodies the core domain-specific functionality, and A_c is delegated with an external management role using parameterized procedures and rules specific to the domain. See Fig. 2 for an illustration of the functional blocks to realize the hierarchical control relationship between A_c and A_p.

2.3 Situational Assessment Feedback

Distinct from the well-known supervisory control methods [4], we incorporate a management-oriented feedback from A_p to a *situational assessment module* (SAM) housed in A_c, for the latter to adjust the control laws employed by A_p. The feedback is a notification about how successful A_p is in realizing the control delegated by A_c. A_p obtains a control reference parameter P_{ref} from A_c, along with domain-specific operating parameters and computational functions (e.g., a rule for changing the packet transmission window size for an RTT-adaptive flow control in TCP). A_p then generates appropriate inputs I to $g(.)$ over multiple invoke-and-observe iterations until the output O of $g(.)$ becomes stable: possibly, with a close

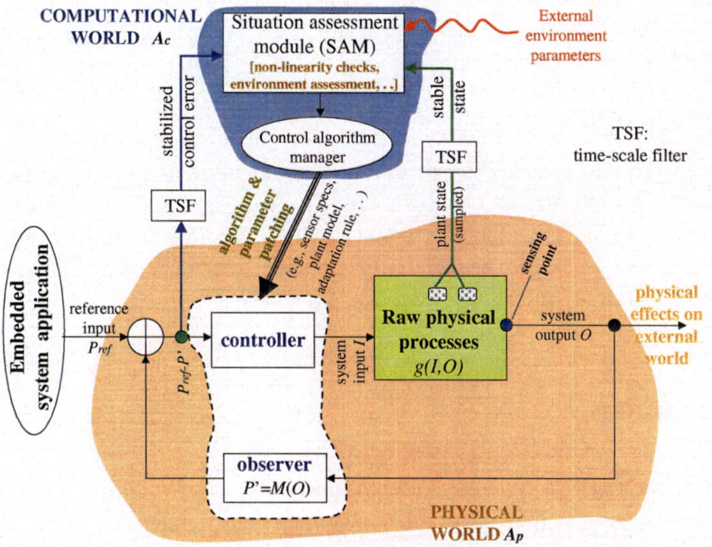

Fig. 2 A meta-level view of hierarchical control in our CPS structure

match to P_{ref}. The computation of I is governed by the domain-specific operating parameters and mapping functions loaded into A_p. Since it may be difficult to model *g(.)* in a closed-form, the local adaptation strategy allows determining the input I needed to attain a stable output O. Any mismatch between O and P_{ref} is notified to A_c for appropriate recovery. The intelligent behavior of A_p is however feasible only over a limited operating region.

If the magnitude of final control error $|P_{ref} - P'|$ indicates that A_p has exceeded its repair capability, A_p seeks the services of A_c for a comprehensive repair, i.e., to bring down the error to a minimum. The comprehensive repair may involve, say, changing the plant and/or the controller parameters—and even the controller algorithm itself. Thus, the adapt/repair cycle is collectively realized by A_p and A_c, with the procedural invocations from A_p occurring infrequently on A_c in comparison to that on $g(I, O)$.

The adaptation cycle executed by A_p is at the machine-level time-scales pertinent to the physical processes. The operations of A_c occur at much slower time-scales. The time-scale separation in the operations of A_c and A_p makes it easier to assert the correctness of application behavior with high confidence.

3 Existing Paradigms for Cyber-Physical Systems

For system specification and analysis purposes, We treat the adaptation and functional behaviors separately. In this light, we broadly categorize the existing works as dealing with:

- Systems engineering for the control-theoretic aspects of adaptation (such as stability and convergence) [6, 7];
- Software engineering for the verification of application requirements (including para-functional ones) [8, 9].

There have also been system-level tools developed to aid these studies: such as probabilistic monitoring and analysis [10, 11], plug-in based model-solvers (e.g., SYSWeaver) [12], and algorithm–level reconfiguration to manage application QoS [5]. In this light, our work falls in a distinct category of model-based engineering of complex embedded systems.

4 Conclusions

Our paper introduced a concrete notion of IPW that removes the explicit hardware-centric boundaries as part of the currently prevalent definitions of an embedded system. The IPW then interacts with an ICW to realize an embedded system application.

The paper described the software engineering issues in orchestrating a harmonious co-existence of the IPW and ICW. A software structural model of CPS aids the study of complex embedded systems applications (such as automotive systems and intelligent data networks) through the prism of ICW–IPW partitioning. The advantages of our CPS-style structure are that it reduces the development cost of distributed control software via software reuse and modular programming. It also enables easier system evolutions in the form of adding and/or modifying the controller functionalities in applications without weakening the software correctness goals. These advantages in turn spurn the discovery/support of newer functionalities in an application that is otherwise difficult with the existing integrated approaches for embedded systems deign.

References

1. Eberhart, R.C., Shi, Y.: Computational intelligence. In: Computational Intelligence: Concepts to Implementations, Chap. 2. Morgan Kaufman Publishers, San Francisco (2007)
2. Kabadayi S., Pridgen A., Julien C.: Virtual sensors: abstracting data from physical sensors. Technical Report 2006-01, University of Texas Austin (2006)
3. Hillier F.S., Lieberman G.J.: Non-linear programming and meta-heuristics. In: Introduction to Operations Research, Chaps. 12, 13, 8th edn., pp. 547–616. McGraw-Hill, NY (2005)
4. Diao Y., Hellerstein J.L., Kaiser G., Parekh S., Phung D.: Self-managing systems: a control theory foundation. IBM Research Report, RC23374 (W0410-080), Oct 2004
5. Ravindran K.: Dynamic protocol-level adaptations for performance and availability of distributed network services. In: Modeling Autonomic Communication Environments, Multicon Lecture Notes-no.6, pp. 191–209, Oct 2007

6. Li, B., Nahrstedt, K.: A control-based middleware framework for quality of service adaptations. IEEE JSAC **17**(9), 1632–1650 (Sept 1999)
7. Lu C., Lu Y., Abdelzaher T.F., Stankovic J.A., Son S.H.: Feedback control architecture and design methodology for service delay guarantees in web servers. IEEE TPDS **17**(7), 1014–1027 (Sept 2006)
8. Schaefer I., Heffter A.P.: Slicing for model reduction in adaptive embedded systems development. Workshop on Software Engineering for Adaptive and Self-managing Systems (SEAMS 2008), May 2008
9. Yi J., Woo H., Browne J.C., Mok A.K., Xie F., Atkins E., Lee C.G.: Incorporating resource safety verification to executable model-based development for embedded systems. IEEE Real-time and Embedded Technology and Applications Symposium, pp. 137–146 (2008)
10. Mikaelian T., Williams B.C., Sachenbacher M.: Probabilistic monitoring from mixed software and hardware specifications. In: Proceedings of ICAP'05 Workshop on Verification and Validation of Model-based Planning and Scheduling Systems (2005)
11. Ravindran K.: Information-theoretic treatment of sensor measurements in network systems. In: Proceedings of Network Operations and Management Symposium (NOMS'10), IEEE–IFIP, Osaka, Japan, April 2010
12. Rowe A., Bhatia G., Rajkumar R.: A model-based design approach for wireless sensor-actuator networks. In: Proceedings of Workshop on Analytic Virtual Integration of Cyber-Physical Systems (AVICPS'10), Nov 2010

Generating Preset Distinguishing Sequences Using SAT

Canan Güniçen, Uraz Cengiz Türker, Hasan Ural and Hüsnü Yenigün

Abstract The preset distinguishing sequence generation problem is converted into a SAT problem to investigate the performance of SAT solvers for generating preset distinguishing sequences. An initial set of experiments are carried out and it is shown that the heuristics of SAT solvers can perform better than brute force algorithms that are used to generate preset distinguishing sequences.

Keywords Finite state machines · Distinguishing sequences · Satisfiability problem

1 Introduction

Finite state machines (FSM) have been widely used as a modeling formalism in areas such as sequential circuits [1], lexical analysis [2], software design [3], communication protocols [4], object-oriented systems [5], web services [6], and in many others. There have been many attempts to automatically generate test

---

C. Güniçen · U. C. Türker · H. Yenigün
Computer Science and Engineering, FENS, Sabancı University Orhanlı Tuzla, 34956 İstanbul, Turkey
e-mail: yenigun@sabanciuniv.edu

C. Güniçen
e-mail: canangunicen@su.sabanciuniv.edu

U. C. Türker
e-mail: urazc@su.sabanciuniv.edu

H. Ural (✉)
SITE, University of Ottawa, 800 King Edward Avenue, Ottawa, ON K1N 6N5, Canada
e-mail: ural@site.uottawa.ca

DOI: 10.1007/978-1-4471-2155-8_62,

sequences from FSM models of systems [3, 7–9]. Among these testing methods, a particular set of methods are based on sequences called *distinguishing sequences* (DS) using which one can identify the current unknown state of an FSM. Although not every FSM may have a DS, it is known that for those machines which have a DS, a polynomial length test case can be constructed [7–11]. This positive result also explains the popularity of DS based testing methods. A DS may be preset (which we will call as PDS from now on) or adaptive. In this paper, we focus on PDS.

It is long known that checking the existence of a PDS for a given FSM is PSPACE-complete [12]. Finding a PDS is therefore obviously a hard problem as well. There is almost no work for designing algorithms to find a PDS, besides the text book algorithms [1, 13, 14]. These algorithms are based on a brute-force construction of a tree with the depth equal to the length of the shortest PDS. Satisfiability problem (SAT) is the first decision problem proven to be NP-complete [15]. However, within the last 15 years, there has been an incredible improvement in the practical performance of SAT solvers using various heuristics in their search algorithms. In this work, we investigate the use of SAT solvers to generate a PDS for a given FSM, by taking advantage of these search heuristics of SAT solvers. To this end, we encode the PDS generation problem as a Boolean formula that is satisfiable if and only if the given FSM has a PDS, and use a standard SAT solver to check the satisfiability of this formula. We also report some initial experiments conducted on a set of randomly generated FSMs.

In Sect. 2, we introduce our notation and give necessary background information formally. The proposed SAT encoding of PDS generation problem is given in Sect. 3. Section 4 presents our initial experimental results. Finally in Sect. 5, we give some concluding remarks and future directions of our work.

2 Preliminaries

An FSM M is a tuple $M = (S, X, Y, \delta, \lambda)$ where $S = \{s_1, s_2, \ldots\}$ is a finite set of states, $X = \{a, b, \ldots\}$ and $Y = \{1, 2, \ldots\}$ are finite set of input and output symbols. Throughout the paper n, p, and q will be used to denote the number of states, the number of input symbols and the number of output symbols, respectively. δ: $S \times X \rightarrow S$ is the transition function, and finally λ: $S \times X \rightarrow Y$ is the output function. Intuitively, if FSM M is at a state $s \in S$, and if an input $x \in X$ is applied, M moves to the state $s' = \delta(s, x)$ and produces the output $\lambda(s, x)$. Such a transition will also be denoted as $(s, x/y, s')$ below. An FSM can be depicted as a directed graph, where the states correspond to the nodes and the transitions correspond to the edges. Furthermore, the edges are labelled by the input and the output of the corresponding transitions. In other words, $\delta(s, x) = s'$ and $\lambda(s, x) = y$, then the edge corresponding to this transition has the label x/y.

We use juxtaposition to denote concatenation. For example, if x_1, x_2, and x_3 are input symbols, $x_1x_2x_3$ is an input sequence. The symbol ε is used to denote empty sequences. The transition function and the output function can be extended to sequences of inputs. By abusing the notation, we will still use δ and λ for the extended functions. These extensions are defined as follows: $\delta(s, x_1x_2 \ldots x_k) = \delta(\delta(s, x_1), x_2x_3 \ldots x_k)$, $\delta(s, \varepsilon) = s$, $\lambda(s, x_1x_2 \ldots x_k) = \lambda(s, x_1)$ $\lambda(\delta(s, x_1), x_2x_3 \ldots x_k)$, and $\lambda(s, \varepsilon) = \varepsilon$. Two states $s, s' \in S$ are said to be *equivalent* if for all input sequences $\alpha \in X^*$, $\lambda(s, \alpha) = \lambda(s', \alpha)$. If there exists an input sequence $\alpha \in X^*$ such that $\lambda(s, \alpha) \neq \lambda(s', \alpha)$, then s and s' are said to be *distinguishable*. An FSM M is minimal if every pair of states of M is distinguishable. M is *completely specified* if both δ and λ are total functions. Since δ and λ are functions (rather than a relation), there can be at most one transition defined from each state with each input symbol. Such machines are called *deterministic*. In this work, we consider only deterministic, completely specified, minimal machines.

A *preset distinguishing sequence* (*PDS*) for an FSM M is an input sequence $\alpha \in X^*$ such that for any two distinct states $s, s' \in S$, $\lambda(s, \alpha) \neq \lambda(s', \alpha)$, i.e. the output produced from each state is unique. Therefore a PDS can be used to identify the state of an FSM. Not every FSM has a PDS, and PDS existence problem is PSPACE-complete [12].

We use common Boolean operators but introduce what we call "exactly-one-OR". We use the symbol ∇ for this operator. It is used as a prefix operator, operating on a set of operands. $\nabla\{r_1, r_2, \ldots, r_k\}$ is true if and only if exactly one of $r_1, r_2, \ldots, r_k$ is true. We will also give below the number of clauses in conjunctive normal form for each formula we introduce. Note that an exactly-one-OR operator having k operands introduces $k(k-1)/2 + 1$ clauses. For each r_i and r_j in $\nabla\{r_1, r_2, \ldots, r_k\}$, we will have the clause $(\neg r_i \vee \neg r_j)$. There are $k(k-1)/2$ such clauses. Finally we have the clause $(r_1 \vee r_2 \vee \ldots \vee r_k)$. For a positive integer k, we use the notation $[k]$ to refer to the set $\{1,2, \ldots, k\}$.

3 SAT Encoding for Preset Distinguishing Sequences

In this work we formulate the constraints of a PDS in the form of Boolean formulae and use a SAT solver to find a PDS. The set of formulae generated depends on the FSM, and also on the length of the PDS that we would like to find. Intuitively, the question asked to the SAT solver is the following: "Is there a PDS of length L?".

Input guessing: We use free Boolean variables to let the SAT solver guess an input sequence. The guessed input sequence is checked to see if it is a PDS or not. If the length we are trying is L, for each $l \in [L]$ and for each input symbol $x \in X$, we generate a Boolean variable $\mathbf{X}_{lx}$. Hence there are pL such variables. The variable $\mathbf{X}_{lx}$ should be set to true only if at step l the input symbol x is used. Obviously, at each step there should be exactly one input symbol being used. Therefore, we generate the following formula: $\varphi_1 = \wedge_{l\in[L]}(\nabla\{\mathbf{X}_{lx} | x \in X\})$. For a step $l \in [L]$, the formula $\nabla\{X_{lx}|x \in X\}$ introduces $p(p-1)/2 + 1$ clauses. Therefore in total, there will be $(p(p-1)/2 + 1)L$ clauses.

State tracing: We need to trace the states that will be visited when the input guessed by formula φ_1 is applied at each state. We use the family of variables $\mathbf{S}_{ijk}$, where $i, k \in [n]$ and $j \in [L]$ for this purpose. Intuitively, $\mathbf{S}_{ijk}$ is set to true only if we are at state s_k at step j when we start from state s_i. There are n^2L such variables. For each state s_i, $\mathbf{S}_{i1i}$ is initialized to true by using the formula: $\varphi_2 = \wedge_{i\in[n]}(\mathbf{S}_{i1i})$. The number of clauses introduced by φ_2 is n. We also make sure that for each starting state, at any step there will always be exactly one current state by using: $\varphi_3 = \wedge_{i\in[n],l\in L}(\nabla\{\mathbf{S}_{ilj}|j \in [n]\})$. Each $(\nabla\{\mathbf{S}_{ilj}|j \in [n]\})$ in φ_3 has n operands. Hence it will generate $n(n-1)/2 + 1$ clauses. There are nL such operators in φ_3, and therefore there will be $nL(n(n-1)/2 + 1)$ clauses for φ_3.

One last thing we have to do for state tracing is to assign the correct variable to true to represent transitions. Let us consider a transition $(s_j, x/y, s_k)$. No matter which state we started from, at a step $l \in [L-1]$, if the current state is s_j and if the input guessed for step l is x, then for the step $l + 1$ the current state should be s_k. The following formula is used to encode the transition information into state tracing variables: $\varphi_4 = \wedge_{i,j\in[n],l\in[L-1],x\in X}((\mathbf{S}_{ilj} \wedge \mathbf{X}_{lx}) \Rightarrow \mathbf{S}_{i(l+1)k})$, where $(s_j, x/y, s_k)$ is a transition in M. $((\mathbf{S}_{ilj} \wedge \mathbf{X}_{lx}) \Rightarrow \mathbf{S}_{ilk}) \equiv (\neg\mathbf{S}_{ilj} \vee \neg\mathbf{X}_{lx} \vee \mathbf{S}_{ilk})$ corresponds to a single clause and there are $n^2(L-1)p$ such clauses.

Output tracing: We use the family of variables $\mathbf{Y}_{ijk}$, where $i \in [n], j \in [L]$, and $k \in Y$ for output tracing. Intuitively, $\mathbf{Y}_{ijk}$ is set to true only if output k is produced at step j if we start at state s_i. There are nLq such variables. For the assignment of these variables, we have two concerns similar to what we had for the state tracing variables. First we have to make sure that, for each starting state and for each step, the guessed input produces exactly one output. Second, we have to make sure that this output produced is the correct output. The first concern can be handled by using the following formula: $\varphi_5 = \wedge_{i\in[n],l\in L}(\nabla\{\mathbf{Y}_{ilk}|k \in [q]\})$. Each $(\nabla\{\mathbf{Y}_{ilk}|k \in [q]\})$ in φ_5 generates $q(q-1)/2 + 1$clauses and therefore in total there will be $nL(q(q-1)/2 + 1)$ clauses due to φ_5.

For the generation of the correct output, we need to use the transition information. Let us consider a transition $(s_j, x/y, s_k)$. No matter which state we started from, at a step $l \in [L]$, if the current state is s_j and if the input guessed for step l is x, then the output produced in step l must be y. The following formula is used to encode this information by using state tracing and input guessing variables: $\varphi_6 =_{i,j\in[n],l\in[L],x\in X} ((\mathbf{S}_{ilj} \wedge \mathbf{X}_{lx}) \Rightarrow \mathbf{Y}_{ily})$, where $(s_j, x/y, s_k)$ is a transition in M $((\mathbf{S}_{ilj} \wedge \mathbf{X}_{lx}) \Rightarrow \mathbf{Y}_{ily}) \equiv (\neg\mathbf{S}_{ilj} \vee \neg\mathbf{X}_{lx} \vee \mathbf{Y}_{ily})$ corresponds to a single clause and there are n^2Lp such clauses.

Output comparison: We introduce a family of Boolean variables $\{\mathbf{E}_{ijl}|i,j \in [n], i<j, l \in [L]\}$ to compare the outputs of pairs of states in each step. Intuitively, $\mathbf{E}_{ijl}$ is set to true only if starting states s_i and s_j produce the same output symbol at step l for the guessed input at step l. There are $Ln(n-1)/2$ such variables. The assignment of these variables is performed according to the following formulae:

$$\varphi_7 = \bigwedge_{i,i\in[n],i<j,l\in[L],k\in[q]}\left((\mathbf{Y}_{ilk} \wedge \mathbf{Y}_{jlk}) \Rightarrow \mathbf{E}_{ijl}\right)$$
$$\varphi_8 = \bigwedge_{i,j\in[n],i<j,l\in[L],k,m\in[q]k\neq m}\left((\mathbf{Y}_{ilk} \wedge \mathbf{Y}_{jlm}) \Rightarrow \neg\mathbf{E}_{ijl}\right)$$

φ_7 is used to set $\mathbf{E}_{ijl}$ to true when starting states s_i and s_j produce the same output at step l, and $\varphi\varphi_8$ is used to set $\mathbf{E}_{ijl}$ to false otherwise. Both $\left((\mathbf{Y}_{ilk} \wedge \mathbf{Y}_{jlk}) \Rightarrow \mathbf{E}_{ijl}\right) \equiv \left(\neg\mathbf{Y}_{ilk} \vee \mathbf{Y}_{jlk} \vee \mathbf{E}_{ijl}\right)$ and $\left((\mathbf{Y}_{ilk} \wedge \mathbf{Y}_{jlm}) \Rightarrow \neg\mathbf{E}_{ijl}\right) \equiv \left(\neg\mathbf{Y}_{ilk} \vee \neg\mathbf{Y}_{jlm} \vee \neg\mathbf{E}_{ijl}\right)$ introduce a single clause. Therefore for φ_7 and φ_8, there are $q^2Ln(n-1)/2$ clauses.

PDS checking: As the final step, we introduce a Boolean formula to decide if each state produces a unique output sequence for the guessed input sequence. Hence for each unordered pair of two distinct states, we check if these two states produce the same output sequence of length L using the following formula:

$$\varphi_9 = \neg\left(\bigvee_{i,j\in[n],i<j}\left(\bigwedge_{l\in[L]}\mathbf{E}_{ijl}\right)\right)$$

In φ_9, we have one conjunct for each unordered pair of states. Therefore φ_9 introduces $n(n-1)/2$ conjuncts. The overall formula for checking the existence of a PDS of length L is

$$\varphi_1 \wedge \varphi_2 \wedge \varphi_3 \wedge \varphi_4 \wedge \varphi_5 \wedge \varphi_6 \wedge \varphi_7 \wedge \varphi_8 \wedge \varphi_9$$

If the SAT solver finds the formula satisfiable, the assignments of the input guessing variables, $\mathbf{X}_{lx}$, give a PDS for the FSM. If the formula is unsatisfiable, then it means the FSM has no PDS of length L.

4 Experiments

We conducted a set of experiments to compare the performance of the classical algorithm (called as CA below) as given in [1, 13, 14] and the SAT based approach (called as SA below) explained in this paper. We constructed 15 sets of FSMs where the numbers of states are 60, 70, 80, 90, and 100, and the numbers of input/output symbols are 10/8, 9/9, and 8/10. There are 100 FSMs in each one of these sets, hence a total of 1,500 FSMs are used. All machines are strongly connected, minimal, completely specified, and have a PDS. The FSMs are generated randomly, by assigning a random output symbol and a random next state for each outgoing transition of each state. The experiments are carried out using MiniSat 2.2.0 running on Linux Ubuntu where the hardware is 2.5 GHz Intel Core2Duo quad-core machine.

For each individual FSM in the test set, we measure the time taken by CA (t_c) and the time taken by SA (t_s), from the start to the end of the tree exploration. We consider the speed up factor of SA over CA, which is simply computed as t_c/t_s. Table 1 summarizes the results of our experimental study. CA finds the length of the shortest PDS. SA checks if there exists a PDS of this shortest length.

Table 1 Speed up factor (MIN/AVG/MAX) of SA over CA (t_c/t_s)

	Number of states				
	60	70	80	90	100
10/8	0.06/3.58/33.73	0.12/5.08/148.5	0.07/4.46/41.95	0.3/4.68/75.45	0.08/3.37/34.82
9/9	0.04/1.05/14	0.04/0.43/7.25	0.03/0.68/5.35	0.01/0.6/4.67	0.02/2.22/25.22
8/10	0.06/0.58/4.8	0.03/0.45/3.5	0.03/0.3/3.53	0.04/0.8/5.53	0.05/0.95/6.63

The experimental results suggest that SA does not necessarily perform better for all the cases, and in fact for some machines CA perform much better. We can see this at MIN rows of Table 1, where the speed up factors are smaller than 1. However, on the average we see that SA performs better and better as the number of input symbols gets higher compared to the number of output symbols. We can see this by comparing the average speed up factors for the cases of input/output symbols 8/10, 9/9, and 10/8, in this order. In the case of 10 input and 8 output symbols, the average speed up factor is consistently greater than 1 for all the state sizes in our test set.

We also wanted to know if there are FSMs for which CA cannot decide if there exists a PDS or not (due to memory limitations), whereas SA can terminate the search. However, please recall that SA only answers the question "Does there exist a PDS of length L?". In our previous experiments, we got L to be tried in the SA runs from the CA runs. This time, such information is not available as CA cannot terminate the search. However, theoretically we know that if there exists a PDS for an FSM with n states and q output symbols, it cannot be shorter than $\lceil \log_q n. \rceil$ To leave a safety margin as well, in our experiments, SA is asked to check if there exists a PDS of length $\lceil 2 \times \log_q n. \rceil$ To find the cases where CA does not terminate, we started from the biggest set of experiments (100 states, 10 inputs and 8 outputs) and increased the number of inputs gradually. When we reached to the machines with the number of inputs 50 (and above), we saw that CA cannot finalize analysis within the heap space bound of 1 GB anymore. Whereas SA terminates successfully for all of our randomly generated machines within 5–6 s using less than 400 MB.

5 Conclusions

In this work, a SAT encoding for the problem of checking if a PDS of length L exists for an FSM is given. Using this encoding method, we investigated and compared the relative performance of classical PDS generation algorithm and a SAT solver. It is observed that SAT based approach performs better than the classical algorithm on the average when the tree that has to be constructed during the search has bigger branching factor. However, the result cannot be generalized easily since it is also observed in our test cases that for some FSMs the classical algorithm performs much better than the SAT based approach.

The results reported here only reflect some initial work on the topic. There are several ways to proceed to extend the investigation. One such path is to reconsider the SAT encoding suggested in this paper. It is possible to have a logarithmic reduction in the number of variables by using standard binary encoding. Our experiments are performed with the standard options of a SAT solver (MiniSat). Different options of this SAT solver or other SAT solvers can be tried out to see how the other heuristics implemented in those variations would affect the performance.

Acknowledgments This work was supported in part by the Natural Sciences and Engineering Research Council of Canada, and the Ontario Centres of Excellence, and by Sabanci University.

References

1. Friedman, A.D., Menon, P.R.: Fault Detection in Digital Circuits. Prentice Hall, Englewood Cliffs (1971)
2. Aho, A.V., Sethi, R., Ullman, J.D.: Compilers: principles, techniques, and tools. In: Reading. Addison-Wesley, MJ (1986)
3. Chow, T.S.: Testing software design modeled by finite state machines. IEEE Trans. Softw. Eng. **SE-4**(3), 178–187 (1978)
4. Holzmann, G.J.: Design and Validation of Protocols. Prentice Hall, Englewood Cliffs (1990)
5. Binder, R.V.: Testing Object-Oriented Systems: Models Patterns and Tools. Addison-Wesley, Boston (1999)
6. Haydar, M., Petrenko, A., Sahraoui, H.: Formal verification of web applications modeled by communicating automata. In: Formal Techniques for Networked and Distributed Systems (FORTE 2004). Springer Lecture Notes in Computer Science, vol. 3235, pp. 115–132. Springer Berlin/Heidelberg, September 2004
7. Hennie, F.C.: Fault-detecting experiments for sequential circuits. In: Proceedings of Fifth Annual Symposium on Switching Circuit Theory and Logical Design, pp. 95–110, Princeton, New Jersey, November 1964
8. Hierons, R.M., Ural, H.: Reduced Length Checking Sequences. IEEE Trans. Comput. **51**(9), 1111–1117 (2002)
9. Hierons, R.M., Ural, H.: Optimizing the length of checking sequences. IEEE Trans. Comput. **55**(5), 618–629 (2006)
10. Gonenc, G.: A method for the design of fault detection experiments. IEEE Trans. Comput. **19**, 551–558 (1970)
11. Moore, E.P.: Gedanken experiments on sequential machines. In: Shannon, C., McCarthy, J. (eds.) Automata Studies. Princeton University Press, Princeton (1956)
12. Lee, D., Yannakakis, M.: Testing finite-state machines: state identification and verification. IEEE Trans. Comput. **43**(3), 306–320 (1994). doi:10.1109/12.272431
13. Gill, A.: Introduction to the Theory of Finite State Machines. McGraw Hill, NY (1962)
14. Kohavi, Z.: Switching and Finite Automata Theory. McGraw Hill, NY (1978)
15. Cook, S.A.: The complexity of theorem proving procedures. In: Proceedings of 3rd Annual ACM Symposium on Theory of Computing, pp. 151–158. ACM, NY (1971)

A Decision-Making Ontology for Analytical Requirements Elicitation

Fahmi Bargui, Hanene Ben-Abdallah and Jamel Feki

Abstract Despite their potential in analytical requirements elicitation, goal-oriented approaches in Data warehouse (DW) projects have not been well exploited. The main difficulties in their adoption is the lack of assistance in goal elicitation and the lack of support in generating suitable information for decision-making, from the defined goals. To address these limitations, in this paper we introduce a decision-making ontology that formalizes the semantic relationships among the decision makers' goals and explicitly relates the decision-making knowledge to the goals. In addition, this ontology is enriched with a set of queries and inference rules that provide for an automatic generation of suitable analytical requirements from the goals fulfilling any business process that the decision maker cares to analyze.

1 Introduction

A data warehouse (DW) is a special type of database dedicated for decision-making support. It organizes data into facts and dimensions based on Multidimensional (MD) modeling. Goal-oriented MD modeling (e.g. [1, 2]) gained an increasing interest in eliciting analytical requirements in DW projects thanks to their graphical languages and supporting tools. However, their widespread usage

F. Bargui (✉) · H. Ben-Abdallah · J. Feki
Mir@cl Laboratory, FSEG, University of Sfax, Po Box 1088 Sfax, Tunisia
e-mail: fahmi.bargui@fsegs.rnu.tn

H. Ben-Abdallah
e-mail: hanene.benabdallah@fsegs.rnu.tn

J. Feki
e-mail: jamel.feki@fsegs.rnu.tn

E. Gelenbe et al. (eds.), *Computer and Information Sciences II*,
DOI: 10.1007/978-1-4471-2155-8_63, © Springer-Verlag London Limited 2012

remains hindered by their lack of assistance in goal elicitation conducted through the decomposition of high level goals into concrete sub-goals. This goal decomposition requires *domain knowledge* mastered only by domain experts. Furthermore, our literature review highlighted that goal-oriented approaches provide little or no assistance in identifying appropriate performance indicators to measure the fulfillment degree of the elicited goals. The same shortage is also present in identifying information (i.e. data to be stored in the DW) that could be analyzed when the elicited goals are not met. In most cases, such information is informally collected and without an explicit link with the goals. Consequently, when decision maker's goals change, it is difficult to trace the parts of the MD model that should be modified.

In this paper, we present a decision-making ontology that provides for the automation of analytical requirements elicitation. The next section, first, describes the ontology and illustrates it for the commercial domain. Section 3 shows how the ontology allows the analyst to overcome the lack of domain knowledge required during requirements elicitation. Finally, Sect. 4 synthesizes our proposal and presents our future work.

2 The Decision-Making Ontology

2.1 Thesaurus Part

Through an interview with decision makers at different hierarchical levels and belonging to different domains, we identified the *terminology* used in the decision-making process. This terminology includes nine concepts (Fig. 1).

A *DecisionMaker* is a person in the enterprise, who evaluates and controls the performance of a *BusinessProcess* composed of a set of related activities that produce a specific service/product for a particular customer. Each business process must fulfill a set of *Goals* during a period of time. A Goal is quantitative and calculated through a *Formula* to evaluate its achievement degree by comparing it to an estimated Target Value. The achievement degree of a goal is given by an *Indicator*. This later can be analyzed by the decision maker through a set of *AnalysisAxes* by aggregating (SUM, MAX, AVG…) the various values produced of the indicator over time.

In addition, an analysis axis is composed of several *AnalysisLevels* each of which represents a granularity echelon to aggregate the indicator values. An analysis level may have additional descriptions (*AnalysisAttribute*) and multiple analysis levels can be organized into an *AnalysisHierarchy*.

Furthermore, the above decision making concepts can be semantically related through the following nine relationships (Fig. 1):

- *Control* (*D*, *P*): The decision maker *D* controls the performance of the business process *P*.

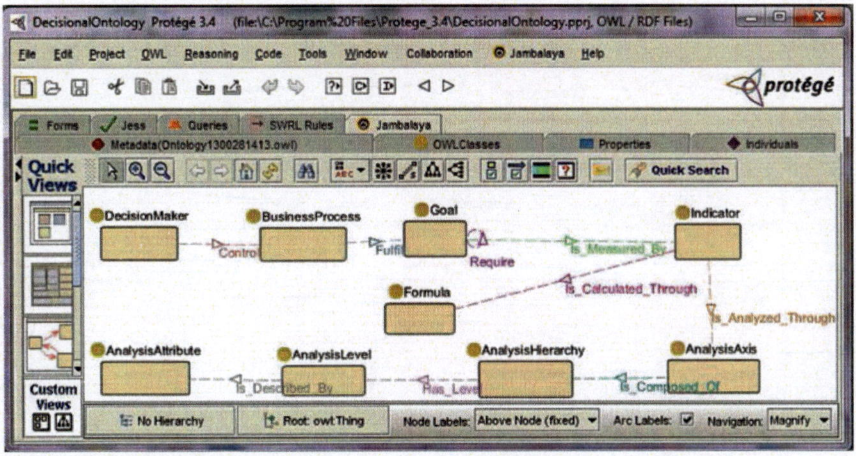

Fig. 1 Decision-making Ontology metamodel in Protégé (thesaurus part)

- *Fulfill* (*P*, *G*): The business process P is defined to fulfill the goal G.
- *Is_Measured_By* (*G*, *I*): The fulfillment degree of the goal *G* is measured by the indicator *I*.
- *Is_Calculated_Through* (*I*, *F*): The value produced by the indicator *I* is calculated through the formula *F*.
- *Is_Analyzed_Through* (*I*, *A*): Indicator *I* is analyzed through the analysis axis *A*.
- *Is_Composed_Of* (*A*, *H*): The analysis axis *A* is composed of a hierarchy *H*.
- *Has_Level* (*H*, *L*): Hierarchy H has level *L*.
- *Is_Described_By* (*L*, *At*): The level *L* is described by the analysis attribute *At*.
- *Require* (*G*, *G'*): Fulfillment of the goal *G* requires the achievement of *G'*.

Figure 2 illustrates an instance of our ontology meta-model for the commercial domain. In this instance, the concept *SalesManager* is defined to *control* the performance of the three business processes: *Order*, *AuctionOrder* and *Delivery*; each of which must *fulfill* some goals. For instance, the process *Order* is defined to fulfill the goal *IncreaseSales*. The realization of this later *requires* the achievement of three goals: *IncreaseCustomers*, *ProductAvailable* and *IncreaseShops*. Note that relationships among goals are represented in the ontology by means of predicates. For example, the realization of the goal *IncreaseSales* requires the availability of the products in the stock, and increasing the number of either customers or shops. This knowledge is represented in the ontology through the predicate: *(Require (IncreaseSales, IncreaseCustomers)* $\vee$ *Require (IncreaseSales, IncreaseShops))* $\wedge$ *Require (IncreaseSales, ProductAvailable)*.

Figure 3, shows various levels organized by analysis hierarchies for the analysis axis *Product*. For example, for the hierarchy *family*, the indicator Turnover could be analyzed according to the level *Category* of a product. This gives the decision maker detailed values of the *Turnover* by category of product. The comparison

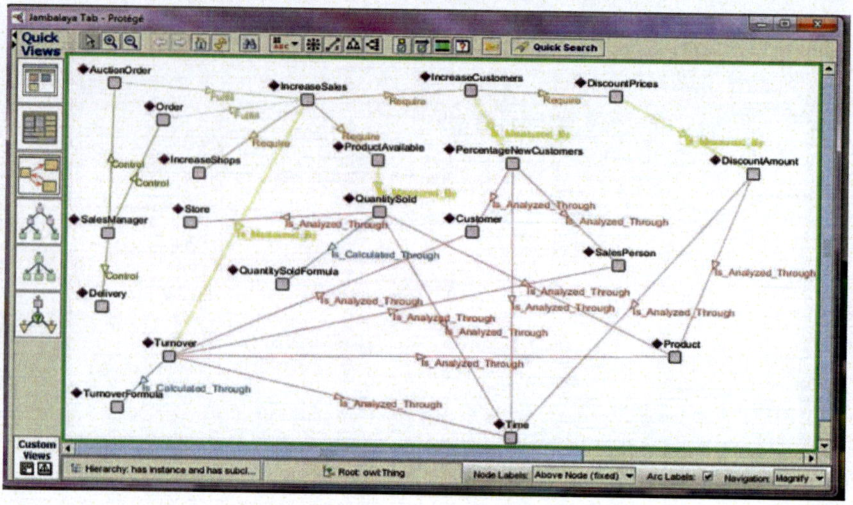

Fig. 2 An extract of the commercial domain ontology (the thesaurus part)

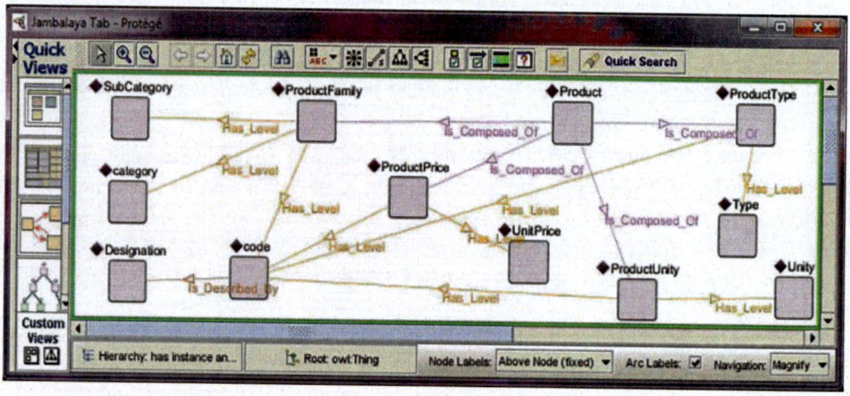

Fig. 3 An extract of the analysis axis concept (product)

of each realized value with an estimated target value allows the decision maker to judge what category of product is not sold as expected.

2.2 *Inference Rules and Queries*

The inference rules are defined by the domain experts and are identified based on the semantic relationships among the domain concepts. The inference rule's premises express constraints on concepts and restrictions on their properties' values and cardinalities. These premises are expressions considered as always true

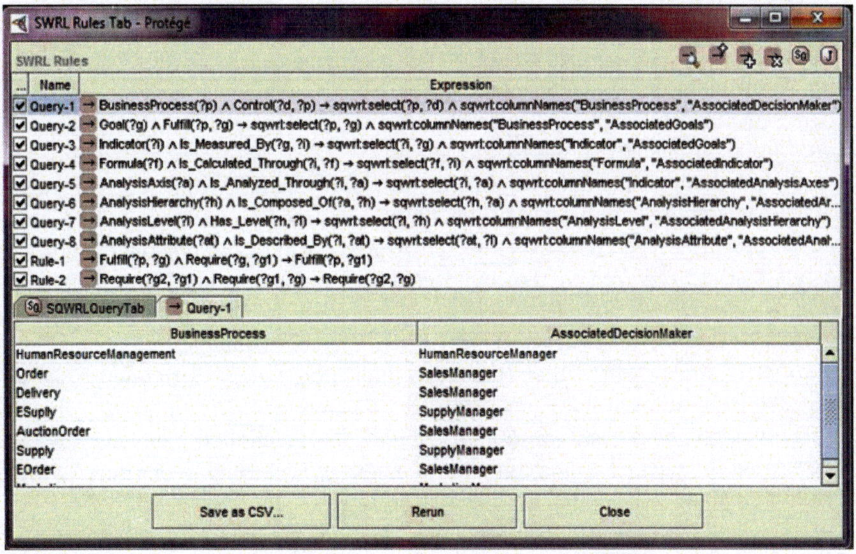

Fig. 4 Inference rules and queries part of the decision-making ontology

by the experts in the domain; that is, they represent knowledge (stored in the ontology) that cannot be contested. When an inference rule is applied, the reasoning engine matches the premises' formulas with the knowledge stored in the ontology to infer a set of new knowledge, as conclusions, not explicitly stored. The deduced information is either retained by the user or can be part of the premises of other rules.

In our context, we have defined a set of inference rules and queries in order to assist the elicitation of the analytical requirement elements and to complete the filling of our natural language based analytical requirements template [3]. The components of this template are determined through an empirical study covering samples of decision-making processes (cf. [4]). Figure 4 shows a sample of eight queries and two inference rules. They are formalized, respectively in the SQWRL and SWRL languages.[1]

3 Analytical Requirements Elicitation

Our requirement elicitation process uses a set of queries and inference rules to extract information from the decision-making ontology (Fig. 4). It is conducted through the following four steps by interaction with the decision maker and the requirements analyst:

[1] http://www.w3.org/Submission/SWRL/ and http://protege.cim3.net/cgi-bin/wiki.pl?SQWRL

- *Business process identification*: Query 1 (Fig. 4) extracts from the ontology *all* business processes *controlled* by the given Decision Maker; this latter selects those business processes of interest.
- *Goal identification*: Query 2 extracts from the ontology all *Goals* that must be *fulfilled* by a business process selected by the *Decision Maker*. In addition, for complex Goals (i.e. might be difficult to fulfill at first); Rule 1 automatically finds any sub-goals indirectly *fulfilled* by the selected business process. Finally, to ensure the completeness of the set of goals, Rule 2 automatically finds all goals that are *required* by the found goals.
- *Indicator retrieval*: For each found goal, Query 3 extracts from the ontology the *Indicator* that *measures* it and Query 4 extracts its *Formula*.
- *Analytical query generation*: For each Indicator, Query 5 through Query 8 extract from the ontology, respectively: its *analysis axis*, their *hierarchies*, their *analysis levels* and *analysis attributes*. The identified elements are then combined to generate a query in natural language according to our query format [5]: < Analyze > + < the > + < Indicator > + < by > + < Analysis axis > + <'s > + < Analysis level > + [< and > + < Analysis Attribute >]. For example, the following analytical query: *Analyze the Turnover by product's code and designation* is generated for the indicator `Turnover` related to the goal `IncreaseSales` and the business process `Order`.

Note that the stakeholders (i.e. decision maker and requirements analyst) are involved in each elicitation step in order to select from/confirm the found results.

4 Conclusion and Future Work

This paper presented an ontology for the decision-making domain. This latter can be used by a requirements analyst to identify *automatically all*: (1) pertinent business processes to analyze, (2) goals and sub-goals that must be fulfilled by the business processes, and (3) the indicators and their associated formulas measuring the achievement degree of the goals. This automatically identified information is then used to formulate automatically a set of analytical queries in natural language. The formulated analytical queries can be used by the decision maker to carry out different analyses when his/her goals are not met. In addition, the ontology helps the analyst to acquire domain knowledge necessary during requirements elicitation. Indeed, our method performs automatic decomposition of the decision makers' goals by referring to the semantic relationships and the inference rules stored in the ontology.

We are currently working on finalizing the development of a toolset supporting our analytical requirements elicitation approach which is based on the presented decision-making ontology. In addition, we plan to examine how the evolution of the decision-making ontology can be handled.

References

1. Giorgini, P., Rizzi, S., Garzetti, M.: GRAnd: A goal-oriented approach to requirement analysis in data warehouses. Decis Support Syst **45**(1), 4–21 (2007)
2. Mazón, J.-N., Pardillo, J., Trujillo, J.: A model-driven goal-oriented requirement engineering approach for data warehouses. ER Workshops, LNCS **4802**, 255–264 (2007)
3. Bargui, F., Feki, J., Ben-Abdallah, H.: A natural language approach for data mart schema design. In the *9th Int*. ACIT, Tunisia (2008)
4. Mard, M., Dunne, R.R., obsborne, E., Rigby, J.S.: Driving your company's value: strategic benchmarking for value. Wiley, New Jersy (2004)
5. Bargui, F., Feki, J., Ben-Abdallah, H.: Vers une expression des besoins décisionnels en langage naturel. In Proc. 3ème Atelier sur les Systèmes Décisionnels, pp. 1–12 (2008)
6. Bargui, F., Ben-Abdallah, H., Feki, J.: Multidimensional Concept Extraction and Validation from OLAP Requirements in NL, pp. 199–206. In Proc IEEE NLP-KE, China (2009)

Part XII
Machine Learning

Spatio-Temporal Pattern and Trend Extraction on Turkish Meteorological Data

Isil Goler, Pinar Senkul and Adnan Yazici

Abstract Due to increasing amount of spatio-temporal data collected from various applications, spatio-temporal data mining has become a demanding and challenging research field requiring development of novel algorithms and techniques for successful analysis of large spatio-temporal databases. In this study, we propose a spatio-temporal mining technique and apply it on meteorological data, which has been collected from various weather stations in Turkey. In addition, we introduce one more mining level on the extracted patterns in order to discover general trends with respect to spatial changes. Generated patterns are investigated under different temporal ranges, in order to monitor the change of the events with respect to temporal changes.

Keywords Spatio-temporal data mining · Trend extraction · Spatio-temporal data

1 Introduction

Automatic extraction of potentially useful information from large spatio-temporal databases requires development of new mining techniques. In contrast to previous studies that focus on either spatial or temporal dimension of data [1–7], in this

I. Goler
Havelsan Inc, Eskişehir Yolu 7.km, 06520 Ankara, Turkey
e-mail: igoler@havelsan.com.tr

P. Senkul (✉) · A. Yazici
Computer Engineering Department, Middle East Technical University,
06531 Ankara, Turkey
e-mail: senkul@ceng.metu.edu.tr

A. Yazici
e-mail: yazici@ceng.metu.edu.tr

E. Gelenbe et al. (eds.), *Computer and Information Sciences II*,
DOI: 10.1007/978-1-4471-2155-8_64,

work, we aim to perform pattern extraction with respect to changes in both spatial and temporal dimensions. In the literature, there are previous studies that consider both spatial and temporal aspects of the data together [8, 9]. Although the techniques proposed in previous studies are theoretically useful for discovering patterns repeating over time and space, they do not present applicability of the techniques with real case studies. Another motivation for this study is the need for extracting more general trends, through the analysis of the extracted spatio-temporal patterns.

In this study, we propose improvements for extracting spatio-temporal association patterns by including a preprocessing step and incorporating sliding window approach into the temporal intervals. In addition, we propose a technique for discovering general trends of events by further analyzing the generated spatio-temporal patterns. The proposed techniques are applied on a real data set, meteorological data, which has been collected from various weather stations in Turkey.

2 The Proposed Approach

2.1 Spatio-Temporal Pattern Discovery

In this study, a data set collected by the Turkish State Meteorological Service [10] is used. The data set covers the measurements taken from 263 major weather stations in Turkey and it includes the monthly averages for temperature, precipitation and the number of snowy days per station from 1970 to 2007. In the preprocessing phase, Turkey map including weather stations is represented in disjoint grid cells in order to define spatial neighboring relationships. Each cell represents a spatial region denoted as (x, y). After mapping to grid representation, the location of weather station is defined in terms of the coordinates of the grid cell it stays in. A neighborhood relation is defined as follows: if the x grid number and y grid number difference of two stations is smaller than or equal to 1, these stations are considered to be neighbors. As an example, $Station_1$ at (2, 3) and $Station_2$ at (3, 4) are assumed to be neighbors since the difference of their x and y grid numbers is one (1). Events are classified on the basis of their intensities, and related event tables are created for each of the event type.

Next phase is the generation of events. For the example given in Fig. 1, there are weather stations in the grid cells (0, 1), (1, 2), (1, 3), (1, 4) and (2, 3) for time partition January 1975. Characters and numbers written in the grid cells stand for the event types. All the events of the first time window (of width 3) for the example given in Fig. 1 are presented in the first column of Fig. 2. The second column shows the events in neighborhood relation for consecutive time partitions, which are marked with rectangles in Fig. 1. Once the events are extracted, consecutive eventsets can be generated. One of the consecutive eventsets generated from the events in Fig. 2 is $\{a, 2, g, b, 2, g\} \rightarrow \{a, 3, g, b, 3, h\}$. The first set in the consecutive eventset includes the events of a selected cell and events of its neighbor cells for January in 1975, and

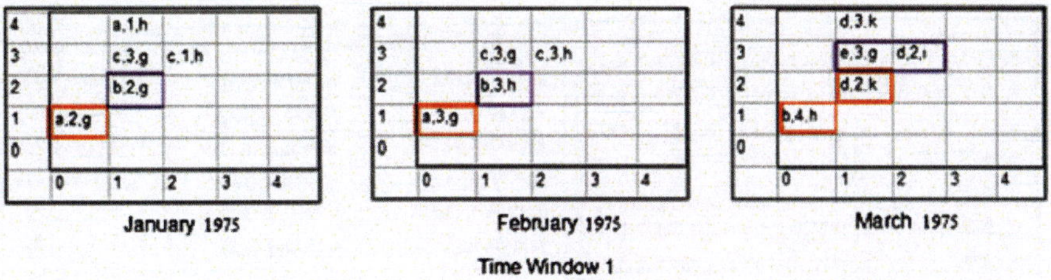

Fig. 1 Events on the grids

the second set includes the events of a cell and events of the neighbor cells in February in 1975. The next step is calculating the frequencies of the events in these sets by searching for these consecutive eventsets in other years. The frequent ones that exceed a given threshold are retained for further processing. All the events in the consecutive eventset might not be frequent. For finding the frequent ones, subsets of the consecutive eventsets are examined.

2.2 Second Level Mining: Mining on the Generated Patterns

This phase is developed in order to find new and higher-level trends from the generated patterns, and it can be called as *second level mining*. As the first step, frequent patterns are classified on the basis of event types, and situation variables are defined in order to express the temporal or spatial relationships in the trends. After this, all of the frequent patterns, which are classified on the basis of event types, are analyzed and events in the consecutive eventsets are compared with each other according to the situation variables. It is assumed that the most frequent situation (the situation variable with the maximum count value) shows the general trend. In meteorological data set, events are grouped according to their event types as snow events, precipitation events and temperature events. For each event type, four situation variables are defined for representing event changes with respect to changes in x dimension and four situation variables are defined for representing event changes with respect to changes in y dimension. As the result of this analysis, trends such as ‘Stations located up north have higher average temperatures compared to the average temperature of their neighbor stations located south to them” can be discovered.

3 Experimental Results

3.1 Interpretation of Patterns Generated for a Sample Region

In this group of experiments, a region in Turkey, which is the area at the coast of Black Sea, namely Black Sea Region, is used. The frequency count is set to 10, which corresponds to about 70% support threshold. Since the data set spans

January	February	March
a, 1975, 1, (0,1)	a, 1975, 2, (0,1)	b, 1975, 3, (0,1)
2, 1975, 1, (0,1)	3, 1975, 2, (0,1)	4, 1975, 3, (0,1)
g, 1975, 1, (0,1)	g, 1975, 2, (0,1)	h, 1975, 3, (0,1)
b, 1975, 1, (1,2)	b, 1975, 2, (1,2)	d, 1975, 3, (1,2)
2, 1975, 1, (1,2)	3, 1975, 2, (1,2)	2, 1975, 3, (1,2)
g, 1975, 1, (1,2)	h, 1975, 2, (1,2)	k, 1975, 3, (1,2)
c, 1975, 1, (1,3)	c, 1975, 2, (1,3)	e, 1975, 3, (1,3)
3, 1975, 1, (1,3)	3, 1975, 2, (1,3)	3, 1975, 3, (1,3)
g, 1975, 1, (1,3)	g, 1975, 2, (1,3)	g, 1975, 3, (1,3)
c, 1975, 1, (2,3)	c, 1975, 2, (2,3)	d, 1975, 3, (2,3)
1, 1975, 1, (2,3)	3, 1975, 2, (2,3)	2, 1975, 3, (2,3)
h, 1975, 1, (2,3)	h, 1975, 2, (2,3)	i, 1975, 3, (2,3)
a, 1975, 1, (1,4)		d, 1975, 3, (1,4)
1, 1975, 1, (1,4)		3, 1975, 3, (1,4)
h, 1975, 1, (1,4)		k, 1975, 3, (1,4)

eventsets of month1	eventsets of month2	eventsets of month3
a, 1975, 1, (0,1)	a, 1975, 2, (0,1)	b, 1975, 3, (0,1)
2, 1975, 1, (0,1)	3, 1975, 2, (0,1)	4, 1975, 3, (0,1)
g, 1975, 1, (0,1)	g, 1975, 2, (0,1)	h, 1975, 3, (0,1)
b, 1975, 1, (1,2)	b, 1975, 2, (1,2)	d, 1975, 3, (1,2)
2, 1975, 1, (1,2)	3, 1975, 2, (1,2)	2, 1975, 3, (1,2)
g, 1975, 1, (1,2)	h, 1975, 2, (1,2)	k, 1975, 3, (1,2)
		e, 1975, 3, (1,3)
		3, 1975, 3, (1,3)
		g, 1975, 3, (1,3)
		d, 1975, 3, (2,3)
		2, 1975, 3, (2,3)
		i, 1975, 3, (2,3)

Fig. 2 Events of the first time window

Fig. 3 Sample frequent events in one time window

```
**************************** Percentage 62.5
month: 1 17022  event is b
month: 1 17070  event is a
month: 2 17070  event is 2
```

through 16 years of recording, if an event occurs in at least 10 years out of 16, it is considered to be frequent. The sample pattern given in Fig. 3, with 62.5% support, corresponds to the following description: *In January when average temperature is [5.0–10.0°C] in Zonguldak (17022) and average temperature is less than 5.0°C in Bolu (17070), in February precipitation is [25.0–50.0 mm] in Bolu (17070).* As a consecutive eventset, this is shown as {b, a} → {2}.

3.2 Experiments on Trend Extraction by Second-level Mining

In order to generate sample higher level patterns and to interpret them, the patterns generated for the coast of Black Sea Region (in the previous group of experiments) are used. The trends extracted for this area are shown in Fig. 4. When the results are inspected, it is seen that the strongest trend (with the highest frequency of 51%) is "*the temperature increases towards east*". Another strong rule with 47% frequency (i.e., it is observed 132 times out of 276 cases involving temperature) says that "*direction the temperature increases towards northern direction (towards the coast)*". The extracted trends are in general parallel to the characteristics of Black Sea Region climate.

Trend	Situation Variable	Frequency
Temperature increases towards north	EventIntencityUp-YGridUp	132 / 276 (47 %)
Temperature decreases towards south	EventIntencityDown-YGridDown	57 / 276 (21 %)
Temperature increases towards east	EventIntencityUp-XGridUp	123 / 241 (51 %)
Temperature decreases towards west	EventIntencityDown-XGridDown	60 / 241 (25 %)

Fig. 4 Sample results for extracted trends

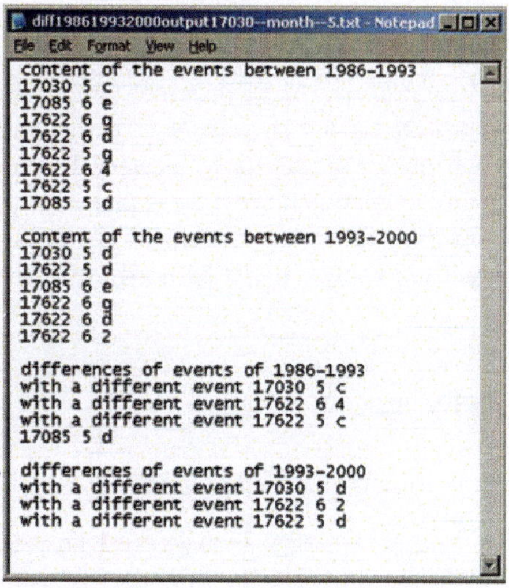

Fig. 5 Pattern changes for different periods

3.3 Experiments on the Climate Change Analysis

Another experiment is conducted on the analysis of climate change over several temporal periods. Three different temporal periods are defined for the years 1986 to 1993, 1993 to 2000 and 2000 to 2007. The patterns discovered for these temporal periods are compared in order to see whether any change in the climate characteristics occurred. To this aim, as the first step, patterns are generated for each of these periods. Afterwards, the patterns are compared to each other and the changes in event for the same weather station and the same time partition (i.e., the same month of the year) are marked. Figure 5 presents a sample set of results for this analysis.

The result set given in Fig. 5 compares the temporal periods 1986–1993 and 1993–2000. As a sample event in this figure, the event *<17030 5 c>* is stated as a climate difference for the patterns of the temporal period 1986–1993. The temporal period 1993–2000 includes the frequent event *<17030 5 g>*, which shows a change of event for the same station and the same month. Event *<17085 5 d>* is also indicated as a difference. When the temporal period 1993–2000 is analyzed, it is seen that there is not any event for <17085 5> (for the same station and the same month). Therefore, this indicates that the starting of a new climate characteristics.

4 Conclusion

Large collections of spatio-temporal data, such as meteorology data, contain possibly interesting information and valuable knowledge. This work proposes improvements for spatio-temporal data mining As a novel feature, a sliding window mechanism is introduced. By using the sliding window mechanism, the loss of correlations of different temporal intervals is prevented. Application of the approach on a real data set, Turkish meteorological data set, is an important contribution of this study, as well. As another novel feature, the generated patterns are mined in one more mining phase, in which the general trends of the events with respect to the location changes are discovered. Experiments demonstrate that the proposed approach can be used for different kinds of analysis effectively.

References

1. Peuquet, D.J.: A conceptual framework and comparison of spatial data models. Cartographica **21**(4), 66–113 (1984)
2. Roddick, J.F., Spiliopoulou, M.: A bibliography of temporal, spatial and spatio-temporal data mining research. ACM SIGKDD, vol. 1, June 1999
3. Yao, X.: Research issues in spatio-temporal data mining. White paper submitted to the University Consortium for Geographic Information Science (UCGIS) workshop on Geospatial Visualization and Knowledge Discovery. Virginia, Nov 18–20 (2003)
4. Koperski, K., Han, J.: Discovery of spatial association rules in geographic information databases. 4th International Symposium Advances in Spatial Databases, vol. 951, pp. 47–66, 6–9 (1995)
5. Mennis, J., Liu, J.: Mining association rules in spatio-temporal data: an analysis of urban socioeconomic and land cover change. Trans. GIS **9**(1), 5–17 (2005). January
6. Subramanyam, R.B., Goswami, A., Prasad, B.: Mining fuzzy temporal patterns from process instances with weighted temporal graphs. Int. J. Data Anal. Tech. Strateg. **1**(1) (2008)
7. S. Shekhar and Y. Huang, Discovery of spatial co-location patterns. International Symposium on Spatial and Temporal Databases (2001)
8. Wang, J., Hsu, W., Li Lee, M., Wang, J.: Finding flow patterns in spatio-temporal databases. In: Proceedings of the 16th IEEE International Conference on Tools with Artificial Intelligence (ICTAI 2004) (2004)
9. Wang, J., Hsu, W., Li Lee, M.: Mining Generalized Spatio-Temporal Patterns. Lecture Notes in Computer Science, LNCS, vol. 3453/2005, pp. 649–661. Springer, Berlin (2005)
10. Turkish state meteorological service. http://www.meteor.gov.tr

Contrastive Learning in Random Neural Networks and its Relation to Gradient-Descent Learning

Alexandre Romariz and Erol Gelenbe

Abstract We apply Contrastive Hebbian Learning to the recurrent Random Neural Network model. Under this learning rule, weight adaptation is performed based on the difference between the network's dynamics when it is input-free and when a teaching signal is imposed. We show that the resulting weight changes are a first order approximation to the gradient-descent algorithm for quadratic error minimization when overall firing rates are constant. The algorithm requires no matrix inversions, and no constraints are placed on network connectivity. A learning result on the XOR problem is presented as an empirical confirmation of these ideas.

1 Introduction

Early models for the neural basis of memory and learning were centred around the idea of connection strength modulation based on activity correlation, in what is generally called Hebbian learning [2]. From the neural-inspired computational

This work was partially supported by a leave of absence from the University of Brasília and a Postdoctoral Scholarship from CNPq, the Brazilian National Council for Science and Technology Development.

A. Romariz (✉)
Integrated Circuits and Devices Lab, Electrical Engineering Department, Universidade de Brasília, Brasília, DF, Brazil
e-mail: romariz@unb.br

E. Gelenbe (✉)
Intelligent Systems and Networks Group, Department of Electrical Engineering, Imperial College London, London, UK
e-mail: e.gelenbe@imperial.ac.uk

E. Gelenbe et al. (eds.), *Computer and Information Sciences II*,
DOI: 10.1007/978-1-4471-2155-8_65,

perspective, more success was obtained with algorithms in which weight adaptation was oriented by an error-reduction goal, as in the perceptron learning rule, and the gradient-descent backpropagation algorithm [8], which later became almost synonymous of neural network learning for application purposes.

Meanwhile, studies on variations of the Hebbian rule continued, highlighting possible connections between the two learning strategies. The Boltzmann machine learning algorithm [6] prescribed weight changes in the form of a difference of correlation of signals between units under two network states: a free-running state, and a forced state, in which a teaching signal clamps a subset of neurons to desired state values. In the framework of deterministic networks, a similar rule was derived directly from the energy-function in the Hopfield network, and called contrastive learning [7].

In those papers and others, an approximate equivalence between gradient-descent on weight space and minimization of the contrastive function is shown, for specific architectures: a single-layer network [7], or a layered network with symmetric back-connections [10], among other approximations and assumptions.

In this paper we claim that the relationship between contrastive and gradient-descent learning can be shown more clearly in the Random Neural Network framework. The Random Neural Network (RNN) [3] is a stochastic model of spiking neuronal activity which provides an effective calculation of the probability distribution of neuron activations, in the form of non-linear fixed-point equations, for arbitrary recurrent networks. It does not require, then, lengthy simulations to calculate equilibrium conditions at each learning step.

A direct comparison between weight changes in this algorithm and the gradient-descent algorithm for the RNN show they are equivalent to first order. No constraints on connectivity are necessary. Even if this does not represent a gain in computational cost, it does make the learning rule operate on mostly local variables, an important aspect for hardware implementation.

This work is organized as follows. The Random Neural Network model is revised in Sect. 2, along with its gradient-descent algorithm. The contrastive learning equations are then presented in Sect. 3, with a demonstration of first-order equivalence to the gradient-descent algorithm. An empirical demonstration is presented along with conclusions in Sect. 4 .

2 Random Neural Network and Learning Algorithm

The RNN [3, 9] models neural activity as a discrete state-space, continuous time stochastic process, mediated by external inputs and by the exchange of signals (spikes) among neurons. The state of each neural cell i is described by its electrical potential value $K_i(t)$ at time $t \geq 0$, a non-negative, integer variable. If excited ($K_i(t) > 0$,) the cell outputs spikes with rate r_i. Spikes may excite another cell j with probability $p^+(i,j)$, or inhibit it with probability $p^-(i,j)$ or yet depart the network without affecting other cells, with probability d_i. Upon arrival of an excitatory spike, the receiving-cell potential is increased by one. It is decreased by one upon arrival of an inhibitory spike, but only if it is not already zero. After sending a spike, the sending cell's electrical potential is also decreased by one. Cells are also affected by external excitatory inputs, represented as poissonian events with rates Λ_i, and inhibitory inputs, represented by λ_i.

In this model, the synaptic weight connection is described by the set of excitatory weights w_{ij}^+ and inhibitory weights w_{ij}^-, satisfying $w_{ij}^+ = r_i p^+(i,j)$ and $w_{ij}^- = r_i p^-(i,j)$. Since all spike outcomes are covered by probabilities p^+, p^- and d, a normalization requirement exists for every cell i. That requirement entails, for the case $d_i = 0$, in which all exiting spikes affect some other neuron,

$$r_i = \sum_j w_{ij}^+ + w_{ij}^- . \tag{1}$$

An important result for this model is that the stationary probability distribution for the state of the network has a product-form solution [3]. Because of this, the probability that cell i is excited, $q_i = P(k_i > 0)$ can be calculated from a set of nonlinear equations given by

$$q_i = \min\left(\frac{\Lambda_i + \sum_j q_j w_{ji}^+}{r_i + \lambda_i + \sum_j q_j w_{ji}^-}, 1 \right). \tag{2}$$

Conditions for existence and uniqueness of the solution are given in [4].

In [4, 5], a gradient-descent learning rule is presented for this recurrent neural network. If t is the neuron for which a desired output value y_t exists, weight changes Δw_{uv}^+ and Δw_{uv}^- are calculated as a step in the opposite direction of the gradient of error E in weight space. So,

$$\Delta w_{uv} = -\eta \frac{\partial E}{\partial w_{uv}} = \eta (y - q_t) \frac{\partial q_t}{\partial w_{uv}} . \tag{3}$$

The derivatives of a cell excitation probability q_t can be stated by the following equations

$$\frac{\partial q_t}{\partial w_{uv}^+} = \left[\sum_i \left(\frac{\partial q_i}{\partial w_{uv}^+} \frac{w_{it}^+ - q_t w_{it}^-}{D_t} \right) \right] + \frac{q_u}{D_t} \delta_{(v=t)} \tag{4}$$

$$\frac{\partial q_t}{\partial w_{uv}^-} = \left[\sum_i \left(\frac{\partial q_i}{\partial w_{uv}^-} \frac{w_{it}^+ - q_t w_{it}^-}{D_t} \right) \right] - \frac{q_t q_u}{D_t} \delta_{(v=t)} , \tag{5}$$

where $D_i = r_i + \lambda_i + \sum_j q_j w_{ji}^-$ is the denominator of Eq. 2, and where $\delta_{[C]}$ equals 1 when condition C holds, and equals 0 otherwise.

In the original work, there were additional terms to be considered on those equations, because of the variation of r_t with changes in weight w_{tv} for any

v. In the analysis that follows, we have noticed that it is convenient to work in a framework where rates do not change with weight updates. This is compatible with an alternative normalization procedure, in which we consider that variations in the weights are accounted for by variations in d_i. This interpretation requires a restriction on weights, namely

$$\sum_j w_{ij}^+ + w_{ij}^- \leq r_i \, . \tag{6}$$

Following Gelenbe [4], and considering the constant-rate case, a closed-form expression can be obtained for the weight adaptation. Equations (4) and (5) can be put in matrix-vector form, as

$$\frac{\partial \overline{q}}{\partial w_{uv}^{\pm}} = \frac{\partial \overline{q}}{\partial w_{uv}^{\pm}} \mathbf{W} + q_u \overline{\gamma}^{\pm}(v) \, , \tag{7}$$

where $\frac{\partial \overline{q}}{\partial w_{uv}^{\pm}}$ is a vector of neuron activation probabilities and matrix $\mathbf{W}$ is defined by

$$[W]_{ij} = \left[\frac{w_{ij}^+ - q_j w_{ij}^-}{D_j} \right]_{ij} \, . \tag{8}$$

Even though the network is recurrent, with arbitrary topology, the update of the weights can be calculated explicitly, and the most complex operation is the matrix inversion implied in Eq. 7.

3 Contrastive Learning and Equivalence

In the contrastive Hebbian learning algorithm [1], weight update Δw_{uv} is a function of network node activities under two different modes: under its free dynamics, and under a teaching (forcing) signal f. For the specific case of the Random Neural Network, a convenient way to impose a forcing signal is by perturbing the neuron's overall input Λ_i. The possibility of using the error as a form of teaching signal in contrastive learning has been pointed out before [1].

To show the correspondence between this learning rule and gradient descent in the context of RNNs, let us calculate the derivative of the neuron activation probability q_t in respect to a perturbation in the overall input of neuron v:

$$\frac{\partial q_t}{\partial \Lambda_v} = \frac{\sum_i \frac{\partial q_i}{\partial \Lambda_v} w_{it}^+}{D_t} + \frac{1}{D_t} \delta_{(t=v)} - q_t \frac{\sum_i \frac{\partial q_i}{\partial \Lambda_v} w_{it}^-}{D_t} \, . \tag{9}$$

Reorganizing terms and multiplying both sides by q_u (from Eq. 8, $q_u \neq 0$ if the gradient rule prescribes any change to Δw_{uv}):

$$q_u \frac{\partial q_t}{\partial \Lambda_v} = \left[\sum_i q_u \frac{\partial q_i}{\partial \Lambda_v} \frac{w_{it}^+ - q_t w_{it}^-}{D_t} \right] + \frac{q_u}{D_t} \delta_{(t=v)}. \tag{10}$$

In vector-matrix form:

$$q_u \frac{\partial \overline{q}}{\partial \Lambda_v} = q_u \frac{\partial \overline{q}}{\partial \Lambda_v} \mathbf{W} + q_u \overline{\gamma}^+(v) \tag{11}$$

and, from Eq. 7, only two possibilities exist. Either:

$$q_u \frac{\partial q_t}{\partial \Lambda_v} = \frac{\partial q_t}{\partial w_{uv}^+} \tag{12}$$

or $(\mathbf{I} - \mathbf{W})$ is singular, which would also prevent the gradient calculation. A similar analysis on the updates for weights w_{uv}^- yields

$$-q_u q_v \frac{\partial q_t}{\partial \Lambda_v} = \frac{\partial q_t}{\partial w_{uv}^-} . \tag{13}$$

The relevance of Eq. 12 is that it allows a very direct relationship between Hebbian contrastive learning and gradient-descent, if we consider the following rule

$$\Delta w_{uv}^+ = q_u (q_t^f - q_t) , \tag{14}$$

where the forced condition q_t^f is obtained by introducing an input $\Lambda_v = \eta(y - q_t)$ on neuron v.

Because η is a small step, we write

$$q_u (q_t^f - q_t) \approx q_u \eta (y - q_t) \frac{\partial q_t}{\partial \Lambda_v} = \eta (y - q_t) \frac{\partial q_t}{\partial w_{uv}^+} , \tag{15}$$

and thus a small step in the direction prescribed by the contrastive rule equals the gradient-learning direction to first order. For each neuron v, then, a signal proportional to the error is added to the input Λ_v, and the new state for the visible neuron q_t^f is used to adapt w_{uv}.

Update of weight w_{uv} can now be performed with information about the state of neuron u, which is locally available, and the activity of the output neuron. This rule is more adequate, then, for a distributed implementation in hardware than the original rule, where the matrix inversion is clearly requires information about all the network connectivity.

4 Empirical Demonstration and Conclusions

As a first empirical test, we performed multiple learning trials of the XOR problem for two algorithms: the modified gradient-descent algorithms for the case of constant rates and constrained weights, and the contrastive learning rule.

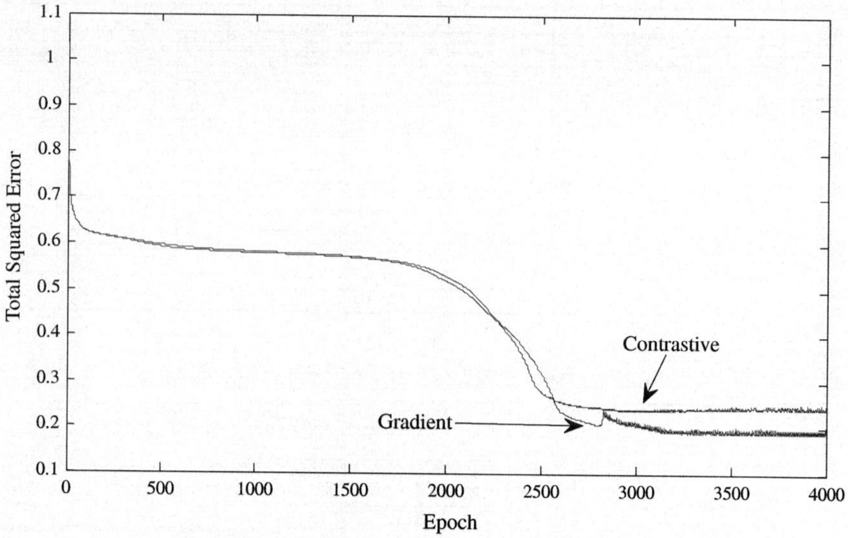

Fig. 1 Evolution of total squared error during supervised learning of the XOR problem

The network has an arbitrary number of seven neurons, four of which receive external inputs (two for each). Connectivity is to be freely learned among the neurons, including recurrent interconnection, but no self-inhibition or self-excitation is allowed. An adaptive bias is present for every neuron, implemented as a connection to a fictitious unit whose q is fixed at 1.

The external input rates were fixed at 0.9 for the higher logical input state and 0.0 for the lower. For the sake of classification, a high logical value was inferred at the neuron output when $q_t > 0.5$. At the end of all learning trials, the correct classification of all 4 input patterns was obtained. Initial conditions were the same for both algorithms in each run, with weights drawn from a uniform distribution with values between 0 and 0.1.

A total of 20 runs were performed. The average number of epochs to get perfect classification was 2760 for the gradient-descent algorithm and 2375 for the contrastive learning algorithm. The difference in averages was not statistically significant ($p = 0.59$). Fig. 1 shows error evolution during one example learning trial for both algorithms. It should be noted that gradient-descent can result in non-monotonic error decrease during learning, depending on the choice for the learning rate.

In conclusion, the idea of Contrastive Hebbian Learning, applied in the framework of Random Neural Networks, yields an algorithm that is equivalent, to first order, to gradient-descent and that does not require matrix inversions. We have also shown that it is possible to keep spikes rates constant during the learning process, through a different interpretation of probability normalization in the model. This simplifies the learning equations.

To address the main problem of the proposed algorithm, which is its need for propagating multiple input perturbations to calculate the effect on the output neuron, we are investigating whether this information can be obtained from a single propagation of information on an adjoint network of back-propagating connections, thus improving on the algorithm's computational cost.

References

1. Baldi, P., Pineda, F.: Contrastive learning and neural oscillations. Neural Comput. **3**, 526–545 (1991)
2. Cooper, S.: Donald O. Hebb's synapse and learning rule: a history and commentary. Neurosci. Biobehav. Rev. **28**, 851–874 (2005)
3. Gelenbe, E.: Random neural networks with negative and positive signals and product form solution. Neural Comput. **1**, 502–510 (1989)
4. Gelenbe, E.: Learning in the recurrent random neural network. Neural Comput. **5**, 154–164 (1993)
5. Gelenbe, E., Hussain, K.: Learning in the multiple class recurrent random neural network. IEEE Trans. Neural Netw. **13**(6), 1257–1267 (2002)
6. Hinton, G., Sejnowski, T.: Learning and relearning in Boltzmann machines. In: Parallel Distributed Processing: Explorations in the Microstructure of Cognition, vol. **1**. The MIT Press, Cambridge (1986)
7. Movellani, J.: Contrastive Hebbian learning in the continuous Hopfield model. In: Proceedings of the 1990 Connectionist Models Summer School, pp. 10–17, Morgan Kaufmann, San Mateo, USA (1990)
8. Rumelhart, D., Hinton, G., Williams, R.: Learning internal representations by error propagation. In: Parallel Distributed Processing: Explorations in the Microstructure of Cognition, vol. **1**. The MIT Press, Cambridge (1986)
9. Timotheau, S.: The random neural network: A survey. Comput. J. **53**(3), 251–267 (2010)
10. Xie, X., Seung, H.: Equivalence of backpropagation and contrastive Hebbian learning in a layered network. Neural Comput. **15**, 441–454 (2003)

A Genetic Algorithms Based Classifier for Object Classification in Images

Turgay Yilmaz, Yakup Yildirim and Adnan Yazici

Abstract Increase in the use of digital images has shown the need for modeling and querying the semantic content, which is usually defined using the objects in the images. In this paper, a Genetic Algorithm (GA) based object classification mechanism is developed for extracting the content of images. Objects are defined by using the Best Representative and Discriminative Feature (BRDF) model, where features are MPEG-7 descriptors. The classifier improves itself in time, with the genetic operations of GA.

1 Introduction

The content of an image can basically be defined as objects and the interaction of these objects. In order to define the content reliably, the categories of the objects should be known. In this context, this study attacks the problem of classification of objects which are extracted from images and develops a Genetic Algorithms (GA) based classifier.

In the literature, studies dealing with objects in images performs either whole image classification or object localization [1, 2]. These studies usually do not

This work is supported in part by a research grant from TÜBİTAK EEEAG with grant number 109E014.

T. Yilmaz · Y. Yildirim · A. Yazici (✉)
Department of Computer Engineering, Middle East Technical University, Ankara, Turkey
e-mail: yazici@ceng.metu.edu.tr

T. Yilmaz
e-mail: turgay@ceng.metu.edu.tr

E. Gelenbe et al. (eds.), *Computer and Information Sciences II*,
DOI: 10.1007/978-1-4471-2155-8_66,

directly deal with the objects, or perform object segmentation and object classification tasks together. Different from the these studies, we propose to separate the object segmentation and classification processes. Keeping in mind that object classification appears more favourable than the object extraction/segmentation process in order to reach the goal of understanding the semantic content; we focus on the object classification problem in this study.

For modelling the objects in images, visual features are the main components. A lot of research has been done in comparing, combining and creating features on different domains, and many different methodologies have been applied. Several studies tend to make improvements on the current features by combining more than one feature [3], making changes in the calculations [4, 5] or making corrections to account for distortions and noise [6] rather than by proposing brand new features. In this study, creating a high-level model that can combine any feature in the literature by means of a GA based classifier is preferred. In addition, the combined features are used in the classifier in a way that each class can be represented with different features. Such ability is performed via the Best Representative and Discriminative Feature (BRDF) model.

In the literature, GA approaches are used in different levels/phases of object/image retrieval: for the spatial segmentation of videos [7], for selecting representative frames of video [8], for object localization in the frame [9], for feature selection [10], for segmentation threshold selection [11], etc. In this study, GA based classifier performs selection of representative images of classes and representative features of images. To the best of our knowledge, no study uses GA for such purposes.

In the developed GA based classifier, object categories are defined with (BRDF) model, where features are MPEG-7 descriptors [12]. The BRDF model developed in this study defines different feature sets that are specific to the categories. During categorization, the decisions of the classifier are calculated using these features and the BRDF model. Due to the nature of GA, the classifier improves itself in time by using genetic operations. Also, the system makes multiple categorization and gives fuzzy decisions on the objects.

2 Our Approach for Object Classification

Two important ideas are considered for the object classification: the BRDF model, to define objects with their features; and a Genetic Algorithm based approach, to classify candidate objects from images. The basic idea is to construct a GA-based classifier that contains representative objects that are modelled with a BRDF mechanism.

The developed system is a supervised learning system and requires a training phase before it is possible to make queries on it. To use the GA approach and genetic operations effectively, the training phase is divided into two parts: one training phase for obtaining the definitions of the training objects (First-Training) and one for making improvements on the definitions by using genetic operations (Second-Training).

2.1 The BRDF Model

The BRDF model is built on the idea that different object types can be represented by different visual features. Using the same features for different domains and types of objects yields unsatisfactory results. For example, shape features are more important than color features for a 'car' object whereas a 'sea' object can be defined with color and texture features.

In [13], Uysal et al. use a single different feature for each object class, which maximizes the correct match in a training set. In this study, we propose to assign weights to the features for each object class. Thus, different from [13], multiple representative and discriminative features can be used to define each class.

2.2 Genetic Algorithms Based Classifier

The core of the genetic algorithm approach is the idea of "survival of the fittest". Using crossovers and mutations during the reproduction will increase the variety and the fitness of offspring solutions. The classifier is given in separate subsections for representation, initialization, fitness function and genetic operations, with the notion of GA. Due to space limitation, the formalism and the algorithms are not included, instead adequate descriptions are given.

Representation and decision-making. Considering a multiple categorization approach, the problem is classifying a candidate object into some categories. In the classifier, each class is represented with a set of chromosomes which are representative images of the object category. Also, each chromosome is represented with a set of genes which are the representative features for that category, in correspondence with the BRDF model.

For the decision making, the genes of each chromosome determine the decision of the chromosome and all chromosomes determine the final decision of a class. The classifier consists of multiple classes and gives a decision on the membership of a candidate object for each of the contained classes. During such calculation, decision of a gene bases on the similarity of candidate object with a representative object in the decision-giving category for a particular feature. Decisions of genes in a chromosome are combined by applying a weighted sum, where the weights are the BRDF indices for the decision-giving category. Decisions of chromosomes are also combined by using a weighted sum in order to obtain the decision of the class. In such calculation, the weights of the chromosomes are the effectiveness values of chromosomes that are updated during the genetic operations according to the fitness of the chromosomes. Initially, effectiveness value of all chromosomes are equal to 1.

Initialization of GA. The initial population is generated by using the First-Training data. In other words, the classes in the above described representation gains chromosomes by identifying new objects in the First-Training phase.

Use of Genetic operations. The genetic operations are applied to the system during Second-Training. For each image obtained during the Second-Training, firstly the decisions of all categories are calculated. Also, fitness of all chromosomes on all categories are computed by using the *Fitness Function.* Then, *Effectiveness Correction*, *Crossover* and *Mutation* operations are performed on the chromosomes of all categories. Thus, each operation is performed for the number of images in the Second-Training.

Fitness Function. The fitness function is used to understand how much a given decision is fit. As the ground truth for the fitness function, the Second-Training data is used. If the class of the chromosome is the same with the candidate object and the calculated chromosomes decision value is larger than a predefined threshold, then the chromosome decision is assigned as the fitness value. Otherwise the fitness value is 0.

Effectiveness correction. To strengthen the mechanism of "survival of the fittest", the effectiveness value of the fit chromosomes is increased by a factor, otherwise it is decreased. This is called "Effectiveness Correction". In the long run, by increasing the effectiveness of the fittest chromosomes and decreasing the effectiveness of others, the chromosomes which vote for incorrect classification lose their existence and fitting chromosomes become more effective on the resulting $Decision_C(S)$ of the class.

Crossovers. During crossover, the mating chromosomes are selected by using a roulette-wheel selection and the probability for participating in mating is directly-proportional with the fitness value of the chromosome. Firstly, an new empty set of chromosomes is created. Until reaching the original number of chromosomes, two chromosomes are selected from the original chromosomes set, gene interchange is performed between them and two new chromosomes are obtained. Effectiveness values of the new chromosomes are assigned as the average of the effectiveness values of the two original chromosomes. Then, fitness of the two new chromosomes are computed. If both of them are better than the original chromosomes, both are included in the new chromosome set, otherwise the best of the four chromosomes (two original, two new) is included only. After filling the chromosome set sufficiently, decision of the new set and the original set are compared. If new set achieves a better decision result, it is replaced with the original set, otherwise discarded.

By using the Crossover Algorithm, it can be stated that the chromosomes which vote for incorrect classification can neither mate nor effect the resulting decision much, and lose their existence in the long run.

Mutation. Different from the traditional GA studies, mutation process is used for learning new information from newly encountered images during the Second-Training. Mutation process is applied similarly as the Crossover. Since each image obtained in the First-Training is imported into the system as a new chromosome, the images in the Second-Training can be thought of outer-chromosomes. From this point of view, the mutation can be assumed to be a crossover between an outer-chromosome and a selected (inner) chromosome. In addition, mutation has a special parameter (mutation factor) for determining interchanging genes. Using this parameter, at which ratio the features of the image are imported into the chromosome is defined.

3 Empirical Study

During the tests, 8 visual features of MPEG-7 [12] in three types are used: Color descriptors (Color Layout, Color Structure, Dominant Color, Scalable Color), Shape descriptors (Contour Shape, Region Shape) and Texture descriptors (Edge Histogram, Homogeneous Texture). As the dataset, the CalTech 101 image dataset [14] is used. It contains pictures of objects belonging to 101 categories. The dataset is divided into three to form First-Training, Second-Training and Test datasets. The maximum of number of images for each one is determined as 30. During First-Training, all 101 categories are used and introduced to the system. For Second-Training and Test. images from 10 categories are selected randomly.

The parameters required for genetic operations are determined as follows: Fitness function threshold is used as the average decision of all chromosomes in each category, in order to accept the chromosomes which are better than the average. The effectiveness correction factor is determined according to the number of images used for each category in the First-Training (1/*number of images in category*). The mutation factor is chosen as 0.5 in order to import half of the genes in each mutation.

In addition to the training data and parameters, a BRDF model based on the expertise knowledge on the image categories is also provided to the system.

3.1 Performance Measurement and Evaluation

During the tests, the retrieval accuracies are measured in five steps: After encountering all First-Training images (S_0), $\frac{1}{4}$, $\frac{2}{4}$, $\frac{3}{4}$ and all of Second-Training images (S_1, S_2, S_3, S_4 respectively).

Figure 1 displays the change of the *ANMRR* and average recall values during test steps. Considering the change between S_0 and S_4 in Fig. 1, all *ANMRR* values decrease. The decreasing values of ANMRR represent improving performance. The improvement on the First-Training set is approximately 34.8%, on the Second-Training set is 58.4% and on the Test set is 46.6%. It is normal to obtain a greater increase in Second-Training, but obtaining an increase in the First-Training and Test sets can be accepted as a success of the GA methodology. Thus, the positive effect of GA is obviously seen. Similar to the evaluation of ANMRR graph, the values in the recall graph tend to increase during the steps. Such increase shows the positive effect of GA methodology.

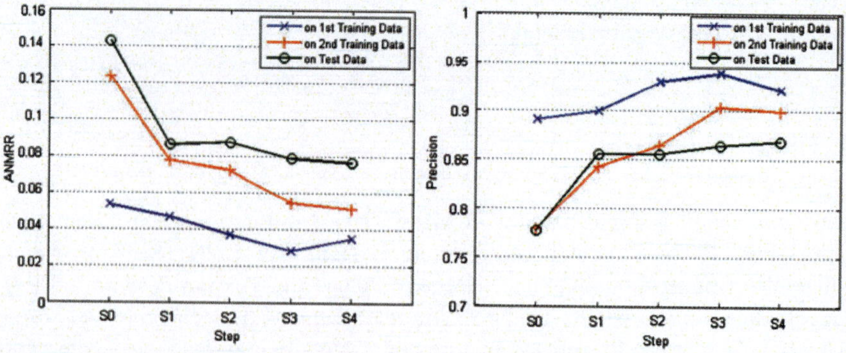

Fig. 1 ANMRR and recall values

Another important evaluation is the to check whether the performance when using multiple features together is better than the performance of each MPEG-7 descriptor individually. In [15], Manjunath et al. calculates the ANMRR values of different color and texture descriptors of MPEG-7. In [15], intervals for ANMRR values of different MPEG-7 descriptors are given as follows: Scalable Color in [0.05, 0.1], Dominant Color in [0.197, 0.252], Color Layout in [0.15, 0.20], Color Structure in [0.046, 0.105] and Edge Histogram in [0.28, 0.36]. The ANMRR values obtained in this study (Fig. 1) are 0.0349, 0.0513 and 0.0764 on First-Training, Second-Training and Test images respectively. Thus, the test results show that combining features is a superior approach than using a single feature.

4 Conclusion

In this study, a GA based object classification mechanism has been developed. Objects were defined with the BRDF model which contains MPEG-7 descriptors, and the classifier makes decisions by using these features and the BRDF model. By using genetic operations of GA, the classifier improves itself in time. Throughout the experiments, the proposed system has achieved much better performances compared to other approaches using single MPEG-7 descriptors as feature and studies utilizing the selection of best single representative feature. Furthermore, the test results have clearly shown the positive effect of the GA model.

References

1. Ferrari, V., Fevrier, L., Jurie, F., Schmid, C.: Groups of adjacent contour segments for object detection. IEEE TPAMI, **30**(1), 36–51 (2008)
2. Datta, R., Li, J., Wang, J.Z.: Content-based image retrieval: approaches and trends of the new age. In: MIR '05, pp. 253–262, ACM, New York, NY, USA (2005)

3. Hadjidemetriou, E., Grossberg, M.D., Nayar, S.K.: Multiresolution histograms and their use for recognition. IEEE TPAMI **26**(7), 831–847 (2004)
4. Bartolini, I., Patella, M.: Warp: accurate retrieval of shapes using phase of fourier descriptors and time warping distance. IEEE TPAMI **27**(1), 142–147 (2005)
5. Latecki, L.J., Lakämper, R.: Shape similarity measure based on correspondence of visual parts. IEEE TPAMI, **22**(10), 1185–1190 (2000)
6. Petrakis, E.G.M., Diplaros, A., Milios, E.: Matching and retrieval of distorted and occluded shapes using dynamic programming. IEEE TPAMI, **24**(11), 1501–1516 (2002)
7. Kim, E.Y., Jung, K.: Object detection and removal using genetic algorithms. In: PRICAI, pp. 411–421 (2004)
8. Doulamis, A., Avrithis, Y., Doulamis, N., Kollias, S.: A genetic algorithm for efficient video content representation. In: Proc. IMACS/IFAC (1998)
9. Swets, D.L., Punch, B., Weng, J.: Genetic algorithms for object recognition in a complex scene. In: Proceedings of the International Conference on Image Processing, Washington DC, vol. 2, 595–598 (1995)
10. Oh, I.S., Lee, J.S., Moon, B.R.: Hybrid genetic algorithms for feature selection. IEEE Trans. Pattern Anal. Mach. Intell. **26**(11), 1424–1437 (2004)
11. Kanungo, P., Nanda, P.K., Samal, U.C.: Image segmentation using thresholding and genetic algorithm. In: Proceedings of SCT 2006, NIT, Rourkela, India (2006)
12. Martínez, J.: Mpeg-7 overview (version 10). Requirements ISO/IEC JTC1 /SC29 /WG11 N6828, International Organisation For Standardisation (2003)
13. Uysal, M., Yarman-Vural, F.T.: Selection of the best representative feature and membership assignment for content-based fuzzy image database. In: CIVR, pp. 141–151 (2003)
14. Fei-Fei, L., Fergus, R., Fergus, R., Fergus, R.: One-shot learning of object categories. IEEE TPAMI, **28**(4), 594 (2006)
15. Manjunath, B.S., Ohm, J.R., Vasudevan, V.V., Yamada, A.: Color and texture descriptors. IEEE Trans. Circuits Syst. Video Technol. **11**(6), 703–715 (2001)

Two Alternate Methods for Information Retrieval from Turkish Radiology Reports

Kerem Hadımlı and Meltem Turhan Yöndem

Abstract Turkish is an highly agglutinative language and this poses problems in information retrieval from Turkish free-texts. In this paper one rule based and one data driven alternate methods for information retrieval from Turkish radiology reports are presented. Both methods do not utilize any medical lexicon or ontology, although inherent medical information within the training reports exist. Performance is measured and evaluated as a retrieval problem.

1 Introduction

Radiology departments work by having a radiologist examine images and describe his/her findings into full-text reports. Extracting structured information from these reports requires knowledge on the field and the language. A vast amount of medical information is contained and shared knowledge between writer and reader is assumed [10]. In spite of this assumption, it has come to our attention that a person with no medical knowledge can still understand the basic relations within Turkish report sentences; even if a small portion of these relations are incorrect due to insufficient medical information.

We propose two methods for retrieval from Turkish radiology reports. Both methods are shaped around the idea of using only Turkish grammar. No medical lexicon or ontologies are used. Although both methods can be used in an extraction scenario, the context of this paper is limited to information retrieval.

Turkish is characterized by a rich agglutinative morphology, free constituent order, and predominantly head-final dependencies [4]. Word structures in Turkish are formed by productive affixations of derivational and inflectional suffixes [7].

K. Hadımlı (✉) · M. T. Yöndem
Department of Computer Engineering, Middle East Technical University,
Ankara, Turkey
e-mail: kerem@ceng.metu.edu.tr

E. Gelenbe et al. (eds.), *Computer and Information Sciences II*,
DOI: 10.1007/978-1-4471-2155-8_67, © Springer-Verlag London Limited 2012

Hemaperitonium ,	mesane	lümen i nde	hematom	ve hava.
Hemoperitoneum ,	urinary-bladder	cavity+ACC+LOC	hematoma	and air.
Hematoma and air in urinary bladder's cavity, hemoperitoneum.				

Fig. 1 A sample report sentence. Note that the locative suffix, *-de*, might extend to *hemoperitoneum* if it was an anatomical location

Morphological features partly determine syntactic relations within a sentence; but nearly half of the words in a sentence are morphologically ambiguous. Moreover, only the final POS tag is not sufficient to completely analyze a sentence; a word's earlier inflectional groups may take part in syntactic relations [8, 9, 11].

Derivational boundaries (DBs) are used to separate inflectional groups (IGs). An IG consists of a single POS tag and some inflectional features. Separation points of IGs are determined by derivational morphemes. In designing the Turkish treebank [9], Oflazer et al. use IGs as the elementary unit between dependency links. Eryiğit and Oflazer [3] and Eryiğit et al. [4] continue using this approach for dependency parsing. Dependency links always emanate from the last IG of a modifier word and land on any IG of head word.

Same medical terms might be spelled differently in different reports as there is no consensus on Turkish spelling of most medical terms. Also, agglutinative nature of Turkish requires morphological analyses of medical terms as well as common words. Correctly identifying and disambiguating alternative morphological analyses is a requirement. (Fig. 1)

2 Rule-Based Method

Our first method uses only local templates and rules. These templates are handcrafted by examining a set of sample reports with a variety of topics.

The algorithm extracts a set of relations by analyzing surface structures. Each relation describes a unique meaning enclosed in the sentence. Relations are mostly between two phrases, but single-phrase or multi-phrase relations are also possible. These relations will be referred as *individual meanings*.

The algorithm is applied in two phases. First phase works on surface forms and morphological analyses, and requires correct morpheme identification. Only the morphemes of last IG are necessary, so a simple morphological analyzer is deemed sufficient. Second phase works on local *individual meanings*, and by use of rules generates non-local ones.

First Phase Morphological analysis is performed using Zemberek library [1]. In our case, morphological analyzer is vulnerable to medical terms that do not exist in its lexicon. To solve this problem, morphemes are made optional in the templates, and an alternative form with common surface representation is created for most of the rules. Use of a morphological analyzer provides us with a priority selection. Templates with surface forms are applied only if the morphological analyzer fails.

Table 1 Sample templates for some individual meanings

Type	Template			Matches to
LOC	..*(a)*..	+LOC	..*(b)*..	lümen+POSS3SG+LOC hematom
	..*(a)*..	+"de"	..*(b)*..	lümeninde hematom
CONJ	..*(c)*..	,	..*(d)*..	hematom , hava
	..*(c)*..	ve	..*(d)*..	hematom ve hava

	Hemaperitonium	,	mesane lümenin	de	hematom	ve	hava
CONJ	....	,	....				
LOC			....	de	....		
CONJ					....	ve	......
	..........(I)............	,	...........(II)..........	de	...(III)...	ve	..(IV)..

Fig. 2 Application of first phase templates. Matching keywords delimit phrases

Table 2 Set of individual meanings after second phase

	Type	Parameter 1	Parameter 2
After	Conjunction	*(left)* Hemaperitonium	*(right)* mesane lümen
1st phase	Location	*(where)* mesane lümen	*(what)* hematom
	Conjunction	*(left)* hematom	*(right)* hava
After	Location	*(where)* mesane lümen	*(what)* hava
2nd phase	Location*	*(where)* Hemaperitonium	*(what)* hematom
	Location*	*(where)* Hemaperitonium	*(what)* hava

Templates match to the target sentence locally, generating an initial set of local *individual meanings*. These describe the relations between consecutive phrases. Boundaries of phrases are determined using keywords of matched templates as delimeters. Sample templates are given in Table 1. If morphological analyzer is successful first template, otherwise second template will match. Note that the blank areas are labeled. Matching keywords are underlined.

After all possible templates align with the input sentence, matching keywords form the delimiters between phrases. This assures that medical terms consisting of multiple words can be grouped together as a single phrase (Fig. 2). Results of template matching on the sample sentence are shown in first part of Table 2.

Second Phase Second phase of our algorithm works on output of first phase. These rules govern the expansion of the initial set with *individual meanings* which exist between nonconsequtive phrases in the sentence.

Rules of second phase specify two input *individual meanings*, and a resultant *individual meaning*. In each iteration, any two *individual meanings* are selected from the input set, *meaning-A* and *meaning-B*. If both *A* and *B* satisfy the rule, then a third *individual meaning*, *meaning-C* is instantiated and added to the set. This process continues until no more expansions are possible.

Resulting set of *individual meanings* after second phase are given in Table 1. Last two *individual meanings* (marked with *) are deduced incorrectly due to missing medical information (*"Hemaperitonium"* is not an anatomical location).

3 Data Driven Method

In order to overcome the problems associated with hand-coded rules, a data driven method is devised. This second method is constructed around dependency links.

Data driven dependency parsing of Turkish is first researched by Eryiğit and Oflazer [3] due to the recent availability of Turkish Treebank [9]. In 2008, Eryiğit et al. [4] developed a classifier based parser for Turkish inside the MaltParser framework [5] and applied their work in CoNLL-X shared task [6].

TRMorph, a freely available two-level morphological analyzer [2] is chosen for morphological analysis. TRMorph is based on a finite state transducer framework, and is able to generate much more detailed morphological analyses than Zemberek. This was a requirement for dependency parsing.

Morphological disambiguation was another requirement. In Turkish, morphemes have a very important role on syntactic relations. Due to the relatively small training set size, purely statistical methods for disambiguation would be inappropriate. Yuret and Türe [11] implement a rule learner for disambiguation of Turkish morphemes, in which a different decision list is learned per morpheme based on surface forms of words. Probabilities of alternative morphological analyses can be calculated and compared using this information. Yuret and Türe also point that as the decision lists can be used as oracles, the disambiguator can be used as a simple morphological analyzer. They achieve fairly good results for this experiment in their dataset. We had to train our own model for disambiguation. The published model was incompatible with TRMorph's output, and our radiology report sentences were very different than the training set of the published model.

MaltParser is used for training an SVM based model. We used the same parameters used in CoNLL-X and CoNNL-07 tasks for Turkish. Public availability of these parameters provided us with the option of training the parser on our own training set. We did not attempt optimizing these parameters because of our small training set size compared to CoNNL tasks.

Training Set Only abdominal reports are selected for creating the training set. Narrowing down the topic is an important requirement because of the need for a sufficiently comprehensive training set. Training set consists of 10 abdominal reports (written by different doctors), 182 sentences and 1,415 words. There are 136 unique sentences and 462 unique morphological analyses.

Extracting a set for use in retrieval Input sentences are parsed for dependency graph. Using this graph, a set of relations between individual words are generated. For each word tuple (w_i, w_j), if there is a directed path from w_i to w_j, tuple (w_i, w_j) is marked as related. There can be other words $w_k, k \neq i, j$ in the path. These

Table 3 Evaluation results

	Method	Match target	Threshold	Recall (%)	Precision (%)	F_1 Score (%)
Set-1	Baseline	Whole report	Single match	96.05	5.15	9.78
	Baseline	Whole report	50	96.05	5.42	10.26
	Baseline	Single sentence	50	96.05	5.46	10.33
	Rule Based	Whole report	Single match	67.10	28.65	40.15
	Rule Based	Whole report	50	65.78	31.84	42.91
	Rule Based	Single sentence	50	65.78	31.84	42.91
Set-2	Rule based	Whole report	Single match	71	66	68
	Dependency based	Whole report	Single match	89	62	73
	Dependency based	Whole report	50	78	81	79

relations are asymmetric. This set of relations may be seen as analogous to the *set of individual meanings* of the first method. Dependency link labels are ignored.

4 Evaluation

In both methods, search operation is performed by set intersection. Search queries are free text, containing 2–4 words. They may be seen as partial sentences, written in similar style with original reports. One of the methods is applied on query and set of relations are extracted. Set intersection is performed with previously extracted sets. For calculation of intersection set, relation comparisons are performed on root lexemes. Intersection can be performed on a single sentence's set, or union of all sets of a report. A threshold of intersection set size vs. query set size is used to decide on which reports to return.

We used two test sets. First test set consists of radiological reports on different topics. There are 53 reports and 100 queries in this set, forming 5,300 data points. Because the set contains multiple topics, only the first method can be measured. Results are compared to a baseline retrieval algorithm which uses individual words as set elements.

Second test set consists of only abdominal radiological reports. There are 10 reports and 28 queries. In this test set, the performance of first and second methods are compared. In second set, there are less words per query than in first set. This is the reason more detailed thresholds can not be tested on second dataset for rule based method.

The baseline method resulted in very high recall and very low precision. The reason may be found in content of radiology reports. Almost all query words exist in all reports. Methods that favor relations result in more balanced recall and precision. Threshold had little effect on rule based method. As the number of words are low in query strings, rule based method generated only 1 or 2 relations per query. (Table 3)

5 Conclusion and Future Work

In this paper we tried to tackle the problem of information retrieval from Turkish radiology reports. Both Turkish language's properties and contents of radiology reports posed very different problems than western counterparts. We've shown a rule based algoritm, an analogical dependency link based data driven method, and comparison of these with a baseline search engine that checks only if the words exists in the reports.

The starting point of our work lies in the observation that a person with no medical knowledge can understand relations in Turkish radiology reports, even if partially. We deliberately avoided using any medical ontology or lexicon. Although current work is limited to information retrieval, we plan to extend this work to information extraction, still with a focus on avoiding medical lexicon and ontologies.

Acknowledgments This study is supported by Turkish Scientific and Technological Research Council (TÜBİTAK) project 3080179.

References

1. Akin, A.A., Akin, M.D.: Zemberek, an open source NLP framework for Turkic Languages (2007)
2. Çöltekin, Ç.: A freely available morphological analyzer for Turkish. In: Proceedings of the Seventh Conference on International Language Resources and Evaluation (LREC'10). European Language Resources Association (ELRA), Valletta, Malta (May 2010)
3. Eryiğit, G., Oflazer, K.: Statistical dependency parsing of Turkish. In: Proceedings of the 11th EACL. pp. 89–96. Trento (3–7 April 2006)
4. Eryiğit, G., Nivre, J., Oflazer, K.: Dependency parsing of Turkish. Comput. Ling. **34**(3), 357–389 (2008)
5. Nivre, J., Hall, J., Nilsson, J., Chanev, A., Eryiğit, G., Kübler, S., Marinov, S., Marsi, E.: Maltparser: a language-independent system for data-driven dependency parsing. Nat. Lang. Eng. J. **13**(2), 99–135 (2007)
6. Nivre, J., Hall, J., Nilsson, J., Eryiğit, G., Marinov, S.: Labeled pseudo-projective dependency parsing with support vector machines. In: Proceedings of the 10th Conference on Computational Natural Language Learning (CoNLL-X). pp. 221–225. New York (8–9 June 2006)
7. Oflazer, K.: Two-level description of Turkish morphology. In: Proceedings of the sixth conference on European chapter of the Association for Computational Linguistics. pp. 472–472. EACL '93, Association for Computational Linguistics, Stroudsburg, PA, USA. http://dx.doi.org/10.3115/976744.976810 (1993)
8. Oflazer, K., Hakkani-Tür, D.Z., Tür, G.: Corpus annotation for parser evaluation. In: Proceedings of the EACL workshop on Linguistically Interpreted Corpora (LINC), pp. 35–41 (1999)
9. Oflazer, K., Say, B., Hakkani-Tür, D.Z., Tür, G.: Building a Turkish Treebank (2003)
10. Taira, R.K., Soderland, S.G., Jakobovits, R.M.: Automatic structuring of radiology free-text reports. Radiographics **21**(1), 237–245 (2001). http://radiographics.rsna.org/content/21/1/237.abstract
11. Yuret, D., Türe, F.: Learning morphological disambiguation rules for Turkish. In: Proceedings of HLT-NAACL (2006)

Scaled On-line Unsupervised Learning Algorithm for a SOM-HMM Hybrid

Christos Ferles, Georgios Siolas and Andreas Stafylopatis

Abstract A hybrid approach combining the Self-Organizing Map (SOM) and the Hidden Markov Model (HMM) is presented. The fusion and synergy of the SOM unsupervised training and the HMM dynamic programming algorithms bring forth a scaled on-line gradient descent unsupervised learning algorithm.

1 Introduction

The present approach analyzes the Self-Organizing Hidden Markov Model Map (SOHMMM). Moreover, a novel scaled on-line unsupervised learning algorithm which is in position to model sequences of arbitrary large lengths, is devised. Since the SOHMMM carries out probabilistic sequence analysis with little or no prior knowledge, it can have a variety of applications in clustering, dimensionality reduction and visualization of large-scale sequence spaces, and also, in sequence discrimination, search and classification.

In the literature, there is a number of approaches based on scenarios involving SOM [1] and HMM modules [2–4]. For instance in [5–7] the training procedures of the two subunits are nearly always disjoint and are carried out independently. All proposed methods can be divided into two classes. According to the first approach a SOM is used as a front-end processor (e.g. vector quantization, preprocessing, feature extraction) for the inputs of a HMM, whereas the second

C. Ferles (✉) · G. Siolas · A. Stafylopatis
Intelligent Systems Laboratory, School of Electrical and Computer Engineering,
National Technical University of Athens, 15780 Zografou, Athens, Greece
e-mail: christos.ferles@gmail.com

G. Siolas
e-mail: gsiolas@islab.ntua.gr

A. Stafylopatis
e-mail: andreas@cs.ntua.gr

E. Gelenbe et al. (eds.), *Computer and Information Sciences II*,
DOI: 10.1007/978-1-4471-2155-8_68,

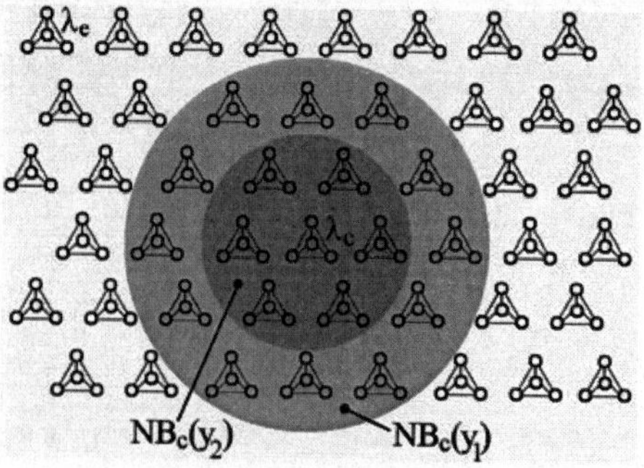

Fig. 1 Paradigm of a SOHMMM lattice and an example of topological neighborhood (y1 < y2)

approach places the SOM on top of the HMM. The SOM(s) are associated with different states (or different groups of states), and subsequently, they are linked via one or several hidden layers (in a multi-layer fashion).

2 The Self-Organizing Hidden Markov Model Map

Studies conducted for many years by a large number of researchers have convincingly shown that the best self-organizing results are obtained if the following two partial processes are implemented in their purest forms [1]:

1. decoding of that neuron that has the best match with the input data pattern (the so-called "winner");
2. adaptive improvement of the match in the neighborhood of neurons centered around the "winner."

Consider Fig. 1, where the SOHMMM defines a mapping from the input observation sequence space onto a two-dimensional array of neurons. In total there are E neurons. With every neuron e, a HMM $\lambda_e \in \Lambda$, also called reference HMM, is associated. The lattice type of the array can be defined to be rectangular, hexagonal or even irregular. In the simplest case, an input observation sequence O is connected to all HMMs in parallel. The input O is compared with all the λ_e and the location of the best match, with respect to the likelihood, is defined as the location of the response. Actually, the input is simply mapped onto this location, like in a set of decoders. Thus, the definition of the best matching HMM (indicated by the subscript c), is given by

$$c = \arg \max_{e} \{\log P(O|\lambda_e)\}. \tag{1}$$

We may then claim that the SOHMMM is a nonlinear projection of the input observation sequence onto the two-dimensional display.

The resulting SOHMMM on-line gradient descent unsupervised learning algorithm is an iterative procedure, where the parameters of interest, namely $W^{(e)}$, $R^{(e)}$ and $U^{(e)}$, are adjusted according to the learning rules

$$w_{ij}^{(e)}(y+1) = w_{ij}^{(e)}(y) + \eta(y)h_{ce}(y) \times \left[P(O|\lambda)^{-1} a_{ij} \sum_{l=1}^{T-1} \left[\alpha_l(i) b_j(o_{l+1}) \beta_{l+1}(j) - \alpha_l(i)\beta_l(i) \right] \Bigg|_{\lambda_e, y} \right], \quad 1 \leq i \leq N, 1 \leq j \leq N \tag{2}$$

$$r_{jt}^{(e)}(y+1) = r_{jt}^{(e)}(y) + \eta(y)h_{ce}(y) \times \left[P(O|\lambda)^{-1} \sum_{l=1}^{T} \left[I\{o_l = t|\lambda\} \alpha_l(j)\beta_l(j) - b_j(t)\alpha_l(j)\beta_l(j) \right] \Bigg|_{\lambda_e, y} \right], \quad 1 \leq j \leq N, 1 \leq t \leq M \tag{3}$$

$$u_j^{(e)}(y+1) = u_j^{(e)}(y) + \eta(y)h_{ce}(y) \times \left[\left[P(O|\lambda)^{-1} \pi_j b_j(o_1) \beta_1(j) - \pi_j \right] \Big|_{\lambda_e, y} \right], 1 \leq j \leq N. \tag{4}$$

In the relaxation process, the function $h_{ce}(y)$ has a very central role: it acts as the neighborhood function, a smoothing kernel defined over the lattice points. For convenience, it is necessary that $h_{ce}(y) \rightarrow 0$ when $y \rightarrow \infty$. Usually $h_{ce}(y) = h(\|\delta_c - \delta_e, y\|)$, where δ_c, $\delta_e \in \Re^2$ are the location vectors of HMMs λ_c and λ_e on the array. With increasing $\|\delta_c - \delta_e\|$, $h_{ce} \rightarrow 0$. The width and form of h_{ce} define the stiffness of the elastic surface to be fitted to the input data.

3 Scaled Unsupervised Learning Algorithm

It can be seen that as t starts to get big (e.g. 10 or more) each term of $a_t(i)$ starts to head exponentially to zero. For sufficiently large t (e.g. 100 or more) the dynamic range of the $a_t(i)$ computation will exceed the precision range of essentially any machine, even in double precision. A similar observation can be made for the backward variables $\beta_t(i)$ as t decreases, they also tend to zero exponentially fast. In general, during the implementation of the SOHMMM algorithm, one has to deal with precision issues. Hence, we need a scaling procedure, where forward and backward variables are scaled during the propagation in order to avoid underflow. The scaled forward variable and the scaled backward variable are defined as

$$\hat{\alpha}_t(i) = P(o_1 o_2 \ldots o_t, q_t = s_i|\lambda) / P(o_1 o_2 \ldots o_t|\lambda) \tag{5}$$

$$\hat{\beta}_t(i) = P(o_{t+1} o_{t+2} \ldots o_T | q_t = s_i, \lambda) / P(o_{t+1} o_{t+2} \ldots o_T | o_1 o_2 \ldots o_t, \lambda). \tag{6}$$

A significant effect the scaling procedure has on the SOHMMM regards the on-line gradient descent learning rules. Therefore, an algorithm for adjusting the SOHMMM's parameters, while simultaneously keeping the computation within precision range, can be derived by integrating (1)–(4) and the scaling procedure.

Table 1 Classification results (namely average accuracy percentages) of the SOHMMM, (relational) NG variants, and k-means on the globin protein family

Model	Overall
SOHMMM	94.13 ± 0.47
Supervised Relational Batch NG	90.0
Relational Batch NG	89.9
Supervised Median Batch NG	89.4
Supervised Relational *k*-Means	88.2

4 Comparative Experiments

The results displayed in Table 1 are obtained from a repeated stratified ten-fold cross-validation averaged over 100 repetitions and have been retrieved form [8]. All reported models (except the SOHMMM) are especially designed for handling relational data (such as pairwise similarities/dissimilarities derived from alignments of biological sequences).

A simple examination of the results reveals that the SOHMMM is superior in terms of performance, not only because it yields the highest accuracy, but also, because it achieves this through a direct unsupervised learning methodology, without requiring any kind of additional information. In support of our claims we mention that the second best algorithm, the supervised relational batch NG, exploits both protein class information and predetermined elaborate evolutionary distances (which are based on prior knowledge or domain information). At the same time, the SOHMMM outperforms all approaches by at least 4.1% without utilizing any kind of class information nor any type of elaborate distances (which must be predetermined by experts). The efficiency of the SOHMMM is justified mainly by its capability to tackle nominal sequences systematically according to a proven probabilistic framework. The SOHMMM manages to capture and decipher information hidden in the proteins' amino acid composition, and also, information distributed across the spatial correlations of the protein monomers.

References

1. Kohonen, T.: Self-Organizing Maps, 3rd edn. Springer, Berlin (2001)
2. Koski, T.: Hidden Markov Models for Bioinformatics. Kluwer Academics Publishers, Dordrecht (2001)
3. Baldi, P., Brunak, S.: Bioinformatics: The Machine Learning Approach, 2nd edn. The MIT Press, Cambridge (2001)
4. Mount, D.W.: Bioinformatics: Sequence and Genome Analysis, 2nd edn. Cold Spring Harbor Laboratory Press, New York (2004)
5. Kang, J., Feng, C.-J., Shao, Q., Hu, H.-Y.: Prediction of chatter in machining process based on hybrid SOM-DHMM architecture. In: Proceedings of the 3rd International Conference on Intelligent Computing, pp. 1004–1013 (2007)

6. Rogovschi, N., Lebbah, M., Bennani, Y.: Learning self-organizing mixture markov models. J. Nonlinear Syst. Appl. **1**, 63–71 (2010)
7. Tsuruta, N., Iuchi, H., Sagheer, A., Tobely, T.: Self-organizing feature maps for HMM based lip-reading. In: Proceedings of the 7th International Conference Knowledge-Based Intelligent Information and Engineering Systems, pp. 162–168 (2003)
8. Hammer, B., Hasenfuss, A.: Relational neural gas. In: Proceedings of the 30th Conference on Artificial Intelligence, pp. 190–204 (2007)

Sequential Pattern Knowledge in Multi-Relational Learning

Carlos Abreu Ferreira, João Gama and Vítor Santos Costa

Abstract In this work we present `XMuSer`, a multi-relational framework suitable to explore temporal patterns available in multi-relational databases. `XMuSer`'s main idea consists of exploiting frequent sequence mining, using an efficient and direct method to learn temporal patterns in the form of sequences. Grounded on a coding methodology and on the efficiency of sequence miners, we find the most interesting sequential patterns available and then map these findings into a new table, which encodes the multi-relational timed data using sequential patterns. In the last step of our framework, we use an ILP algorithm to learn a theory on the enlarged relational database that consists on the original multi-relational database and the new sequence relation. We evaluate our framework by addressing three classification problems.

1 Introduction

Multi-relational databases are widely used to represent and store data. A multi-relational database is often composed by a *target* table and by a number of *fact* tables. The target table will represent the main objects of interest (say, patients in a medical domain); fact tables will represent the information being accumulated about the entities in the target table (say, medical visits or drug usage in the medical domain). We expect target tables to be relatively stable or to grow slowly over time; in contrast, fact tables may grow quickly. Moreover, quite often

C. A. Ferreira (✉)
LIAAD-INESC and ISEP, Polytechnic Institute of Porto, Porto, Portugal
e-mail: cgf@isep.ipp.pt

J. Gama
LIAAD-INESC and Faculty of Economics, University of Porto, Porto, Portugal

V. S. Costa
CRACS-INESC and Faculty of Sciences, University of Porto, Porto, Portugal

E. Gelenbe et al. (eds.), *Computer and Information Sciences II*,
DOI: 10.1007/978-1-4471-2155-8_69,

the information stored in fact tables is time-based and consists of *sequences* that reflect the evolution of a phenomenon of interest.

The main goal of this work is to exploit heterogeneous temporal information stored in the multi-relational sequences of events. To do so, we propose the `XMuSer` (eXtended MUlti-relational SEquential patteRn knowledge learning) framework that encodes *timed data*, stored in one or several fact tables, into a separate sequence relation, uses an optimized sequence learner to find the most interesting such sequences, maps back the sequences to the relational database, and then learns a theory on the extended multi-relational database. The extended database thus contains all primitive relations and an additional relation that stores temporal patterns for each example.

This methodology allows us to explore multi-relational datasets that have different types of timed data, either sequence data or time-series data. On the one hand, we can benefit from computationally efficient sequence miners such as PrefixSpan [5] to find the most remarkable sequential patterns. On the other hand, we still have access to the original data and can take advantage of the flexibility of Inductive Logic Programming (ILP) [6] to learn in the extended multi-relational dataset. Indeed, we argue that the first step provides a good insight into the search space, and may enable `XMuSer` to perform better than classical ILP based strategies or than ILP based algorithms that define logical temporal formalisms in order to explore temporal sequences [1]. We observe that the sequence miner and ILP learning algorithm are decoupled: we can use other sequence miners such as SPIRIT [4], that finds a constrained set of sequential patterns.

We experimentally evaluate our methodology with two datasets, addressing three classification problems.

In the following section we introduce and evaluate the performance of the `XMuSer` algorithm. In the last section we present some conclusions.

Algorithm 1. `XMuSer` pseudo-code

---

input a multi-relational database $\mathbf{r}$; three thresholds λ, δ and k
output a classifier model
Data Coding
1: $\mathbf{s} \leftarrow$ SequenceCoding ($\mathbf{r}$)
Frequent Sequential Patterns
2: $\mathbf{s}_1, \ldots, \mathbf{s}_k \leftarrow$ partition ($\mathbf{s}$)
3: **for** $i = 1$ **to** k **do**
4: $\quad sf_i \leftarrow$ SequenceMiner ($\mathbf{s}_i$, λ)
5: **end for**
Filtering
6: $S_{discriminative} \leftarrow$ discriminate($sf_1, \ldots, sf_k$)
7: $S_{interesting} \leftarrow$ chi-square ($S_{discriminative}$, δ, k)
Mapping
8: $\mathbf{r}_{sr} \leftarrow$ Mapping ($S_{interesting}$, $\mathbf{r}_{targetExamples}$)
9: $\mathbf{E_r} \leftarrow \mathbf{r} \cup \mathbf{r}_{sr}$
Learn a Theory with an ILP algorithm
10: ILP Algorithm ($\mathbf{E_r}$)

---

2 The `XMuSer` Algorithm

Our framework has five main steps. In the first phase we code the temporal data into a sequence database. In a second phase, we run a sequence miner to find all sequential patterns in each class partition. In a third phase, we apply two filters to select the most discriminative and class related patterns. In a fourth phase, for each example in the target table, we build a relation where the example is characterized by presence or absence of the most interesting sequential patterns. Last, we learn a theory on the enlarged database, where the enlarged database is the union of the original database with the new *sequence relation*.

Algorithm 1 presents the pseudo-code for `XMuSer`. Next, we explain each one of the major components of our framework. Throughout, we follow an illustrative example, a classification problem, at Fig. 1. The example has three tables registering the follow-up of two patients. One of the tables is the target table, named *Patient*, where each record describes each patient, identified by a masked ID, and registers the class of each patient. The other two tables are fact tables registering timed blood analysis and urinalysis examinations.

Data coding. The algorithm that we present next takes a multi-relational dataset as input, represented as a database of Prolog facts. To explore the richness of this representation, and namely temporal patterns, we introduce a strategy that converts multi-relational timed data into an amenable sequence database. First, we find all relations that have temporal records. Second, we sort the records in these relations by time order. We thus obtain a chronological sequence of multiple events for each example. Figure 2 shows patient one event sequence. The sequence includes a sequence of blood and urine analysis. Third, we build a temporal attribute-value sequence for each example. In this new sequence each itemset corresponds to all records registered at a given date/time. We have `(plt = high) (RBC = high, WBC = normal, alb = normal, ttp = normal) (alb = normal) (RBC = high, WBC = high)` for patient one. We then define a one-to-one coding map $f : \mathit{Attributes} \times \mathit{Values} \longrightarrow \mathbb{N}$. This mapping associates an *unique* number to each attribute-value pair. In the example, we use the map to code the attribute-value sequence into an integer number sequence. The definition of this map is done according to the type of attributes in each database relation. The mapping assumes discrete attributes (continuous attributes will be discretized beforehand). Table 1 shows the transformed sequence database: each sequence tuple corresponds to an example in the target table and each subsequence corresponds to all one-time events. In the example, we define the one-to-one map as $f(rbc, low) = 1, f(rbc, normal) = 2, f(rbc, high) = 3, f(wbc, low) = 4, f(wbc, normal) = 5, f(wbc, high) = 6, f(alb, normal) = 7, f(ttp, normal) = 8, f(plt, high) = 9$ and patient one sequence of events is thus coded as (9) (3 5 7 8) (7) (3 6).

This stage obtains a sequence suitable to be processed by a sequence miner. In Table 1 we present a sequence database registering the coded sequence of patients one and two.

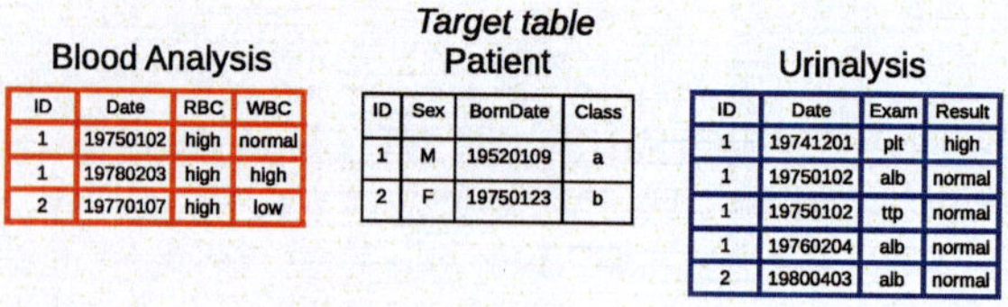
Blood Analysis

ID	Date	RBC	WBC
1	19750102	high	normal
1	19780203	high	high
2	19770107	high	low

Target table
Patient

ID	Sex	BornDate	Class
1	M	19520109	a
2	F	19750123	b

Urinalysis

ID	Date	Exam	Result
1	19741201	plt	high
1	19750102	alb	normal
1	19750102	ttp	normal
1	19760204	alb	normal
2	19800403	alb	normal

Fig. 1 Database relations. ID is the patient ID, date is in numeric format, RBC and WBC are blood parameters, and we show alb, plt and ttp urine exams

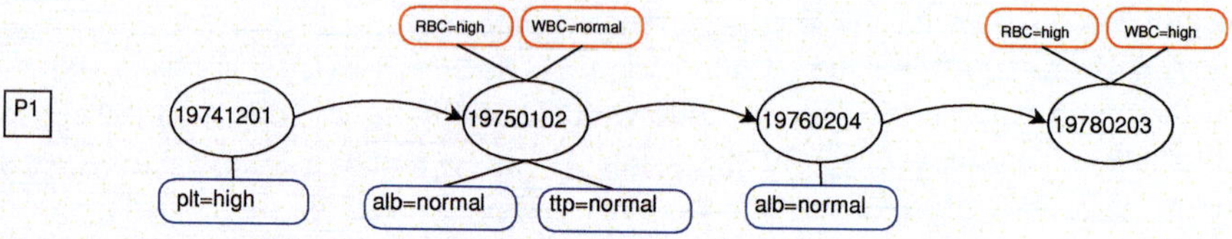

Fig. 2 Temporal patient events for patient one

Table 1 Transformed event database

ID	Sequence	Class
1	(9) (3 5 7 8) (7) (3 6)	a
2	(3 4) (7)	b

Finding Frequent Sequential Patterns. We run a sequence miner in each class/partition of the sequence database in order to find *frequent* sequences, that is, having a support equal or higher than a user defined threshold. Thus, for each partition and each class we obtain all frequent patterns, the sf_i set.

Filtering. The previous step usually obtains a large number of findings. We would like to retain highly discriminative, class correlated, patterns and drop uninteresting and redundant patterns [7]. To do so, we introduce a *Discriminative Filter* that selects the set of discriminative sequential patterns, $S_{discriminative}$, that have support above a user defined value, the λ parameter, in one and only one partition. The filter is implemented by using matching.

Moreover, we use a chi-square filter and the respective δ significance threshold to select at most top-k class-related features. The sequential composition of the two filters thus results in the set of interesting features/sequences that we will use to build the final classification model. For instance, in our example we get two discriminative and class-related features, S_1, or (3 4), and S_8, or (5) (7).

Mapping back interesting sequences. We have coded the timed data as a sequence database. We now want to build a relation where each example is characterized by the most interesting sequences. Table 2 shows the key idea in our approach. We construct a new table, the *sequence relation*, with an entry per example and with $k+1$ attributes: the top-k most interesting features and the example ID. If the sequence associated with the new attribute is a subsequence of the example sequence at the sequence database, the new attribute takes value one. Otherwise, it takes the value zero.

Table 2 Sequence relation

ID	S_1	S_8
1	0	1
2	1	0

Learning a Theory. Last, we add the new *sequence relation* to the initial tables and use an ILP algorithm, such as Aleph [8], to learn a first-order classifier model consisting of a set of clauses. One of the learned clauses is:

```
patient_info(A,B,C,a):- blood_analysis(A,D,high,normal),
urinalysis(A,E,plt,high), sequence_relation(A,0,1).
```

This clause calls the predicate *blood_analysis*, the predicate *urinalysis* and the predicate associated with the sequence relation, predicate *sequence_relation*. The clause explains (or covers) patient number one, a patient from class **a**.

2.1 Experimental Evaluation

We evaluated our algorithm in two real-life datasets obtained from the PKDD data-mining competitions, the Hepatitis dataset and the Financial dataset (see http://lisp.vse.cz/challenge/CURRENT/). We experimented three classification problems: Hepatitis Subtype and Hepatitis Fibrosis Degree, in the Hepatitis dataset, and the prediction of successful loans, in the Financial dataset.

Throughout, we used PrefixSpan algorithm to find all frequent sequences and Aleph algorithm to learn a logical theory. We applied the following predefined parameter configuration: $\delta = 0.10$ and $k = 30$. We test two different λ values, 90 and 80%. For comparison purposes we run a reference algorithm, in this case we run the stand-alone Aleph algorithm, to solve each one of the three problems. We evaluate our framework using a ten-fold cross-validation procedure and compute: the mean generalization accuracy and standard deviation. We also compute the Wilcoxon hypothesis test p-value to measure how significantly our algorithm differs from the reference algorithm. We use the same background knowledge and bias when running the Aleph algorithm, either as the last component of `XMuSer` or as stand-alone. The major difference is that in `XMuSer` we introduce an extra relation, the *sequence relation*. The bias is relatively simple, relying on the predefined tables in the database.

To address these unbalanced multi-relational datasets, recording a large amount of data and where each table has a wide number of attributes, discrete or continuous ones, most previous works performs preprocessing steps, such as balancing the datasets, that makes impossible to make a fair comparison. In Table 3 we present experimental results for $\lambda = 90\%$ and $\lambda = 80\%$. Lower values of λ cause a memory explosion. Further experiments with other values of k do not show significant changes in performance, although larger k might provide more interesting patterns and slightly accuracy gains.

Table 3 Mean generalization accuracy: XMuSer against the stand-alone Aleph algorithm

	λ	XMuSer (With Aleph)	Stand-alone Aleph	Wilcoxon p-value
Hepatitis subtype	0.9	0.79 (0.10)	0.78 (0.11)	0.269
	0.8	0.80 (0.10)		0.073
Hepatitis fibrosis	0.9	0.64 (0.06)	0.58 (0.09)	0.097
	0.8	0.61 (0.10)		0.407
Financial	0.9	0.71 (0.07)	0.71 (0.07)	0.854
	0.8	0.73 (0.04)		0.138

We believe that we obtained very good results. In all three classification problems, `XMuSer` obtained better or equal results than the stand-alone Aleph algorithm. The best accuracy gains were obtained in the Hepatitis Fibrosis task. In this problem our mean accuracy gains range from 3 to 6%. In the Hepatitis Fibrosis problem `XMuSer`, with a support value of $\lambda = 90\%$, is significantly better than the stand-alone Aleph algorithm, at a confidence level 10%. In Subtype we obtained a significant win for a support value $\lambda = 80\%$. These results are among the best that we could find in related work that addresses this problem (see [3]).

3 Conclusions

In this work, we presented `XMuSer`, a multi-relational framework that explores temporal information stored in a database. The methodology is an effective alternative to previous ILP-based approaches that incorporate timed data in the final first-order theory by either using time aggregation strategies [2], like time window aggregation, or by refining ILP clauses or predicates that explicit explore time. Differently from these approaches, our methodology, can take advantage of the strengths of sequence miners to explore efficiently any kind of timed data.

Moreover, we can say that we develop a new methodology to translate any multi-relational timed database into a sequence database and that we develop a new strategy to explore efficiently timed patterns in an ILP framework. Our ILP based framework gains both from the descriptive power of the ILP algorithms and the efficiency of the sequence miners.

Acknowledgments This work was supported by the Portuguese Foundation for Science and Technology (FCT) under the projects KDUS (PTDC/EIA-EIA/098355/2008) and HORUS (PTDC/EIA-EIA/100897/2008). Carlos Abreu Ferreira was financially supported by the Portuguese Polytechnic Institute of Porto (ISEP/IPP).

References

1. Esposito, F., Di Mauro, N., Basile, T.M.A., Ferilli, S.: Multi-dimensional relational sequence mining. Fundamenta Informaticae **89**(1), 23–43 (2009)

2. Ferreira, C.A., Gama, J., Costa, V.S.: RUSE-WARMR: rule selection for classifier induction in multi-relational data-sets. In: 20th IEEE International Conference on Tools with Artificial Intelligence. IEEE Computer Society, Dayton, vol. 1, pp. 379–386 (2008)
3. Ferreira, C.A., Gama, J., Costa, V.S.: Sequential pattern mining in multi-relational datasets. In: Proceedings of the 13th Conference of the Spanish Association for Artificial Intelligence, LNCS 5988. Springer, Seville, pp. 121–130 (2010)
4. Garofalakis, M., Rastogi, R., Shim, K.: Mining sequential patterns with regular expression constraints. IEEE Trans. on Know. and Data Eng. **14**(3), 530–552 (2002)
5. Jian, P., Han, J., Mortazavi-asl, B., Pinto, H., Chen, Q., Dayal, U., Hsu, M.: Prefixspan: mining sequential patterns efficiently by prefix-projected pattern growth. In: Proceedings of the 17th International Conference on Data Engineering. IEEE Computer Society, Heidelberg, pp. 215–224 (2001)
6. Muggleton, S.: Inductive logic programming, vol 38. Academic Press, London, (1992)
7. Novak, P., Lavrac, N., Webb, G.: Supervised descriptive rule discovery: a unifying survey of contrast set, emerging pattern and subgroup mining. Mach. Learn. Res. **10**, 377–403 (2009)
8. Srinivasan, A.: The Aleph Manual, http://www.comlab.ox.ac.uk/activities/machinelearning/Aleph/aleph.html (2003)

Part XIII
Algorithms

Zoom-In/Zoom-Out Algorithms for FCA with Attribute Granularity

Radim Belohlavek, Bernard De Baets and Jan Konecny

Abstract We present a method that makes it possible to control, in a parameterized way, the output of FCA (formal concept analysis) by varying the granularity of the input attributes. Depending on the user's need, the method makes it possible to re-analyze the data by using finer or coarser attributes, resulting in a different hierarchy that contains finer or coarser formal concepts. In particular, we present algorithms that make it possible to compute the corresponding zoom-in and zoom-out of the original hierarchy. In addition, we present illustrative example.

1 Introduction

Problem Setting FCA proved to be useful in various areas such as organization of web search results into a hierarchical structure of concepts based on common topics [2], information retrieval [3, 8], hierarchical analysis of software code [4, 9, 11], visualization in software engineering [5, 10], to name just a few. Further references to applications of FCA can be found in [2, 6]. In the basic setting of FCA, it is assumed that the input data is in the form of a binary table describing which attributes (columns) apply to which objects (rows). The main output of FCA is a partially ordered set, called a concept lattice, consisting of particular biclusters, called formal concepts, in the data. The choice of attributes is a fundamental

R. Belohlavek (✉) · J. Konecny
Palacký University, Olomouc, Czech Republic
e-mail: radim.belohlavek@acm.org

J. Konecny (✉)
e-mail: Konecnyj@inf.upol.cz

B. D. Baets
Ghent University, Ghent, Belgium

E. Gelenbe et al. (eds.), *Computer and Information Sciences II*,
DOI: 10.1007/978-1-4471-2155-8_70, © Springer-Verlag London Limited 2012

step in FCA because it determines the extracted formal concepts. In the basic setting of FCA, this choice is done at the beginning and is considered fixed. In [1], the authors proposed a way to treat a granularity of attributes. In this paper, we continue this line of research in that we provide algorithm that enable the user to compute the different concept lattices corresponding to different choices of attribute granularity, i.e., in a sense, to zoom-in and zoom-out in the data and the corresponding structure concept lattice.

Preliminaries from Formal Concept Analysis For a comprehensive background on formal concept analysis (FCA), we refer to [6]. The input data to FCA is represented by a triplet $\langle X, Y, I \rangle$, called *formal context*, in which I is a binary relation between X (objects) and Y (attributes) and $\langle x, y \rangle \in I$ means that object x has attribute y. For every set $A \subseteq X$ of objects from X denote by $A^{\uparrow}$ a subset of Y consisting of all of attributes shared by all objects from A, i.e., $A^{\uparrow} = \{y|$ for each $x \in A : \langle x, y \rangle \in I\}$. Similarly, for $B \subseteq Y$ denote by $B^{\downarrow}$ a subset of X consisting of all objects sharing all attributes from B. A *formal concept* of $\langle X, Y, I \rangle$ is a pair $\langle A, B \rangle$ of $A \subseteq X$ (extent) and $B \subseteq Y$ (intent) satisfying $A^{\uparrow} = B$ and $B^{\downarrow} = A$. The set $\mathcal{B}(X, Y, I) = \{\langle A, B \rangle | A^{\uparrow} = B, B^{\downarrow} = A\}$ of all formal concepts of $\langle X, Y, I \rangle$, called a *concept lattice of I*, can be equipped with a partial order $\leq$ defined by

$$\langle A_1, B_1 \rangle \leq \langle A_2, B_2 \rangle \text{ iff } A_1 \subseteq A_2 (\text{or, equivalently,} B_2 \subseteq B_1).$$

$\mathcal{B}(X, Y, I)$ happens to be a concept lattice whose structure is described by the so-called basic theorem of concept lattices [6].

2 Granularity of Attributes

Suppose now a user wants to use a concept lattice to visualize the structure of car accidents in a certain area. He might want to use attributes of accidents such as "technical cause", "driver's fault" (describing the cause of the accident), "before noon", "after noon" (describing when the accident happened), etc. The concept lattice over such attributes may not contain interesting patterns because the selected attributes are too coarse. If one uses attributes with a higher level of granularity, such as "alcohol", "skid", ..., "early morning", "late evening", etc., the concept lattice may reveal some interesting patterns. The capability to change a level of attribute granularity in FCA to capture relevant patterns in data is therefore a natural requirement. The intitial steps in emplying granularity levels to FCA were done in [1] and we recall the basic notions below.

Definition 1 *Let X be a set of objects. A g-tree (granularity tree) for attribute y is a rooted tree with the following properties:*

- each node of the tree is labeled by attribute name; the root is labeled by y;
- to each label z of a node there is associated a set $\{z\}^{\downarrow} \subseteq X$; objects from $\{z\}^{\downarrow}$ are considered as objects to which attribute z applies;

- if nodes labeled by $z_1,\ldots,z_n$ are the successors of a node labeled by z, then $\{\{z_1\}^{\downarrow},\ldots,\{z_n\}^{\downarrow}\}$ is a partition of $\{z\}^{\downarrow}$.

A selection of an appropriate level of granularity can be described by the following notion of a cut in a g-tree.

Definition 2 *A cut in a g-tree for y is a set* $C=\{y_1,\ldots,y_n\}$ *of node labels of the g-tree such that for each leaf node u, there exists exactly one node v on the path from the root to u such that the label of v belongs to C.*

A cut represents a level of granularity which can be obtained by moving down the paths of the tree, starting from the root. Moving down from a node labeled by y to its successors labeled by $y_1,\ldots,y_n$ represents a replacement of a coarser attribute y by finer attributes $y_1,\ldots,y_n$. The relation of a refinement induces a partial order on the set of all cuts of a given g-tree by putting for cuts $C_1=\{y_1,\ldots,y_n\}$ and $C_2=\{z_1,\ldots,z_m\}$, $C_1\leq C_2$ *if and only if* $\{\{y_1\}^{\downarrow},\ldots,\{y_n\}^{\downarrow}\}$ *is a subpartition of* $\{\{z_1\}^{\downarrow},\ldots,\{z_m\}^{\downarrow}\}$.

Let now $\langle X,Y,I\rangle$ be an input data table. Suppose that we have for each attribute $y\in Y$ a g-tree T_y for y. Let for each $y\in Y$, C_y be a cut in T_y and denote by

$$\mathcal{C}=\{C_y|y\in Y\}$$

the collection of all these cuts. Each such a collection $\mathcal{C}$ induces a data table $\langle X,Y_{\mathcal{C}},I_{\mathcal{C}}\rangle$ such that

$$Y_{\mathcal{C}}=\bigcup_{y\in Y}C_y$$

and we put for each $z\in Y_{\mathcal{C}}$, $\langle x,z\rangle\in I_{\mathcal{C}}\text{iff}x\in\{z\}^{\downarrow}$. That is, $\langle X,Y_{\mathcal{C}},I_{\mathcal{C}}\rangle$ results from $\langle X,Y,I\rangle$ by replacing each attribute $y\in Y$ by the corresponding collection C_y of attributes (refinements of y). Clearly, putting $\mathcal{C}=\{\{y\}|y\in Y\}$, we have $\langle X,Y,I\rangle=\langle X,Y_{\mathcal{C}},I_{\mathcal{C}}\rangle$.

Denote the concept lattice corresponding to $\langle X,Y_{\mathcal{C}},I_{\mathcal{C}}\rangle$ by $\mathcal{B}(X,Y_{\mathcal{C}},I_{\mathcal{C}})$ or simply by $\mathcal{B}_{\mathcal{C}}$. Given two granularity levels, $\mathcal{C}_1$ and $\mathcal{C}_2$, what is the relationship between the corresponding concept lattices $\mathcal{B}(X,Y_{\mathcal{C}_1},I_{\mathcal{C}_1})$ and $\mathcal{B}(X,Y_{\mathcal{C}_2},I_{\mathcal{C}_2})$? We restrict ourselves to the practically most important case, namely when $\mathcal{C}_1$ is a refinement of $\mathcal{C}_2$, denoted by $\mathcal{C}_1\leq\mathcal{C}_2$, by which we mean that for each $y\in Y$ we have $C_{1y}\leq C_{2y}$ for the corresponding cuts $C_{1y}\in\mathcal{C}_1$ and $C_{2y}\in\mathcal{C}_2$. Replacing $\mathcal{C}_2$ by $\mathcal{C}_1$ and going from $\mathcal{B}(X,Y_{\mathcal{C}_2},I_{\mathcal{C}_2})$ to $\mathcal{B}(X,Y_{\mathcal{C}_1},I_{\mathcal{C}_1})$ and *vice versa* may be seen as *zooming in* and *zooming out*.

Theorem 1 If $\mathcal{C}_1\leq\mathcal{C}_2$ *then for each formal concept* $\langle A,B\rangle\in\mathcal{B}(X,Y_{\mathcal{C}_2},I_{\mathcal{C}_2})$ *there exist unique formal concepts* $\langle A_k,B_k\rangle\in\mathcal{B}(X,Y_{\mathcal{C}_1},I_{\mathcal{C}_1})$, $k\in K$, *such that* $A=\bigcup_{k\in K}A_k$.

Theorem 1 says that if we refine our attributes then the extent (cluster of objects) of each formal concept from the concept lattice of the "rougher attributes" is a union of extents (clusters of objects) of the concept lattice of the "finer attributes".

3 Zoom-in/Zoom-out Algorithms

In this section, we describe an algorithm that enables us to compute $\mathcal{B}(X, Y_{\mathcal{C}_2}, I_{\mathcal{C}_2})$ from $\mathcal{B}(X, Y_{\mathcal{C}_1}, I_{\mathcal{C}_1})$ and *vice versa*. $\mathcal{B}(X, Y_{\mathcal{C}_2}, I_{\mathcal{C}_2})$ or $\mathcal{B}(X, Y_{\mathcal{C}_1}, I_{\mathcal{C}_1})$ can be computed by the available algorithms for computing concept lattices, see [7]. However, in a natural scenario of data exploration utilizing the change in attribute granularity, a user explores $\mathcal{B}(X, Y_{\mathcal{C}_1}, I_{\mathcal{C}_1})$ and if desirable, zooms out and explores $\mathcal{B}(X, Y_{\mathcal{C}_2}, I_{\mathcal{C}_2})$ instead. Therefore, a question arises of how to compute $\mathcal{B}(X, Y_{\mathcal{C}_2}, I_{\mathcal{C}_2})$ directly from $\mathcal{B}(X, Y_{\mathcal{C}_1}, I_{\mathcal{C}_1})$ using the information described by the g-trees.

Particularly, the AddAttribute algorithm computes a new concept lattice from an original concept lattice $\mathcal{B}$ and a new attribute w. Removing an attribute using RemoveAttribute algorithm is a reverse process. We use extended versions of presented algorithms which exploit disjunct character of cuts in g-trees and enable to process more attributes at once. Description and explanation of the algorithms is left for full version of the paper.

Procedure AddAttribute($\mathcal{B}, w, \{w\}^{\downarrow}$)

```
M := {⟨∅↓, ∅↓↑⟩ ∈ 𝓑};
while M ≠ ∅ do
    C := maximal element of M;
    M := M \ {C}
    if Ext(C) ⊆ {w}↓ then
        Int(C) := Int(C) ∪ {w}
        M := M ∪ Succ(C)
        Child(C) := C
        foreach minimal P ∈ (𝒰(C) \ C) with a Child do
            add edge (Child(P), C);
        end
    end
    else if exists D ∈ Succ(C) s.t. Ext(C) ∩ {w}↓ ⊆ Ext(D) then
        M := M ∪ {D}
    end
    else
        M := M ∪ Succ(C)
        create a new node C_new := ⟨Ext(C) ∩ {w}↓, Int(C) ∪ {w}⟩;
        Child(C) := C_new;
        foreach S ∈ Succ(C) s.t. Ext(S) ⊆ Ext(C) ∩ {w}↓ do
            add edge (C_new, S);
            remove edge (C, S);
        end
        add edge (C, C_new);
        foreach minimal P ∈ (𝒰(C) \ C) with a Child do
            add edge (Child(P), C);
        end
    end
end
```

4 Examples and Experimental Evaluation

Car accidents example. The data in Table 1 describes a part of a real dataset. We are interested in whether FCA can reveal some interesting patterns in car accidents. We consider the following g-tree for the attributes cause and time: "cause" is split to "driver" and "car defect", "driver" is split to "alcohol" and "priority", "car defect" is split to "steering" and "brakes"; "time" is split to "morning", "afternoon", "night", "morning" is split to "5 AM", ..., "12 AM", "afternoon" is split to "1 PM", ..., "7 PM", "night" is split to "8 PM", ..., "4 AM". We now illustrate how the selection of appropriate cuts in the g-trees influences the patterns revealed by formal concept analysis in the car accident data. Let us now consider the granularity of attributes corresponding to the following cuts:

$$C_{\text{cause}} = \text{alcohol, priority, steering, brakes}\}, \tag{1}$$

$$C_{\text{time}} = \text{morning, afternoon, night}. \tag{2}$$

Procedure RemoveAttribute($\mathcal{B}, w$)

$M := \{\langle \emptyset^{\uparrow\downarrow}, \emptyset^{\uparrow}\rangle \in \mathcal{B}\}$
while $M \neq \emptyset$ **do**
  $C :=$ minimal element of M;
  $M := M \setminus \{C\}$
  if exists $P \in \text{Pred}(C)$ **s.t.** $\text{Int}(P) = \text{Int}(C) \setminus \{w\}$ **then**
    foreach $S \in \text{Succ}(C)$ **s.t.** *no* P' *exists with* $P' \in \text{Succ}(P) \setminus \{C\}$ *and* $P' \subseteq S$ **do**
      add edge (S, P)
    end
    $M := M \cup (\text{Pred}(C) \setminus \{P\})$;
    remove C **from** $\mathcal{B}$ and all edges inciding with C too
  end
  else
    $M := M \cup \{P' \in \text{Pred}(C) \mid w \in \text{Int}(P')\}$;
    $\text{Int}(C) := \text{Int}(C) \setminus \{w\}$;
  end
end

Such a granularity may be suggested by a person who expects to see interesting patterns when looking at the particular causes of accidents and the different parts of the day. The corresponding concept lattice $\mathcal{B}(X, Y_C, I_C)$ is depicted in Fig. 1. The size of the nodes and thickness of the lines are set in a way to emphasize possibly interesting concepts (based on the size of extents, details are left out due to the limited scope). For example, the large size of the node representing formal concept No. 2 indicates that there is a significant number of "night accidents caused by alcohol"; similarly for concept No. 6. These two types of accidents may thus be regarded as typical accidents.

Table 1 Car accidents

Ref. No.	Driver name	Cause of accident	Time of accident
029	Lech	Steering	6 AM
082	Schwartz	Priority	6 AM
103	Zulle	Alcohol	10 PM
105	Kiril	Brakes	9 PM
⋮	⋮	⋮	⋮

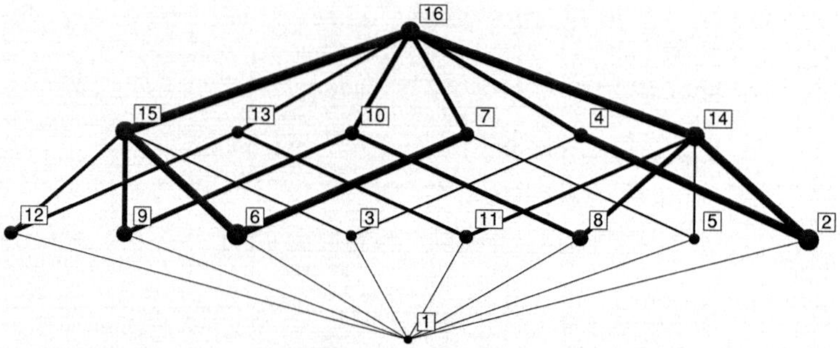

Fig. 1 The concept lattice of the car accident data corresponding to the cuts given by (1) and (2)

Experimental evaluation of zoom-in/zoom-out algorithms. We performed experiments which show that using the zoom-in/zoom-out algorithms is significantly more efficient than computing the concept lattices in question (corresponding to newly selected granularity) from scratch. Details will be presented in the full version of this paper.

Acknowledgments Supported by the Bilateral Scientific Cooperation Flanders–Czech Republic (Kontakt 1–2006–33; Special Research Fund of Ghent University, project nr. 011S01106), and by ESF Grant No. CZ.1.07/2.3.00/20.0059.

References

1. Belohlavek, R., Sklenar, V.: Formal concept analysis over attributes with levels of granularity. Proc. CIMCA, IEEE CS, Los Alamitos, CA, pp. 619–624 (2005)
2. Carpineto, C., Romano, G.: Concept data analysis theory and applications. Wiley, NY (2004)
3. Cho, W.C., Richards, W.: Improvement of precision and recall for information retrieval in a narrow domain: reuse of concepts by formal concept analysis. IEEE/WIC/ACM WI, pp. 370–376 (2004)
4. Dekel, U., Gil, Y.: Visualizing class interfaces with formal concept analysis. In: ACM SIGPLAN OOPSLA03, Anaheim, California, pp. 288–289 (2003)
5. Ganapathy, V., King, D., Jaeger, T., Jha, S.: Mining security-sensitive operations in legacy code using concept analysis. ICSE, pp. 458–467 (2007)

6. Ganter, B., Wille, R.: Formal concept analysis mathematical foundations. Springer, Berlin (1999)
7. Kuznetsov, S., Obiedkov, S.: Comparing performance of algorithms for generating concept lattices. J. Exp. Theor. Artif. Intell. **14**(2–3), 189–216 (2002)
8. Okubo, Y., Haraguchi, M.: Finding conceptual document clusters with improved top-n formal concept search. IEEE/WIC/ACM WI, pp. 347–351 (2006)
9. Snelting, G., Tip, F.: Understanding class hierarchies using concept analysis. ACM Trans. Program. Lang. Sys. **22**, 540–582 (2000)
10. Streckenbach, M., Snelting, G.: Refactoring class hierarchies with KABA. In: ACM SIGPLAN OOPSLA 2004, pp. 315–330 (2004)
11. Tonella, P.: Using a concept lattice of decomposition slices for program understanding and impact analysis. IEEE Trans. Softw. Eng. **29**, 495–509 (2003)

Author Biographies

Radim Belohlavek (SM'07) graduated in Theoretical Cybernetics and Informatics and Systems Science from Palacky University (Czech Republic) and obtained his MSc *summa cum laude* in 1994. He obtained a PhD in Computer Science in 1998 and a PhD in Mathematics in 2001. In 2005, he was appointed a Full Professor of Computer Science by the President of the Czech Republic. In 2008, he obtained a DSc in Computer Science and Cybernetics from the Academy of Sciences of the Czech Republic. R. Belohlavek's academic interests are in the areas of uncertainty and information, logic and algebra, and relational data analysis. R. Belohlavek was a principal investigator of numerous grants in these areas. He authored 2 monographs (Kluwer, Springer) and over 130 papers in conference proceedings and journals including J. Computer and System Sciences, Ann. Pure Appl. Logic, IEEE Trans. Systems, Man, and Cybernetics, Neural Computation, J. Logic and Computation, J. Math. Anal. Appl., Fundamenta Informaticae, Information Sciences, J. Exp. Theor. Artificial Intelligence, Math. Log. Quart. Radim Belohlavek is a Senior Member of IEEE and a Member of ACM and AMS. From 2001 to 2007, he was a Head of Department of Computer Science, Palacky University (Czech Republic). Since 2007, he is a Professor of Systems Science at the Binghamton University—State University of New York.

Bernard De Baets holds an M.Sc. in Maths (1988), a Postgraduate degree in Knowledge Technology (1991) and a Ph.D. in Maths (1995), all *summa cum laude* from Ghent University (Belgium), and is a Government of Canada Award holder (1988). He is a Full Professor in Applied Mathematics (2008) at Ghent University, where he is leading KERMIT, the research unit *Knowledge-Based Systems*. He is an Honorary Professor of Budapest Tech (2006). His publications comprise more than 170 papers in international journals and about 40 book chapters. He acted as supervisor of 20 Ph.D. students. He serves on the Editorial Boards of various international journals, in particular as co-editor-in-chief of Fuzzy Sets and Systems. B. De Baets coordinates EUROFUSE, the EURO Working Group on Fuzzy Sets, and is member of the Board of Directors of EUSFLAT, the Technical Committee on Artificial Intelligence and Expert Systems of IASTED, and of the Administrative Board of the Belgian OR Society.

Jan Konecny graduated in Computer Sciience from Palacky University (Czech Republic) and obtained his MSc *summa cum laude* in 2005. He is a PhD student at the State Univerisity of New York in Binghamton. His academic interests are in the area of data analysis and management of uncertainty. He published several papers in these areas.

A Hyper-Heuristic Based on Random Gradient, Greedy and Dominance

Ender Özcan and Ahmed Kheiri

Abstract Hyper-heuristics have emerged as effective general methodologies that are motivated by the goal of building or selecting heuristics automatically to solve a range of hard computational search problems with less development cost. HyFlex is a publicly available hyper-heuristic tool for rapid development and research which currently provides an interface to four problem domains along with relevant low level heuristics. A multistage hyper-heuristic based on random gradient and greedy with dominance heuristic selection methods is introduced in this study. This hyper-heuristic is implemented as an extension to HyFlex. The empirical results show that our approach performs better than some previously proposed hyper-heuristics over the given problem domains.

1 Introduction

Hyper-heuristics represent a class of search methodologies which explore the space of heuristics rather than the space of solutions, directly. There are two main types of hyper-heuristic methodologies in literature: methodologies to *select* or *generate* heuristics [1]. The main goal in this line of research is to raise the level of generality by providing hyper-heuristic solution methodologies that are applicable to different problem domains without requiring any additional development cost. A selection hyper-heuristic is a high level problem solving framework that can select and apply an appropriate low-level heuristic at each decision point, given

E. Özcan (✉) · A. Kheiri
School of Computer Science, University of Nottingham,
Jubilee Campus, Wollaton Road, Nottingham NG8 1BB, UK
e-mail: exo@cs.nott.ac.uk

A. Kheiri (✉)
e-mail: axk@cs.nott.ac.uk

E. Gelenbe et al. (eds.), *Computer and Information Sciences II*,
DOI: 10.1007/978-1-4471-2155-8_71,

a particular problem instance and a number of low-level heuristics. This study focusses on the selection hyper-heuristics.

Cowling et al. [2] presented a variety of selection hyper-heuristics embedding simple heuristic selection methods, such as *Random Descent* (Gradient), *Greedy* and a sophisticated one, namely *Choice Function*. Random Descent selects a low level heuristic randomly and applies it as long as the solution is improved. If the solution worsens, then another low level heuristic is selected and the same process is repeated. Greedy applies all low level heuristics to the candidate solution and selects a heuristic, hence a solution which provides the best quality. The new solution could be still worse than the current solution. The authors reported the success of a learning hyper-heuristic; namely, Choice Function. Greedy also showed some potential. The *peckish* heuristic selection methods attempt to reduce the number of low level heuristics either using an online or an offline mechanism. Cowling et al. [3] suggested a Tabu Search based hyper-heuristic that utilised a list to disallow the use of low level heuristics generating worsening results. More on hyper-heuristics can be found in [1].

To the best knowledge of authors, there are two publically available hyper-heuristic tools: Hyperion [4] and Hyflex [5]. Hyperion provides a general recursive framework for the development of hyper-heuristics (or metaheuristics), supporting the selection hyper-heuristic frameworks provided in [6]. Hyflex provides reusable hyper-heuristic (meta-heuristic) components, having a support for the problem domains of Boolean Satisfiability (MAX-SAT), One Dimensional Bin Packing, Permutation Flow Shop (PFS) and Personnel Scheduling (PS) each with ten different instances and a set of low-level heuristics. Burke et al. [7] investigated the performance of a range of selection hyper-heuristics implemented as part of HyFlex. This was a proof of concept study for CHeSC: Cross-Domain Heuristic Search Competition.[1] The best selection hyper-heuristic will be determined among CHeSC competitors which generalises the best across a set of problem instances from different problem domains. Burke et al. [7] reported that the best performing hyper-heuristic was an iterated local search approach.

This study describes a multistage selection hyper-heuristic. The proposed hyper-heuristic is implemented based on HyFlex. Either Greedy or Random Gradient heuristic selection method is used as a heuristic selection method at any stage. Therefore, each stage will be referred to as Greedy or Random Gradient stage depending on the heuristic selection method used. Heuristic selection is followed by a Naïve move acceptance (NV) strategy [7] to decide whether to accept or reject the new solution considering its quality. The performance of the proposed hyper-heuristic is tested over these problem domains and compared against some previous approaches.

---

[1] http://www.asap.cs.nott.ac.uk/chesc2011/

2 Methodology

Figure 1 provides the pseudocode of the proposed hyper-heuristic. The multistage selection hyper-heuristic mechanism starts with a Greedy stage. The Greedy heuristic selection method allows all the low level heuristics to process a given candidate solution successively for a number of steps to build a *List of Active Heuristics* (LAH) in the Greedy stage. LAH is a list of the low level heuristics that are expected to perform relatively well. This is an opposite strategy employed by the Tabu Search based hyper-heuristic [3] which utilises a list to disallow the use of low level heuristics generating worsening results. In the first step of the Greedy stage, LAH contains all low level heuristics. The Greedy heuristic selection method combined with a dominance based strategy is used to reduce the number of active heuristics for the next stage. The Greedy stage is always followed by a Random Gradient stage. The best solution found during the Greedy stage is used as the current solution to be processed by the Random Gradient stage. In this stage, Random Gradient heuristic selection method picks a low level heuristic from LAH randomly and applies it to the solution in hand repeatedly until there is no improvement. In the case of obtaining a non-improving solution, the hyper-heuristic will go into the Random Gradient stage again without accepting the new solution with a probability of P_s; or it will go into the Greedy stage for updating the list of active heuristics with a probability of P_u; or it will accept the non-improving solution with a probability of $(1 - P_s - P_u)$ and continue with the Random Gradient stage. The following parts explain how the stages interact in more details.

In the Greedy stage, the Greedy heuristic selection method is employed for n successive steps. The best performing heuristics are determined using a strategy inspired from the concept of Pareto Front [8] in multi-objective optimisation. Given a set of k low level heuristic points $LLH = \{LLH_1, LLH_2, \ldots, LLH_k\}$ in 2-dimensional space, each represented by its x (Step) and y (Fitness) coordinates. At each step, the fitness of each solution generated by the corresponding low level heuristic is calculated. Well performing low level heuristics that have the potential to improve within the n steps are those points that are not dominated by any other point. A low level heuristic may make a small improvement in the solution taking a short time and performance-wise this is as good as a heuristic which improves a solution more taking a longer time. Assuming a minimisation problem where we are seeking for the low level heuristics that generate minimum fitness, a point LLH_i is considered to be dominated by point LLH_j if and only if $(LLH_{i.x} \geq LLH_{j.x})$ and $(LLH_{i.y} \geq LLH_{j.y})$.

Figure 2 shows an example on how to build the list of active heuristics for $(k = 5)$ low level heuristics. Note that in the figure, LLH_2 has been added three times to the list and that makes this heuristic to be selected with higher probability. Note also that in Step1, LLH_2 and LLH_3 have been both added to the list, since they have the same fitness value and they are not dominated by any other point.

As there is a limited time to find the best fitness, the value of n depends on the time of applying Greedy method in one step. n decreases when the time needed

```
Framework(Parameter: S_initial; P_s and P_u ∈ [0,1])
  S = S_initial; f = Evaluate(S); L = BuildLAH(); LLH = SelectRandomlyFrom(L)
  while (TimeLimitNotReached)
      S' = ApplyHeuristic(LLH, S)
      f' = Evaluate(S')
      if (f' does not improve f) then
          r = GenerateUniformRandomNumberIn(0, 1)
          if (r < P_s)
              LLH = SelectRandomlyFrom(L); P_a = 0
          else if (r < P_s + P_u)
              L = UpdateLAH(); LLH = SelectRandomlyFrom(L); P_a = 0
          else P_a=1
      S = NaiveMoveAcceptance(P_a, S, S')
  return {S}
```

Fig. 1 Pseudocode of the dominance based selection hyper-heuristic

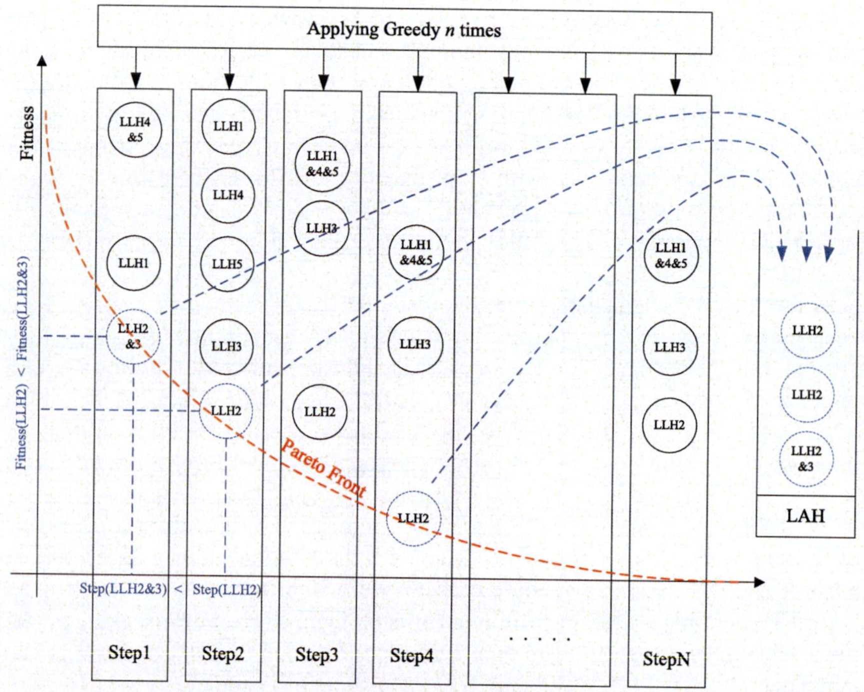

Fig. 2 An illustration showing how the list of active heuristics is built

to apply the Greedy method was high. An exponential function has been considered to find n, where zero is a possible value. However, in case of n equals zero, then the list of active heuristics will contain the whole k low level heuristics. The value of n will be calculated before starting the main loop and it should be an integer value. $n = Ae^{-f(t)}$ where A is the maximum possible value of n. $f(t)$ is the total time

Table 1 Comparisons of the different hyper-heuristics over each domain based on Formula1 scores

Domain	HH1	HH2	HH3	HH4	HH5	HH6	HH7	HH8	PHH
MAX-SAT	37.0	67.5	47.0	35.5	0.0	64.0	42.0	2.0	**85.0**
1D Bin Packing	37.0	38.0	80.0	63.0	0.0	59.0	21.0	0.0	**82.0**
Personnel Scheduling	**63.0**	61.0	14.0	54.0	36.5	0.0	61.0	33.5	57.0
Permutation Flow Shop	33.5	28.0	24.5	74.0	7.0	**84.0**	30.0	74.0	25.0
Overall	170.5	194.5	165.5	226.5	43.5	207.0	154.0	109.5	**249.0**

The best values are highlighted in bold

of applying Greedy method for one step divided by the limited time that required to complete the whole process; the total time of applying Greedy method for one step equals the summation of the time required to apply each low level heuristic on a given candidate solution: The goal of this stage is to improve the solution at hand as much as possible turning the framework into a hill climber. A low level heuristic is selected randomly from the list of active heuristics, created during the Greedy stage, and applied repeatedly until no improvement is achieved. The Naïve move acceptance [7] is used as the move acceptance strategy which accepts all improving moves. In case of non-improving move ($P_a = 1$), the solution accepted with a probability of $(1 - P_s - P_u)$; otherwise, the solution remains unchanged ($P_a = 0$).

3 Empirical Results

The proposed hyper-heuristic performance is compared to the performances of eight different hyper-heuristics (HH1–HH8) as provided at the competition website and in [7]. It is put into a mock competition against these eight hyper-heuristics based on the Formula1 scoring system. The best hyper-heuristic gets 10 points, the second gets 8, and then 6,5,4,3,2,1 and then all the remaining get no point. These points are accumulated as a score for a hyper-heuristic over all instances from four problem domains each with ten instances.

The parameter values are chosen as $A = 9$, $P_s = 0.50$ and $P_u = 0.25$. These values are decided after a set of exhaustive experiments using different combinations of values which is not reported in this paper due to space requirements. A single run is performed using each problem instance. The experiments were performed on an i3 CPU M330 at 2.13 GHz with a memory of 4.00 GB. A run terminates after 946 s as the competition requires. This value is obtained using the benchmarking tool provided at the competition website

In the MAX-SAT problem domain, our hyper-heuristic produces the best result in 3 out of 10 instances and there is a tie in 2 instances. It is the best hyper-heuristic in this domain. In the bin packing problem domain, our hyper-heuristic performs still well, but in the personnel scheduling and permutation flow shop problems, its performance is not as good as expected. It is observed that mostly, hill climbing heuristics are chosen. Table 1 summarises the results for each problem domain. The proposed hyper-heuristic performs better than the previously proposed hyper-heuristics in the Max SAT and 1D Bin Packing problem domains and worse in the rest of the domains. In the overall, our hyper-heuristic ranks the first with a score of 249.0.

4 Conclusion

In this study, a multistage selection hyper-heuristic combining two heuristic selection methods and a Naïve move acceptance method is described. Greedy with dominance and Random Gradient are used as the heuristic selection methods in an alternating manner at successive stages. Greedy aims to detect the low level heuristics with good performance and maintains a list of active heuristics considering the trade-off between the change (improvement) in the solution quality and the number of steps taken. If a heuristic takes a large number of successive steps and generating a large improvement in the solution quality, the performance of this heuristic is considered to be similar to the one which takes less number of successive steps and improves the solution quality less as well. Random Gradient selects from the (possibly) reduced set of low level heuristics to improve the solution in hand at each step. Whenever the search by Random Gradient stagnates, then the Greedy stage may restart for detecting new list of active heuristics. This is a viable strategy considering that at different points during the search, different heuristics may be performing well. The experimental results show that our hyper-heuristic is a general methodology and performs better than eight previously proposed hyper-heuristics based on their overall rankings considering all problem instances from four different domains.

References

1. Burke, E.K., Hyde, M., Kendall, G., Ochoa, G., Ozcan, E., Qu, R.: Hyper-heuristics: a survey of the state of the art. School of Computer Science and Information Technology, University of Nottingham, Computer Science Technical Report No. NOTTCS-TR-SUB-0906241418-2747 (2010)
2. Cowling, P., Kendall, G., Soubeiga, E.: A hyperheuristic approach to scheduling a sales summit. In: Selected Papers from the Third International Conference on Practice and Theory of Automated Timetabling, pp. 176–190. Springer-Verlag, London (2001)
3. Cowling, P., Chakhlevitch, K. (2003) Hyperheuristics for managing a large collection of low level heuristics to schedule personnel. In: Proceedings of the 2003 Congress on Evolutionary Computation, pp. 1214–1221
4. Swan, J., Özcan, E., Kendall, G.: Hyperion—a recursive hyper-heuristic framework. In: Coello, C.A.C. (ed.) Learning and Intelligent Optimization, 5th International Conference, LION 5 (2011)
5. Burke, E., Curtois, T., Hyde, M., Kendall, G., Ochoa, G., Petrovic, S., Vazquez-Rodriguez, J.: Hyflex: a flexible framework for the design and analysis of hyper-heuristics. In: Proceedings of the Multidisciplinary International Scheduling Conference (MISTA09), pp. 790–797 (2009)

6. Özcan, E., Bilgin, B., Korkmaz, E.: A comprehensive analysis of hyper-heuristics. Intell. Data Anal. **12**(1), 3–23 (2008)
7. Burke, E.K., Curtois, T., Hyde, M.R., Kendall, G., Ochoa, G., Petrovic, S., Rodríguez, J.A.V., Gendreau, M.: Iterated local search vs. hyper-heuristics: towards general-purpose search algorithms. In: IEEE Congress on Evolutionary Computation, pp. 1–8 (2010)
8. Deb, K.: Multi-objective genetic algorithms: problem difficulties and construction of test problems. Evol. Comput. **7**, 205–230 (1999)

A Class of Methods Combining L-BFGS and Truncated Newton

Lennart Frimannslund and Trond Steihaug

Abstract We present a class of hybrid methods for large-scale nonlinear optimization which allows for a smooth transition between the limited memory BFGS and truncated Newton methods. Numerical experiments suggest that the resulting methods in the class are more robust than the parent methods, in the sense that the methods solve more of the problems tested, at a moderate computational cost.

1 Introduction

We consider the unconstrained optimization problem $\min_{x\in\mathbb{R}^n} f(x)$, where $f: \mathbb{R}^n \to \mathbb{R}$ is twice continuously differentiable, and where n is relatively large, so that limited memory methods must be used. If Hessian information is available, one may use the Truncated Newton method (TN) [3, 4] which approximately solves the Newton equation using a conjugate gradient method:

$$\nabla^2 f(x_k)\, p_k = -\nabla f(x_k) + r_k, \quad \|r_k\| \leq \eta_k \|\nabla f(x_k)\|, \tag{1}$$

where the subscript k denotes iteration number and $0 \leq \eta_k \leq \eta < 1$. The parameter η_k can be taken to be, for instance,

$$\eta_k = \min\{1/k, \|\nabla f(x_k)\|\}, \tag{2}$$

which was introduced in [4].

L. Frimannslund · T. Steihaug (✉)
Department of Informatics, University of Bergen, Bergen, Norway
e-mail: Trond.Steihaug@ii.uib.no

L. Frimannslund
e-mail: lennart@ii.uib.no

E. Gelenbe et al. (eds.), *Computer and Information Sciences II*,
DOI: 10.1007/978-1-4471-2155-8_72,

If only gradients are available, one can use limited-memory quasi-Newton methods, such as the L-BFGS method, [5, 8, 13]. Numerical testing [11] suggests that the TN method performs better than L-BFGS if the problem is near-quadratic, and that L-BFGS performs better than TN when the problem is highly nonlinear.

Several attempts have been made to create a method which combines the properties of the truncated Newton method and the L-BFGS method. It has been proposed to use the difference pairs accumulated by L-BFGS as a preconditioner for the iterative solution of the step equation [12], using difference pairs from within an inexact conjugate gradient (CG) solution of (1) to provide fresh difference pairs for the L-BFGS method while alternating between the two methods [2, 9], and using inexact or exact Hessian information to create an incomplete or modified Cholesky factorization used as the the initial Hessian approximation, H_0^k, in the L-BFGS method [7, 16]. Constructing H_0^k from second-derivative information is also discussed in [15].

In this paper we present a novel class of methods which bridges the gap between the truncated Newton method and the L-BFGS method, in a smooth way. We determine the initial Hessian approximation H_0^k using Hessian information in conjunction with the conjugate gradient method (CG). The reason we consider the L-BFGS method in particular is that it has the advantage over other limited memory quasi-Newton methods that the product involving H_0^k is isolated from the rest of the computations [1].

We numerically test the performance of several members of the class. This paper is organized as follows. In Sect. 2 we give necessary details about existing methods. In Sect. 3 we outline the hybrid approach, give numerical results in Sect. 4 and give some concluding remarks in Sect. 5.

2 Preliminaries

BFGS and Limited Memory BFGS: Limited-memory BFGS or L-BFGS [8, 13] is a modification of the BFGS method. BFGS usually works by maintaining an approximation to the inverse of the Hessian. It performs a rank-two update at each iteration. Specifically, given at iteration k the approximation $H_k \approx \left(\nabla^2 f(x_k)\right)^{-1}$, and given $y_k = \nabla f(x_{k+1}) - \nabla f(x_k)$ and $s_k = x_{k+1} - x_k$, then H_{k+1} is taken to be

$$H_{k+1} = (I - \rho_k s_k y_k^T) H_k (I - \rho_k y_k s_k^T) + \rho_k s_k s_k^T, \tag{3}$$

where $\rho_k = \frac{1}{y_k^T s_k}$. The initial approximation H_0 to the inverse of the Hessian is up to the user, as long as H_0 is symmetric and positive definite.

L-BFGS Update Formula: L-BFGS, unlike BFGS, only has available information from iterations $k-1, \ldots, \max\{0, k-m\}$, the most recent iteration listed first, where m is a user-defined parameter. The two-loop procedure for computing $r = H_k v$ for $v = -\nabla f(x_k)$ is given in Table 1. (For simplicity we use the notation

Table 1 L-BFGS matrix-vector product procedure [13]

Given H_0^k and v, computes $r = H_k v$ using $\min\{k, m\}$ stored vector pairs

$q = v$	$r = H_0^k q$
for $i = k-1$ **step** -1 **until** $k-m$	**for** $i = k-m$ **step** 1 **until** $k-1$
$\quad \mu_i = \rho_i \cdot s_i^T q$	$\quad \beta = \rho_i \cdot r^T y_i$
$\quad q = q - \mu_i y_i$	$\quad r = r - (\beta - \mu_i) s_i$
end	**end**

μ_i for the scalar corresponding to iteration $i \in [\max\{1, k-m\}, k-1]$.) Since the initial approximation to the inverse of the Hessian appears only in the matrix-vector product between the two for-loops, it may be different at each iteration. Consequently, we denote this matrix by H_0^k rather than H_0. The choice of matrix H_0^k is up to the user. One may choose $H_0^k = I$ to avoid an extra matrix-vector product, but a choice which has proved to be effective is to set

$$H_0^k = \frac{s_{k-1}^T y_{k-1}}{y_{k-1}^T y_{k-1}} \cdot I, \tag{4}$$

Globalization: An approach to make Newton-based methods globally convergent is to use line searches. We use line searches which satisfy the strict Wolfe conditions (see e.g. [14], Chap. 3.1). Given a descent direction p_k at iteration k, then the next iterate x_{k+1} is taken to be $x_{k+1} = x_k + \alpha_k p_k$, for α_k satisfying

$$f(x + \alpha_k p_k) \leq f(x_k) + c_1 \alpha_k \nabla f(x_k)^T p_k, \tag{5}$$

and

$$|\nabla f(x_k + \alpha_k p_k)^T p_k| \leq c_2 |\nabla f(x_k)^T p_k|, \tag{6}$$

with $c_1 = 10^{-4}$ and $c_2 = 0.9$ [8, 13]. In our experiments we will use a line search procedure based on the algorithm described in [10].

3 Hybrid Approach

Consider the assignment

$$r = H_0^k q, \tag{7}$$

between the two for-loops in Table 1. If H_0^k is the inverse of the Hessian at x_k, then (7) would correspond to solving for r in the linear equation $\nabla^2 f(x_k)\, r = q$. Solving this system approximately using CG gives an error in the equation that will be required to satisfy

$$\|\nabla^2 f(x_k)\, r - q\| \leq \eta_k \|q\|. \tag{8}$$

Table 2 Pseudocode for hybrid class

```
Given x_0, k = 0.
while  ||∇f(x_k)|| > tolerance:
    Set v = −∇f(x_k).
    Perform first L-BFGS for-loop, resulting in vector q.
    Solve ∇²f(x_k)r = q with CG, terminating when ||∇²f(x_k)r − q|| ≤ η_k||q||.
    Perform second L-BFGS loop, resulting in search direction p_k.
    Find α_k satisfying (72.5) and (72.6), and set x_{k+1} ← x_k + α_k p_k.
    Discard difference pair (or pairs) (s_i, y_i), i ≤ k − m.
    Compute difference pair (s_k, y_k).    Set k ← k + 1.
end
```

We combine TN with L-BFGS by replacing r in (7) by the r obtained by applying CG to $\nabla^2 f(x_k)\, r = q$ so that r will satisfy (8). For L-BFGS to produce descent directions the matrix H_0^k must be positive definite, so using CG on the system $\nabla^2 f(x_k)\, r = q$ must be equivalent to multiplying q with some positive definite matrix. This is addressed in the following theorem.

Theorem 1 *Let A be a symmetric but not necessarily positive definite matrix. Let x_k be an inexact solution to the equation $Ax = b$, returned from CG starting with $x_0 = 0$ after iteration k, before negative curvature is encountered. Then there exists a symmetric and positive definite matrix B such that $Bx_k = b$.*

Proof The proof is by construction. Let p_i be the ith search direction employed by CG, and let α_i be the corresponding step length. Then we have $x_k = \sum_{i=1}^{k} \alpha_i p_i$. Furthermore, in CG with $x_0 = 0$ the identity $\alpha_i = (p_i^T b)/(p_i^T A p_i)$, holds, and consequently we have $x_k = \sum_{i=1}^{k} (p_i^T b)/(p_i^T A p_i p_i)$. Now, let $\hat{P}_k = \left[(p_1)/(\sqrt{p_1^T A p_1}) \right| \cdots \left| (p_k)/(\sqrt{p_k^T A p_k}) \right]$. Then we have $\hat{P}_k \hat{P}_k^T b = x_k$. The matrix $\hat{P}_k$ is in general not square, however. If we add $n - k$ columns of unit length that are mutually orthogonal and orthogonal to the first k columns, say $q_{k+1}, \ldots, q_n$ so that the matrix is square and invertible, then the relation still holds. That is, let $P_k = [\, \hat{P}_k \quad q_{k+1} \quad \cdots \quad q_n \,]$. Then, $P_k P_k^T b = x_k$. B is then the inverse of this matrix, $B = (P_k P_k^T)^{-1}$. □

If CG encounters negative curvature it should be terminated, since otherwise the CG method may become unstable and it is not guaranteed that it produces a direction of descent. Pseudocode for the hybrid method class is given in Table 2 . It is a parent method class of both L-BFGS and TN, since, if $m = 0$ (and (8) is solved to, say, tolerance (2)), then the algorithm reduces to truncated Newton. If no iterations are performed in (8), and $m > 0$, then the code becomes L-BFGS using $H_0^k = I$, which easily can be replaced with, say, (4).

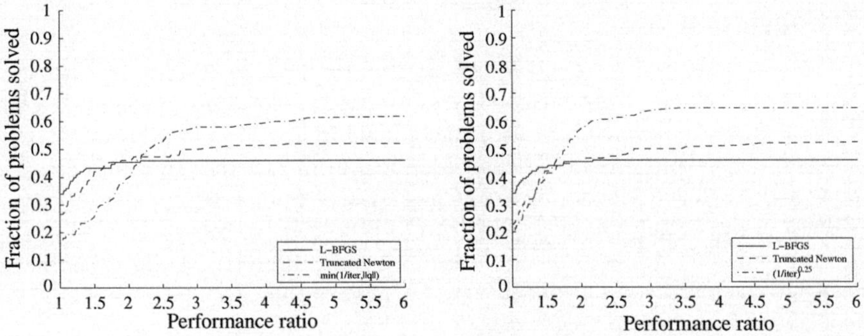

Fig. 1 *Left* Results for the forcing sequence $\eta_k = \min\{1/k, \|q\|\}$ compared to pure L-BFGS and pure truncated Newton. *Right* Results for the forcing sequence $\eta_k = (1/k)^{0.25}$ compared to pure L-BFGS and pure truncated Newton

4 Numerical Testing

We test the methods on 150 different problems from the CUTEr collection [6], with dimensions from 2 to 10,000. The problems are most of the unconstrained problems in the collection classified either as quadratic, least squares or otherwise nonlinear. For all experiments the convergence criterion is $\|\nabla f(x)\| \leq 10^{-7}$. First we test pure L-BFGS, pure TN and a member of the hybrid method class, with the following parameters: For L-BFGS and the hybrid method, $m = 6$. For TN, we use (2) as the forcing sequence, and for the hybrid method, we use $\eta_k = \min\{1/k, \|q\|\}$. The cost of one Hessian-vector product is equal one gradient evaluation. For all three methods the maximum number of iterations is set to 200.

A performance profile for these settings is given in the left half of Fig. 1. The performance profiles in Fig. 1 show the fraction of problems solved on the vertical axis, and the cost compared to the best method for each problem on the horizontal axis. The best method may be different from problem to problem. The very left of horizontal axis (performance ratio equal 1) shows the fraction of problems for which the method is best. At value p of the performance ratio on the horizontal axis the vertical axis displays the fraction of problems solved with at most p times the cost of the best method for each problem.

The figures show that the hybrid method solves more problems than either of its two parents with about 60%, but the fraction of problems where it is the most efficient is smaller than for the other two methods. This suggests that the hybrid approach is more stable. The figure to the left indicates that the forcing sequence might be too strict. For this reason we have tried many variations of η_k as forcing sequences. The most successful of the sequences in our experiments has been the sequence $\eta_k = (1/k)^{0.25}$, and a performace profile for this sequence is given to the right in Fig. 1. As one can see from this figure, the hybrid method solves more problems than with the sequence shown on the left in Fig. 1, and the fraction of problems for which it is the best method is roughly equal to pure truncated Newton.

5 Conclusion

We have presented a hybrid method class for unconstrained optimization which is a parent class of truncated Newton methods and Limited Memory BFGS. The class is flexible in such a way that it can be tailored to favor either gradient computations and the linear algebra of L-BFGS, Hessian-vector products and the linear algebra of the conjugate gradient method, or anywhere in betweeen.

Although only a few members of the class have been tested, our experiments indicate that the hybrid methods may be more stable than their parents, at a moderate computational cost.

References

1. Byrd, R.H., Nocedal, J., Schnabel, R.B.: Representations of quasi-Newton matrices and their use in limited memory methods. Math. Program. **63**, 129–156 (1994)
2. Byrd, R.H., Nocedal, J., Zhu C.: Towards a discrete Newton method with memory for large-scale optimization. In: Pillo, G.D., Giannessi, F. (eds.) Nonlinear Optimization and Applications, pp. 1–12. Plenum Press, New York (1996)
3. Dembo, R., Eisenstat, S., Steihaug, T.: Inexact Newton methods. SIAM J. Numer. Anal. **19**(2), 400–408 (1982)
4. Dembo, R.S., Steihaug, T.: Truncated-Newton algorithms for large-scale unconstrained optimization. Math. Program. **26**, 190–212 (1983)
5. Gilbert, J.C., Lemaréchal, C.: Some numerical experiments with variable-storage quasi-Newton algorithms. Math. Program. **45**, 407–435 (1989)
6. Gould, N.I.M., Orban, D., Toint, P.L.: CUTEr and SifDec: A constrained and unconstrained testing environment, revisited. ACM Trans. Math. Softw. **29**, 373–394 (2003)
7. Jiang, L., Byrd, R.H., Eskow, E., Schnabel, R.B.: A preconditioned L-BFGS algorithm with application to molecular energy minimization. Technical Report CU-CS-982-04. Department of Computer Science, University of Colorado, Boulder, Colorado 80309 (2004)
8. Liu, D.C., Nocedal, J.: On the limited memory BFGS method for large scale optimization. Math. Program. **45**, 503–528 (1989)
9. Morales, J.L., Nocedal, J.: Enriched methods for large-scale unconstrained optimization. Comput. Optim. Appl. **21**, 143–154 (2002)
10. Moré, J.J., Thuente, D.J.: Line search algorithms with guaranteed sufficient decrease. ACM Trans. Math. Softw. **20**(3), 286–307 (1994)
11. Nash, S., Nocedal, J.: A numerical study of the limited memory BFGS method and the Truncated-Newton method for large scale optimization. SIAM J. Optim. **1**(3), 358–372 (1991)
12. Nash, S.G.: Preconditioning of truncated-Newton methods. SIAM J. Sci. Stat. Comput. **6**(3), 599–616 (1985)
13. Nocedal, J.: Updating quasi-Newton matrices with limited storage. Math. Comput. **35**(151), 773–782 (1980)
14. Nocedal, J., Wright S.J.: Numerical Optimization. Springer, New York (1999). ISBN 0-387-98793-2
15. Schlick, T.: Optimization methods in computational chemistry. Rev. Comput. Chem. **3**, 1–71 (1992)
16. Xie, D., Schlick, T.: Efficient implementation of the truncated Newton method for large scale chemistry applications. SIAM J. Optim. **10**(1), 132–154 (1999)

Adaptation and Fine-Tuning of the Weighted Sum Method on Personnel Assignment Problem with Hierarchical Ordering and Team Constraints

Yılmaz Arslanoglu and Ismail Hakki Toroslu

Abstract This work deals with the multi-objective optimization problem, Personnel Assignment Problem with Hierarchical Ordering and Team Constraints, which appears in personnel assignment of large hierarchical organizations, such as military. Weighted Sum Method, which is one of the pioneering approaches towards multi-objective optimization, is considered as the means of solution to the problem. The adaptation of this intuitive and successful method to the problem, and an analysis on fine-tuning of its various parameters are discussed in this paper.

1 Introduction

Personnel Assignment Problem with Hierarchical Ordering and Team Constraints is a real life problem, which appears in the personnel assignment of large hierarchical organizations, such as military. In such hierarchical organizations, the personnel to be assigned form a level graph reflecting the rank structure of personnel, and the positions to be filled form a forest, where each tree in the forest represents a set of hierarchically related positions. The matching should satisfy the hierarchical ordering constraints represented on two partitions, which means a person assigned to a superior position should have a higher rank. Also, the nodes of both partitions can form mutually exclusive sets, where all members of a set in

Y. Arslanoglu
Marine Environmental Technology, SINTEF Materials and Chemistry,
7465 Trondheim, Norway
e-mail: yilmaz.arslanoglu@sintef.no

I. H. Toroslu (✉)
Department of Computer Engineering, Middle East Technical University,
06531 Ankara, Turkey
e-mail: toroslu@ceng.metu.edu.tr

E. Gelenbe et al. (eds.), *Computer and Information Sciences II*,
DOI: 10.1007/978-1-4471-2155-8_73,

the personnel partition are expected to be matched with the members from the same set in the positions partition. On the personnel partition, these sets may represent families that must be assigned to positions at the same location, and, on the positions partition, these sets may represent the positions that lie in the same location. Finally, the weights on the edges connecting the personnel to the positions reflect the personnel preferences or their expected performance on jobs. Besides the objective of maximizing the sum of weights assigned to the edges of the bipartite graph, the hierarchical ordering and team *constraints* are also treated as *objectives* which are subject to minimization.

Due to the problem's complexity and the fact that it was proven to be NP-Hard; the Genetic Algorithm (GA) based *Weighted Sum Method*, which is a classical, yet a promising approach, is preferred as a solution [7]. Since one of the main challenges in GAs is calibrating the various parameters depending on the nature of a given problem, especially when multi-criteria optimization is of concern, the main contribution of this paper stands out as a systematic way of searching for such parameters that will best suit for the problem in subject.

2 The Problem

In personnel assignment problem, the possible ways of assigning a set of personnel to a set of positions with the same cardinality are sought, while taking also into account the weights each person is assigned with respect to each possible position. The weight value of a $\langle$ *personnel, position* $\rangle$ tuple may keep information about the eligibility or the preference of the person for the position in subject. The aim is to match each person from the personnel set with one of the elements from the positions set in such a way that, the sum of the corresponding weights of all edges resulting from this assignment will be the maximum.

However, it is natural to expect some constraints on possible matchings. For example, in a military organization the assignment is done while respecting the *hierarchical ordering constraint*. According to this variation of the problem, there is a hierarchy among the elements of both the set of personnel and the set of positions, as it can be seen in Fig. 1.

As the *level graph* on the left of Fig. 2 depicts, military organizations are composed of levels, where every person belongs to exactly one of the levels (ranks) in the level graph, and a member of an upper level *dominates* all members in lower levels. The right part of the figure illustrates the hierarchical structure among the elements of a set of positions. Intuitionally, this may be interpreted as a hierarchy in the case of a general manager, managers and their sub-workers at a typical organization. However, in the case of positional hierarchy, the hierarchical structure is represented as a *tree*, where a position *only* dominates the positions that could be reached by following its branches.

When a person is assigned to a position, relocation could be required, and there could also be married couples or families in the personnel set. Therefore, it turns

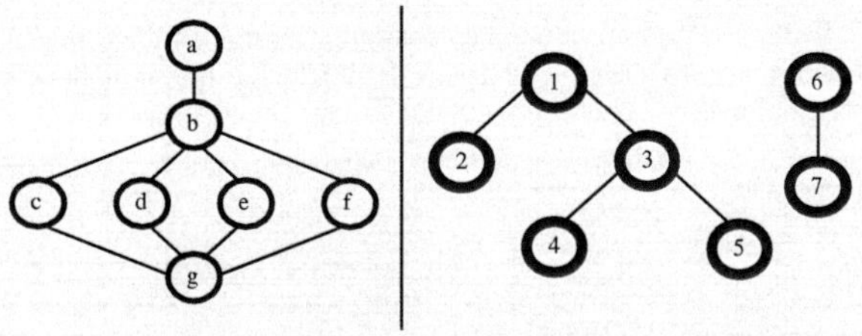

Fig. 1 Personnel and positional hierarchies

```
fitness =

- teamWeight        x ( teamViol      / maxPossibleTeamViol      )
- hierarchyWeight   x ( hierarchyViol / maxPossibleHierarchyViol )
+ weightTotalWeight x ( weightTotal   / maxPossibleWeightTotal   )
```

Fig. 2 Aggregated fitness function

out to be necessary to consider this issue in the assignment task. This version of the problem is called the *Assignment Problem with Hierarchical Ordering and Team Constraints*. The new constraint is called *the team constraint*, generalizing the concept of couples or families into teams. According to this new constraint, all members of a set from the personnel partition should be matched with positions which are members of the same set from the positions partition.

3 Adaptation of the Weighted Sum Method

The classical Weighted Sum Method is picked up as the appropriate approach among some other multi-objective evolutionary algorithm alternatives [7]. Being successfully used for solving different multi-objective optimization problems [3, 6], this simple evolutionary approach mainly deals with aggregating multiple objectives into a single one by assigning a weight to each of the objective functions, depending on their importance.

The aggregated fitness function given in Fig. 2 is used. Since team and hierarchy violations are subject to minimization, their contributions are negated in the fitness function, as the duality principle suggests [5]. Furthermore, the effect of each objective is normalized by dividing the actual value by its associated maximum possible value, in order to abstain from biasing the solution towards an individual objective. The coefficients *teamWeight*, *hierarchyWeight*, and *weightTotalWeight* reflect the degrees of importance assigned to their corresponding objectives, and these coefficients must sum up to 1.0.

The core of the assignment problem is matching a set of personnel to a set of positions, where the cardinality of these two sets are equal. Since each person can be assigned to exactly one position, a potential solution could be represented as an

Table 1 Problem parameter configurations: ELIT.X% means that selection is done among the top X% individuals. TOURN.X% means that tournament selection size is the X% of the population size

	Crossover rate	Mutation rate	Crossover method	Selection method
Values	0.2	0.000	PMX	ELIT.20%
		0.002		
	0.4	0.004		ELIT.40%
		0.006		ELIT.50%
	0.5	0.008	CX	ELIT.60%
		0.010		ELIT.80%
	0.6	0.020		R. WHEEL
		0.040	OX	TOURN.2%
	0.8	0.060		TOURN.4%
		0.080		TOURN.6%
		0.100		TOURN.8%

ordering of personnel (ordering of positions is also possible). From this perspective, it seems best to use a pure permutation representation because of the nature of the problem and the simplicity of the representation.

Traditional techniques such as one-point crossover cannot be applied directly on permutation-based chromosomes, since they cannot preserve the chromosome structure. For this reason, either the permutation representation needs to be encoded into a specific binary form [8], or special crossover techniques which could directly be applied on permutation-based chromosomes need to be utilized. In this work, the second option is followed, and the well known three permutation-based crossover techniques are considered: PMX (Partially Mapped Crossover) [2], CX (Cycle Crossover) [4], and OX (Order Crossover) [1]. For mutation, simply two gene positions are selected randomly, and their values are swapped.

The standard GA template is adopted. The crossover rate, mutation rate, selection mechanism, reproduction mechanism, and mutation mechanism can be chosen among different alternatives, as it is the case in all GAs. The success of the GA on a problem usually varies depending on the values chosen, and it is not a trivial task to determine the parameters that would best fit to a given problem. Since the number of possible combinations can be extremely large, available mechanisms and parameter values that are considered in this work are kept restricted with a subset.

4 Results and Conclusions

In this work, potential parameters that could directly affect the performance of the algorithm are considered are *crossover rate, mutation rate, selection mechanism,* and *reproduction mechanism.* These parameters altogether form a *parameter configuration.* In order to find out the best parameter configuration, 1,650 random combinations of the parameter values given in Table 1 were analyzed.

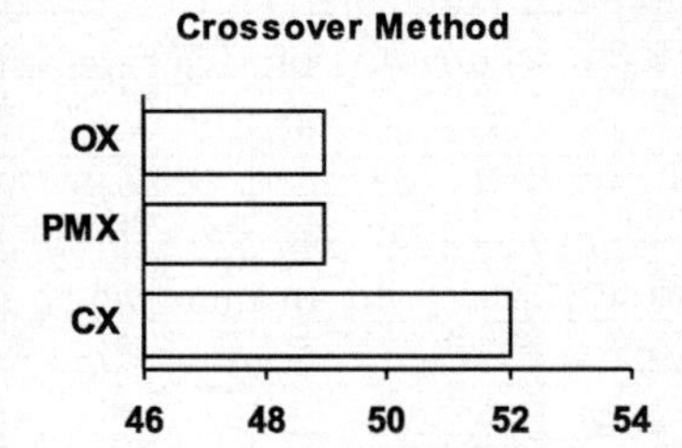	Each crossover technique has almost the same effect. This possibly results from the sensitivity to changes on permutation based chromosomes. Even simple changes on chromosomes may lead to a high number of violations or possibly, in a much better state. Moreover, positive changes in one objective may affect the others negatively. Because of the nature of the problem and the chromosome structure, none of these crossover methods outperform the others.
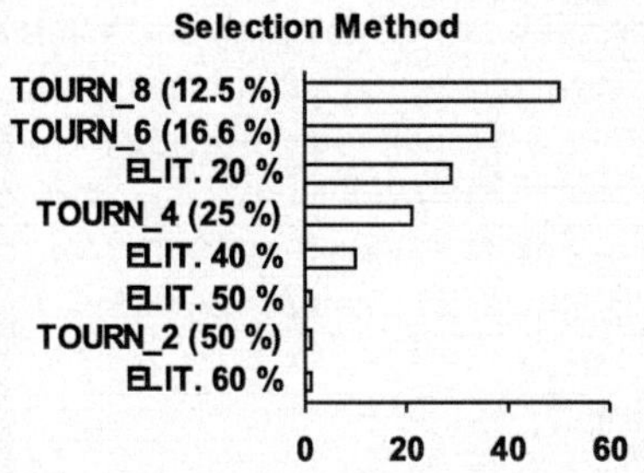	The concept of elitism has a drastic effect on the results. According to this figure, tournament selection with size 8 performs the best, which corresponds to elitism with 12.5% selection range. It is followed by tournament selection with size 6, which corresponds to an elitism selection with roughly top 16.6%. It is followed by elitism selection with top 20% selection range, and the other values follow in a consistent manner. To sum up, as the top selection range decreases, performance increases.
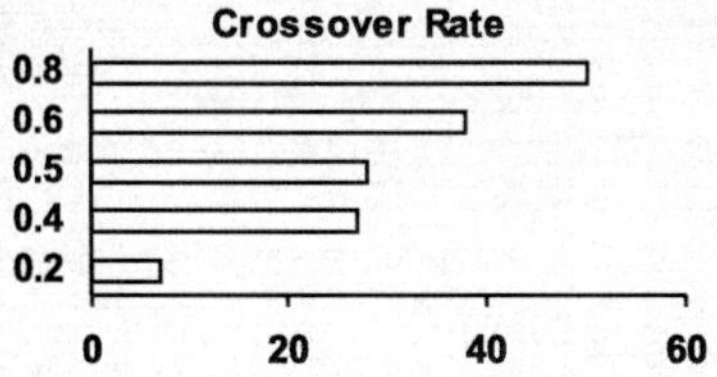	It can be concluded that, the higher the crossover rate is, the better the results are. This means that, it gets better if newer offspring, thus new combinations of genetic content, are produced instead of transferring the existing individuals to the next generation. In further tests with larger crossover rates, even better solutions were obtained.
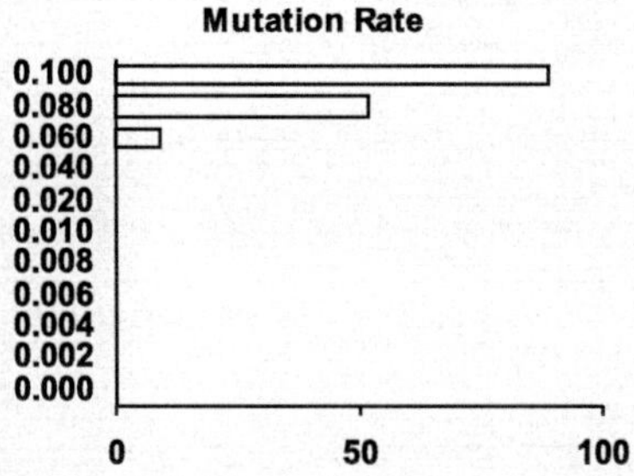	High mutation rates simply lead to better results. In further tests, even higher mutation rates such as 100% (1.000) were applied, leading to further better results. Please note that the mutation rates are expressed per genotype.

Fig. 3 Parameter analysis results

Three different problems, each with a chromosome length of 100 (cardinality of the position and the personnel set is 100) were considered, running exhaustively with 1,650 different parameter configurations, each of which is a random combination of the options given in Table 1. A population of size 100 was used and the algorithms were run 200 generations.

After obtaining the results, the best 150 parameter configurations (roughly the top 10%) among the 1,650 were considered and the occurrences of each parameter value in this top 150 parameter configuration set were counted *independent* of each other, due to no apparent dependency that could be detected between the parameters. The same approach was followed for each of the three problem instances, and the average of these occurrences was calculated. In simulations, all the three objectives were assigned equal importance. Figure 3 summarizes the results.

According to the results given in Fig. 3, it turns out to be the best to employ high rates of mutation and crossover, and also very high rates of elitism at the same time. This may possibly be due to the extreme landscape of the problem. While it tries to widely explore the landscape with high crossover and mutation rates, a balance between exploration and exploitation is also sought by favoring the best solutions, making use of the elitism concept. Through the employment of a parameter configuration in this direction, considerably better results can be obtained, compared to commonly adopted GA parameter configurations.

References

1. Davis, L.: Applying adaptive algorithms to epistatic domains. In: Proceedings of the International Joint Conference on Artificial Intelligence, pp. 162–164 (1985)
2. Goldberg, D. E., Linge, Jr.R.: Alleles, loci and the TSP. In: Proceedings of the 1st International Conference Genetic Algorithms and Their Applications, Hillsdale, NJ, pp. 154-159 (1985)
3. Liu, X., Begg, D., Fishwick, R.J.: Genetic approach to optimal topology/controller design of adaptive structures. Int. J. Numer. Method. Eng. **41**, 815–830 (1998)
4. Oliver, I.M., Smith, D.J., Holland, J.R.C.: A study of permutation crossover operators on the traveling salesman problem. In: Proceedings of the second International Conference on Genetic Algorithms, 224–230 (1987)
5. Rao, S.S.: Optimization: theory and applications. Wiley, New York (1984)
6. Rubenstein-Montano, B., Malaga, R.A.: A weighted sum genetic algorithm to support multiple-party multiple-objective negotiations. IEEE Trans. Evol. Comput. **6**(4), 366–377 (2002)
7. Toroslu, I.H., Arslanoglu, Y.: Genetic algorithm for the personnel assignment problem with multiple objectives. Inf. Sci. **177**(3), 787–803 (2007)
8. Ucoluk, G.: Genetic algorithm solution of the TSP avoiding special crossover and mutation. Intell. Autom. Soft Comput. 3(8), TSI Press, 265–272 (2002)

An Option Pricing Model Calibration Using Algorithmic Differentiation

Emmanuel M. Tadjouddine and Yi Cao

Abstract We study the application of gradient-based optimization methods for calibrating a stochastic volatility model used for option pricing. To this end, we derived and analyzed Monte Carlo estimators for computing the gradient of a certain payoff function using Finite Differencing and Algorithmic Differentiation. We have assessed the accuracy and efficiency of both methods and their impacts into the optimization algorithm. Numerical results are presented and discussed. This work can benefit investors in financial products with the need for fast and more precise predictions of future market data.

Keywords Calibration · Optimization · Stochastic pricing models · Algorithmic differentiation · Monte Carlo simulation

1 Introduction

Pricing goods is an essential element of trading in a market based economics. Goods may be a commodity or a contract with some financial returns such as an option. We may sell goods through auctions wherein a seller or a buyer need to price or value goods as reflected by the market so as to make informed decisions on the bidding process. Goods may have a *reserve price* (under which the item cannot be sold) unknown to the bidders; the seller may not be aware of the bidders' valuations of the item. Finding the 'right' price for a good on sale in such an uncertain environment is challenging but can be carried out using a model dependent on certain parameters. Ideally, we would like the value predicted by such

E. M. Tadjouddine (✉) · Y. Cao
Department of Computer Science and Software Engineering, Xian Jiaotong-Liverpool University, SIP, Suzhou, Jiangsu Province, China
e-mail: emmanuel.tadjouddine@xjtlu.edu.cn; etadjouddine@gmail.com

E. Gelenbe et al. (eds.), *Computer and Information Sciences II*,
DOI: 10.1007/978-1-4471-2155-8_74,

a computational model to be as close as possible to the winning bid value. Since the future is uncertain, we would like the model to be at least consistent with historical market data. This is the aim of model *calibration*. In finance for example, engineers have to recalibrate model parameters to intraday market prices almost in real time.

In this paper, we study the problem of calibrating a pricing model in a risk neutral world, where no *arbitrage* opportunities exist. In economics, an arbitrage is a concept allowing for a positive chance to make a positive profit and a zero chance for a loss. To account for price volatility, we have considered the family of financial stochastic volatility models known as Vasicek models [1] and *European call options* to be calibrated to market data. A European call option is a contract giving its holder the right but not the obligation to buy a product for a prescribed price at a prescribed time in the future. Calibrating such a model amounts to optimizing an objective function representing the misfit between the predicted and observed data. This is caried out by a gradient descent method. In a Monte Carlo simulation (a simulation based on pseudo-random numbers) of the pricing model, we have derived estimators of the gradient by two different methods: *FD (Finite Differencing)* and *AD (Algorithmic Differentiation)*, see for example [2]. FD is the traditional numerical formula that approximates the derivative of a mathematical function. AD is a set of algorithms allowing for the evaluation of a numerical function represented by a computer program. Numerical results from within Matlab showed a sublinear convergence of the minimizer using FD supplied derivatives but a nearly linear convergence of the minimizer when using AD, therefore matching the theoretical convergence of the minimizer.

Related Work. A platform long used for trading goods is that of auctions. It is predicted that future agent mediated e-commerce will involve open systems of financially-motivated agents moving between different auction houses, learning about auction rules, and devising bidding strategies. This 'mobile bidder model' as termed in [3] is increasingly plausible within web-based internet auctions wherein software agents can interact, enter and leave auctions at will. [3] studied the performance of networked auctions and evaluated the income generated by the seller for a well-defined mathematical model of the network. As for model calibration, related work includes that of [4] wherein stochastic volatility models are calibrated using a Malliavin gradient method, which combines gradient method with Malliavin calculus, for the gradient calculation. [5] solved the calibration problem for the Heston's model by an Excel solver using Monte-Carlo simulation, finite difference method and closed-form solution of the model. In our paper, we advocate the use of the Algorithmic Differentiation technique to obtain the gradient and discuss its impact on the convergence of the minimizer.

2 Model Calibration Scheme

Our approach to the calibration process is as follows. We consider the Vasicek model, a 2-dimensional stochastic differential equations (SDEs),

$$dS_t = S_t(\sqrt{1-\gamma_1^2}\sigma_t dW_t^1 + \gamma_1\sigma_t dW_t^2),$$
$$d\sigma_t = \kappa(\theta - \sigma_t)dt + \gamma_2 dW_t^2, \tag{1}$$

where dW^1 and dW^2 are correlated Brownian motion processes with constant cross correlation ρ. Equations 1 are discretized using the Euler scheme along with a European call option. We have also found the analytical solution of the 2-dimensional SDEs. Subsequently, the option price is estimated within a Monte Carlo simulation by either the Euler scheme on the system of SDEs or the analytical solution. Afterwards, we formed the objective function in terms of the parameters $(\kappa, \theta, \gamma_1, \gamma_2)$, which is minimized within MATLAB by a gradient method using both Finite Differencing (FD) and Algorithmic Differentiation (AD) for evaluating sensitivities.

3 Numerical Results

Using Matlab R2009a on a PC with an Intel core i7 CPU, we have simulated option values $\mathbf{v}_{\theta,\kappa,\gamma_1,\gamma_2}(S_T)$ for the European call option price S_T with the model parameters $(\kappa, \theta, \gamma_1, \gamma_2) = (0.5, 0.2, -0.2, 0.1)$, the initial price $S_0 = 1$, the initial volatility $\sigma_0 = 0.2$ and maturity time $T = 1$. In this Monte Carlo simulation, we used 5×10^4 simulation paths and a time discretization $\Delta t = 0.01$. Then, we evaluated the gradient $\frac{\partial v_{\theta,\kappa,\gamma_1,\gamma_2}(S_T)}{\partial(\theta,\kappa,\gamma_1,\gamma_2)}$ by using the FD and AD (using the Mad's forward mode AD [6]) methods. Table 1 shows the numerical results as well as the runtime performance of each method. In the "Deviation" column of the Table 1, we chose the gradient results by AD as the standard value and calculate the maximum difference between it and that obtained by FD. CPU($\mathbf{v}$) represents the average CPU time of a Matlab function $\mathbf{v}$ over 1000 runs. The average elapsed CPU time for the pricing function was about 1.16 ms.

From Table 1 , we can see that the FD method gave results with a deviation about 10^{-5}. This may due to the fact that the variance of the estimated derivatives is inversely proportional with the step size h and linearly dependent on its bias [7]. Furthermore, FD (using the central difference formula) showed a runtime ratio of 7.04, almost in line with its theoretical complexity of 8 function evaluations. We then used Matlab's `fmincon` function to minimize the objective function subject to the constraints $\hat{\Theta} = [\kappa \geq 0, \theta \geq 0, \gamma_1 \leq 0, \gamma_2 \geq 0]$. We set the initial guess $\hat{\Theta}_0 = [0.8, 0.5, -0.5, 0.3]$ for 1000 simulation paths with a time-discretization of $\Delta t = 0.01$. Finally, we monitored the convergence rate of the solver for different iteration numbers. The simulation results for this gradient-based optimization approach are shown in Fig. 1 in a logarithmic scale. Figure 1 shows the maximum absolute error against the number of iterations by the optimization solver. Theoretically, the number of iterations required for $\hat{\Theta}_k - \Theta \leq \xi$ is $O(1/\sqrt{\xi})$ [8]. From this result, we have added the theoretical convergence curve for a typical

Table 1 Evaluation of the Sensitivities

Method	Deviation	CPU(**v**′)/CPU(**v**)
Forward mode AD	—	18.78
Finite Difference	1.62×10^{-5}	7.04

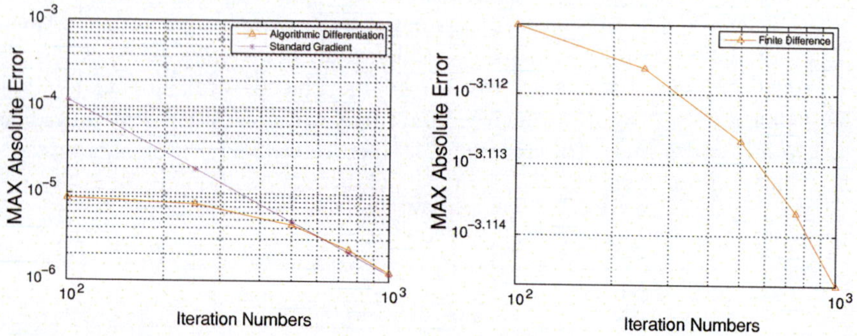

Fig. 1 Convergence of the Optimizer using a supplied gradient obtained using AD, and FD

gradient method in Fig. 1. Moreover, constrained quasi-Newton methods in the Matlab Optimization Toolbox guarantee linear convergence by using a quasi-Newton updating procedure.

From the numerical results shown in Fig. 1, we can see that: (i) The left picture of Fig. 1 shows that for a small number of iterations, the forward mode AD has reached more precise results than expected by the theoretical curve. As the number of iterations increases, the gradient method using AD derivatives matched the expected linear convergence. (ii) In the the right picture of Fig. 1, we have omitted the expected convergence curve since its inclusion makes the curves by FD nearly invisible from the graph. The obtained curve has shown a sublinear convergence.

4 Concluding Remarks

In this paper, we have explored the idea of AD-based calibration method for a class of stochastic pricing models. We have used an optimization algorithm using a gradient obtained via FD and AD. Numerical results have shown that accurate derivatives by the forward mode AD yielded high precision calibration results while FD displayed a low convergence rate. In finance for example, there is a need for complex stochastic models in order to exactly predict future market prices. As a result, financial engineers have to recalibrate model parameters to intraday market data almost in real time. The methods we have presented here aim to partially respond to this kind of needs for low latency computation.

References

1. Vasicek, O.: An equilibrium characterisation of the term structure. J. Finan. Econ. **5**, 177–188 (1977)
2. Tadjouddine, E.M.: Vertex-ordering algorithms for automatic differentiation of computer codes. Comput. J. **51**(6), 688–699 (2008)
3. Gelenbe, E.: Analysis of single and networked auctions. ACM Trans. Internet Techn. **9**(2), 1–24 (2009)
4. Ewald, C.O., Zhang, A.: A new technique for calibrating stochastic volatility models: the malliavin gradient method. Quant. Finan. **6**(2), 147–158 (2006)
5. Mikhailov, S., Nögel, U.: Heston's stochastic volatility model: Implementation, calibration and some extensions. Wilmott magazine. 74–79 (2003)
6. Forth, S.A.: An efficient overloaded implementation of forward mode automatic differentiation in MATLAB. ACM Trans. Math. Softw. **32**(2), 195–222 (2006)
7. Glasserman, P.: Monte Carlo Methods in Financial Engineering.. Springer, New York (2004)
8. Polyak, B.: Introduction to Optimization. Optimization Software. Inc Publ. Division, New York (1987)

Author Index

E. Gelenbe et al. (eds.), *Computer and Information Sciences II*,
DOI: 10.1007/978-1-4471-2155-8,